C. MASiello

D0141736

INTRODUCTION TO LINEAR ALGEBRA

INTRODUCTION TO LINEAR ALGEBRA

Philip Gillett
UNIVERSITY OF WISCONSIN CENTER SYSTEM

HOUGHTON MIFFLIN COMPANY • BOSTON
Atlanta • Dallas • Geneva, Illinois
Hopewell, New Jersey • Palo Alto • London

FOR DOOIE

Grateful acknowledgment is due Prindle, Weber, and Schmidt, Inc., publishers of my book *Linear Mathematics,* for permission to adapt certain material from that book for use in this one.

Printed in the United States of America.

Library of Congress Catalog Card Number: 74-12899

ISBN: 0-395-18574-2

Contents

(Sections marked † are optional.)

PREFACE ix

CHAPTER 1 · CONCRETE VECTOR SPACES 1

1.1 Vectors in the Plane 1
1.2 The Dimension of $\mathcal{R}^2$ 6
1.3 Real and Complex *n*-Space 14
Review Quiz 22

CHAPTER 2 · ABSTRACT VECTOR SPACES 25

2.1 Definitions and Preliminary Theorems 25
2.2 Linear Independence in a Vector Space 33
†2.3 Linear Independence of Functions 37
2.4 Dimension of a Vector Space 42
2.5 The Minimax Principle 50
2.6 Subspaces of a Vector Space 53
Review Quiz 60

CHAPTER 3 · LINEAR MAPS 63

3.1 Definitions and Examples 63
3.2 The Law of Nullity 74
3.3 Invertible Linear Maps 83
3.4 Concrete Spaces Revisited 91
3.5 The Vector Space of Linear Maps 97
Review Quiz 104

CHAPTER 4 · MATRICES 107

4.1 The Matrix Representation of a Linear Map 107
4.2 The Isomorphism between Maps and Matrices 117
4.3 The Transpose of a Matrix 122
Review Quiz 128

CHAPTER 5 · MULTIPLICATION OF MAPS AND MATRICES 131

5.1 Composition of Linear Maps 131
5.2 The Inverse of a Linear Map 136
5.3 Multiplication of Matrices 142
5.4 The Inverse of a Matrix 153
Review Quiz 164

CHAPTER 6 · EQUIVALENCE AND SIMILARITY 167

6.1 Matrix Representations Revisited 167
6.2 Similar Matrices 174
6.3 Change of Basis 178
Review Quiz 188

CHAPTER 7 · LINEAR SYSTEMS 191

7.1 Systems of Linear Equations 191
7.2 Row Echelon Form 199
7.3 Row Operations 207
7.4 Column Operations 219
7.5 Rank 223
7.6 Theory of Linear Systems 231
Review Quiz 235

CHAPTER 8 · DETERMINANTS 239

8.1 Definitions 239
8.2 Expansion by a Row 244
8.3 Expansion by a Column 252
8.4 Uniqueness of Determinants 254
†8.5 Cramer's Rule 257
Review Quiz 264

CHAPTER 9 · EIGENVALUES 267

9.1 Diagonal Matrices 267
9.2 Eigenvalues of a Linear Operator 271
9.3 Eigenvalues of a Matrix 278
9.4 Diagonalization of a Matrix 285
†9.5 An Application to Differential Equations 295
†9.6 The Cayley-Hamilton Theorem 297
Review Quiz 300

CHAPTER 10 · INNER PRODUCTS 303

10.1 Geometry in *n*-Space 304
10.2 Euclidean and Unitary Spaces 313
10.3 The Cauchy-Schwarz Inequality 320
10.4 Orthogonality 324
10.5 Parseval's Identity 334
†10.6 Gram-Schmidt Orthogonalization 345
Review Quiz 352

CHAPTER 11 · THE PROJECTION THEOREM 355

11.1 Sums of Subspaces 355
11.2 Orthogonal Subspaces of a Unitary Space 362
11.3 The Projection Theorem 369
†11.4 Applications to Geometry and Analysis 375
Review Quiz 384

CHAPTER 12 · LINEAR OPERATORS ON A UNITARY SPACE 387

12.1 The Adjoint of a Linear Operator 387
12.2 Orthogonal and Unitary Operators 396
†12.3 Orthogonal Transformations of the Plane 406
†12.4 An Example in Hilbert Space 410
12.5 Another Characterization of Unitary Operators 413
12.6 Self-adjoint Operators 417
Review Quiz 425

CHAPTER 13 · SPECTRAL THEORY 429

13.1 Spectral Theorems for Self-adjoint Operators 429
†13.2 An Application to Geometry 438
13.3 Normal Operators and the Spectral Theorem 443
Review Quiz 451

APPENDIX 1 453

A1.1 The Dual Space of a Vector Space 453
A1.2 The Representation Theorem 455
A1.3 Orthogonality Revisited 459

APPENDIX 2 464

A2.1 Triangulation of a Matrix 464
A2.2 The Cayley-Hamilton Theorem 471

BIBLIOGRAPHY 474

SELECTED ANSWERS AND HINTS 476

INDEX 518

Preface

Linear algebra has become established in the undergraduate curriculum as the ideal place for students fresh out of calculus to start taking mathematics seriously. Not that calculus should be treated lightly, but in some ways it is too difficult to present in a logical order; it is an unusual student who emerges from a calculus class with much more than an intuitive grasp of what must often seem like a subject enveloped in fog.

Linear algebra clears the air. It starts with simple ideas and goes a long way before any deep theorems have to be borrowed. The only thing in this book that is just flatly stated without any argument, for example, is the fundamental theorem of algebra. Most students (like mathematicians before Gauss) have learned to live without a proof of that.

Of course linear algebra is significant as well as lucid. It is a body of ideas that runs all through mathematics and its applications; one can hardly overestimate its value to the serious student. Moreover, it is a delight to teach and a pleasure to learn.

But why another book on the subject? I doubt if this is ever a real question in the mind of an author until he starts apologizing in a preface. Obviously he thinks he can build a better mousetrap or he wouldn't take three years out of his life to try. In my case I labor under the impression that I have learned how to communicate with undergraduates. (After twenty years of teaching and writing, one acquires these illusions.) And I have long maintained, despite the evidence to the contrary, that there is nothing incompatible between good mathematics and a decent literary style. I have tried to make this book readable, even occasionally enjoyable, although I am sure that much of it is still as ponderous as only a mathematics textbook can be. Some will find the style verbose; I will defend that on the grounds that a first course should err on the side of too much explanation rather than too little. The book is not written for my colleagues, but for their students.

Another principle that I tried to follow is that it is better to do some things well rather than many things badly. The book is not an exhaustive treatise: great gobs of classical material are left out. Neither is it a superficial survey. It is simply a book to get undergraduates into algebra in a substantial and coherent way.

The level of abstraction is supposed to be somewhere between a good calculus course and higher algebra; thus the climb is Alpine but not Himalayan. It starts with vectors in the plane and ends with spectral theorems for symmetric and normal operators on finite-dimensional inner product spaces. Anyone who has taught linear algebra to undergraduates knows what a rewarding trip this can be. One wants an ice axe from time to time, but nobody needs any oxygen if he is willing to put one foot down after the other in good order.

The first thing worthy of note in this book is the time spent on concrete vector spaces as a preparation for abstraction. Chapter 1 is as intuitive as I could make it, starting in $\mathcal{R}^2$ with an easygoing discussion of vector algebra and then repeating these ideas in the more formal setting of real and complex *n*-space. It is essential, incidentally, to include complex space; many of the jewels of linear algebra lose their sparkle in the light of $\mathcal{R}$ alone.

The second chapter begins with axioms for a vector space suggested by experience with *n*-tuples. An example of a space which is not finite-dimensional is presented right away, however, both to keep things honest and to prepare for things to come. The standard theorems about finite-dimensional spaces are developed with as little pain as possible. Any instructors who have conducted a class of undergraduates through vector spaces will recognize how many obstacles have been removed when they read this chapter; yet I believe they will not find any bad mathematics. The theorems are all proved, but not with a sledge hammer.

At this point instructors who want to take up inner product spaces as a logical continuation of the first two chapters may go directly to Chaps. 10 and 11, which are designed to be largely independent of the intervening material. Others, who want to get off the ground early, should move on to Chap. 3. This chapter, on linear maps, opens with a variety of important illustrations. After the basic ideas of kernel, range, and invertibility are developed, there is a section entitled Concrete Spaces Revisited in which I have taken special pains to exhibit the connection between abstract finite-dimensional space and *n*-space (and to embellish the concept of isomorphism).

Chapter 4 uses linear maps to motivate the introduction of matrices, and begins the traverse from one to the other which is such a fascinating part of linear algebra. In the old argument concerning the relative importance of matrices and linear maps, I have tried to be neutral. Maps come first in this book, as they do historically, because they lead in a natural way to matrices. But the usual isomorphisms are exhibited; the controversy expires, it seems to me, when each domain is seen from the standpoint of the other. And of course students are enriched by seeing both.

Chapter 5, on multiplication of maps and matrices, is fundamental. One could follow it with Chap. 7 and call the result a short course in linear algebra. But if eigenvalues are going to be covered, one needs Chap. 6 on equivalence and similarity. This is a short, almost lyrical interlude; when students recognize that a linear map and an equivalence class of matrices are essentially the same thing, they may begin to perceive what mathematicians mean when they speak of beauty.

Linear systems (Chap. 7) come late in this book. My reason for not starting with them, as many writers do, is that I think the business of row operations, rank, and echelon form is really very cluttered. I have seen too many students (not to mention instructors) repelled by it, so I have tried to minimize the pain both by delay and by omission. (The existence and uniqueness of row echelon form, for example, are stated without proof, not because it is deep but because it is dull.)

Determinants (Chap. 8) are also given little encouragement to proliferate. Their treatment is deliberately artificial, and brief, the idea being that anything is preferable to permutation theory—particularly in a course on linear algebra. Students will not know what is going on until they are halfway through it, but in the end they will have acquired the necessary artillery without permanent alienation of their affection for mathematics. My own practice (designed to save time for more important things) is to spend only one day on this material in the classroom, merely announcing the main results and assigning the rest for outside reading.

Only the first four sections of Chap. 9 (on eigenvalues) are intended for the standard linear algebra course. The fifth (an application of eigenvalues to linear systems of differential equations) is worth doing if there is time; the last (on the Cayley-Hamilton theorem) is suitable for an honors class or a special report (particularly in connection with its sequel in Appendix 2). Of course every graduate department of mathematics should have an inscription over the door reading "Let no one ignorant of the Cayley-Hamilton theorem enter here." One should certainly tell the undergraduate what the theorem says.

Chapters 10 and 11, on geometry in n-space and its generalization to abstract inner product spaces, can be taken up any time after Chap. 2. The theory in these chapters is abundantly illustrated by examples and problems drawn both from geometry and from analysis. Of particular interest are the points of contact with Fourier analysis, a connection which is important not only because the subject developed that way historically, but because students need to see nontrivial applications. I am not aware of any other book in which this is done on such an elementary level. Students are not expected to know anything at all about Fourier series, nor are any substantial demands made of their knowledge of calculus. But if they were frightened by an integral sign at an early age, they may need encouragement to read it.

The last two chapters cover the part of linear algebra I enjoy teaching more than any other. The application of virtually all the preceding material to linear operators on unitary spaces (culminating in the several versions of the spectral theorem) is such a lovely episode in mathematics that it is hard not to get emotional. I will say no more about it, except that I have tried to avoid spoiling it by too much abstraction, jargon, and notation.

Two appendices at the end of the book are intended to provide additional material for honors sections (and to put the finishing touches on one or two incomplete results in the main body of the text). The answer section contains all numerical answers called for in the problems (of which there are more than 1200), together with many hints and sketches for proofs. Following each chapter is a review quiz consisting of true-false questions. These are not intended to provide any really substantial review, but simply to help students discover where it is needed.

I would like to thank Associate Professor Richard E. Goodrick, of California State College at Hayward, and Professor Warren S. Loud, of the University of Minnesota, for their criticisms and reviews of the manuscript. It is also a pleasure to acknowledge the exceptionally helpful suggestions of Professor Richard Dowds, whose prepublication reviews were of unusual depth and insight. The final form of the book owes much to his well-organized criticism.

I must also mention the late Hilaire Belloc, from whose classic *The Path to Rome* I borrowed the idea for Auctor-Lector conversations. No instructor should pay much attention to these; they are intended to evoke an occasional reaction from readers, who may need to know that their difficulties and objections are understood by the writer.

INTRODUCTION TO LINEAR ALGEBRA

CHAPTER 1

Concrete Vector Spaces

The purpose of this chapter is to introduce the fundamental concept of ''vector space'' by exhibiting several concrete instances and examining their properties. You may be pleased to discover how much of this material you already know. Much of the chapter is nothing but the adoption of a certain point of view regarding familiar objects and operations of mathematics. Ordinary geometry and a working acquaintance with real and complex numbers are the only tools required.

1.1 Vectors in the Plane

If $\mathcal{R}$ is the set of real numbers, then

$$\mathcal{R}^2 = \{(x_1,x_2)\,|\,x_1 \in \mathcal{R},\ x_2 \in \mathcal{R}\}$$

is the set of ordered pairs of real numbers, commonly known as the ''cartesian plane'' in honor of René Descartes (1596–1650), who made the identification of ordered pairs with points in a coordinate plane. Sometimes these pairs are also identified with arrows, as indicated in Fig. 1.1. The agreement here is

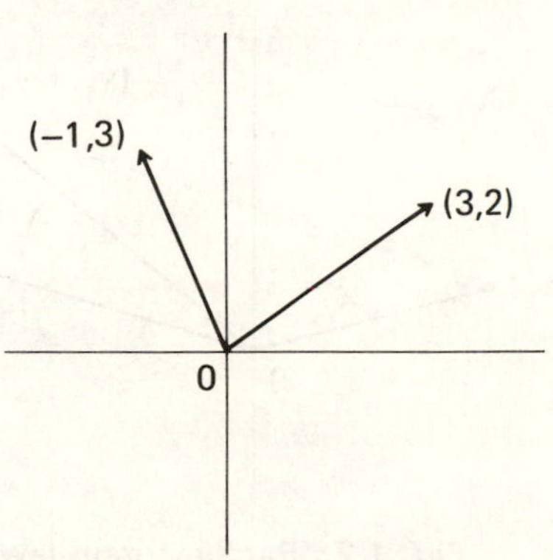

Fig. 1.1 Ordered pairs interpreted as arrows

that the arrow associated with (x_1,x_2) begins at the origin and terminates at the point with coordinates (x_1,x_2).

These three ways of thinking of the elements of $\mathcal{R}^2$ (as ordered pairs, points, or arrows) are usually regarded as interchangeable. Strictly speaking, there are three sets involved, but since each set can be placed in one-to-one correspondence with the others, we shall not distinguish between them.

The set $\mathcal{R}^2$ acquires algebraic interest when we impose operations on it. Let us agree that the *sum* of two elements (x_1,x_2) and (y_1,y_2) of $\mathcal{R}^2$ is the element $(x_1 + y_1, x_2 + y_2)$ of $\mathcal{R}^2$:

†ADDITION IN $\mathcal{R}^2$

$$(x_1,x_2) + (y_1,y_2) = (x_1 + y_1, x_2 + y_2)$$

Note that this agreement (like all definitions) describes a new idea in terms of an old one, which brings up the question of where we choose to start.

For the present, we are taking as given the usual algebraic apparatus in $\mathcal{R}$, the set of real numbers. Later we'll assume that $\mathcal{C}$, the set of complex numbers, is also familiar (although its properties will be reviewed). There are only one or two places in the book where we need anything deep about these systems; on the whole our attitude concerning their basic properties will be offhand. The foundations of mathematics recede into a bottomless abyss when they are sought out! We are moving onto the slope with an idea of climbing instead. (We are not going to worry about base camp, but about how the ropes are attached!)

Geometrically we may interpret addition on $\mathcal{R}^2$ by the usual parallelogram law (Fig. 1.2). If **X** and **Y** are the arrows corresponding to the pairs (x_1,x_2)

† *Whenever a box appears in this book, a definition is being given Not all new terminology and notation is displayed in boxes, but most is.*

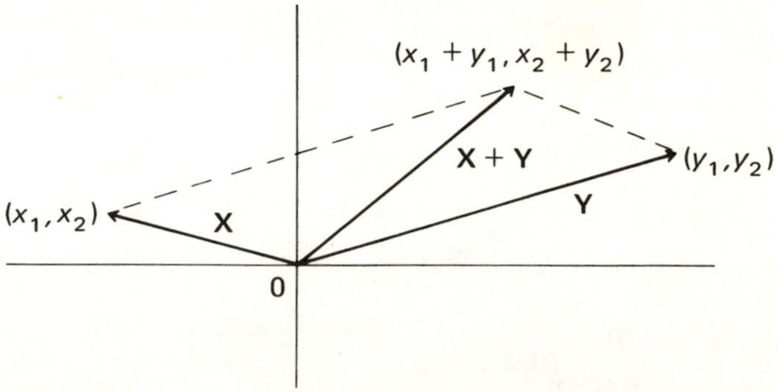

Fig. 1.2 Parallelogram law of addition

and (y_1,y_2), then $\mathbf{X} + \mathbf{Y}$ is the arrow indicated in Fig. 1.2. Of course in some cases no parallelogram is formed, as when $\mathbf{X}$ and $\mathbf{Y}$ are parallel or when either $\mathbf{X}$ or $\mathbf{Y}$ is zero.† There is no need to worry about this; the algebraic definition is clear in any case. This fact is sometimes formally stated as the *closure law:*

> If $\mathbf{X}$ and $\mathbf{Y}$ are elements of $\mathcal{R}^2$, then $\mathbf{X} + \mathbf{Y}$ is a uniquely defined element of $\mathcal{R}^2$.

Here is our first encounter with abstraction! (Note that it is possible to prove the closure law, unnecessary as it may seem to do so.) If $\mathbf{X} = (x_1,x_2)$ and $\mathbf{Y} = (y_1,y_2)$ are elements of $\mathcal{R}^2$, we have defined $\mathbf{X} + \mathbf{Y} = (x_1 + y_1, x_2 + y_2)$. The closure law in $\mathcal{R}$ (one of the properties of real numbers we are taking for granted) says that $z_1 = x_1 + y_1$ and $z_2 = x_2 + y_2$ are uniquely defined real numbers. Hence $(z_1,z_2) = (x_1 + y_1, x_2 + y_2)$ is a uniquely defined element of $\mathcal{R}^2$.‡

Other laws of addition in $\mathcal{R}^2$ can be proved in the same way, by referring back to properties of $\mathcal{R}$ with which we assume you are familiar. (See the problems.)

The second operation we impose on $\mathcal{R}^2$ is not internal like addition, but involves both elements of $\mathcal{R}^2$ and elements of $\mathcal{R}$. If $c \in \mathcal{R}$ and $(x_1,x_2) \in \mathcal{R}^2$, the *product* of c and (x_1,x_2) is the element (cx_1,cx_2) of $\mathcal{R}^2$:

SCALAR MULTIPLICATION IN $\mathcal{R}^2$

$$c(x_1,x_2) = (cx_1,cx_2)$$

This operation is called "scalar multiplication" because the elements of $\mathcal{R}$ ("scalars") can be represented by points on a number scale.

The geometric interpretation of scalar multiplication (Fig. 1.3) is that if $\mathbf{X}$ is the arrow corresponding to (x_1,x_2), then $c\mathbf{X}$ is the arrow whose length is $|c|$ times the length of $\mathbf{X}$ and whose direction is the same as or opposite to the direction of $\mathbf{X}$, depending on whether $c > 0$ or $c < 0$. If $c = 0$, no direction is assigned to $c\mathbf{X}$, but this is not surprising in view of the fact that

$$0\mathbf{X} = 0(x_1,x_2) = (0,0)$$

which is an "arrow" of length 0.

† The zero element in $\mathcal{R}^2$ is $\mathbf{0} = (0,0)$, the "arrow" beginning and ending at the origin. See Prob. 3.

‡ The equality of pairs $(z_1,z_2) = (x_1 + y_1, x_2 + y_2)$ follows from the agreement, implicit in the definition of $\mathcal{R}^2$, that (a_1,a_2) and (b_1,b_2) are distinct if and only if $a_1 \neq b_1$ or $a_2 \neq b_2$. Sometimes this is called the *definition of equality* in $\mathcal{R}^2$ and is stated in the equivalent form

$$(a_1,a_2) = (b_1,b_2) \quad \Leftrightarrow \quad a_1 = b_1 \quad \text{and} \quad a_2 = b_2$$

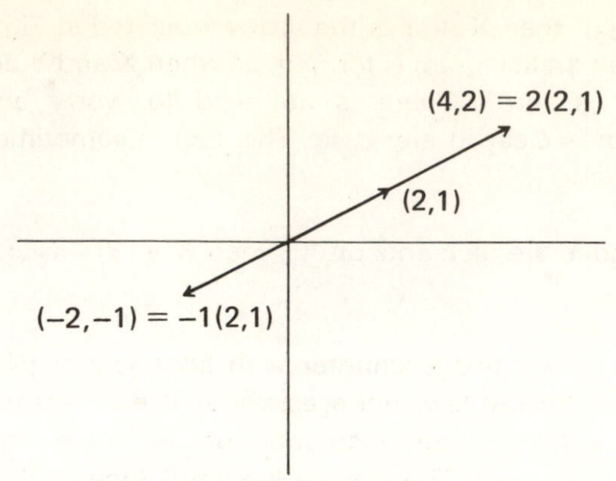

Fig. 1.3 Scalar multiplication

Laws of scalar multiplication can be derived from the properties of $\mathcal{R}$ in the same way as those for addition. For example, if a and b are real numbers and $\mathbf{X} \in \mathcal{R}^2$, then

$$(a + b)\mathbf{X} = a\mathbf{X} + b\mathbf{X}$$

Suppose that $\mathbf{X} = (x_1,x_2)$. Then

$$\begin{aligned}(a + b)\mathbf{X} &= (a + b)(x_1,x_2)\\ &= [(a + b)x_1, (a + b)x_2]\\ &= (ax_1 + bx_1, ax_2 + bx_2) \qquad \text{(distributive law in } \mathcal{R})\\ &= (ax_1,ax_2) + (bx_1,bx_2)\\ &= a(x_1,x_2) + b(x_1,x_2)\\ &= a\mathbf{X} + b\mathbf{X}\end{aligned}$$

In the problems we'll ask you to discuss a variety of properties of addition and scalar multiplication, with the end in view that $\mathcal{R}^2$ is a ''vector space.'' This term simply summarizes certain essential properties concisely; we'll have more to say about it later.

VECTOR SPACE PROPERTIES OF $\mathcal{R}^2$

1. $\mathcal{R}^2$ is closed relative to addition: $\mathbf{X} \in \mathcal{R}^2$ and $\mathbf{Y} \in \mathcal{R}^2 \Rightarrow \mathbf{X} + \mathbf{Y} \in \mathcal{R}^2$
2. Addition is associative: $(\mathbf{X} + \mathbf{Y}) + \mathbf{Z} = \mathbf{X} + (\mathbf{Y} + \mathbf{Z})$
3. $\mathcal{R}^2$ contains an element $\mathbf{0}$ with the property that $\mathbf{X} + \mathbf{0} = \mathbf{0} + \mathbf{X} = \mathbf{X}$ for every $\mathbf{X}$.
4. For each $\mathbf{X} \in \mathcal{R}^2$ there is an element $\mathbf{Y} \in \mathcal{R}^2$ such that $\mathbf{X} + \mathbf{Y} = \mathbf{Y} + \mathbf{X} = \mathbf{0}$.
5. Addition is commutative: $\mathbf{X} + \mathbf{Y} = \mathbf{Y} + \mathbf{X}$

6. $\mathcal{R}^2$ is closed relative to scalar multiplication: $c \in \mathcal{R}$ and $\mathbf{X} \in \mathcal{R}^2$ $\Rightarrow c\mathbf{X} \in \mathcal{R}^2$
7. Scalar multiplication is associative: $(ab)\mathbf{X} = a(b\mathbf{X})$
8. Scalar multiplication is distributive:

$$(a + b)\mathbf{X} = a\mathbf{X} + b\mathbf{X} \qquad \text{and} \qquad a(\mathbf{X} + \mathbf{Y}) = a\mathbf{X} + a\mathbf{Y}$$

9. For each $\mathbf{X} \in \mathcal{R}^2$, $1\mathbf{X} = \mathbf{X}$.

Problems

1. Suppose that **X** and **Y** are nonzero, noncollinear arrows in $\mathcal{R}^2$. Explain why $\mathbf{X} + \mathbf{Y}$ is the diagonal of a parallelogram having **X** and **Y** as sides (as in Fig. 1.2).

2. Derive the associative law of addition in $\mathcal{R}^2$:

$$(\mathbf{X} + \mathbf{Y}) + \mathbf{Z} = \mathbf{X} + (\mathbf{Y} + \mathbf{Z})$$

3. The identity law of addition says that $\mathcal{R}^2$ contains an element **0** with the property that if $\mathbf{X} \in \mathcal{R}^2$, then $\mathbf{X} + \mathbf{0} = \mathbf{0} + \mathbf{X} = \mathbf{X}$. Obviously (0,0) qualifies as such an element; prove that it is the only element that can properly be called "zero"; that is, if **Z** has the property that $\mathbf{X} + \mathbf{Z} = \mathbf{Z} + \mathbf{X} = \mathbf{X}$ for every **X**, then $\mathbf{Z} = (0,0)$.

4. The inverse law of addition says that for each $\mathbf{X} \in \mathcal{R}^2$ there is an element $\mathbf{Y} \in \mathcal{R}^2$ such that $\mathbf{X} + \mathbf{Y} = \mathbf{Y} + \mathbf{X} = \mathbf{0}$. Given $\mathbf{X} \in \mathcal{R}^2$, show that $-1\mathbf{X}$ (usually designated $-\mathbf{X}$) qualifies as such an element.

5. Given that $\mathbf{X} \in \mathcal{R}^2$, explain why $-\mathbf{X}$ (defined in Prob. 4) is the only element of $\mathcal{R}^2$ that can properly be called the "inverse" of **X**; that is, if $\mathbf{X} + \mathbf{Y} = \mathbf{Y} + \mathbf{X} = \mathbf{0}$, then $\mathbf{Y} = -\mathbf{X}$.

6. Prove the commutative law of addition in $\mathcal{R}^2$: $\mathbf{X} + \mathbf{Y} = \mathbf{Y} + \mathbf{X}$.

7. If **X** and **Y** are elements of $\mathcal{R}^2$, designate the sum $\mathbf{X} + (-\mathbf{Y})$ by the symbol $\mathbf{X} - \mathbf{Y}$.

a. Letting $\mathbf{X} = (x_1,x_2)$ and $\mathbf{Y} = (y_1,y_2)$, show that $\mathbf{X} - \mathbf{Y} = (x_1 - y_1, x_2 - y_2)$.

b. Draw a picture illustrating the geometric interpretation of $\mathbf{X} - \mathbf{Y}$.

8. Prove the closure law of scalar multiplication: If $c \in \mathcal{R}$ and $\mathbf{X} \in \mathcal{R}^2$, then $c\mathbf{X}$ is a uniquely defined element of $\mathcal{R}^2$. What law in $\mathcal{R}$ did you appeal to?

9. Prove the associative law of scalar multiplication:

$$(ab)\mathbf{X} = a(b\mathbf{X})$$

where a and b are scalars and $\mathbf{X} \in \mathcal{R}^2$.

10. We derived the distributive formula $(a + b)\mathbf{X} = a\mathbf{X} + b\mathbf{X}$ in the text. Prove the other distributive law,

$$a(\mathbf{X} + \mathbf{Y}) = a\mathbf{X} + a\mathbf{Y}$$

where $a \in \mathcal{R}$ and $\mathbf{X}$ and $\mathbf{Y}$ are in $\mathcal{R}^2$.

11. Explain why $1\mathbf{X} = \mathbf{X}$ for every $\mathbf{X} \in \mathcal{R}^2$.

12. Let $c \in \mathcal{R}$ and $\mathbf{X} \in \mathcal{R}^2$. Show that

$$c\mathbf{X} = \mathbf{0} \quad \Leftrightarrow \quad c = 0 \text{ or } \mathbf{X} = \mathbf{0}$$

as in ordinary algebra. [Note the different usages of zero. In $c\mathbf{X} = \mathbf{0}$ and $\mathbf{X} = \mathbf{0}$ the "zero" is (0,0), an element of $\mathcal{R}^2$, whereas $c = 0$ is in $\mathcal{R}$.]

13. Assuming that $c \neq 0$ and $\mathbf{X} \neq \mathbf{0}$, explain why $c\mathbf{X}$ is an arrow whose length is $|c|$ times the length of $\mathbf{X}$ and whose direction is the same as or opposite to the direction of $\mathbf{X}$, depending on whether $c > 0$ or $c < 0$.

14. So far we have been concerned with the cartesian plane $\mathcal{R}^2$, but everything we have said can be specialized to $\mathcal{R}^1$, the cartesian "line." Thus $\mathcal{R}^1$ can be regarded as the set $\mathcal{R}$ of real numbers, the set of points on a number scale, or the set of arrows beginning at the origin and terminating at points on the number scale. What do the various laws we have discussed in $\mathcal{R}^2$ reduce to in $\mathcal{R}^1$?

15. Cartesian "space" is

$$\mathcal{R}^3 = \{(x_1,x_2,x_3) \,|\, x_1 \in \mathcal{R},\ x_2 \in \mathcal{R},\ x_3 \in \mathcal{R}\}$$

Run through in your mind what would be involved in generalizing the properties of $\mathcal{R}^2$ to $\mathcal{R}^3$. If necessary, write out a few proofs.

1.2 The Dimension of $\mathcal{R}^2$

Any set on which operations of addition and scalar multiplication are defined that satisfy the algebraic laws we have discussed is called a "vector space." (We'll be more precise later on.) Thus we may summarize what we have done so far by saying that $\mathcal{R}^1$, $\mathcal{R}^2$, and $\mathcal{R}^3$ are vector spaces.† Everybody recognizes that they are different spaces; we now wish to address ourselves to the question of what it is that makes them different. Of course there is no mystery about it: the elements of $\mathcal{R}^1$ are single real numbers, the elements of $\mathcal{R}^2$ are pairs of real numbers, and the elements of $\mathcal{R}^3$ are triples of real numbers. The geometric interpretations are a line, a plane, and cartesian space. While $\mathcal{R}^1$, $\mathcal{R}^2$, and $\mathcal{R}^3$ behave alike as far as the properties of addition and scalar multiplication are concerned, they differ in "dimension."

† For definitions of $\mathcal{R}^1$ and $\mathcal{R}^3$, see Probs. 14 and 15 in Sec. 1.1.

Even though this idea is transparent, we want to say more about it, for we shall soon be dealing with vector spaces in which the concept of dimension is not an easy one, and for which no obvious geometric interpretation exists.

Let us agree to denote the elements (1,0) and (0,1) of $\mathcal{R}^2$ by $\mathbf{E}_1$ and $\mathbf{E}_2$, respectively. Then (as you may recall from calculus, where the symbols **i** and **j** are often used instead) every element of $\mathcal{R}^2$ can be written as a "linear combination" of $\mathbf{E}_1$ and $\mathbf{E}_2$: if $\mathbf{X} = (x_1,x_2) \in \mathcal{R}^2$, then

$$\mathbf{X} = x_1(1,0) + x_2(0,1) = x_1\mathbf{E}_1 + x_2\mathbf{E}_2$$

(See Fig. 1.4.) Moreover, the coefficients in this expression are unique, for if there are scalars c_1 and c_2 such that $\mathbf{X} = c_1\mathbf{E}_1 + c_2\mathbf{E}_2$, then

$$(x_1,x_2) = c_1(1,0) + c_2(0,1) = (c_1,c_2)$$

from which $c_1 = x_1$ and $c_2 = x_2$. (Why?)

Now, it is a remarkable fact that these properties of $\mathbf{E}_1$ and $\mathbf{E}_2$ cannot be duplicated by using *fewer* than two elements or *more* than two elements of $\mathcal{R}^2$. To see what we mean, look at the elements

$$\mathbf{U}_1 = (1,1) \qquad \mathbf{U}_2 = (-1,1) \qquad \mathbf{U}_3 = (0,-1)$$

Every element of $\mathcal{R}^2$ can be expressed as a linear combination of $\mathbf{U}_1$, $\mathbf{U}_2$, $\mathbf{U}_3$, for if $\mathbf{X} = (x_1,x_2)$ is given, the equation

$$\mathbf{X} = c_1\mathbf{U}_1 + c_2\mathbf{U}_2 + c_3\mathbf{U}_3$$

or

$$(x_1,x_2) = c_1(1,1) + c_2(-1,1) + c_3(0,-1)$$

is equivalent to the system

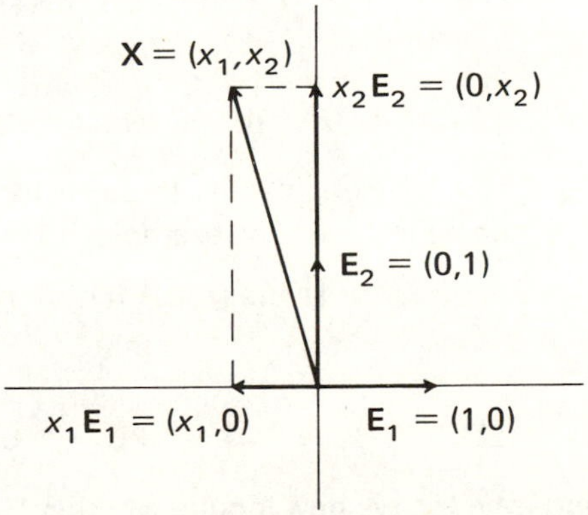

Fig. 1.4 Linear combination of $\mathbf{E}_1$ and $\mathbf{E}_2$

$$c_1 - c_2 = x_1$$
$$c_1 + c_2 - c_3 = x_2$$

To find c_1, c_2, c_3 satisfying this system, we can take c_2 to be arbitrary, say $c_2 = t$, and write

$$c_1 = t + x_1 \qquad c_2 = t \qquad c_3 = 2t + x_1 - x_2$$

(Confirm this result!) The simplest solution is obtained by letting $t = 0$, which yields $c_1 = x_1$, $c_2 = 0$, and $c_3 = x_1 - x_2$; that is,

$$\mathbf{X} = x_1\mathbf{U}_1 + 0\mathbf{U}_2 + (x_1 - x_2)\mathbf{U}_3$$

We say that $\mathbf{U}_1$, $\mathbf{U}_2$, $\mathbf{U}_3$ *span* $\mathcal{R}^2$, because everything in $\mathcal{R}^2$ is dependent on them, that is, can be expressed in terms of them. (Think of the load-bearing girders of a bridge!)Thus they share this property with $\mathbf{E}_1$ and $\mathbf{E}_2$. But the above expression for an arbitrary element of $\mathcal{R}^2$ in terms of $\mathbf{U}_1$, $\mathbf{U}_2$, $\mathbf{U}_3$ is not unique. We could just as well take $t = x_2$ (say) and write

$$\mathbf{X} = c_1\mathbf{U}_1 + c_2\mathbf{U}_2 + c_3\mathbf{U}_3$$

where $c_1 = x_1 + x_2$, $c_2 = x_2$, and $c_3 = x_1 + x_2$. (Confirm this!) For example, the element $(3,-1) \in \mathcal{R}^2$ can be written in at least two ways as a linear combination of $\mathbf{U}_1$, $\mathbf{U}_2$, $\mathbf{U}_3$:

$$\begin{aligned}(3,-1) &= 3(1,1) + 0(-1,1) + 4(0,-1)\\ &= 2(1,1) - (-1,1) + 2(0,-1)\end{aligned}$$

The reason for this failure of uniqueness is that we are suffering from an embarrassment of riches. It doesn't take three elements to span $\mathcal{R}^2$, but only two. (As we have already seen, $\mathbf{E}_1$ and $\mathbf{E}_2$ span $\mathcal{R}^2$.) This suggests that we could remove one of the three and get along with the others; indeed, it suggests that one of the three elements is dependent on the others. To see why, take $(x_1,x_2) = (0,0)$ in the relation

$$(x_1,x_2) = (t + x_1)(1,1) + t(-1,1) + (2t + x_1 - x_2)(0,-1)$$

and obtain

$$\mathbf{0} = t\mathbf{U}_1 + t\mathbf{U}_2 + 2t\mathbf{U}_3$$

Any nonzero choice of t enables us to solve for one of the vectors $\mathbf{U}_1$, $\mathbf{U}_2$, $\mathbf{U}_3$ in terms of the other two. For example, $t = 1$ yields

$$\mathbf{U}_1 + \mathbf{U}_2 + 2\mathbf{U}_3 = \mathbf{0}$$

from which

$$\mathbf{U}_1 = -\mathbf{U}_2 - 2\mathbf{U}_3 \qquad \mathbf{U}_2 = -\mathbf{U}_1 - 2\mathbf{U}_3 \qquad \text{or} \qquad \mathbf{U}_3 = -\tfrac{1}{2}\mathbf{U}_1 - \tfrac{1}{2}\mathbf{U}_2$$

Discarding (say) $\mathbf{U}_3$, we now inquire whether $\mathbf{U}_1$ and $\mathbf{U}_2$ span $\mathcal{R}^2$. If $\mathbf{X} \in \mathcal{R}^2$, do scalars c_1 and c_2 exist such that $\mathbf{X} = c_1\mathbf{U}_1 + c_2\mathbf{U}_2$? This question is equivalent to asking for a solution of the system

$$c_1 - c_2 = x_1$$
$$c_1 + c_2 = x_2$$

where $\mathbf{X} = (x_1, x_2)$. You can easily solve this system to find

$$c_1 = \tfrac{1}{2}(x_1 + x_2) \qquad c_2 = \tfrac{1}{2}(x_2 - x_1)$$

from which

$$\mathbf{X} = \tfrac{1}{2}(x_1 + x_2)\mathbf{U}_1 + \tfrac{1}{2}(x_2 - x_1)\mathbf{U}_2$$

For example,

$$(3, -1) = (1, 1) - 2(-1, 1)$$

Moreover, this expression is unique—because the above system of equations has only one solution. (See Fig. 1.5.)

Thus $\mathbf{E}_1$ and $\mathbf{E}_2$ are not the only two elements that span $\mathbb{R}^2$ without redundancy; $\mathbf{U}_1$ and $\mathbf{U}_2$ serve the same purpose, though not as conveniently. Can we go one step further and drop, say, $\mathbf{U}_2$? The answer is no! The single element $\mathbf{U}_1$ is not sufficient to span $\mathbb{R}^2$, for there are elements of $\mathbb{R}^2$ that cannot be written in the form $\mathbf{X} = c_1\mathbf{U}_1$—for example, $(-1, 1)$ cannot be.

This fact is apparent geometrically, since the set of scalar multiples of $\mathbf{U}_1 = (1, 1)$, considered as a set of points in the plane, is the straight line containing the points (0,0) and (1,1). Any point not on this line cannot be expressed as a multiple of $\mathbf{U}_1$. Equivalently, any arrow which is not parallel to the arrow $\mathbf{U}_1$ from (0,0) to (1,1) is not a multiple of $\mathbf{U}_1$. (See Fig. 1.6.)

We may summarize all this by saying that $\mathbf{U}_1$, $\mathbf{U}_2$, $\mathbf{U}_3$ span $\mathbb{R}^2$, but they span it redundantly; $\mathbf{U}_1$ and $\mathbf{U}_2$ (like $\mathbf{E}_1$ and $\mathbf{E}_2$) span $\mathbb{R}^2$ without redundancy; $\mathbf{U}_1$ does not span $\mathbb{R}^2$ at all. (It spans a 1-dimensional subspace of $\mathbb{R}^2$, namely the straight line shown in Fig. 1.6. We'll discuss subspaces of a vector space in Chap. 2.)

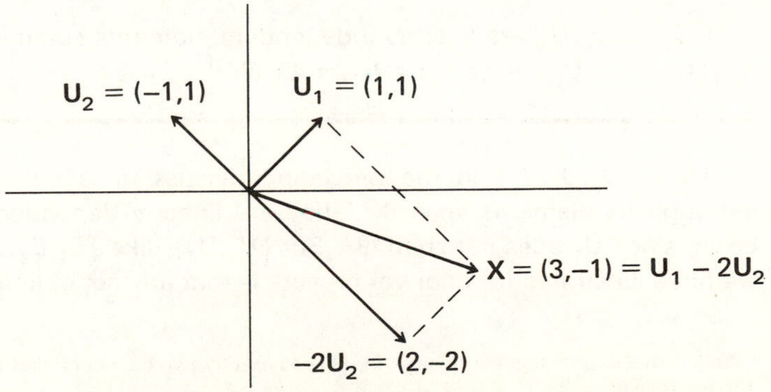

Fig. 1.5 Linear combination of $\mathbf{U}_1$ and $\mathbf{U}_2$

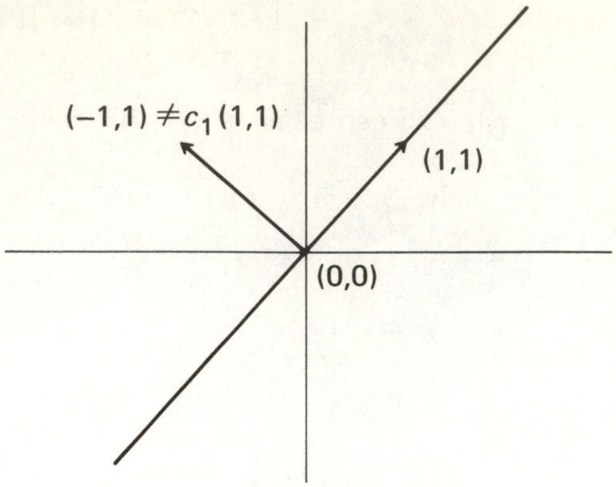

Fig. 1.6 $\mathcal{R}^2$ **cannot be spanned by** $\mathbf{U}_1$.

Alternative terminology, to replace the awkward phrase "without redundancy," is as follows.

LINEAR INDEPENDENCE IN $\mathcal{R}^2$

If no element of $\mathcal{R}^2$ can be written as a linear combination of $\mathbf{U}_1, \ldots, \mathbf{U}_m$ in more than one way, we say that $\mathbf{U}_1, \ldots, \mathbf{U}_m$ are *linearly independent*.

Elements that are not linearly independent are called *linearly dependent*, the terminology being suggested by what we wrote before, when we found that $\mathbf{U}_1 + \mathbf{U}_2 + 2\mathbf{U}_3 = \mathbf{0}$. (If the motivation for this language is not clear to you, don't worry about it. We intend to say more about linear independence and dependence as we go along.)

BASIS FOR $\mathcal{R}^2$

If $\mathbf{U}_1, \ldots, \mathbf{U}_m$ are linearly independent elements spanning $\mathcal{R}^2$, the set $\{\mathbf{U}_1, \ldots, \mathbf{U}_m\}$ is called a *basis* for $\mathcal{R}^2$.†

Thus $\{\mathbf{U}_1,\mathbf{U}_2,\mathbf{U}_3\}$ in the preceding discussion is not a basis, because although its elements span $\mathcal{R}^2$, they are linearly dependent. Nor is $\{\mathbf{U}_1\}$ a basis, since $\mathbf{U}_1$ does not span $\mathcal{R}^2$. But $\{\mathbf{U}_1,\mathbf{U}_2\}$, like $\{\mathbf{E}_1,\mathbf{E}_2\}$, is a basis. What we have illustrated (but not yet proved) is that any set of linearly independent

† We'll prove shortly that every basis for $\mathcal{R}^2$ has exactly two elements; that is, the definition itself implies that $m = 2$.

elements spanning $\mathcal{R}^2$ has exactly two members. This is the algebraic version of the intuitively obvious notion that $\mathcal{R}^2$ is 2-dimensional. The *dimension* is the number of elements in a basis!

If you are interested in a proof of the above, read on; you will get some idea of the problems involved. This material is not essential, however, since it is a special case of a more general statement in Chap. 2.

THEOREM 1.1

If $\{\mathbf{U}_1, \ldots, \mathbf{U}_m\}$ is a basis for $\mathcal{R}^2$, then $m = 2$.

Proof

First we observe that $m > 1$, since $\mathbf{U}_1$ alone cannot span $\mathcal{R}^2$. Otherwise we could find scalars a and b such that

$$\mathbf{E}_1 = a\mathbf{U}_1 \qquad \text{and} \qquad \mathbf{E}_2 = b\mathbf{U}_1$$

If $\mathbf{U}_1 = (x,y)$, we have

$$a(x,y) = (1,0) \quad \Rightarrow \quad ax = 1 \quad \Rightarrow \quad x \neq 0$$

and

$$b(x,y) = (0,1) \quad \Rightarrow \quad bx = 0 \quad \Rightarrow \quad b = 0$$

But this means that $\mathbf{E}_2 = b\mathbf{U}_1 = \mathbf{0}$, whereas $\mathbf{E}_2 \neq \mathbf{0}$.

Thus m is at least 2. Since $\mathbf{U}_1, \ldots, \mathbf{U}_m$ are linearly independent, we know there is only one way to write $\mathbf{0}$ in terms of them, namely, $\mathbf{0} = 0\mathbf{U}_1 + \cdots + 0\mathbf{U}_m$. The implications

$$\begin{aligned} x_1\mathbf{U}_1 + x_2\mathbf{U}_2 = \mathbf{0} \quad &\Rightarrow \quad x_1\mathbf{U}_1 + x_2\mathbf{U}_2 + 0\mathbf{U}_3 + \cdots + 0\mathbf{U}_m = \mathbf{0} \\ &\Rightarrow \quad \text{each coefficient is } 0 \\ &\Rightarrow \quad x_1 = x_2 = 0 \end{aligned}$$

are therefore valid. Letting $\mathbf{U}_1 = (a_{11},a_{21})$ and $\mathbf{U}_2 = (a_{12},a_{22})$, we may write these in the form

$$x_1(a_{11},a_{21}) + x_2(a_{12},a_{22}) = (0,0) \quad \Rightarrow \quad x_1 = x_2 = 0$$

But this is equivalent to saying that the system

$$\begin{aligned} a_{11}x_1 + a_{12}x_2 &= 0 \\ a_{21}x_1 + a_{22}x_2 &= 0 \end{aligned}$$

has only the trivial solution $x_1 = x_2 = 0$. It follows that the determinant of the system, $a_{11}a_{22} - a_{12}a_{21}$, is not 0.†

† If you have never heard of determinants or are unaware of the theorem quoted here, see Prob. 28.

We claim that $\mathbf{U}_1$ and $\mathbf{U}_2$ span $\mathcal{R}^2$, for if $\mathbf{B} = (b_1,b_2)$ is any element of $\mathcal{R}^2$, we can find scalars x_1 and x_2 such that $\mathbf{B} = x_1\mathbf{U}_1 + x_2\mathbf{U}_2$ by solving the system

$$\begin{aligned} a_{11}x_1 + a_{12}x_2 &= b_1 \\ a_{21}x_1 + a_{22}x_2 &= b_2 \end{aligned}$$

for x_1 and x_2. A solution is guaranteed because $a_{11}a_{22} - a_{12}a_{22} \neq 0$! (See Prob. 29.)

Now suppose that $m > 2$. Since $\mathbf{U}_1$ and $\mathbf{U}_2$ span $\mathcal{R}^2$, we can write $\mathbf{U}_m$ as a linear combination of them, say, $\mathbf{U}_m = c_1\mathbf{U}_1 + c_2\mathbf{U}_2$. This equation can be put in the form

$$\mathbf{0} = c_1\mathbf{U}_1 + c_2\mathbf{U}_2 + 0\mathbf{U}_3 + \cdots + 0\mathbf{U}_{m-1} + (-1)\mathbf{U}_m$$

which expresses $\mathbf{0}$ as a linear combination of $\mathbf{U}_1, \ldots, \mathbf{U}_m$. Since the last coefficient is -1, we have arrived at a contradiction. (As observed earlier, there is no way to express $\mathbf{0}$ as a linear combination of $\mathbf{U}_1, \ldots, \mathbf{U}_m$ except by writing $\mathbf{0} = 0\mathbf{U}_1 + \cdots + 0\mathbf{U}_m$ because $\mathbf{U}_1, \ldots, \mathbf{U}_m$ are linearly independent.) Therefore $m = 2$. ▮†

You can see that the same argument might be extended to $\mathcal{R}^3$, provided that we could establish the necessary facts about systems of three linear equations in three unknowns. But this is not easy. We prefer to postpone such questions; the dimension of $\mathcal{R}^3$ (and of higher dimensional spaces) will drop out as a corollary of a different argument in the next chapter.

Problems

16. Show that $\mathbf{U}_1 = (1,1)$ and $\mathbf{U}_2 = (0,1)$ constitute a basis for $\mathcal{R}^2$. That is, prove that they span $\mathcal{R}^2$ and that no element of $\mathcal{R}^2$ can be written as a linear combination of $\mathbf{U}_1$ and $\mathbf{U}_2$ in more than one way.

17. Explain why $\mathbf{U}_1 = (1,-2)$ and $\mathbf{U}_2 = (-2,4)$ do not constitute a basis for $\mathcal{R}^2$. In other words, either find an element of $\mathcal{R}^2$ that cannot be written as a linear combination of $\mathbf{U}_1$ and $\mathbf{U}_2$, or name one that can be expressed in terms of $\mathbf{U}_1$ and $\mathbf{U}_2$ in more than one way.

18. If $\mathbf{U}_1$ and $\mathbf{U}_2$ are linearly independent, why can we say the following?

$$c_1\mathbf{U}_1 + c_2\mathbf{U}_2 = \mathbf{0} \quad \Rightarrow \quad c_1 = c_2 = 0$$

19. Suppose that $\mathbf{U}_1$ and $\mathbf{U}_2$ have the property that

$$c_1\mathbf{U}_1 + c_2\mathbf{U}_2 = \mathbf{0} \quad \Rightarrow \quad c_1 = c_2 = 0$$

† The symbol ▮ means "The proof is finished."

Why does it follow that $\mathbf{U}_1$ and $\mathbf{U}_2$ are linearly independent? *Hint:* You must show that no element of $\mathcal{R}^2$ can be written as a linear combination of $\mathbf{U}_1$ and $\mathbf{U}_2$ in more than one way, that is,

$$\mathbf{X} = a_1\mathbf{U}_1 + a_2\mathbf{U}_2 \quad \text{and} \quad \mathbf{X} = b_1\mathbf{U}_1 + b_2\mathbf{U}_2 \Rightarrow a_1 = b_1 \text{ and } a_2 = b_2$$

Problems 18 and 19 show that linear independence of $\mathbf{U}_1$ and $\mathbf{U}_2$ is equivalent to the statement that $c_1\mathbf{U}_1 + c_2\mathbf{U}_2 = \mathbf{0} \Rightarrow c_1 = c_2 = 0$. This statement is often given as the *definition* of linear independence (of two elements).

20. If $\mathbf{U}_1$ and $\mathbf{U}_2$ are not linearly independent, we call them linearly dependent. Explain why this is equivalent to saying that there are scalars c_1 and c_2, not both zero, such that $c_1\mathbf{U}_1 + c_2\mathbf{U}_2 = \mathbf{0}$.

21. Show that $\mathbf{U}_1 = (1,-2)$ and $\mathbf{U}_2 = (-2,4)$ in Prob. 17 are linearly dependent.

22. Prove that two elements of $\mathcal{R}^2$ are linearly dependent if and only if one is a scalar multiple of the other.†

23. If $\mathbf{U}_1$ and $\mathbf{U}_2$ span $\mathcal{R}^2$ and are linearly independent, then by definition $\{\mathbf{U}_1,\mathbf{U}_2\}$ is a basis. Explain why either of these properties is sufficient, that is, explain why:

a. If $\mathbf{U}_1$ and $\mathbf{U}_2$ span $\mathcal{R}^2$, they are necessarily independent, and hence $\{\mathbf{U}_1,\mathbf{U}_2\}$ is a basis.

b. If $\mathbf{U}_1$ and $\mathbf{U}_2$ are linearly independent, they necessarily span $\mathcal{R}^2$, and hence $\{\mathbf{U}_1,\mathbf{U}_2\}$ is a basis.

24. If $m > 2$, the elements $\mathbf{U}_1, \ldots, \mathbf{U}_m$ of $\mathcal{R}^2$ cannot be linearly independent. Why?

25. According to Prob. 24, the elements $\mathbf{U}_1 = (1,2)$, $\mathbf{U}_2 = (-1,1)$, $\mathbf{U}_3 = (-1,0)$ must be linearly dependent. Name scalars c_1, c_2, c_3, not all zero, such that $c_1\mathbf{U}_1 + c_2\mathbf{U}_2 + c_3\mathbf{U}_3 = \mathbf{0}$. Then write one of the elements as a linear combination of the others.

26. Show that $\mathbf{U}_1$, $\mathbf{U}_2$, $\mathbf{U}_3$ in Prob. 25 span $\mathcal{R}^2$. Then demonstrate that each element of $\mathcal{R}^2$ can be written in more than one way as a linear combination of $\mathbf{U}_1$, $\mathbf{U}_2$, $\mathbf{U}_3$.

27. Reduce $\{(1,-1), (2,0), (-3,5), (-2,2)\}$ to a basis for $\mathcal{R}^2$ by omitting some of its members. In how many ways can this be done?

★ **28.** Let a, b, c, d be elements of $\mathcal{R}$. Prove that the system

† The phrase *if and only if* is a translation of $\Leftrightarrow$. Note that the "if" part is $\Leftarrow$ and the "only if" part is $\Rightarrow$. Many people have a tendency to get these turned around.

★ A problem marked with a star is not necessarily to be worked out, nor is it necessarily harder than other problems. But the problem should be noted, either because it is interesting in its own right or because it will be useful later.

$$ax + by = 0$$
$$cx + dy = 0$$

has a unique solution if and only if $ad - bc$ (the determinant of coefficients) is not zero. *Hint:* The "if" part is easy; just solve the system. To prove the "only if" part, assume that $ad - bc = 0$ and show there are solutions other than the trivial $(x,y) = (0,0)$.

★ **29.** Prove that if $ad - bc \neq 0$, the system

$$ax + by = e$$
$$cx + dy = f$$

has a unique solution.

30. Specializing our discussion to $\mathcal{R}^1$, let $\mathbf{E}_1 = 1$. Prove that every element of $\mathcal{R}^1$ can be written in exactly one way as a multiple of $\mathbf{E}_1$. (Thus $\{\mathbf{E}_1\}$ is a basis for $\mathcal{R}^1$. If every other basis has just one member, we may define the dimension of $\mathcal{R}^1$ to be 1.)

31. Generalize to $\mathcal{R}^3$ by showing that $\mathbf{E}_1 = (1,0,0)$, $\mathbf{E}_2 = (0,1,0)$, $\mathbf{E}_3 = (0,0,1)$ constitute a basis for $\mathcal{R}^3$. (If every other basis turns out to have three members, we can say that the dimension of $\mathcal{R}^3$ is 3.)

1.3 Real and Complex *n*-Space

Our discussion of $\mathcal{R}^2$ in the preceding sections has been mostly algebraic, flavored with a little geometry to serve as motivation. As you have seen in the problems, it can be specialized to $\mathcal{R}^1$ and generalized to $\mathcal{R}^3$, in both cases without losing its geometric appeal.

Now we are going to abandon geometry in order to talk about ordered *n*-tuples of numbers, where *n* is any positive integer. Moreover, we are going to interpret "scalar" to mean not only "element of $\mathcal{R}$" (as in the preceding sections) but also "element of $\mathcal{C}$," where $\mathcal{C}$ is the set of complex numbers.† Thus $\mathcal{C}^3$, for example, means the set of ordered triples of complex numbers; a typical element is $(1 + i, 2 - 3i, 5)$.

The type of scalar allowed in a given discussion will be indicated by writing $\mathcal{R}^n$ or $\mathcal{C}^n$. In the former case only real numbers qualify, while in the latter the scalars may be any complex (including real) numbers. More generally, $\mathcal{F}^n$ will mean the set of ordered *n*-tuples of elements of $\mathcal{F}$ (where $\mathcal{F}$ is $\mathcal{R}$ or $\mathcal{C}$ depending on the context). We impose two operations on $\mathcal{F}^n$, exactly as in $\mathcal{R}^2$:

† These are numbers of the form $a + bi$, where a and b are real and $i = \sqrt{-1}$. See the problems for a review of their properties.

REAL AND COMPLEX n-SPACE

Let n be any positive integer and suppose that $\mathfrak{F}$ is either $\mathfrak{R}$ or $\mathfrak{C}$. Then

$$\mathfrak{F}^n = \{(x_1, \ldots, x_n) \mid x_j \in \mathfrak{F}, j = 1, \ldots, n\}$$

If $\mathbf{X} = (x_1, \ldots, x_n)$ and $\mathbf{Y} = (y_1, \ldots, y_n)$ are elements of $\mathfrak{F}^n$ and if $c \in \mathfrak{F}$, then

$$\mathbf{X} + \mathbf{Y} = (x_1 + y_1, \ldots, x_n + y_n)$$
$$c\mathbf{X} = (cx_1, \ldots, cx_n)$$

Note particularly that scalar multiplication in $\mathfrak{F}^n$ involves scalars from $\mathfrak{F}$. We do not, on the one hand, allow scalars from outside $\mathfrak{R}$ when dealing with $\mathfrak{R}^2$; nor, on the other hand, do we restrict the scalars to $\mathfrak{R}$ when talking about $\mathfrak{C}^2$. Thus $i(1,3) = (i,3i)$ makes sense in $\mathfrak{C}^2$, but not in $\mathfrak{R}^2$, while $2(1,3) = (2,6)$ is legitimate in either context. In the latter case we might write $2(1,3) \in \mathfrak{F}^2$ to indicate that we are indifferent.

THEOREM 1.2

$\mathfrak{F}^n$ is a vector space.

Proof

Although we have not stated precisely what a vector space is, we have indicated in Sec. 1.1 what we have in mind. The laws that need checking are the closure, associative, identity, inverse, and commutative properties of addition, and the closure, associative, distributive, and identity properties of scalar multiplication. Since the arguments in $\mathfrak{R}^2$ generalize in a trivial fashion to $\mathfrak{F}^n$, we spare you the tedium of going through them again. ▮†

LINEAR INDEPENDENCE IN n-SPACE

If $\mathbf{U}_1, \ldots, \mathbf{U}_m$ are elements of $\mathfrak{F}^n$ with the property that

$$c_1\mathbf{U}_1 + \cdots + c_m\mathbf{U}_m = \mathbf{0} \quad \Rightarrow \quad c_1 = \cdots = c_m = 0$$

we say that $\mathbf{U}_1, \ldots, \mathbf{U}_m$ are *linearly independent*. But if there are scalars $c_1, \ldots, c_m$, not all zero,‡ such that $c_1\mathbf{U}_1 + \cdots + c_m\mathbf{U}_m = \mathbf{0}$, then $\mathbf{U}_1, \ldots, \mathbf{U}_m$ are said to be *linearly dependent*.

† In an earlier footnote we said the symbol ▮ indicates completion of the proof. In this case we didn't prove anything, but just waved our hands. Nevertheless, the symbol is useful; we'll try to reserve it for more respectable arguments in the future.

‡ Don't misread this expression as saying "all not zero"! The point is that *at least one* (not necessarily all) of the coefficients is nonzero.

These definitions are not the same as those given in $\mathcal{R}^2$. However, they are equivalent, as we shall show. (Also see Probs. 18 and 19, Sec. 1.2.)

EXAMPLE 1.1

The elements $\mathbf{U}_1 = (2,-1,1)$ and $\mathbf{U}_2 = (5,-1,3)$ are linearly independent in $\mathcal{R}^3$ because

$$c_1\mathbf{U}_1 + c_2\mathbf{U}_2 = \mathbf{0} \quad \Rightarrow \quad \begin{aligned} 2c_1 + 5c_2 &= 0 \\ -c_1 - c_2 &= 0 \\ c_1 + 3c_2 &= 0 \end{aligned} \quad \Rightarrow \quad c_1 = c_2 = 0$$

In other words, the expression of $\mathbf{0}$ as a linear combination of $\mathbf{U}_1$ and $\mathbf{U}_2$ ($\mathbf{0} = 0\mathbf{U}_1 + 0\mathbf{U}_2$) is unique. It is an interesting, perhaps even surprising, fact that this uniqueness implies that the expression of *any* element in terms of $\mathbf{U}_1$ and $\mathbf{U}_2$ is unique; for suppose that $\mathbf{X} = a_1\mathbf{U}_1 + a_2\mathbf{U}_2$ and also that $\mathbf{X} = b_1\mathbf{U}_1 + b_2\mathbf{U}_2$. Then

$$(a_1 - b_1)\mathbf{U}_1 + (a_2 - b_2)\mathbf{U}_2 = \mathbf{0}$$

which implies (as we have just seen) that $a_1 - b_1 = 0$ and $a_2 - b_2 = 0$. Hence $a_1 = b_1$ and $a_2 = b_2$.

Note that this does not mean that $\mathbf{U}_1$ and $\mathbf{U}_2$ span $\mathcal{R}^3$. We are not asserting that every element can be written as a linear combination of $\mathbf{U}_1$ and $\mathbf{U}_2$, but merely that no element can be expressed in terms of $\mathbf{U}_1$ and $\mathbf{U}_2$ in more than one way. In the problems we'll ask you to show that $\mathbf{U}_1$ and $\mathbf{U}_2$ span a plane in $\mathcal{R}^3$. (See Fig. 1.7.) It is impossible to write any vector not in this plane as a linear combination of $\mathbf{U}_1$ and $\mathbf{U}_2$.

EXAMPLE 1.2

The elements $\mathbf{U}_1 = (1,0,3)$, $\mathbf{U}_2 = (-2,1,4)$, $\mathbf{U}_3 = (4,-1,2)$ are linearly dependent in $\mathcal{R}^3$. This fact is not apparent, however. What we have to find out is what kind of scalars x_1, x_2, x_3 satisfy $x_1\mathbf{U}_1 + x_2\mathbf{U}_2 + x_3\mathbf{U}_3 = \mathbf{0}$. Such a triple of scalars must be a solution of the system

$$\begin{aligned} x_1 - 2x_2 + 4x_3 &= 0 \\ x_2 - x_3 &= 0 \\ 3x_1 + 4x_2 + 2x_3 &= 0 \end{aligned}$$

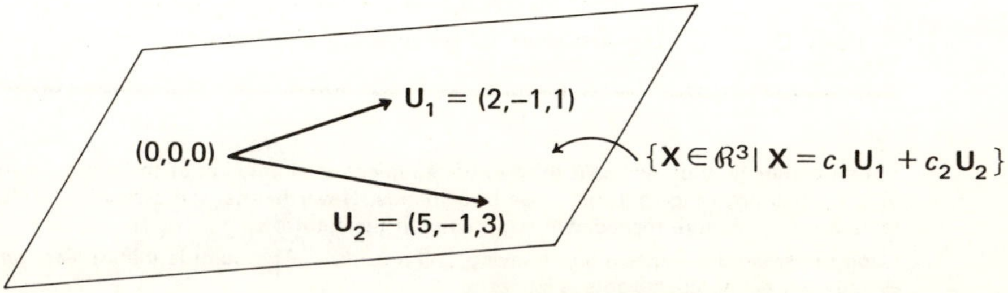

Fig. 1.7 Plane spanned by $\mathbf{U}_1$ and $\mathbf{U}_2$

Of course the triple (0,0,0) satisfies this system, but that is not the question. Is there a nontrivial solution? In other words, are there scalars x_1, x_2, x_3, not all zero, satisfying the system? If so, then $\mathbf{U}_1$, $\mathbf{U}_2$, $\mathbf{U}_3$ are linearly dependent.

Your experience with this sort of thing may be limited. As a matter of fact, the subject of linear systems of equations is one of the main topics covered in this book. For the present, let's proceed by the seat of our pants. The second equation tells us that $x_3 = x_2$; substitution of this result in the first and third equations yields the equivalent system†

$$\begin{aligned} x_1 + 2x_2 &= 0 \\ x_2 - x_3 &= 0 \\ 3x_1 + 6x_2 &= 0 \end{aligned}$$

But this system is clearly "dependent" in the sense of that word as used in elementary algebra; the third equation is just a multiple of the first. Working with only the first two equations, we see that x_2 may be taken to be any scalar whatever, say $x_2 = t$. Then $x_1 = -2x_2 = -2t$ and $x_3 = x_2 = t$; that is, solutions of the system are triples (x_1,x_2,x_3) given in terms of the "parameter" t by

$$x_1 = -2t \qquad x_2 = t \qquad x_3 = t$$

A nontrivial solution of the system certainly exists! For example, we may take $t = 1$ and find the solution $(-2,1,1)$. Thus we have obtained a relation of linear dependence between $\mathbf{U}_1$, $\mathbf{U}_2$, $\mathbf{U}_3$, namely

$$-2\mathbf{U}_1 + \mathbf{U}_2 + \mathbf{U}_3 = \mathbf{0}$$

BASIS AND STANDARD BASIS IN *n*-SPACE

If $\mathbf{U}_1, \ldots, \mathbf{U}_m$ are linearly independent elements spanning $\mathfrak{F}^n$, we call $\{\mathbf{U}_1, \ldots, \mathbf{U}_m\}$ a *basis* for $\mathfrak{F}^n$.

Let j be any one of the integers $1, \ldots, n$. The symbol $\mathbf{E}_j$ designates the element of $\mathfrak{F}^n$ whose jth entry is 1 and whose other entries are 0. We call $\{\mathbf{E}_1, \ldots, \mathbf{E}_n\}$ the *standard basis* for $\mathfrak{F}^n$.‡

EXAMPLE 1.3

The standard basis for $\mathcal{R}^4$ (or $\mathcal{C}^4$) is

$$\{(1,0,0,0), (0,1,0,0), (0,0,1,0), (0,0,0,1)\}$$

To check that it is, in fact, a basis, we note that every 4-tuple $\mathbf{X} = (x_1,x_2,x_3,x_4)$ can be written as a linear combination of $\mathbf{E}_1$, $\mathbf{E}_2$, $\mathbf{E}_3$, $\mathbf{E}_4$:

† Two systems of equations are *equivalent* if every solution of one is a solution of the other, that is, if their solution sets are the same.

‡ See Theorem 1.3 for a justification of this terminology.

$$\begin{aligned}\mathbf{X} &= (x_1,0,0,0) + (0,x_2,0,0) + (0,0,x_3,0) + (0,0,0,x_4)\\ &= x_1(1,0,0,0) + x_2(0,1,0,0) + x_3(0,0,1,0) + x_4(0,0,0,1)\end{aligned}$$

Moreover, the coefficients in this expression are unique, for if c_1, c_2, c_3, c_4 are *any* scalars such that

$$\mathbf{X} = c_1\mathbf{E}_1 + c_2\mathbf{E}_2 + c_3\mathbf{E}_3 + c_4\mathbf{E}_4$$

it follows that

$$\begin{aligned}(x_1,x_2,x_3,x_4) &= (c_1,0,0,0) + (0,c_2,0,0) + (0,0,c_3,0) + (0,0,0,c_4)\\ &= (c_1,c_2,c_3,c_4)\end{aligned}$$

and hence

$$c_1 = x_1 \qquad c_2 = x_2 \qquad c_3 = x_3 \qquad c_4 = x_4$$

As you can see, this argument would be messy if we were dealing with $\mathfrak{F}^{100}$. To facilitate discussions involving standard basis elements, we now introduce a notation you may not like—but it turns out to be handy.

---

†KRONECKER'S DELTA

Let i and j be any positive integers. We define the symbol δ_{ij} by writing

$$\delta_{ij} = \begin{cases} 1 & \text{if } i = j \\ 0 & \text{if } i \neq j \end{cases}$$

For example, $\delta_{13} = 0$ and $\delta_{22} = 1$. Kronecker's delta is merely an "on-off" function of two variables ("on" when they are the same, "off" when they are different). Like the absolute value function

$$|x| = \begin{cases} x & \text{if } x \geq 0 \\ -x & \text{if } x < 0 \end{cases}$$

it should not be any cause for alarm; it is simply a convenience.

EXAMPLE 1.4

The standard basis elements of $\mathfrak{F}^n$ are

$$\begin{aligned}\mathbf{E}_1 &= (\delta_{11},\delta_{21}, \ldots ,\delta_{n1})\\ \mathbf{E}_2 &= (\delta_{12},\delta_{22}, \ldots ,\delta_{n2})\\ &\vdots\\ \mathbf{E}_n &= (\delta_{1n},\delta_{2n}, \ldots ,\delta_{nn})\end{aligned}$$

† Leopold Kronecker (1823–1891) was one of the precursors of the modern idea that every number system should be constructed using the positive integers as building blocks.

For each $j = 1, \ldots, n$, we have

$$\mathbf{E}_j = (\delta_{1j}, \delta_{2j}, \ldots, \delta_{nj})$$

This notation corresponds to our earlier verbal description of $\mathbf{E}_j$ as the element of $\mathfrak{F}^n$ whose jth entry (δ_{jj}) is 1 and whose other entries (δ_{ij}, where $i \neq j$) are 0.

LECTOR *You're right, I don't like it.*
AUCTOR *Look at the proof of the next theorem before you decide.*
LECTOR *I don't think I like this fellow Kronecker either.*
AUCTOR *Well, he said something I rather enjoy.*
LECTOR *OK, I'll ask. What did he say?*
AUCTOR *"The natural numbers come from God; everything else is the work of man."*
LECTOR *Next time I won't ask.*

THEOREM 1.3

$\{\mathbf{E}_1, \ldots, \mathbf{E}_n\}$ is a basis for $\mathfrak{F}^n$.

Proof

Let $\mathbf{X} = (x_1, \ldots, x_n)$ be any element of $\mathfrak{F}^n$. The existence and uniqueness of scalars $c_1, \ldots, c_n$ such that $\mathbf{X} = c_1\mathbf{E}_1 + \cdots + c_n\mathbf{E}_n$ may be established by noting that if $c_1, \ldots, c_n$ are *any* scalars, then

$$\begin{aligned} c_1\mathbf{E}_1 + \cdots + c_n\mathbf{E}_n &= \sum_{j=1}^{n} c_j\mathbf{E}_j \\ &= \sum_{j=1}^{n} c_j(\delta_{1j}, \ldots, \delta_{nj}) && \text{(definition of } \mathbf{E}_j\text{)} \\ &= \sum_{j=1}^{n} (c_j\delta_{1j}, \ldots, c_j\delta_{nj}) && \text{(scalar multiplication)} \\ &= \left(\sum_{j=1}^{n} c_j\delta_{1j}, \ldots, \sum_{j=1}^{n} c_j\delta_{nj}\right) && \text{(addition)} \end{aligned}$$

The ith entry of this n-tuple (where i is any one of the integers $1, \ldots, n$) is

$$\sum_{j=1}^{n} c_j\delta_{ij} = c_i\delta_{ii} = c_i \qquad \text{(why?)}$$

so we have

$$c_1\mathbf{E}_1 + \cdots + c_n\mathbf{E}_n = (c_1, \ldots, c_n)$$

This equals the n-tuple $(x_1, \ldots, x_n)$ if and only if we choose $c_j = x_j$, $j = 1, \ldots, n$; that is, $\mathbf{X}$ can be written as a linear combination of $\mathbf{E}_1, \ldots, \mathbf{E}_n$ in one and only one way, namely,

$$\mathbf{X} = x_1\mathbf{E}_1 + \cdots + x_n\mathbf{E}_n$$

Hence $\{\mathbf{E}_1, \ldots, \mathbf{E}_n\}$ is a basis. ▮†

One is tempted at this point to say that $\mathfrak{F}^n$ is n-dimensional, just as we called $\mathcal{R}^2$ 2-dimensional because $\mathbf{E}_1 = (1,0)$ and $\mathbf{E}_2 = (0,1)$ serve as basis elements (and because every basis has two elements). But it is conceivable, unless we prove otherwise, that there are subsets of $\mathfrak{F}^n$ with fewer than n members, or with more than n members, which satisfy the definition of basis as well as $\{\mathbf{E}_1, \ldots, \mathbf{E}_n\}$ does. What has to be shown (but we won't get around to it until the next chapter) is that although there may be many subsets of $\mathfrak{F}^n$ with the properties of a basis, they all have n members. Then it will make sense to call n the dimension of $\mathfrak{F}^n$.

Problems

32. Suppose that scalar multiplication in $\mathcal{R}^2$ were allowed with complex scalars, so that, for example, $i(1,-2) = (i,-2i)$. What property of a vector space would be violated?

33. Let $S = \{(x_1,x_2) \mid x_1 \in \mathcal{C}, x_2 \in \mathcal{C}\}$. Defining addition as in $\mathcal{C}^2$, but restricting scalar multiplication to products of the form $c(x_1,x_2)$, where $c \in \mathcal{R}$, explain why S is a vector space.‡

34. Let $\mathbf{U}_1 = (2,-1,0)$, $\mathbf{U}_2 = (1,0,1)$, $\mathbf{U}_3 = (0,0,2)$, $\mathbf{U}_4 = (-1,0,3)$.
a. Write $\mathbf{U}_4$ as a linear combination of $\mathbf{U}_1$, $\mathbf{U}_2$, $\mathbf{U}_3$.
b. Show that $\mathbf{U}_1$ cannot be expressed as a linear combination of $\mathbf{U}_2$, $\mathbf{U}_3$, $\mathbf{U}_4$.
c. Show that $\mathbf{U}_1$, $\mathbf{U}_2$, $\mathbf{U}_3$, $\mathbf{U}_4$ span $\mathcal{R}^3$.
d. Show that $\mathbf{U}_4$ is not needed in part *c* because $\mathbf{U}_1$, $\mathbf{U}_2$, $\mathbf{U}_3$ alone span $\mathcal{R}^3$.

† Did we prove what we set out to prove? Our definition of basis requires $\mathbf{E}_1, \ldots, \mathbf{E}_n$ to be linearly independent elements spanning $\mathfrak{F}^n$. We proved that they span $\mathfrak{F}^n$ by demonstrating the existence of scalars $c_1, \ldots, c_n$ such that $\mathbf{X} = c_1\mathbf{E}_1 + \cdots + c_n\mathbf{E}_n$ (where $\mathbf{X}$ is any element of $\mathfrak{F}^n$). But instead of proving that

$$c_1\mathbf{E}_1 + \cdots + c_n\mathbf{E}_n = \mathbf{0} \Rightarrow c_1 = \cdots = c_n = 0$$

(required by our definition of linear independence), we showed that the scalars in $\mathbf{X} = c_1\mathbf{E}_1 + \cdots + c_n\mathbf{E}_n$ are unique. Do you see why this uniqueness implies linear independence? (See Prob. 37 for more on this point.)

‡ This space, although legitimate, should not be confused with $\mathcal{C}^2$, in which scalar multiplication is unrestricted. When the symbol $\mathfrak{F}^n$ is used, the agreement is that the underlying set of scalars is $\mathfrak{F}$. We'll see in Chap. 2 how different S and $\mathcal{C}^2$ are.

e. Show that $\mathbf{U}_1$ and $\mathbf{U}_2$ do not span $\mathcal{R}^3$.
f. Which of the sets $\{\mathbf{U}_1,\mathbf{U}_2,\mathbf{U}_3,\mathbf{U}_4\}$, $\{\mathbf{U}_1,\mathbf{U}_2,\mathbf{U}_3\}$, $\{\mathbf{U}_2,\mathbf{U}_3,\mathbf{U}_4\}$, $\{\mathbf{U}_1,\mathbf{U}_2\}$ is a basis for $\mathcal{R}^3$? Defend your answer in each case!

35. What part of Prob. 34 shows that $\mathbf{U}_1$, $\mathbf{U}_2$, $\mathbf{U}_3$, $\mathbf{U}_4$ are linearly dependent? Name scalars c_1, c_2, c_3, c_4, not all zero, such that $c_1\mathbf{U}_1 + c_2\mathbf{U}_2 + c_3\mathbf{U}_3 + c_4\mathbf{U}_4 = \mathbf{0}$.

36. Prove that $\mathbf{U}_1$, $\mathbf{U}_2$, $\mathbf{U}_3$ in Prob. 34 are linearly independent.

37. Prove that $\mathbf{U}_1, \ldots, \mathbf{U}_m$ are linearly independent if and only if no element of $\mathcal{F}^n$ can be written as a linear combination of $\mathbf{U}_1, \ldots, \mathbf{U}_m$ in more than one way. *Note:* This establishes the equivalence of our definitions of linear independence in Sec. 1.2 and in this section.

38. Let $\mathbf{U}_1 = (2,-1,1)$ and $\mathbf{U}_2 = (5,-1,3)$. The set of all linear combinations of $\mathbf{U}_1$ and $\mathbf{U}_2$ in $\mathcal{R}^3$ is $T = \{\mathbf{X} \mid \mathbf{X} = c_1\mathbf{U}_1 + c_2\mathbf{U}_2,\ c_1 \in \mathcal{R},\ c_2 \in \mathcal{R}\}$.
a. Convince yourself that T is a vector space. (It is a "subspace" of $\mathcal{R}^3$, an idea that we'll discuss in detail in Chap. 2.)
b. Letting $\mathbf{X} = (x,y,z)$, write the condition $(x,y,z) = c_1\mathbf{U}_1 + c_2\mathbf{U}_2$ as a system of parametric equations giving x, y, z in terms of the parameters c_1 and c_2.
c. Eliminate the parameters to obtain the equation $2x + y - 3z = 0$. (This shows that T is a plane in 3-space. It is the plane spanned by $\mathbf{U}_1$ and $\mathbf{U}_2$, which serve as basis elements for T. It seems reasonable, both on intuitive geometric grounds and because $\mathbf{U}_1$ and $\mathbf{U}_2$ constitute a basis, to say that T is 2-dimensional.)

39. Let $\mathbf{U}_1 = \mathbf{E}_1$, $\mathbf{U}_2 = \mathbf{E}_2$, $\mathbf{U}_3 = \mathbf{E}_3$, $\mathbf{U}_4 = \mathbf{E}_1 + \mathbf{E}_2 + \mathbf{E}_3$ in $\mathcal{F}^4$. Explain why $\{\mathbf{U}_1,\mathbf{U}_2,\mathbf{U}_3,\mathbf{U}_4\}$ is not a basis for $\mathcal{F}^4$.

40. Let $\mathbf{U}_1 = \mathbf{E}_1$, $\mathbf{U}_2 = \mathbf{E}_2$, $\mathbf{U}_3 = \mathbf{E}_3$, $\mathbf{U}_4 = \mathbf{E}_1 + \mathbf{E}_2 + \mathbf{E}_3 + \mathbf{E}_4$ in $\mathcal{F}^4$. Prove that $\{\mathbf{U}_1,\mathbf{U}_2,\mathbf{U}_3,\mathbf{U}_4\}$ is a basis for $\mathcal{F}^4$.

41. When $\mathbf{U}_1, \ldots, \mathbf{U}_m$ are linearly independent elements spanning $\mathcal{F}^n$, we call the set $\{\mathbf{U}_1, \ldots, \mathbf{U}_m\}$ a basis. There is a subtle difficulty here, for if any of the elements $\mathbf{U}_1, \ldots, \mathbf{U}_m$ were the same, we could not write $\{\mathbf{U}_1, \ldots, \mathbf{U}_m\}$.† Show that linear independence precludes this possibility; that is, if any two of the elements $\mathbf{U}_1, \ldots, \mathbf{U}_m$ are the same, then $\mathbf{U}_1, \ldots, \mathbf{U}_m$ must be linearly dependent.

★ **42.** Let $\mathcal{C}$ be the set of numbers of the form $a + bi$, where a and b are real and $i = \sqrt{-1}$.‡ Confirm that the definitions of addition and multiplication in $\mathcal{C}$

† Set notation is understood to involve no duplications; that is, the elements listed are supposed to be distinct.

‡ In Chap. 5 (Prob. 40, Sec. 5.3) we'll show you a way of describing $\mathcal{C}$ in terms of real numbers, without assuming that $\sqrt{-1}$ makes sense. For the present, we assume that you have learned to live with i.

$$(a + bi) + (c + di) = (a + c) + (b + d)i$$
$$(a + bi)(c + di) = (ac - bd) + (ad + bc)i$$

are consistent with the assumption that the familiar algebraic laws in $\mathcal{R}$ also apply to $\mathcal{C}$, provided that i^2 is replaced by -1 whenever it occurs.

★ **43.** The *conjugate* of a complex number $z = a + bi$ is defined to be $\bar{z} = a - bi$; the *absolute value* of z is $|z| = \sqrt{a^2 + b^2}$.

a. Show that $z \in R \Leftrightarrow \bar{z} = z$. What does $|z|$ reduce to in this case?

b. Explain why $|\bar{z}| = |z|$.

c. Prove that $z\bar{z} = |z|^2$. Is this the same as $|z^2|$?

d. Show that $z + \bar{z} \leq 2|z|$, with equality holding if and only if z is a nonnegative real number.

e. Show that if u and v are complex numbers, then

$$\overline{u + v} = \bar{u} + \bar{v} \qquad \text{and} \qquad \overline{uv} = \bar{u}\,\bar{v}$$

(In words: The conjugate of a sum is the sum of the conjugates, and the conjugate of a product is the product of the conjugates.)

f. Suppose that u and v are complex numbers and $v \neq 0$. Explain why u/v may be found by multiplying numerator and denominator by $\bar{v}$. Carry out this operation in the case $u = 2 + i$ and $v = 2 - i$.

44. Show that $\mathbf{U}_1 = (1 - i, 1)$ and $\mathbf{U}_2 = (i,i)$ constitute a basis for $\mathcal{C}^2$.

45. Show that $\mathbf{U}_1 = (i,0)$, $\mathbf{U}_2 = (0,i)$, $\mathbf{U}_3 = (1,0)$ span $\mathcal{C}^2$. Then show that $\mathbf{U}_3$ is unnecessary, that $\mathbf{U}_1$ and $\mathbf{U}_2$ alone span $\mathcal{C}^2$. Is $\{\mathbf{U}_1, \mathbf{U}_2\}$ a basis for $\mathcal{C}^2$?

†Review Quiz

True or False?

1. The vectors $\mathbf{U}_1, \ldots, \mathbf{U}_m$ are linearly independent if $\sum_{j=1}^{m} c_j\mathbf{U}_j = \mathbf{0}$ implies that $c_j = 0$, $j = 1, \ldots, m$.

2. It is possible to write $\mathbf{X} = (2,-3)$ as a linear combination of $\mathbf{U}_1 = (1,1)$ and $\mathbf{U}_2 = (-1,1)$ in two ways.

3. The vectors $\mathbf{U}_1 = (1,2,-1)$, $\mathbf{U}_2 = (0,-4,1)$, $\mathbf{U}_3 = (1,-2,0)$ span $\mathcal{R}^3$.

4. If $\mathbf{U}_1$ and $\mathbf{U}_2$ are linearly dependent, then each is a scalar multiple of the other.

5. The vectors $\mathbf{U}_1 = (1,0)$, $\mathbf{U}_2 = (0,1)$, $\mathbf{U}_3 = (3,-5)$ are linearly independent.

† By deciding whether each of the statements in this list is true or false (try this without looking through the book), you can get some idea of how well you understand the ideas of the chapter. The statements are not in any particular order; they range over the whole chapter. Answers are in the back of the book.

6. Every set of elements spanning $\mathcal{R}^2$ has at least two members.
7. If $\mathbf{U}_1 + 0\mathbf{U}_2 + 0\mathbf{U}_3 = \mathbf{0}$, then $\mathbf{U}_1$, $\mathbf{U}_2$, $\mathbf{U}_3$ are linearly dependent.
8. In $\mathcal{R}^3$ it is correct to write $\mathbf{E}_2 = (\delta_{21},\delta_{22},\delta_{23})$.
9. If $c_1, \ldots, c_n$ are scalars and i is one of the integers $1, \ldots, n$, then $\Sigma_{j=1}^{n} c_j\delta_{ij} = c_i$.
10. The vector $-2(1,-2)$ is twice as long as $(1,-2)$.
11. René Descartes was a famous French actress.
12. The set of all linear combinations of $\mathbf{U}_1 = (1,i)$ and $\mathbf{U}_2 = (i,-1)$ is $\mathcal{C}^2$.
13. If no element of $\mathcal{F}^n$ can be written as a linear combination of $\mathbf{U}_1, \ldots, \mathbf{U}_m$ in more than one way, then $\mathbf{U}_1, \ldots, \mathbf{U}_m$ are linearly independent.
14. $\mathbf{U}_1, \ldots, \mathbf{U}_m$ are linearly independent if and only if the $\mathbf{U}_i$ are distinct.
15. If $\mathbf{U}_1 = \mathbf{U}_2$, then $\mathbf{U}_1$ and $\mathbf{U}_2$ are linearly dependent.
16. In translating the statement $P \Leftrightarrow Q$ by "P if and only if Q," the "if" part is $P \Rightarrow Q$.
17. Any two linearly independent elements of $\mathcal{R}^2$ constitute a basis for $\mathcal{R}^2$.
18. $\{2\}$ is a basis for $\mathcal{R}^1$.
19. If $z \in \mathcal{C}$, then $z\bar{z} = |z|$.
20. $2i/(1 + i) = 1 + i$
21. If u and v are complex numbers, the conjugate of their sum is the sum of their conjugates.

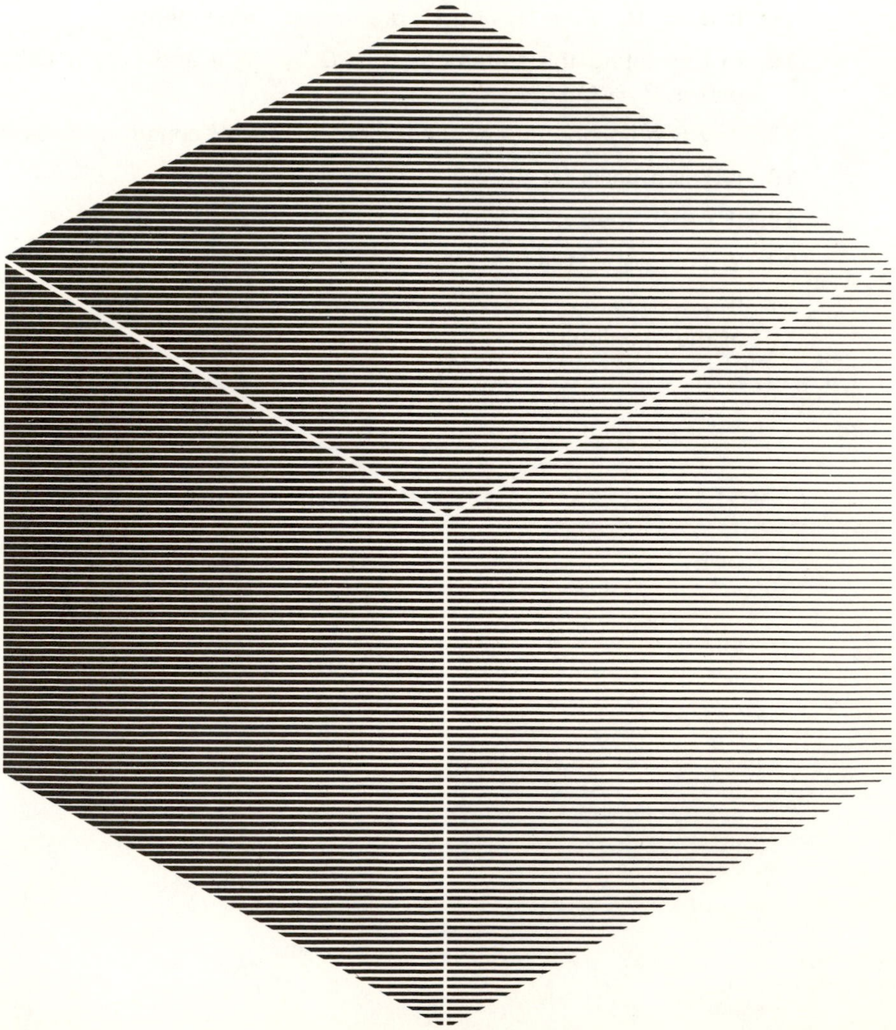

CHAPTER 2

Abstract Vector Spaces

We are now going to forget about *n*-tuples and talk about vector spaces in general. There are many mathematical systems with the same properties of addition and scalar multiplication we discussed in Chap. 1; the elements of such systems may not resemble "vectors" at all (in the sense of "arrows in space"). So you should not read anything into the word "vector" as we use it in this chapter beyond what the definitions themselves say about it.

2.1 Definitions and Preliminary Theorems

VECTOR SPACE

Let $\mathfrak{F}$ be either $\mathfrak{R}$ or $\mathfrak{C}$ and let S be a nonempty set on which operations of "addition" and "scalar multiplication" have been defined satisfying the following laws.

ADDITION

Closure (*Law 1*): If x and y are elements of S, then $x + y$ is a uniquely defined element of S.

Associative (*Law 2*): If x, y, z are elements of S, then $(x + y) + z = x + (y + z)$.

Identity (*Law 3*): S contains an element 0 with the property that if $x \in S$, then $x + 0 = 0 + x = x$.

Inverse (*Law 4*): For each $x \in S$, there is an element $y \in S$ such that $x + y = y + x = 0$.

Commutative (*Law 5*): If x and y are elements of S, then $x + y = y + x$.

SCALAR MULTIPLICATION

Closure (*Law 6*): If $c \in \mathfrak{F}$ and $x \in S$, then cx is a uniquely defined element of S.

Associative (*Law 7*): If a and b are elements of $\mathfrak{F}$ and $x \in S$, then $(ab)x = a(bx)$.

Distributive (*Law 8*): If a and b are elements of $\mathfrak{F}$ and x and y are elements of S, then $(a + b)x = ax + bx$ and $a(x + y) = ax + ay$.

Identity (*Law 9*): If $x \in S$, then $1x = x$.

In these circumstances S is called a *vector space over* $\mathfrak{F}$ (also *linear space*). Its elements are called *vectors;* the elements of $\mathfrak{F}$ are called *scalars*.

LECTOR *That's a* definition?

AUCTOR *Yes, why?*

LECTOR *I think a good rule to follow is that a definition should be stated in 25 words or less.*

AUCTOR *It took you 19 just to say that.*

LECTOR *It took you 285 to state what a vector space is.*

AUCTOR *Well, the whole book is about vector spaces. You wouldn't ask Shakespeare to describe Hamlet in a couplet, would you?*

LECTOR *"But I have that within which passeth show;*
These but the trappings and the suits of woe."

In more abstract treatments of vector spaces $\mathfrak{F}$ is not restricted to be $\mathcal{R}$ or $\mathcal{C}$, but is any "field." We then refer to a "vector space over a field." For our purposes in this book, however, it is unnecessary to know precisely what a field is (although we shall eventually define it). Our scalars will always be real or complex numbers.

It is worth mentioning that any set with a binary operation (not necessarily addition) satisfying the closure, associative, identity, and inverse laws is called a *group* (a *commutative* or *abelian*† group if the commutative law is also satisfied). With this terminology we may shorten our original definition of vector space as follows: a *vector space* is an abelian group relative to addition, with a scalar multiplication satisfying the closure, associative, distributive, and identity laws. The theory of groups, while not important in this book, is a major topic in higher algebra.

Another matter to note is that we have used lowercase italic letters

† Named for Niels Abel (1802–1829).

$(x,y,z, \ldots)$ for the elements of a vector space, whereas in Chap. 1 we used boldface capital letters $(\mathbf{X},\mathbf{Y},\mathbf{Z}, \ldots)$. Use of the former is intended to indicate that we make no assumption about the elements; they may be n-tuples or they may be something quite unrelated (as we shall see). On the other hand, whenever we *are* dealing with n-tuples, we shall revert to the boldface capital letters of Chap. 1.

EXAMPLE 2.1

Suppose that $S = \{(x,y) \mid x \in \mathcal{C}, y \in \mathcal{C}\}$ and $\mathfrak{F} = \mathcal{R}$, with addition and scalar multiplication defined by

$$(x_1,y_1) + (x_2,y_2) = (x_1 + x_2, y_1 + y_2)$$
$$c(x,y) = (cx,cy)$$

It is easy to check out the properties listed in the above definition (see Prob. 33, Sec. 1.3); we conclude that S is a vector space over $\mathcal{R}$. To appreciate the force of the phrase "over $\mathcal{R}$," note that the vector space S is not the same as $\mathcal{C}^2$ (although its elements are the same). The agreement in Sec. 1.3 is that scalar multiplication in $\mathfrak{F}^n$ involves scalars from $\mathfrak{F}$; in the language of our present definition, $\mathcal{R}^n$ is a vector space over $\mathcal{R}$ and $\mathcal{C}^n$ is a vector space over $\mathcal{C}$.

To understand the distinction more clearly, recall from Sec. 1.3 that $\mathbf{E}_1 = (1,0)$ and $\mathbf{E}_2 = (0,1)$ constitute a basis for $\mathcal{C}^2$. But they don't even span S. The element $(1,i)$, for example, cannot be expressed as a linear combination of $\mathbf{E}_1$ and $\mathbf{E}_2$ except by writing

$$(1,i) = 1(1,0) + i(0,1)$$

This expression is legitimate in $\mathcal{C}^2$, but since $i \notin \mathcal{R}$, it is not allowed in S; scalars are confined to real numbers.

A basis for S is $\{\mathbf{U}_1,\mathbf{U}_2,\mathbf{U}_3,\mathbf{U}_4\}$, where

$$\mathbf{U}_1 = (1,0) \qquad \mathbf{U}_2 = (i,0) \qquad \mathbf{U}_3 = (0,1) \qquad \mathbf{U}_4 = (0,i)$$

for suppose (x,y) is any element of S, where $x = a + bi$ and $y = c + di$ $(a,b,c,d \in \mathcal{R})$. Then

$$(x,y) = c_1\mathbf{U}_1 + c_2\mathbf{U}_2 + c_3\mathbf{U}_3 + c_4\mathbf{U}_4$$

where the coefficients are given (uniquely) by

$$c_1 = a \qquad c_2 = b \qquad c_3 = c \qquad c_4 = d$$

(Confirm!) For example,

$$(1,i) = 1\mathbf{U}_1 + 0\mathbf{U}_2 + 0\mathbf{U}_3 + 1\mathbf{U}_4$$

This time the coefficients are real.

Anticipating the results of Sec. 2.4 (in which we'll prove that every basis for a vector space has the same number of elements, and define the "dimension'' to be this number), we may say that $\mathcal{C}^2$ is 2-dimensional and S is 4-dimensional. Yet the only distinction between them is that $\mathcal{C}^2$ is a vector space over $\mathcal{C}$ and S is a vector space over $\mathcal{R}$.

Many students who encounter the definition of vector space after studying $\mathfrak{F}^n$ persist in thinking that the elements of a vector space have to be n-tuples of numbers. To cure this tendency, we offer the following illustrations.

EXAMPLE 2.2

A 2×2 ''matrix'' is an array of four numbers in two rows and two columns, say

$$A = \begin{bmatrix} a & b \\ c & d \end{bmatrix}$$

(We'll have good reasons later for introducing matrices, but now our only purpose is to show you a certain vector space.) Matrices are added by summing corresponding entries:

$$\begin{bmatrix} a & b \\ c & d \end{bmatrix} + \begin{bmatrix} e & f \\ g & h \end{bmatrix} = \begin{bmatrix} a+e & b+f \\ c+g & d+h \end{bmatrix}$$

Multiplication of a matrix by a scalar is defined by

$$k\begin{bmatrix} a & b \\ c & d \end{bmatrix} = \begin{bmatrix} ka & kb \\ kc & kd \end{bmatrix}$$

It is tedious to check that the set S of 2×2 matrices (with entries in $\mathfrak{F}$) is a vector space over $\mathfrak{F}$, so we'll skip it. The point we are making here is that there exist vector spaces whose elements are not ordinarily referred to as ''vectors.''

Of course it may have occurred to you that if we label the elements differently, writing (a,b,c,d) instead of

$$\begin{bmatrix} a & b \\ c & d \end{bmatrix}$$

then they would become elements of $\mathfrak{F}^4$. This sort of thing will be exploited later (Sec. 3.4), when we talk about vector spaces that are in some sense identical to the n-spaces already discussed. Example 2.3, however, shows that this is not always the case.

EXAMPLE 2.3

Let S be the set of real-valued functions differentiable on the interval I, with addition and scalar multiplication defined by

$$f \in S,\ g \in S \Rightarrow f + g \in S \qquad \text{where } (f+g)(x) = f(x) + g(x) \text{ for all } x \in I$$
$$c \in \mathcal{R},\ f \in S \Rightarrow cf \in S \qquad \text{where } (cf)(x) = cf(x) \text{ for all } x \in I$$

Note that $f + g$ and cf are real-valued and differentiable because f and g are (and $c \in \mathcal{R}$); hence the closure properties $f + g \in S$ and $cf \in S$ are correct. The zero element of S is the function $0: I \to \mathcal{R}$ whose rule of correspondence is $0(x) = 0$ for all $x \in I$; this is real-valued and differentiable, too.† The other vector space properties are easy to verify.

Thus S is a vector space over $\mathcal{R}$, but its elements are hardly the kind of thing we would normally call "vectors." Moreover (unlike the vector space of matrices in Example 2.2), it is impossible to relabel these elements somehow so that they will correspond to the elements of a space of n-tuples. To see why, consider the functions $\phi_1, \phi_2, \phi_3, \ldots$ defined by

$$\phi_1(x) = 1 \qquad \phi_2(x) = x \qquad \phi_3(x) = x^2 \qquad \cdots$$

Let m be a positive integer. We claim that $\phi_1, \ldots, \phi_m$ are linearly independent elements of S,‡ for suppose that

$$c_1\phi_1 + \cdots + c_m\phi_m = 0 \qquad \text{(the zero function)}$$

Then for every $x \in I$ we have

$$c_1\phi_1(x) + \cdots + c_m\phi_m(x) = 0 \qquad \text{(the number 0)}$$

that is,

$$c_1 + c_2x + \cdots + c_mx^{m-1} = 0 \qquad \text{for all } x \in I$$

If one or more of the coefficients $c_1, \ldots, c_m$ were different from 0, we would have a polynomial of degree $\leq m - 1$ with infinitely many roots, which we know to be impossible.¶ Hence all the coefficients must be 0, as required by the definition of linear independence.

Now suppose that S were really just a space of n-tuples in disguise, so that it could be spanned by n elements. Taking $m > n$ in the above argument, we see that S has a subset containing more than n linearly independent elements, a situation which should offend your intuition (if you have understood what we said about n-space in Chap. 1) and which, in any case, we shall soon prove is impossible in *any* vector space spanned by n elements. (See Theorem 2.8.) Thus S is a space which is fundamentally different from the n-tuple spaces of Chap. 1. It cannot be assigned a finite dimension because it has subsets of arbitrarily large size consisting of linearly independent elements. Some people refer to it as an "infinite-dimensional" space.

Having convinced you (we trust) that vector spaces come in a variety of forms, and that therefore you cannot assume anything about them except what

† We presume that you are familiar with the notation $f: A \to B$, which indicates a function whose domain is A and whose values are in B, that is, for each $x \in A$, $f(x) \in B$.

‡ Linear independence has been defined only in n-space, but the meaning in any vector space is the same. See Sec. 2.2.

¶ Recall from elementary algebra that a polynomial has a factor $x - r$ corresponding to each root r. There cannot be more than $m - 1$ factors. Why?

is contained in the definition, we are ready to develop some basic ideas. The theorems in this section are trivial in appearance, representing nothing more than what you have known for years about numbers. Keep in mind, however, that the elements of a vector space are not necessarily familiar objects. The fact that they behave decently is reassuring, but the proofs are not so simple (as you will see) that we can call them obvious.

In the rest of this section we assume that S is a vector space over $\mathfrak{F}$, where $\mathfrak{F}$ is $\mathcal{R}$ or $\mathcal{C}$.

THEOREM 2.1

The zero element in S is unique.

Proof

Suppose that z is an element of S with the property that $x + z = z + x = x$ for every $x \in S$. Then (taking $x = 0$) it must be true that $0 + z = 0$. But $0 + z = z$; hence $z = 0$. ∎

See Prob. 3, Sec. 1.1, where you proved this theorem in $\mathcal{R}^2$. However, there you may have used the properties of ordered pairs of real numbers, which we cannot do in the present context because we don't know what the elements of S are.

THEOREM 2.2

Each element of S has exactly one inverse.

Proof

Given $x \in S$, the inverse law of addition says that there is an element $y \in S$ such that $x + y = y + x = 0$. Suppose that z is *any* element with this property; the objective (which we leave for the problems) is to prove that $z = y$.

ADDITIVE INVERSE AND SUBTRACTION

If $x \in S$, the unique inverse guaranteed by Theorem 2.2 is denoted by $-x$; that is, if $x + y = y + x = 0$, then $y = -x$.

If x and y are elements of S, then

$$x - y = x + (-y)^\dagger$$

† This expression defines subtraction in terms of addition of an inverse, as in elementary algebra. See Prob. 7, Sec. 1.1, for the same idea in $\mathcal{R}^2$.

The next theorem identifies the additive inverse of x with the scalar multiple $-1x$ (as in Prob. 4, Sec. 1.1).

THEOREM 2.3

If $x \in S$, then $-1x = -x$.

Proof

The idea is to show that $x + (-1x) = 0$; for if this is true, then Theorem 2.2 says that $-1x$ is the inverse of x, that is, $-1x = -x$:

$$\begin{aligned} x + (-1x) &= 1x + (-1)x \\ &= [1 + (-1)]x \\ &= 0x \\ &= 0 \qquad \text{(by Theorem 2.4)} \ ▮ \end{aligned}$$

LECTOR *Whoa.*
AUCTOR *I know I haven't proved Theorem 2.4.*
LECTOR *You haven't even stated it.*
AUCTOR *Well, look ahead.*
LECTOR *What if you prove it using Theorem 2.3?*
AUCTOR *I'll turn in my union card.*

THEOREM 2.4

If $c \in \mathfrak{F}$ and $x \in S$, then $cx = 0 \Leftrightarrow c = 0$ or $x = 0$.

Proof

To show that $c = 0$ or $x = 0 \Rightarrow cx = 0$, we write

$$\begin{aligned} c = 0 \quad &\Rightarrow \quad x + cx = x + 0x = 1x + 0x = (1 + 0)x = 1x = x \\ &\Rightarrow \quad cx = 0 \qquad \text{(Theorem 2.1)} \end{aligned}$$

and

$$\begin{aligned} x = 0 \quad &\Rightarrow \quad cx + c0 = c0 + c0 = c(0 + 0) = c0 \\ &\Rightarrow \quad cx = 0 \qquad \text{(again by Theorem 2.1)} \end{aligned}$$

Conversely, suppose that $cx = 0$. If $c = 0$, there is nothing to prove; so assume that $c \neq 0$. Then

$$\begin{aligned} cx = 0 \quad &\Rightarrow \quad c^{-1}(cx) = c^{-1}0 \\ &\Rightarrow \quad (c^{-1}c)x = 0 \qquad (c^{-1}0 = 0, \text{ by the part of the theorem just proved}) \\ &\Rightarrow \quad 1x = 0 \\ &\Rightarrow \quad x = 0 \ ▮ \end{aligned}$$

Note that we did not use Theorem 2.3 in this argument!

Problems

1. Explain why $\mathcal{C}$ is a vector space over $\mathcal{R}$. What is a basis for this space?[†]
2. In the definition of vector space, suppose that S is taken to be $\mathfrak{F}$, that is, consider $\mathfrak{F}$ as a vector space over itself.
   a. Why is this legitimate?
   b. What is a basis for this space?
   c. Explain why $\mathfrak{F}$ as a vector space over $\mathfrak{F}$ is the same as $\mathfrak{F}^1$ (defined in Sec. 1.3).
   d. If $\mathfrak{F} = \mathcal{C}$, we are saying that $\mathcal{C}$ is a vector space over $\mathcal{C}$ (and that this is the same as $\mathcal{C}^1$). How does this differ from $\mathcal{C}$ as a vector space over $\mathcal{R}$ (Prob. 1)?
3. Convince yourself that the set S of 2×2 matrices with entries in $\mathfrak{F}$ (Example 2.2) is a vector space over $\mathfrak{F}$. What is the zero element of S? Given a 2×2 matrix, say

$$\begin{bmatrix} 2 & -1 \\ 0 & 5 \end{bmatrix}$$

   what is its additive inverse?
4. Name a basis for the vector space of 2×2 matrices.
5. Let T be the set of 2×2 matrices of the form

$$\begin{bmatrix} a & b \\ -b & a \end{bmatrix}$$

   where a and b are elements of $\mathfrak{F}$. Prove that T is a vector space over $\mathfrak{F}$. (Thus it is a ''subspace'' of the space in Example 2.2.) Name a basis for this space.
6. Finish the proof of Theorem 2.2 by showing that if $x + y = 0$ and $x + z = 0$, then $z = y$. *Hint:* Add y to each side of $x + z = 0$.
7. Go through the proofs of Theorems 2.1 through 2.4 and see if you can identify the places where the various properties of a vector space are used. (Note that every property is used, at least implicitly. The point is that there are no superfluous axioms in the definition of a vector space, at least not if these proofs are given. The definition has been carefully pruned by mathematicians, so that it contains the bare essentials.)
8. Convince yourself that the set of integers $\{0, \pm 1, \pm 2, \ldots\}$ is an abelian group relative to addition. What is the identity element? Given an integer, what is its inverse? Why doesn't the set of nonnegative integers comprise a group relative to addition?

† ''Basis'' for an abstract vector space has not yet been defined, but the idea is the same as in Chap. 1.

9. Convince yourself that the set of nonzero rational numbers is an abelian group relative to multiplication. What is the identity element? Given a nonzero rational number, what is its inverse? Why isn't the set of positive integers a group relative to multiplication?

10. Explain why the set of complex numbers $\{1, -1, i, -i\}$ is an abelian group relative to multiplication. Find the inverse of each element.

2.2 Linear Independence in a Vector Space

The basic ideas in this section are repetitious, having been introduced in one form or another in Chap. 1. However, we want to broaden the definitions slightly and apply them not only in n-space but in other vector spaces as well. Throughout this section, S is understood to be a vector space over $\mathfrak{F}$, where $\mathfrak{F}$ is $\mathfrak{R}$ or $\mathfrak{C}$.

SPANNING

A *linear combination* of elements $u_1, \ldots, u_m$ of S is an element of the form

$$c_1u_1 + \cdots + c_mu_m = \sum_{j=1}^{m} c_ju_j$$

where the c_j are scalars.

If every element of S can be expressed as a linear combination of $u_1, \ldots, u_m$, we say that $u_1, \ldots, u_m$ *span* S and that S is *finite-dimensional*. If the u_j are distinct, we also call $\{u_1, \ldots, u_m\}$ a *spanning set* for S.†

Note that we have not yet defined "dimension"; that will come in Sec. 2.4. Also note that the term "finite-dimensional" is not empty of content, as it would be if every space could be spanned. In Example 2.3 we showed you a vector space which is not finite-dimensional.‡

† We require $u_1, \ldots, u_m$ to be distinct in this last definition because no duplication is allowed in set notation.

‡ All we actually showed you is that the vector space of real-valued functions differentiable on I has subsets of arbitrarily large size which consist of linearly independent elements. In Sec. 2.4 we'll prove that this cannot happen in a finite-dimensional space.

BASIS

If $u_1, \ldots, u_m$ span S and the expression of each element of S as a linear combination of $u_1, \ldots, u_m$ is unique, we call $\{u_1, \ldots, u_m\}$ a *basis* for S.

In Sec. 1.3 we said that a basis is a set of linearly independent elements spanning the space. The definitions are equivalent, as we'll show in Theorem 2.5.

LINEAR INDEPENDENCE AND DEPENDENCE

If $u_1, \ldots, u_m$ are elements of S with the property that

$$c_1u_1 + \cdots + c_mu_m = 0 \quad \Rightarrow \quad c_1 = \cdots = c_m = 0$$

we say that they are *linearly independent*. But if there are scalars $c_1, \ldots, c_m$, not all zero, such that

$$c_1u_1 + \cdots + c_mu_m = 0$$

then $u_1, \ldots, u_m$ are called *linearly dependent*.

If the u_j are distinct, the *set* $\{u_1, \ldots, u_m\}$ is called linearly independent (or dependent) when its elements are.

As we noted before, the reason we require $u_1, \ldots, u_m$ to be distinct in the last part of this definition is that redundancy is not allowed in set notation; it is understood that each element is listed only once. In the case of linear independence, there is no problem, since $u_1, \ldots, u_m$ cannot be independent if any two of the u_j are the same. (Why?) But linear dependence of $u_1, \ldots, u_m$ and linear dependence of the set consisting of the distinct elements in the list are not equivalent. (See Prob. 14.)

THEOREM 2.5

The elements $u_1, \ldots, u_m$ of S are linearly independent if and only if no element of S can be written as a linear combination of $u_1, \ldots, u_m$ in more than one way.

Proof

If no element of S can be written as a linear combination of $u_1, \ldots, u_m$ in more than one way, then, in particular, 0 cannot be. The only possibility is $0 = 0u_1 + \cdots + 0u_m$, which means that

$$c_1u_1 + \cdots + c_mu_m = 0 \quad \Rightarrow \quad c_1 = \cdots = c_m = 0$$

that is, $u_1, \ldots, u_m$ are linearly independent. Conversely, suppose that $u_1, \ldots, u_m$ are linearly independent. Then

$$x = \sum_{j=1}^{m} a_j u_j \text{ and } x = \sum_{j=1}^{m} b_j u_j \quad \Rightarrow \quad \sum_{j=1}^{m} (a_j - b_j) u_j = 0$$
$$\Rightarrow \quad a_j - b_j = 0 \qquad j = 1, \ldots, m$$
$$\Rightarrow \quad a_j = b_j \qquad j = 1, \ldots, m$$

Hence no element of S can be written as a linear combination of $u_1, \ldots, u_m$ in more than one way. ▮

Note that Theorem 2.5 says nothing about $u_1, \ldots, u_m$ spanning S; there may be elements of S which cannot be written in terms of $u_1, \ldots, u_m$ at all. (As we'll see, this is certainly the case if S is infinite-dimensional or if it has a finite dimension larger than m.) What the theorem says is that linear independence and uniqueness of expression amount to the same thing. When either of these equivalent conditions is combined with the requirement that the vectors in question span S, we are dealing with a basis, as stated in Corollary 2.5a.

COROLLARY 2.5a

$\{u_1, \ldots, u_m\}$ is a basis for S if and only if it is a linearly independent spanning set.

Many writers use this corollary to define basis (as in fact we already did in Sec. 1.3). It is worth remembering.

THEOREM 2.6

The elements $u_1, \ldots, u_m$ of S are linearly dependent if and only if one of them is a linear combination of the others.

Proof

If one of the u_j is a linear combination of the others, we may relabel (if necessary) so that this element is u_m. Then

$$u_m = \sum_{j=1}^{m-1} c_j u_j$$

from which we have

$$c_1 u_1 + \cdots + c_{m-1} u_{m-1} + (-1) u_m = 0$$

This is a linear dependence relation because at least one of the coefficients (namely, -1) is different from zero.

Conversely, suppose that $u_1, \ldots, u_m$ are linearly dependent. Then there are scalars $c_1, \ldots, c_m$, not all zero, such that

$$c_1u_1 + \cdots + c_mu_m = 0$$

By relabeling (if necessary) we can arrange it so that c_m is nonzero. Then, solving for u_m, we have

$$u_m = a_1u_1 + \cdots + a_{m-1}u_{m-1} \qquad \text{where } a_j = -c_j/c_m,\ j = 1, \ldots, m-1$$

Hence one of the u_j (whichever one we relabeled to be u_m) is a linear combination of the others. ∎

LECTOR *What if $m = 1$?*

AUCTOR [*inaudible*]

LECTOR *You can't write $u_m = \sum_{j=1}^{m-1} c_ju_j$ if there's only one element.*

AUCTOR *I admit I was assuming that $m > 1$. But if $m = 1$, we have linear dependence $\Leftrightarrow c_1u_1 = 0$ for some nonzero $c_1 \Leftrightarrow u_1 = 0$.*

LECTOR *The theorem says that "one of them is a linear combination of the others."*

AUCTOR *It's vacuous; there aren't any "others" to worry about.*

LECTOR *So you admit it's false.*

AUCTOR *It doesn't say anything. How can it be false?*

LECTOR *How can it be true?*

AUCTOR *Let's don't argue. Put in $m > 1$ if it makes you feel better.*

COROLLARY 2.6a

Two elements of S are linearly dependent if and only if one is a scalar multiple of the other.

COROLLARY 2.6b

Let $u_1, \ldots, u_m$ be elements of S. If one of them is 0, or if any two are the same, they are linearly dependent.

Theorem 2.6 should go a long way toward explaining the terminology of linear independence and dependence (if you have not already figured it out). Linear independence of $u_1, \ldots, u_m$ means that any dependence relation

$$c_1u_1 + \cdots + c_mu_m = 0$$

is only apparent; the coefficients must all be zero, and hence it is impossible to solve for any u_j in terms of the others. Linear dependence, on the other hand, is equivalent to this possibility.

As a special case of these ideas, look at the element u_1 by itself (assume that u_1 is nonzero). The idea of "linear combination equal to 0" now reduces to the equation $c_1u_1 = 0$, which implies that $c_1 = 0$ because $u_1 \neq 0$. Hence, according to the definition of linear independence of $u_1, \ldots, u_m$ (if we take

$m = 1$), $\{u_1\}$ is a linearly independent subset of S. On the other hand, $\{0\}$ is a linearly dependent subset (as pointed out in the above conversation). These observations sound strange, but that is because we are dealing with a special case.

We end this section by remarking that a vector space consisting of a zero element alone is called a *zero space*. For example, the subset of $\mathcal{R}^2$ whose only element is $\mathbf{0} = (0,0)$ is a zero space. Such spaces are of no great interest, but they cause trouble. Since $\{0\}$ is always a linearly dependent subset of its parent space, the concept of linear independence is meaningless in a zero space. Hence the zero space cannot have a basis.† Nevertheless it is finite-dimensional, because it can be spanned—by the zero element!

Problems

11. Explain why a nonzero vector space must have a linearly independent subset.

12. Why is $\{0\}$ a linearly dependent subset of its parent space?

13. In our definition of basis we did not specify that $u_1, \ldots, u_m$ must be distinct, even though we used the notation $\{u_1, \ldots, u_m\}$. Explain why this is not an oversight.

14. Suppose that $u_1 = u_2$ in the vector space S.
a. Explain why u_1 and u_2 are linearly dependent.
b. Why is it nonsense to say that $\{u_1,u_2\}$ is linearly dependent?

The point of this problem is that linear dependence of the *elements* $u_1, \ldots, u_m$ and of the *set* $\{u_1, \ldots, u_m\}$ are equivalent ideas only when the u_j are distinct.

15. Show that any (nonempty) subset of a linearly independent set is linearly independent.

16. Prove Corollaries 2.6a and 2.6b. *Note:* You have already done these proofs in Chap. 1 (Prob. 22, Sec. 1.2, and Prob. 41, Sec. 1.3). This time, however, show why they are consequences of Theorem 2.6.

‡2.3 Linear Independence of Functions

Linear independence is an important concept in connection with vector spaces whose elements are functions. The following examples will indicate some of its ramifications.

† In Sec. 2.4 we'll define the dimension of a zero space to be 0. Some writers, in order to fit this definition into the idea that dimension is the number of elements in a basis, adopt the convention that the empty set is a basis for a zero space. This convention is convenient for some purposes, but we prefer to avoid it.

‡ This section is not essential for continuity, except for an occasional reference to the examples. If you are short on time, give it a quick reading and move on to the next section.

EXAMPLE 2.4

Let $S(I)$ be the vector space of real-valued functions differentiable on the interval I.† If $I = (-\pi,\pi)$, the functions ϕ_1 and ϕ_2 defined by $\phi_1(x) = \cos x$ and $\phi_2(x) = \sin x$ are linearly independent elements of $S(I)$ for if $c_1\phi_1 + c_2\phi_2 = 0$ (the zero function), then for all $x \in I$ we have

$$c_1 \cos x + c_2 \sin x = 0 \qquad \text{(the scalar zero)}$$

In particular this condition holds when $x = 0$, which means that $c_1(1) + c_2(0) = 0$ and hence that $c_1 = 0$. The condition also holds when $x = \pi/2$, which yields $c_2 = 0$. Thus

$$c_1\phi_1 + c_2\phi_2 = 0 \quad \Rightarrow \quad c_1 = c_2 = 0$$

as required by the definition of linear independence.

EXAMPLE 2.5

The argument in Example 2.4 is awkward. Moreover, it breaks down if I does not happen to contain the numbers 0 and $\pi/2$. We could use other numbers instead, but in the absence of more explicit information about I, we need a more subtle approach.

Observe that if $c_1\phi_1 + c_2\phi_2 = 0$, then $c_1\phi_1' + c_2\phi_2' = 0$ (where the primes, as usual, mean differentiation). Choosing $x \in I$ (but not worrying about its value), we can write

$$\begin{aligned}(\cos x)c_1 + (\sin x)c_2 &= 0\\ (-\sin x)c_1 + (\cos x)c_2 &= 0\end{aligned}$$

which is a system of two linear equations in c_1 and c_2. The determinant of coefficients (Prob. 28, Sec. 1.2) is

$$\begin{vmatrix} \cos x & \sin x \\ -\sin x & \cos x \end{vmatrix} = \cos^2 x + \sin^2 x = 1$$

Since this value is not zero, the system has only one solution—namely, $c_1 = c_2 = 0$. Thus $c_1\phi_1 + c_2\phi_2 = 0 \Rightarrow c_1 = c_2 = 0$.

The mathematician who discovered this trick was Jozef Wronski (1778–1853); for that reason the function

$$W = \begin{vmatrix} \phi_1 & \phi_2 \\ \phi_1' & \phi_2' \end{vmatrix}$$

is called the "Wronskian" of ϕ_1 *and* ϕ_2. The above argument can be generalized to show that if ϕ_1 and ϕ_2 are any elements of $S(I)$ whose Wronskian is not the zero function, then ϕ_1 and ϕ_2 are linearly independent.

†We use the notation $S(I)$ instead of S (Example 2.3) because the class of functions we are talking about depends on I. For example, if $I = (-\infty,\infty)$ and $J = (0,\infty)$, the function $\phi(x) = \sqrt{x}$ is an element of $S(J)$ but not of $S(I)$.

LECTOR *I can prove that in two steps.*

AUCTOR *I yield to the distinguished Lector for his comments.*

LECTOR *If ϕ_1 and ϕ_2 are linearly dependent, one is a scalar multiple of the other.*

AUCTOR *That is correct.*

LECTOR *Then their Wronskian is the zero function. As they say in the trade,* ∎.

AUCTOR *Let the record show that no exception is taken to this line of reasoning.*

LECTOR *Let it also show that prolonged applause from the gallery was ended only by the Chairman's calling a recess.*

EXAMPLE 2.6

Let $I = (-\infty,\infty)$ and suppose that $\phi_1(x) = x^2$ and $\phi_2(x) = x\,|x|$.[†] The Wronskian of ϕ_1 and ϕ_2 is defined by

$$W(x) = \begin{vmatrix} x^2 & x|x| \\ 2x & 2|x| \end{vmatrix} = 2x^2|x| - 2x^2|x| = 0$$

which means that W is the zero function. Does it follow that ϕ_1 and ϕ_2 are linearly dependent? The answer is no, as we'll ask you to show in the problems. The explanation, of course, is that the theorem we just quoted,

$$W \neq 0 \quad \Rightarrow \quad \phi_1 \text{ and } \phi_2 \text{ are linearly independent}$$

cannot be read backward; its converse is false.

EXAMPLE 2.7

Change I to $J = (0,\infty)$ in Example 2.6. Then $\phi_1(x) = x^2$ and $\phi_2(x) = x|x|$ are identical, hence linearly dependent in $S(J)$. This seeming discrepancy does not contradict Example 2.6; recall from calculus that a function is not properly described by a rule alone, but requires a domain. The functions ϕ_1 and ϕ_2 in Example 2.6 are not the same as ϕ_1 and ϕ_2 here! (See Figs. 2.1 and 2.2.)

From our present point of view, we may say this differently by observing that the vector spaces in Examples 2.6 and 2.7 are not the same. There is no reason to be surprised by linear independence of ϕ_1 and ϕ_2 in $S(I)$ and linear dependence of (different) functions ϕ_1 and ϕ_2 in $S(J)$. It just happens that ϕ_1 and ϕ_2 are defined by the same formulas in both cases.

Having said this, we confess it is clumsy. Books on differential equations (in which linear independence is usually discussed without reference to vector spaces) describe the situation more simply by calling x^2 and $x|x|$ linearly independent "on $(-\infty,\infty)$" and linearly dependent "on $(0,\infty)$." This convention is all right to use if it does not reinforce the idea that functions are defined merely by formulas. We'll adopt the terminology when it is convenient, trusting you to remember what it means.

[†] Note that ϕ_2 is differentiable (even at $x = 0$) although the absolute value function is not. To see why, use the definition of derivative to compute $\phi_2'(0)$.

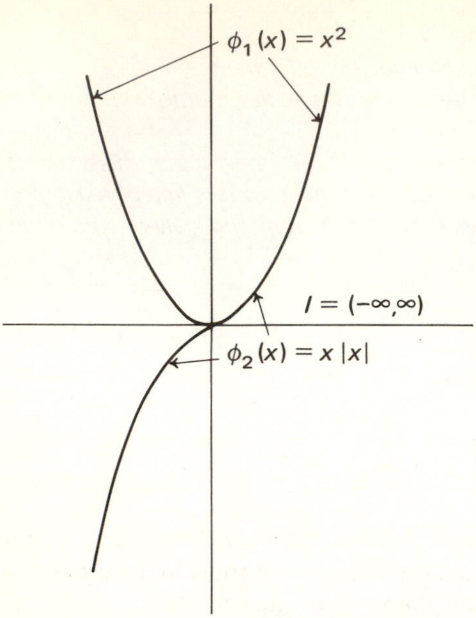

Fig. 2.1 ϕ_1 **and** ϕ_2 **as elements of** $S(I)$

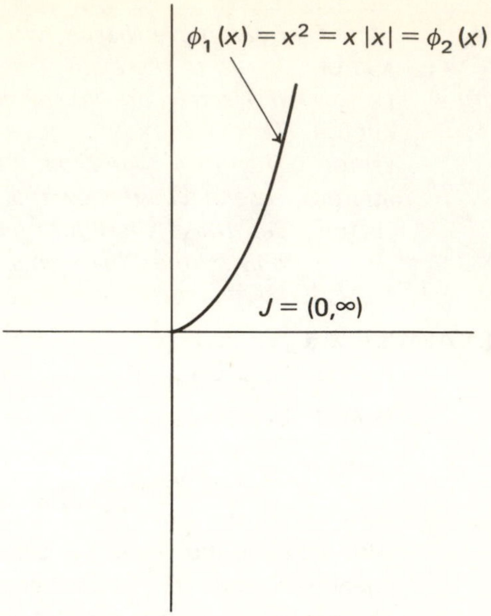

Fig. 2.2 ϕ_1 **and** ϕ_2 **as elements of** $S(J)$

EXAMPLE 2.8

Let S be the set of solutions on $I = (-\infty,\infty)$ of the differential equation $y'' + y = 0$.† Since the zero function is a solution, and since S is closed relative to addition and scalar multiplication (why?), it is a vector space over $\mathcal{R}$. (The other vector space properties are trivial, as you can see by checking them out.)

The functions $\phi_1(x) = \cos x$ and $\phi_2(x) = \sin x$ are clearly elements of S; hence so are all linear combinations of the form $\phi(x) = c_1 \cos x + c_2 \sin x$. (Why?) But one can show more than this. The linear independence of ϕ_1 and ϕ_2 on I (Example 2.5) implies that *every* solution is a linear combination of these two.‡ In other words, $\{\phi_1,\phi_2\}$ is a basis for S. Anticipating Sec. 2.4, we conclude that S is 2-dimensional.

More generally, one learns in differential equations that the set of solutions of an nth-order linear homogeneous equation is an n-dimensional vector space. This elegant result is hard to communicate without the language of vector spaces; it is one of the reasons linear algebra is worth studying.

† A "solution on I" is a real-valued function ϕ (with domain I) having the property that $\phi'' + \phi = 0$. If you have studied differential equations, you will recognize $y'' + y = 0$ as a second-order linear homogeneous equation with constant coefficients, whose "general solution" is $\phi(x) = c_1 \cos x + c_2 \sin x$.

‡ This claim is not obvious! It is usually proved in books on differential equations by appealing to a theorem guaranteeing the uniqueness of solutions satisfying certain initial conditions. See Philip Gillett, *Linear Mathematics,* Boston: Prindle, Weber and Schmidt, 1970.

Problems

17. Let S be the vector space of real-valued functions differentiable on $I = (-\infty, \infty)$.

a. Show that the functions ϕ_1, ϕ_2, ϕ_3, defined by $\phi_1(x) = 1$, $\phi_2(x) = x$, $\phi_3(x) = x^2$, are linearly independent elements of S. *Note:* We have already done this in Example 2.6. See if you can give a different argument.

b. The *hyperbolic cosine* and *hyperbolic sine* are defined in calculus by

$$\cosh x = \tfrac{1}{2}(e^x + e^{-x}) \qquad \text{and} \qquad \sinh x = \tfrac{1}{2}(e^x - e^{-x})$$

Explain why the functions

$$\phi_1(x) = e^x \qquad \phi_2(x) = e^{-x} \qquad \phi_3(x) = \cosh x \qquad \phi_4(x) = \sinh x$$

are linearly dependent in S. Name scalars c_1, c_2, c_3, c_4, not all zero, such that

$$c_1\phi_1 + c_2\phi_2 + c_3\phi_3 + c_4\phi_4 = 0$$

18. Let ϕ_1 and ϕ_2 be elements of $S(I)$, the vector space of real-valued functions differentiable on I.

a. Generalize the argument in Example 2.5 to show that if the Wronskian of ϕ_1 and ϕ_2 is not the zero function, then ϕ_1 and ϕ_2 are linearly independent.

b. If ϕ_1 and ϕ_2 are linearly dependent, we know that one is a scalar multiple of the other. Compute the Wronskian in this case, showing that it is the zero function.

c. Explain why part *b* constitutes another proof of part *a*.

19. Use the Wronskian to prove that if r_1 and r_2 are distinct real numbers, then the functions $\phi_1(x) = e^{r_1 x}$ and $\phi_2(x) = e^{r_2 x}$ are linearly independent on I (where I is any interval).

20. The Wronskian of three functions ϕ_1, ϕ_2, ϕ_3 (twice differentiable on I) is[†]

$$W = \begin{vmatrix} \phi_1 & \phi_2 & \phi_3 \\ \phi_1' & \phi_2' & \phi_3' \\ \phi_1'' & \phi_2'' & \phi_3'' \end{vmatrix}$$

Explain why it is reasonable to expect that

$$W \neq 0 \quad \Rightarrow \quad \phi_1, \phi_2, \phi_3 \text{ are linearly independent on } I$$

Note: You need not prove it! We merely want you to determine which theorem about systems of linear equations is involved.

[†] If you haven't learned what a third-order determinant is, skip Probs. 20 and 21.

21. Assuming that the theorem in Prob. 20 is true, use the Wronskian to confirm linear independence of the following functions on $I = (-\infty,\infty)$.

a. $\phi_1(x) = 1$, $\phi_2(x) = x$, $\phi_3(x) = x^2$

b. $\phi_1(x) = e^{r_1 x}$, $\phi_2(x) = e^{r_2 x}$, $\phi_3(x) = e^{r_3 x}$, where r_1, r_2, r_3 are distinct real numbers. *Hint:* It is sufficient to show that $W(0) \neq 0$. (Why?) To do this, prove that $W(0) = (r_1 - r_2)(r_2 - r_3)(r_3 - r_1)$.†

22. Let S be the set of solutions on $(-\infty,\infty)$ of the differential equation $y'' + y = 0$ (Example 2.8).

a. Explain why the zero function is an element of S.

b. Prove that S is closed relative to addition and scalar multiplication.

c. We stated in the text that $\phi_1(x) = \cos x$ and $\phi_2(x) = \sin x$ constitute a basis for S. Assuming that this is true, find the solution ϕ of $y'' + y = 0$ that satisfies the initial conditions $\phi(0) = 2$ and $\phi'(0) = -1$.

23. Let S be the vector space of solutions on $(-\infty,\infty)$ of the differential equation $y'' - y = 0$.

a. Confirm that cosh and sinh are linearly independent elements of S. *Hint:* Recall from calculus that $\cosh' = \sinh$ and $\sinh' = \cosh$.

b. It can be shown that any two linearly independent solutions constitute a basis for S. Use this fact to find (in terms of $\cosh x$ and $\sinh x$) the solution ϕ satisfying the initial conditions $\phi(0) = 1$ and $\phi'(0) = 3$.

c. Confirm that the solution found in part *b* can be written in the form

$$\phi(x) = 2e^{x} - e^{-x}$$

d. Confirm that e^x and e^{-x} are linearly independent solutions. Use them in part *b* instead of $\cosh x$ and $\sinh x$, and come up with the same answer as in part *c*.

24. Confirm that the function ϕ defined by $\phi(x) = x^2 - 4$ is a solution of $y'' + y = x^2 - 2$ on $I = (-\infty,\infty)$, but that 2ϕ is not a solution. Why does this fact show that the set of solutions on I is not a vector space? (Note that the given equation is nonhomogeneous; that is, the function on the right-hand side is not zero.)

2.4 Dimension of a Vector Space

The dimension of a vector space, as we have suggested a number of times, is the number of elements in a basis. But except in the case of $\mathcal{R}^2$ (Sec. 1.2) we have not given you any good reason to believe this definition makes sense. We need to prove two major theorems, one to tell us which vector spaces have a basis and another to guarantee that every basis for a given space has the same number of elements.

† $W(0)$ is the 3×3 case of a well-known expression called *Vandermonde's* (1735–1796) *determinant*.

Each of these theorems takes a while to prove, yet the main ideas of this section (and the next) are simple. A basis for a vector space (assuming that one exists) is a linearly independent spanning set. To construct such a set, we may proceed in either of two ways. One approach is to start with a spanning set. If it is linearly independent, then nothing need be done, for it is already a basis. If it is not linearly independent, then it is too large; the problem is to weed out its superfluous elements—those which are linear combinations of others and whose omission, therefore, does not affect the spanning power of the set. In this way we reduce the set to a basis.

The other approach is to start with a linearly independent set. If its elements span the space, we are done. If they do not, the set is too small; the problem is to build it up to a basis by finding enough new elements (not dependent on the others) to span the space.

We are going to adopt the second of these methods in our first theorem. Before stating and proving the theorem, however, we give an example of how the idea works.

EXAMPLE 2.9

Let $\mathbf{U}_1 = (0,0,0,2,0)$, $\mathbf{U}_2 = (1,1,0,0,0)$, $\mathbf{U}_3 = (1,0,0,0,0)$. Then $\{\mathbf{U}_1,\mathbf{U}_2,\mathbf{U}_3\}$ is a linearly independent subset of $\mathcal{R}^5$. (Confirm!) We propose to build it up to a basis by using appropriate elements of the spanning set $\{\mathbf{E}_1, \ldots, \mathbf{E}_5\}$, the standard basis for $\mathcal{R}^5$.† (See Sec. 1.3.) To illustrate the notation in the proof of Theorem 2.7, we call this spanning set $\{\mathbf{V}_1, \ldots, \mathbf{V}_5\}$ instead.

The first thing that might occur to us (if we were stone blind, that is) is to enlarge the set $\{\mathbf{U}_1,\mathbf{U}_2,\mathbf{U}_3\}$ by including $\mathbf{V}_1$. However, $\mathbf{V}_1 = (1,0,0,0,0) = \mathbf{U}_3$ and since its inclusion duplicates an element already present, there is no point in considering it. The situation is similar with $\mathbf{V}_2 = (0,1,0,0,0) = \mathbf{U}_2 - \mathbf{U}_3$; the set $\{\mathbf{U}_1,\mathbf{U}_2,\mathbf{U}_3,\mathbf{V}_2\}$ is also of no help.

The first $\mathbf{V}_j$ which is not a linear combination of $\mathbf{U}_1$, $\mathbf{U}_2$, $\mathbf{U}_3$ is $\mathbf{V}_3$, as you can check. Relabel the $\mathbf{V}_j$ by writing

$$\mathbf{V}_1 = \mathbf{E}_3 \qquad \mathbf{V}_2 = \mathbf{E}_4 \qquad \mathbf{V}_3 = \mathbf{E}_5 \qquad \mathbf{V}_4 = \mathbf{E}_1 \qquad \mathbf{V}_5 = \mathbf{E}_2{}^{\ddagger}$$

Then, using the new labels, we have constructed the linearly independent set $\{\mathbf{U}_1,\mathbf{U}_2,\mathbf{U}_3,\mathbf{V}_1\}$. If this were a basis, we would stop. Since it isn't, we look at the remaining elements of our (relabeled) spanning set, $\mathbf{V}_2$, $\mathbf{V}_3$, $\mathbf{V}_4$, $\mathbf{V}_5$, and try to find one that is not a linear combination of $\mathbf{U}_1$, $\mathbf{U}_2$, $\mathbf{U}_3$, $\mathbf{V}_1$. Since $\mathbf{V}_2 = \mathbf{E}_4 = \frac{1}{2}\mathbf{U}_1$, it does not qualify; the first element that does is $\mathbf{V}_3 = \mathbf{E}_5$. (Confirm this!)

† It may seem silly to worry about building a basis when we already have one. The point of the example, however, is not to demonstrate the existence of a basis, but to prepare the ground for Theorem 2.7.

‡ Note that we relegated the useless elements $\mathbf{E}_1$ and $\mathbf{E}_2$ to the end of the list. The reason for relabeling is to illustrate what goes on in the proof of Theorem 2.7, where we try to avoid becoming bogged down in notation.

Now relabel the $\mathbf{V}_j$ again by writing

$$\mathbf{V}_2 = \mathbf{E}_5 \qquad \mathbf{V}_3 = \mathbf{E}_4 \qquad \mathbf{V}_4 = \mathbf{E}_1 \qquad \mathbf{V}_5 = \mathbf{E}_2$$

(leaving $\mathbf{V}_1 = \mathbf{E}_3$ as is). Then $\{\mathbf{U}_1,\mathbf{U}_2,\mathbf{U}_3,\mathbf{V}_1,\mathbf{V}_2\}$ is linearly independent. It is, in fact, a basis, as you can check, and so we are done.

Note that the idea is straightforward. All we do to build up $\{\mathbf{U}_1,\mathbf{U}_2,\mathbf{U}_3\}$ to a basis is to include $\mathbf{E}_3$ and $\mathbf{E}_5$ from the spanning set $\{\mathbf{E}_1, \ldots, \mathbf{E}_5\}$. In general (when the given linearly independent set may have, say, 2 elements and the spanning set, 100), we need a systematic procedure that can be described in reasonable notation. As you work through the following theorem, refer back to this example to see what we mean.

THEOREM 2.7

A linearly independent subset of a nonzero finite-dimensional space S either is a basis for S or can be built up to a basis.

Proof

Let $\{u_1, \ldots, u_m\}$ be the linearly independent subset. If it is already a basis, there is nothing to prove; so assume that it isn't. By Corollary 2.5a, its failure to be a basis is due to the fact that $u_1, \ldots, u_m$ do not span S. Yet S can be spanned, say by the set $\{v_1, \ldots, v_n\}$, because it is finite-dimensional. We propose to use appropriate elements of this spanning set to build up $\{u_1, \ldots, u_m\}$ to a basis.

Our first step is to go through the list $v_1, \ldots, v_n$ until we come to the first v_j which is not a linear combination of $u_1, \ldots, u_m$. (This search is bound to succeed, because if *every* v_j is a linear combination of $u_1, \ldots, u_m$, then the u_i span S, which contradicts our opening assumption. See the problems for elucidation of this point.)

Relabel the elements $v_1, \ldots, v_n$ (if necessary) so that the v_j found above is called v_1. Then the elements $u_1, \ldots, u_m, v_1$ are linearly independent; for suppose that

$$a_1u_1 + \cdots + a_mu_m + b_1v_1 = 0$$

If $b_1 \neq 0$, we can solve for v_1 as a linear combination of $u_1, \ldots, u_m$, contradicting the definition of v_1. Hence $b_1 = 0$. But then we have

$$a_1u_1 + \cdots + a_mu_m = 0$$

which implies that $a_1 = \cdots = a_m = 0$ because $u_1, \ldots, u_m$ are linearly independent.

Thus we have built up $\{u_1, \ldots, u_m\}$ to the linearly independent set $\{u_1, \ldots, u_m, v_1\}$. We are finished if this set is a basis; if it is not, its elements do not span S, and we repeat our search-and-screening act.

To be convinced that this process may be continued to a successful conclusion, we need only look at the general step. Suppose that we have arrived after p steps (each one involving a possible relabeling of elements) at the linearly independent set $\{u_1, \ldots, u_m, v_1, \ldots, v_p\}$. If this set is a basis, we are done. If it is not a basis, then $p < n$, for if $p = n$ our set spans S because $v_1, \ldots, v_n$ alone do. Look through the list $v_{p+1}, \ldots, v_n$ until the first v_j is reached which is not a linear combination of $u_1, \ldots, u_m, v_1, \ldots, v_p$. (Why is this search bound to succeed?) Call this element v_{p+1} (by relabeling, if necessary). Then $u_1, \ldots, u_m, v_1, \ldots, v_{p+1}$ are linearly independent, because

$$
\begin{aligned}
a_1u_1 + \cdots + a_mu_m + b_1v_1 + \cdots + b_{p+1}v_{p+1} = 0 \\
&\Rightarrow\ b_{p+1} = 0 \qquad \text{(Why?)} \\
&\Rightarrow\ a_1u_1 + \cdots + a_mu_m + b_1v_1 + \cdots + b_pv_p = 0 \\
&\Rightarrow\ \text{each coefficient is } 0 \qquad \text{(Why?)}
\end{aligned}
$$

Thus the completion of the pth step (without reaching a basis) guarantees the completion of the $p + 1$ step. After at most n steps, we are done. ∎†

COROLLARY 2.7a

A vector space has a basis if and only if it is nonzero and finite-dimensional.

This is the first of the two major results we announced at the beginning of this section. Note that both conditions are essential. A zero space does not have a basis, as we explained at the end of Sec. 2.2. Nor does a space which is not finite-dimensional have a basis, since by definition such a space cannot be spanned.‡

THEOREM 2.8

A vector space spanned by m elements cannot have a linearly independent subset containing more than m elements.

Proof

Suppose that $\{u_1, \ldots, u_m\}$ is a spanning set for S, let n be any integer greater than m, and let $v_1, \ldots, v_n$ be any n (distinct) elements of S. The problem is to prove that $v_1, \ldots, v_n$ are linearly dependent.

Suppose that they are linearly independent. The idea of the proof is to span S by using m of them. (We'll see in a moment how to do this.) Relabeling (if necessary), we can denote this spanning set by $\{v_1, \ldots, v_m\}$. But there are some v_j left over, because $n > m$; more precisely, v_n is certainly not

† We'll see shortly that the process cannot exhaust the v_j, but must stop before $p = n$. However, for the present argument it is unnecessary to take this fact into account.

‡ In more advanced treatments, the concept of "infinite basis" is introduced. See Example 2.3, in which we might consider the space spanned by the infinite class of linearly independent functions $\phi_1(x) = 1, \phi_2(x) = x, \phi_3(x) = x^2, \cdots$. (The elements of this space would be polynomials of arbitrary degree, together with the zero polynomial.) In this book, however, the term "basis" refers to finite sets.

included. Writing v_n as a linear combination of $v_1, \ldots, v_m$ (therefore also as a linear combination of $v_1, \ldots, v_{n-1}$), we see by Theorem 2.6 that $v_1, \ldots, v_n$ are linearly *dependent,* which contradicts our assumption and completes the proof.

To construct the desired spanning set, we begin by observing that since the u_i span S, we can write

$$v_1 = a_1u_1 + \cdots + a_mu_m$$

However, $v_1 \neq 0$. (Remember that we are assuming the linear independence of $v_1, \ldots, v_n$.) Hence not all the a_i are 0; by relabeling (if necessary) we may assume that $a_1 \neq 0$. Then we can solve the above equation for u_1 in terms of $v_1, u_2, \ldots, u_m$; we claim that these elements still span S. (Confirm!) Of course if $m = 1$, the list stops with v_1, and we already have the desired spanning set. But if $m > 1$, there are some u_i left, and we must repeat the argument.

After carrying out this procedure p times (possibly relabeling each time), we arrive at the stage where $u_1, \ldots, u_p$ have been replaced by $v_1, \ldots, v_p$. If $p = m$, the desired spanning set $\{v_1, \ldots, v_m\}$ has been constructed, and so assume that $p < m$. Then the spanning set is $\{v_1, \ldots, v_p, u_{p+1}, \ldots, u_m\}$, and we write

$$v_{p+1} = a_1v_1 + \cdots + a_pv_p + a_{p+1}u_{p+1} + \cdots + a_mu_m$$

Not all the coefficients $a_{p+1}, \ldots, a_m$ are 0, since otherwise we have

$$v_{p+1} = a_1v_1 + \cdots + a_pv_p$$

which contradicts the linear independence of $v_1, \ldots, v_{p+1}$. By relabeling (if necessary) arrange it so that $a_{p+1} \neq 0$, and solve for u_{p+1} in terms of $v_1, \ldots, v_{p+1}$ and the remaining u_i (if any). This solution shows that $v_1, \ldots, v_{p+1}$ and the remaining u_i span S, thus completing the $p + 1$ step. Obviously the process ends after m steps (when the u_i are exhausted). ▮

COROLLARY 2.8a

Any two bases for a vector space contain the same number of elements.

Proof

This proof is left for the problems. It is a beautiful argument and can be done in three or four lines.

This is the second major result we announced at the beginning of the section. We now know two crucial things:

1. Every nonzero finite-dimensional vector space has a basis.

2. Every basis for such a space has the same number of elements.

Hence we can make the following definition:

DIMENSION

Let S be a finite-dimensional space. The *dimension* of S, denoted by $\dim S$, is the number of elements in a basis (unless S is a zero space, in which case its dimension is defined to be 0).

The most immediate application of this idea is to fill in a gap from Chap. 1, where we kept saying that n-space is n-dimensional without proving it (except in the case of $\mathcal{R}^2$, which we did in Sec. 1.2).

THEOREM 2.9

$\dim \mathfrak{F}^n = n$

Proof

Look at the standard basis.

You should compare this triviality with the argument for $\dim \mathcal{R}^2$ in Sec. 1.2 (or with the similar argument for higher dimensions suggested there). Our venture into abstraction has not been altogether fruitless!

THEOREM 2.10

If $\dim S = r$, any spanning set must contain at least r elements.

Proof

This proof is left for the problems.

Just for the record, go back now to the proof of Theorem 2.7, where we built up the linearly independent set $\{u_1, \ldots, u_m\}$ to a basis by using certain elements of the spanning set $\{v_1, \ldots, v_n\}$. We said that the buildup stops in at most n steps. Actually it stops in fewer steps, for if $\dim S = r$, and our built-up basis is $\{u_1, \ldots, u_m, v_1, \ldots, v_p\}$, the number of steps is p, where $m + p = r$.† Since m is a positive integer, it follows that $p < r$. But $\{v_1, \ldots, v_n\}$ is a spanning set, so $r \leq n$ by Theorem 2.10. Hence $p < n$.

† If $\{u_1, \ldots, u_m\}$ is already a basis, so that no buildup is required, then $p = 0$.

Problems

25. Find the dimension of each of the following spaces:

a. $\mathcal{C}$ as a vector space over $\mathcal{R}$. (See Prob. 1, Sec. 2.1.)

b. $\mathcal{C}$ as a vector space over $\mathcal{C}$. (See Prob. 2, Sec. 2.1.)

c. The vector space of 2×2 matrices with entries in $\mathcal{F}$. (See Prob. 4, Sec. 2.1.)

d. The vector space of 2×2 matrices of the form

$$\begin{bmatrix} a & b \\ -b & a \end{bmatrix}$$

where a and b are elements of $\mathcal{F}$. (See Prob. 5, Sec. 2.1.)

e. The vector space consisting of all linear combinations of ϕ_1 and ϕ_2, where $\phi_1(x) = e^{2x}$ and $\phi_2(x) = \phi_1'(x)$ for all $x \in \mathcal{R}$. (Before answering, convince yourself that this set of functions is a vector space! Otherwise it wouldn't even have a dimension.)

26. Let $\mathbf{U}_1 = (-1,2,1)$, $\mathbf{U}_2 = (0,1,1)$, $\mathbf{V}_1 = (1,0,1)$, $\mathbf{V}_2 = (-2,5,3)$, $\mathbf{V}_3 = (-1,0,1)$, and $\mathbf{V}_4 = (-1,3,2)$.

a. Confirm that $\{\mathbf{U}_1,\mathbf{U}_2\}$ is linearly independent in $\mathcal{R}^3$. Why isn't it a spanning set?

b. Confirm that $\{\mathbf{V}_1,\mathbf{V}_2,\mathbf{V}_3,\mathbf{V}_4\}$ is a spanning set for $\mathcal{R}^3$. Why isn't it linearly independent?

c. Which of the $\mathbf{V}_j$ would you use to build up $\{\mathbf{U}_1,\mathbf{U}_2\}$ to a basis for $\mathcal{R}^3$?

27. In the first step of the proof of Theorem 2.7 we assume $u_1, \ldots, u_m$ to be linearly independent elements not spanning S, and go through the spanning elements $v_1, \ldots, v_n$ to find the first one which is not a linear combination of $u_1, \ldots, u_m$. What if this search fails because every v_j is a linear combination of $u_1, \ldots, u_m$? Show that this leads to a contradiction, as follows.

a. Every $x \in S$ can be written in the form $x = \sum_{j=1}^{n} x_j v_j$, where the x_j are scalars. Why?

b. If each v_j is a linear combination of $u_1, \ldots, u_m$, we can write

$$v_j = \sum_{i=1}^{m} a_{ij} u_i \qquad j = 1, \ldots, n$$

Why are double subscripts needed on the scalars?

c. It follows that

$$x = \sum_{j=1}^{n} x_j \left(\sum_{i=1}^{m} a_{ij} u_i \right)$$

By writing this out and collecting terms, convince yourself that

$$x = \sum_{i=1}^{m} \left(\sum_{j=1}^{n} a_{ij} x_j \right) u_i$$

d. Let

$$b_i = \sum_{j=1}^{n} a_{ij} x_j \qquad i = 1, \ldots, m$$

so that

$$x = \sum_{i=1}^{m} b_i u_i$$

What does this result imply and why is it a contradiction?

28. Referring to the argument in Prob. 27, how many scalars a_{ij} are there? (It is irresistible to arrange them in a rectangular array

$$\begin{bmatrix} a_{11} & a_{12} & \cdots & a_{1n} \\ a_{21} & a_{22} & \cdots & a_{2n} \\ \cdots & \cdots & \cdots & \cdots \\ a_{m1} & a_{m2} & \cdots & a_{mn} \end{bmatrix}$$

in which case the counting is easy. This is the first time in the book that a matrix has jumped out at us unbidden! It was in such situations as this that matrices were invented.)

29. Suppose that $\{v_1, \ldots, v_n\}$ is a spanning set for the vector space S. Explain why any larger (finite) set containing $v_1, \ldots, v_n$ is also a spanning set. Where did we use this fact in the proof of Theorem 2.7?

30. Prove that a vector space has a basis if and only if it is nonzero and finite-dimensional (Corollary 2.7a).

31. Where in the proof of Theorem 2.8 did we use the fact that any (nonempty) subset of a linearly independent set is linearly independent? (See Prob. 15, Sec. 2.2.)

32. Explain why a vector space having linearly independent subsets of arbitrarily large size cannot be finite-dimensional. (This fact justifies our remarks about the vector space of functions in Example 2.3.)

33. Prove that any two bases for a vector space contain the same number of elements (Corollary 2.8a).

34. Prove that if $\dim S = r$, then any spanning set must contain at least r elements (Theorem 2.10).

2.5 The Minimax Principle

In the last section we described two approaches to the problem of constructing a basis for a vector space. One approach is to reduce a spanning set to a basis by omitting its superfluous elements (if any); the other is to build up a linearly independent set to a basis by adding enough new elements to span the space (if any are needed). This suggests that a basis is simultaneously a *minimal spanning set* and a *maximal linearly independent set,* an idea known as the "minimax principle." The purpose of this section is to be precise about the details.

THEOREM 2.11

A set of r elements in a vector space of dimension $r > 0$ is linearly independent if and only if it is a spanning set.

Proof

Let $\{u_1, \ldots, u_r\}$ be the set and suppose that it is a spanning set. If it were linearly dependent, one of its elements would be a linear combination of the others, and these others would constitute a spanning set containing $r - 1$ elements.† This is a contradiction of Theorem 2.10; hence the set must be linearly independent.

Conversely, suppose that the set is linearly independent. Then by Theorem 2.7 either it is a basis or it can be built up to a basis. Since every basis contains r elements, no buildup is possible; the set is a basis already. Hence it must be a spanning set. ∎

COROLLARY 2.11a

A set of r elements in a vector space of dimension $r > 0$ is a basis if it is either linearly independent or a spanning set.

This result is important. Every spanning set must have at least r elements (Theorem 2.10). This corollary says that if the set has the minimum number, then it is automatically a basis. Similarly, every linearly independent set has at most r elements (Theorem 2.8); if it has the maximum number, it is automatically a basis. Thus, if we can find a set of r elements having one of the properties of a basis, we need not check the other property—it follows automatically (provided that we already know that $\dim S = r$). For example, the vectors $(2,-1)$ and $(5,3)$ are linearly independent in $\mathcal{R}^2$; since $\dim \mathcal{R}^2 = 2$,

† If $r = 1$, this argument breaks down, but then the theorem is trivial.

we may conclude immediately that they constitute a basis (without checking whether they span $\mathcal{R}^2$).[†]

MAXIMAL AND MINIMAL SETS

Let S be a nonzero vector space. A linearly independent set $\{u_1, \ldots, u_m\}$ is called *maximal* if, for every $x \in S$, $u_1, \ldots, u_m, x$ are linearly dependent. In other words, the set cannot be enlarged without destroying its linear independence.

A spanning set $\{v_1, \ldots, v_n\}$ is called *minimal* if it no longer spans S when any of its elements are left out. In other words, the set cannot be reduced without losing its spanning power.

THEOREM 2.12 (The Minimax Principle)

Let S be a nonzero vector space.

a. A linearly independent subset of S is a basis if and only if it is maximal.

b. A spanning set for S is a basis if and only if it is minimal.

Proof

For part *a*, suppose that $\{u_1, \ldots, u_m\}$ is a maximal linearly independent set. We claim that it is a spanning set (hence a basis). If x is any element of S, we know that $u_1, \ldots, u_m, x$ are linearly dependent (by the definition of "maximal"). Hence there are scalars $c_1, \ldots, c_m, c$, not all zero, such that

$$c_1u_1 + \cdots + c_mu_m + cx = 0$$

Since $c \neq 0$ (otherwise $c_1u_1 + \cdots + c_mu_m = 0$, and all the scalars are zero), we can solve for x as a linear combination of $u_1, \ldots, u_m$.

We leave the converse (the "only if" part) for the problems.

For part *b* suppose that $\{v_1, \ldots, v_n\}$ is a minimal spanning set. We claim it is linearly independent (hence a basis). For suppose that $c_1v_1 + \cdots + c_nv_n = 0$. If one of the c_j were different from zero, we could solve for the corresponding v_j in terms of the others. These others would then constitute a spanning set, contradicting the minimal property of $\{v_1, \ldots, v_n\}$. Hence all the c_j are zero.[‡]

The "only if" part is left for the problems. ∎

[†] We proved this conclusion a long time ago in $\mathcal{R}^2$: see Prob. 23, Sec. 1.2. Now we can do the same thing in any finite-dimensional space.

[‡] This argument breaks down if $n = 1$. What should be done in that case?

THEOREM 2.13

Any spanning set for a nonzero finite-dimensional space S either is a basis for S or can be reduced to a basis.

Proof

Let $r = \dim S$ and suppose that $\{v_1, \ldots, v_n\}$ is a spanning set. If it is minimal, then by Theorem 2.12*b* it is already a basis. If it is not minimal, then by definition of "minimal" at least one of its elements can be omitted without destroying its spanning power. Reduce the set by omitting such an element; then repeat the argument. Since $n \geq r$ (Theorem 2.10), the set is eventually reduced to a spanning set with r elements. Corollary 2.11a then implies that the set is a basis. ▮

Theorem 2.13 completes our discussion of dimension, since our original idea (explained at the beginning of Sec. 2.4) is now justified. A basis can be constructed by reducing a spanning set or building up a linearly independent set. Either way we get a linearly independent set with r elements, where $r = \dim S$.

Problems

35. Why does our proof of Theorem 2.11 misfire if $r = 1$? Explain why the theorem is true in this case.

36. Prove that a set of r elements in a vector space of dimension $r > 0$ is a basis if it is either linearly independent or a spanning set (Corollary 2.11a).

37. Explain why a nonzero vector space is infinite-dimensional if and only if it does not have a maximal linearly independent subset.

38. Prove that a basis for a vector space is a maximal linearly independent set (the "only if" part of Theorem 2.12*a*).

39. Why does our proof of the "if" part of Theorem 2.12*b* break down if $n = 1$? Give an argument that works in this case.

40. Prove that a basis for a vector space is a minimal spanning set (the "only if" part of Theorem 2.12*b*).

41. Which of the following subsets of $\mathcal{C}^2$ are maximal linearly independent sets? Which are minimal spanning sets?

a. $\{(1,i), (i,-1)\}$

b. $\{(1,1), (-1,1)\}$

c. $\{(1,0), (i,0), (0,1), (0,i)\}$

42. Reduce $\{(1,0,1,0), (0,1,0,1), (1,1,1,1), (1,1,0,0), (0,0,1,1), (2,2,1,0)\}$ to a basis for $\mathcal{R}^4$.

2.6 Subspaces of a Vector Space

SUBSPACE

Let S be a vector space over $\mathfrak{F}$ and suppose that T is a (nonempty) subset of S which is itself a vector space over $\mathfrak{F}$ (using the same operations of addition and scalar multiplication as in S). Then T is called a *subspace* of S.†

EXAMPLE 2.10

Let $S = \mathcal{R}^2$ and suppose that

$$T = \{(x,y) \in \mathcal{R}^2 \mid (x,y) = t(-1,1),\ t \in \mathcal{R}\}$$

is the set of scalar multiples of $(-1,1)$. Thus

$$T = \{(x,y) \in \mathcal{R}^2 \mid y = -x\}$$

which represents the straight line whose equation in cartesian coordinates is $y = -x$. We claim that it is a subspace of $\mathcal{R}^2$.

To justify this statement, we have to show that T has all the properties listed in the definition of vector space in Sec. 2.1 (with S and $\mathfrak{F}$ replaced by T and $\mathcal{R}$, respectively, throughout). Five of these—namely the associative and commutative laws of addition, and the associative, distributive, and identity laws of scalar multiplication—are trivial; they apply in the case of T because we already know that they hold in $\mathcal{R}^2$. This might be called the "heredity principle": any subset of a vector space S "inherits" these properties from S.

The remaining four properties need verification in T:

1. Is T closed relative to addition? In other words, if we are given that $(-a,a)$ and $(-b,b)$ are elements of T, can we conclude that their sum is also an element of T? The answer is yes, for if we let $c = a + b$, then

   $$(-a,a) + (-b,b) = (-c,c) \in T$$

   Of course this is geometrically apparent! The sum of two arrows lying in the line T is another arrow in T. (See Fig. 2.3.)

2. Does T contain a zero element? We already know that $\mathcal{R}^2$ does, namely $\mathbf{0} = (0,0)$. But does $\mathbf{0} \in T$? Obviously yes!

3. Given that $\mathbf{X} \in T$, does T contain the inverse of $\mathbf{X}$? We know that

†Note the distinction between "subspace" and "subset." Arbitrary subsets of S do not necessarily have any coherent algebraic structure. But a subspace is a subset which has all the properties of a vector space.

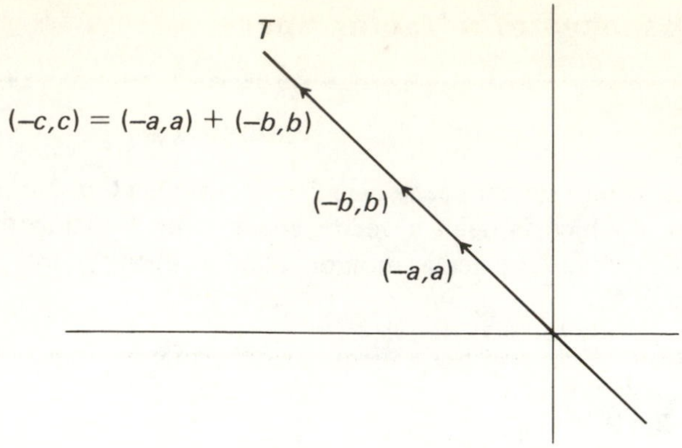

Fig. 2.3 Closure of *T* relative to addition

$-\mathbf{X} \in \mathcal{R}^2$, but does $-\mathbf{X} \in T$? Again the answer is yes, for if (x_1,x_2) satisfies the equation $y = -x$, so does $(-x_1,-x_2)$.

4. Is T closed relative to scalar multiplication? That is, if $(-t,t) \in T$ and $c \in \mathcal{R}$, does it follow that $c(-t,t) \in T$? Yes.

Thus T is a subspace of $\mathcal{R}^2$. Since $\mathbf{U}_1 = (-1,1)$ is a nonzero element which spans T, a basis for T is $\{\mathbf{U}_1\}$ and dim $T = 1$. Note how this purely algebraic statement fits our geometric intuition that the dimension of a straight line is 1. On the other hand, one must not be carried away by geometry. The straight line $\{(x,y) \in \mathcal{R}^2 \mid y = x + 2\}$ is not even a subspace of $\mathcal{R}^2$. (Why?) While we might call it 1-dimensional in another context, it does not have a dimension in the sense we have defined.

In practice it is unnecessary to be tedious as we were in Example 2.10. When a subset T of S is proposed as a candidate for a subspace, the heredity principle accounts for five of the properties of a vector space automatically. Of the remaining four properties, only the closure laws need checking; the identity and inverse laws of addition turn out to be consequences of these, as we now show.

THEOREM 2.14

A nonempty subset of a vector space is a subspace if and only if it is closed relative to addition and scalar multiplication.

Proof

Let S be the space and T the subset. If T is a subspace, all nine properties hold; thus the ''only if'' part of the theorem is trivial. Conversely, suppose that T is closed relative to addition and scalar multiplication. To show that T is a subspace of S, we need only argue that the identity and inverse laws of addition hold:

1. Choose some $x \in T$. Then $0x \in T$ because T is closed relative to scalar multiplication. By Theorem 2.4, $0x = 0$. Hence $0 \in T$.
2. If $x \in T$, then $-1x \in T$. However, since $-1x = -x$ by Theorem 2.3, we conclude that $-x \in T$. ∎

COROLLARY 2.14a

A nonempty subset T of a vector space is a subspace if and only if $ax + by \in T$ for all scalars a and b and all x and y in T.

This single test is often easier to apply than the two closure criteria of Theorem 2.14.

EXAMPLE 2.11

Let T be the subset of $\mathcal{R}^3$ consisting of all linear combinations of $\mathbf{U}_1 = (2,-1,1)$ and $\mathbf{U}_2 = (5,-1,3)$:

$$T = \{\mathbf{X} \in \mathcal{R}^3 | \mathbf{X} = c_1\mathbf{U}_1 + c_2\mathbf{U}_2,\ c_1 \in \mathcal{R},\ c_2 \in \mathcal{R}\}$$

Letting $\mathbf{X} = (x,y,z)$, we may write the condition $(x,y,z) = c_1\mathbf{U}_1 + c_2\mathbf{U}_2$ as a system of parametric equations giving x, y, z in terms of the parameters c_1 and c_2:

$$\begin{aligned} x &= 2c_1 + 5c_2 \\ y &= -c_1 - c_2 \\ z &= c_1 + 3c_2 \end{aligned}$$

Adding the second and third of these to eliminate c_1, we obtain

$$y + z = 2c_2$$

Eliminating c_1 from the first and second equations by adding twice the second to the first, we find that

$$x + 2y = 3c_2$$

Hence (solving for c_2) we have

$$\tfrac{1}{2}(y + z) = \tfrac{1}{3}(x + 2y)$$

or simply

$$2x + y - 3z = 0$$

This is the cartesian equation representing T; hence T is a plane in 3-space. (See Example 1.1 and Prob. 38, Sec. 1.3.) This conclusion is obvious geometrically, since linear combinations of the arrows $\mathbf{U}_1$ and $\mathbf{U}_2$ in $\mathcal{R}^3$ must lie in the plane determined by $\mathbf{U}_1$ and $\mathbf{U}_2$.†

† We consider $\mathbf{U}_1$ and $\mathbf{U}_2$ here as arrows beginning at the origin. Another way of saying it is that T is the plane containing the *points* $\mathbf{0} = (0,0,0)$, $\mathbf{U}_1 = (2,-1,1)$, $\mathbf{U}_2 = (5,-1,3)$. See Sec. 1.1 for remarks about alternative interpretations of the elements of $\mathcal{R}^2$ (also applicable in $\mathcal{R}^3$).

To prove that T is a subspace of $\mathcal{R}^3$ (without depending on intuition), we need only apply Corollary 2.14a. Let $\mathbf{X} = c_1\mathbf{U}_1 + c_2\mathbf{U}_2$ and $\mathbf{Y} = d_1\mathbf{U}_1 + d_2\mathbf{U}_2$ be any elements of T (linear combinations of $\mathbf{U}_1$ and $\mathbf{U}_2$) and suppose that a and b are scalars. Then

$$\begin{aligned} a\mathbf{X} + b\mathbf{Y} &= a(c_1\mathbf{U}_1 + c_2\mathbf{U}_2) + b(d_1\mathbf{U}_1 + d_2\mathbf{U}_2) \\ &= (ac_1 + bd_1)\mathbf{U}_1 + (ac_2 + bd_2)\mathbf{U}_2 \end{aligned}$$

which is also an element of T.

Since T is a plane, it ought to be 2-dimensional. We confirm this by observing that $\mathbf{U}_1$ and $\mathbf{U}_2$ are linearly independent elements spanning T. A basis for T is $\{\mathbf{U}_1,\mathbf{U}_2\}$ and hence $\dim T = 2$.

Note that our description of T in Example 2.11 was "the subset of $\mathcal{R}^3$ consisting of all linear combinations of $\mathbf{U}_1$ and $\mathbf{U}_2$." Subspaces of a vector space are so often characterized this way that a special notation has been adopted to represent them:

If $u_1, \ldots, u_m$ are elements of a vector space, the set of all linear combinations $\sum_{j=1}^{m} c_j u_j$ is written as $\langle u_1, \ldots, u_m \rangle$.†

THEOREM 2.15

If $u_1, \ldots, u_m$ are elements of a vector space S, then $T = \langle u_1, \ldots, u_m \rangle$ is a subspace of S.

Proof

Let $x = \sum_{j=1}^{m} c_j u_j$ and $y = \sum_{j=1}^{m} d_j u_j$ be any elements of T and suppose that a and b are scalars. Then

$$ax + by = a\sum_{j=1}^{m} c_j u_j + b\sum_{j=1}^{m} d_j u_j$$

If you are familiar with the properties of the "sigma notation" for sums, you won't have any trouble seeing why the above equation is the same as

$$ax + by = \sum_{j=1}^{m} ac_j u_j + \sum_{j=1}^{m} bd_j u_j = \sum_{j=1}^{m} (ac_j + bd_j)u_j$$

† Do not confuse this set with $\{u_1, \ldots, u_m\}$, which is the (finite) set of "generators" of $\langle u_1, \ldots, u_m \rangle$. In Example 2.11 the set $\langle \mathbf{U}_1,\mathbf{U}_2 \rangle$ is the whole plane spanned by $\mathbf{U}_1$ and $\mathbf{U}_2$. But $\{\mathbf{U}_1,\mathbf{U}_2\}$ contains only two elements.

If it is not familiar to you, write out the sums and convince yourself that this sort of thing is legitimate. In any case, $ax + by \in T$.

LECTOR *What happened to ▮?*

AUCTOR *The proof isn't finished.*

LECTOR . . .

AUCTOR *Reread Corollary 2.14a.*

LECTOR *I did, and I must say it is unlikely to make the best-seller list.*

AUCTOR *What if T were empty?*

LECTOR *Ah, so.*

AUCTOR *Since* $0u_1 + \cdots + 0u_m = 0 \in T$, *the point is moot, but in general it is worth remembering. It is an embarrassing experience to laboriously prove that a set is closed relative to addition and scalar multiplication only to discover that the argument is vacuous.*

LECTOR *I should think it would be more embarrassing to split an infinitive and misuse ''moot,'' all in two sentences.*

THEOREM 2.16

If T is a subspace of the finite-dimensional space S, then T is finite-dimensional and $\dim T \leq \dim S$.

Proof

If T is the zero subspace of S, the theorem is clearly true, so assume that $T \neq \{0\}$. Then T has a linearly independent subset. (Why?) If $n = \dim S$, then S cannot have a linearly independent subset containing more than n elements (Theorem 2.8). Hence neither can T (being a subset of S). But then T must have a maximal linearly independent subset, say $\{u_1, \ldots, u_m\}$, where $1 \leq m \leq n$. By the minimax principle this is a basis for T, so T is finite-dimensional and $\dim T = m \leq n = \dim S$. ▮

LECTOR *Since T has an independent subset, why not just build it up to a basis for T and build this up to a basis for S? Then* $\dim T \leq \dim S$.

AUCTOR *It's a good idea, but it doesn't work.*

LECTOR *Why not?*

AUCTOR *Reread Theorem 2.7.*

THEOREM 2.17

Let T be a subspace of the finite-dimensional space S. Then $T = S \Leftrightarrow \dim T = \dim S$.

Proof

Obviously $T = S \Rightarrow \dim T = \dim S$. Conversely, suppose that $\dim T = \dim S = n$. If $n = 0$, it is clear that $T = S$, so assume that $n > 0$. To prove that $T = S$, we must establish the two inclusions $T \subset S$ and $S \subset T$.† The first of these is, of course, true, so look at the second, which is true if $x \in S \Rightarrow x \in T$. Let x be any element of S. Since T has a basis consisting of n linearly independent elements ($\dim T = n$), these elements must span S (Theorem 2.11). Hence x is a linear combination of them. But T is a subspace, thus closed relative to addition and scalar multiplication. Therefore $x \in T$. ∎

Problems

43. Describe each of the following subsets of $\mathcal{R}^2$ or $\mathcal{R}^3$ geometrically. Which ones are subspaces? Name a basis for each one that is a subspace (if possible).

a. $\{(x,y) \mid y = 2x - 1\}$
b. $\{(x,y,z) \mid 2x - y + z = 0\}$
c. $\{(x,y) \mid y = x\}$
d. $\{(x,y,z) \mid x^2 + y^2 + z^2 = 0\}$
e. $\{(x,y,z) \mid x = t,\ y = 2t,\ z = 0,\ t \in \mathcal{R}\}$
f. $\langle (2,-1), (-4,2) \rangle$

44. Explain why the only subspaces of $\mathcal{R}^2$ are the zero space, straight lines containing the origin, and the whole plane. What are the possibilities in $\mathcal{R}^3$?

45. Let S be the vector space of 2×2 matrices with entries in $\mathfrak{F}$, and let T be the subset of S consisting of "diagonal" 2×2 matrices, that is, matrices of the form

$$\begin{bmatrix} a & 0 \\ 0 & b \end{bmatrix}$$

Show that T is a 2-dimensional subspace of S. (See Prob. 25*c*, Sec. 2.4, where you found that $\dim S = 4$.)

46. Let S be the vector space of real-valued functions differentiable on $(-\infty,\infty)$. If ϕ_1 and ϕ_2 are the elements of S defined by $\phi_1(x) = e^{2x}$ and $\phi_2(x) = \phi_1'(x)$, what is the dimension of the subspace $\langle \phi_1, \phi_2 \rangle$? What does Theorem 2.16 say about this?

47. Where in the proof of Theorem 2.14 did we use the fact that T is nonempty?

† Some writers use $\subseteq$, reserving $\subset$ for "proper" subset. In this book we make no distinction.

48. Prove that a nonempty subset T of a vector space is a subspace if and only if $ax + by \in T$ for all scalars a and b and all x and y in T (Corollary 2.14a).

★ **49.** Prove that the intersection of two subspaces of a vector space is a subspace. Then show that this is not necessarily true of the union of two subspaces.

50. Explain why the dimension of $\langle u_1, \ldots, u_m \rangle$ is not necessarily m.

51. Explain why Lector's "proof" of Theorem 2.16 is invalid.

52. Use Theorem 2.17 to explain why $\langle (3,1), (1,-1) \rangle = \mathcal{R}^2$.

★ **53.** Let S be $\mathcal{R}^2$ or $\mathcal{R}^3$. If we adopt the convention that the zero vector (being directionless) is perpendicular ("orthogonal") to every vector, then

$$\mathbf{X} \text{ and } \mathbf{Y} \text{ are perpendicular} \quad \Leftrightarrow \quad \mathbf{X} \cdot \mathbf{Y} = 0$$

where $\mathbf{X} \cdot \mathbf{Y}$ is the "dot product" of $\mathbf{X}$ and $\mathbf{Y}$.† Let U be a (nonempty) subset of S and define the *orthogonal complement* of U to be

$$U^\perp = \{\mathbf{Y} \in S \mid \mathbf{X} \cdot \mathbf{Y} = 0 \text{ for every } \mathbf{X} \in U\}$$

the set of all vectors perpendicular to every vector in U.

a. What is $U^\perp$ if U is the zero subspace of S?

b. If $S = \mathcal{R}^3$ and U consists of the vector $(1,-1,2)$ alone, show that

$$U^\perp = \{(x,y,z) \mid x - y + 2z = 0\}$$

Note that this is the plane through the origin perpendicular to $(1,-1,2)$. In analytic geometry $(1,-1,2)$ is called a "normal" to the plane; its components are the coefficients in the equation $x - y + 2z = 0$.

c. What is $U^\perp$ if $U = \langle (1,-1,2) \rangle$? Note that now U is the straight line through the origin containing $(1,-1,2)$ and is itself a subspace.

d. If $S = \mathcal{R}^2$ and $U = \langle (3,-2) \rangle$, show that $U^\perp = \{(x,y) \mid 3x - 2y = 0\}$. Note that U and $U^\perp$ are perpendicular lines through the origin.

e. If $U = S$, what is $U^\perp$?

f. Prove that, regardless of what U is, $U^\perp$ is a subspace of S. *Hint:* Use the fact that

$$\mathbf{X} \cdot (a\mathbf{Y} + b\mathbf{Z}) = a(\mathbf{X} \cdot \mathbf{Y}) + b(\mathbf{X} \cdot \mathbf{Z})$$

for all scalars a and b and vectors $\mathbf{X}$, $\mathbf{Y}$, and $\mathbf{Z}$.

g. In parts *a*, *c*, *d*, and *e* confirm that $\dim U + \dim U^\perp = \dim S$. (This is a special case of the *projection theorem,* to be proved in Chap. 11.)

† Recall from calculus that if $\mathbf{X} = (x_1,x_2)$ and $\mathbf{Y} = (y_1,y_2)$, then $\mathbf{X} \cdot \mathbf{Y} = x_1y_1 + x_2y_2$, while in $\mathcal{R}^3$ the formula is $\mathbf{X} \cdot \mathbf{Y} = x_1y_1 + x_2y_2 + x_3y_3$.

Review Quiz

True or false?

1. A linearly independent subset of a nonzero vector space either is a basis or can be built up to a basis.
2. The union of two subspaces of a vector space is a subspace.
3. $\mathcal{C}$ is a 2-dimensional vector space over $\mathcal{R}$.
4. The vector space of real-valued functions continuous on an interval I is finite-dimensional.
5. If T is a nonempty subset of the vector space S and linear combinations of elements of T are always in T, then T is a subspace of S.
6. If T is a subspace of S and $\dim T = \dim S$, then $T = S$.
7. Every nonzero vector space has a basis.
8. In a vector space of dimension $r > 0$, any set of r elements spanning the space is a basis.
9. $\{(x,y) \mid y = 3x - 1\}$ is a 1-dimensional subspace of the plane.
10. If $T = \langle u_1, \ldots, u_m \rangle$, then $\dim T = m$.
11. The vector space of real-valued functions differentiable on $(-\infty,\infty)$ has a maximal linearly independent subset.
12. If T is a subspace of the vector space S, the zero element of S is the zero element of T.
13. If the Wronskian of two functions ϕ_1 and ϕ_2 is the zero function on the interval I, then ϕ_1 and ϕ_2 are linearly dependent on I.
14. The set of solutions on $(-\infty,\infty)$ of the differential equation $y'' - y = 0$ is a vector space.
15. If a spanning set for the space S loses its spanning power when any of its elements are left out, it is a basis for S.
16. If $u_1, \ldots, u_m$ are linearly dependent elements of a vector space S, any (finite) subset of S containing these elements is linearly dependent.
17. $\{(1,0,1), (0,1,0), (2,1,1), (-1,2,1)\}$ is a linearly independent subset of $\mathcal{R}^3$.
18. $\{(1,i), (i,-1), (-1,-i)\}$ can be reduced to a basis for $\mathcal{C}^2$.
19. $\langle (1,-1,1), (1,0,1) \rangle$ is a plane in $\mathcal{R}^3$.
20. The intersection of two subspaces of a vector space cannot be empty.
21. $\{(x,y,z) \mid x^2 + y^2 + z^2 \leq 1\}$ is a subspace of $\mathcal{R}^3$.
22. $\mathcal{R}$ is a 1-dimensional vector space over $\mathcal{R}$.
23. The set of positive integers is a group relative to multiplication.

24. If $u_1, \ldots, u_m$ are linearly dependent, the set consisting of the distinct elements in the list is linearly dependent.

25. Linear independence of $u_1, \ldots, u_m$ implies that $u_j \neq 0$, $j = 1, \ldots, m$.

26. If u_1, u_2, u_3 are linearly independent, so are u_1 and u_2.

27. $\{(1,1,1), (0,1,0)\}$ is a maximal linearly independent subset of $\mathbb{R}^3$.

CHAPTER 3

Linear Maps

One of the fundamental ideas of mathematics is the concept of ''function.'' And at the heart of linear algebra is the notion of ''linear function'' from one vector space to another (more often called ''linear transformation,'' or as we shall call it, ''linear map''). Having spent two chapters providing an introduction to vector spaces, we are now ready to begin the real business of this book.

LECTOR *Did you say* begin?
AUCTOR *Well, we're past timberline.*
LECTOR *I'm not sure I appreciate these references to mountains.*
AUCTOR *The best part of the climb is ahead.*
LECTOR *In Mordor where the shadows are.*
AUCTOR *In Lórien where dwells Galadriel.*

3.1 Definitions and Examples

Calculus students usually understand the word *function* to mean a rule by which a unique real number is associated with each element of a domain D in $\mathcal{R}^n$. At first $n = 1$, but when functions of several variables are studied, $n \geq 2$. For example, the function f defined on

$$D = \{(x,y) \mid x^2 + y^2 \leq 1\}$$

by the equation

$$f(x,y) = \sqrt{1 - x^2 - y^2}$$

associates the real number $z = \sqrt{1 - x^2 - y^2}$ with each element $(x,y) \in D$. The range of this function is the subset of $\mathcal{R}$ given by

$$f(D) = \{z \mid z = f(x,y),\ (x,\ y) \in D\}$$

which in this case is the interval $[0,1]$.†

† The symbol $f(D)$, where D is the domain, stands for a *set,* namely the set of functional values corresponding to elements of the domain.

There is no reason why this idea should be confined to a domain in $\mathcal{R}^n$ or a range in $\mathcal{R}$. When it is generalized to arbitrary sets, however, mathematicians tend to use the word *mapping* (or *map*) instead of *function*. To be precise, we have the following definition.

MAPPING

Let A and B be any (nonempty) sets. A function f with domain A and range

$$f(A) = \{y \in B \mid y = f(x),\ x \in A\}$$

is called a *mapping* (or *map*) from A to B, denoted by

$$f: A \to B$$

Given that $x \in A$, the element $y = f(x) \in B$ is called the *image* of x under f, and f is said to *map x into y*.

Of course this is not much of a definition, since it merely substitutes the word *mapping* for *function,* presupposing that you already know what a function is.† It is designed to reinforce your acquaintance with an old friend, together with words like *domain* and *range* and notation like $f(A)$ and $f: A \to B$.

As a frivolous example, let A be the set of American Presidents serving before 1976 and let B be the set of nonnegative integers. For each $x \in A$, define $f(x)$ to be the number of times x was elected President. Then

$$f(\text{Washington}) = 2 \qquad f(\text{Fillmore}) = 0 \qquad f(\text{Truman}) = 1$$

and so on. Evidently f is a mapping from A to B; its range is the subset of B consisting of the "images" of the Presidents, namely

$$f(A) = \{0,1,2,4\}$$

This map is not of much interest because A has no algebraic structure. We cannot, for example, inquire whether

$$f(\text{Washington} + \text{Fillmore}) = f(\text{Washington}) + f(\text{Fillmore})$$

for although the right-hand side is $2 + 0$, the left-hand side makes no sense. In this chapter (and throughout the book) we'll confine our attention to mappings *from a vector space to a vector space,* so that we can pose meaningful algebraic questions.

†On the other hand, we do this all the time! When a "set" is described as a "collection," what has been accomplished? Naturally one tries to keep to a minimum the technical terms that are "defined" this way, but it is impossible to define everything. A famous example is Samuel Johnson's definition of a *net* as "anything reticulated, with interstices between the intersections." When one looks up *reticulated* in his dictionary one finds "formed into, or like, a net." Such circularity is inherent in language and its use in logic.

EXAMPLE 3.1

Define $f\colon \mathcal{R}^1 \to \mathcal{R}^2$ by $f(t) = (t,3t)$.† Since both $\mathcal{R}^1$ and $\mathcal{R}^2$ are vector spaces over $\mathcal{R}$, we can talk about the effect of f on sums and scalar multiples of the elements of $\mathcal{R}^1$. Observe that for all x and y in $\mathcal{R}^1$ we have

$$f(x + y) = [x + y, 3(x + y)] = (x,3x) + (y,3y) = f(x) + f(y)$$

while if k is any scalar,

$$f(kx) = (kx,3kx) = k(x,3x) = kf(x)$$

If we let $\mathbf{X} = f(x)$ and $\mathbf{Y} = f(y)$, this amounts to

$$f(x + y) = \mathbf{X} + \mathbf{Y} \qquad \text{and} \qquad f(kx) = k\mathbf{X}$$

Addition of x and y in $\mathcal{R}^1$ is therefore duplicated in $\mathcal{R}^2$ by addition of their images $\mathbf{X}$ and $\mathbf{Y}$; scalar multiplication of x in $\mathcal{R}^1$ is duplicated in $\mathcal{R}^2$ by the same scalar multiplication of $\mathbf{X}$.

These properties of addition and scalar multiplication are not shared by functions in general. For example, the mapping $f\colon \mathcal{R}^1 \to \mathcal{R}^1$ defined by $f(x) = x^2$ satisfies neither:

$$f(x + y) = (x + y)^2 \neq f(x) + f(y) \qquad \text{(unless } x \text{ or } y \text{ is 0)}$$

$$f(kx) = (kx)^2 \neq kf(x) \qquad \text{(unless } x = 0 \text{ or } k \text{ is 0 or 1)}$$

But the class of mappings from one vector space to another which have these properties is a fundamental one, and we single it out by the following definition.

LINEAR MAP

Let S and T be vector spaces over $\mathfrak{F}$ (where $\mathfrak{F}$ is $\mathcal{R}$ or $\mathcal{C}$) and suppose that $f\colon S \to T$ is a mapping from S to T with the properties

$$f(x + y) = f(x) + f(y) \qquad \text{for all } x \text{ and } y \text{ in } S$$
$$f(kx) = kf(x) \qquad \text{for all } k \in \mathfrak{F} \text{ and } x \in S$$

Then f is called a *linear map* (also *linear transformation*). If $T = S$ (so that f maps S into itself), then f is also called a *linear operator*.

It is important to understand this definition. One way of looking at it is to let $u = f(x)$ and $v = f(y)$ and to note that the linearity properties

$$f(x + y) = f(x) + f(y) = u + v \qquad \text{and} \qquad f(kx) = kf(x) = ku$$

† In calculus this is called a "vector-valued" function, since each $t \in \mathcal{R}$ determines a vector $(t,3t)$ in $\mathcal{R}^2$. Such a function would arise, for example, in the study of a bug whose position at time t is $(x,y) = (t,3t)$. The path of the bug is the straight line with parametric equations $x = t$ and $y = 3t$, or simply $y = 3x$ in cartesian coordinates.

are equivalent to the single statement

$$f(ax + by) = af(x) + bf(y) = au + bv$$

where a and b are scalars and x and y are in S. This criterion is convenient to use in some circumstances and is worth stating in words:

A map is linear if and only if (for all x and y in its domain) it sends every linear combination of x and y into the same linear combination of their images. In other words, for all scalars a and b, $ax + by$ is mapped into $au + bv$ (where u and v are the images of x and y, respectively).

Thus the algebraic structure of S is carried over into T by the map f; that is, f "preserves sums and scalar products." What goes on in one space is duplicated in the other. Of course there may be considerable constriction in the process; we do not mean to imply that a linear map from S to T makes two vector spaces alike when in fact they are not. An extreme example is the following.

EXAMPLE 3.2

Let S and T be vector spaces over $\mathfrak{F}$. The "zero map" from S to T is the function 0: $S \to T$ defined by $0(x) = 0$ for every $x \in S$. This sends every element of S into the one element $0 \in T$ (Fig. 3.1) and is a very dull function. Nevertheless it is linear, as you can check.

EXAMPLE 3.3

Let S be a vector space over $\mathfrak{F}$. The "identity map" on S is the linear operator i: $S \to S$ defined by $i(x) = x$ for every $x \in S$. When we want to distinguish this from the identity map on another space, we use the symbol i_S. For example, the identity map on $S = \mathfrak{R}^2$ is

$$i_S(x,y) = (x,y)$$

while on $T = \mathfrak{R}^3$ it is

$$i_T(x,y,z) = (x,y,z)$$

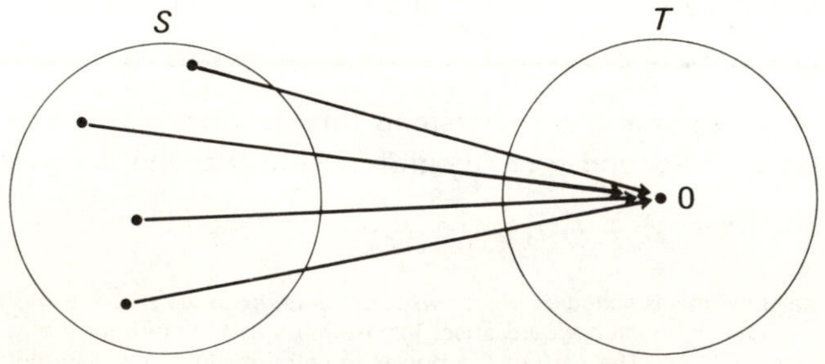

Fig. 3.1 The zero map

EXAMPLE 3.4

Define f: $\mathcal{R}^2 \to \mathcal{R}^3$ by $f(x,y) = (x + y, 2x, x - y)$. Then f is linear, for if $\mathbf{X}_1 = (x_1,y_1)$ and $\mathbf{X}_2 = (x_2,y_2)$ are elements of $\mathcal{R}^2$ and a_1 and a_2 are scalars, we have

$$\begin{aligned}
&f(a_1\mathbf{X}_1 + a_2\mathbf{X}_2)\\
&\quad = f(a_1x_1 + a_2x_2, a_1y_1 + a_2y_2)\\
&\quad = (a_1x_1 + a_2x_2 + a_1y_1 + a_2y_2, 2a_1x_1 + 2a_2x_2, a_1x_1 + a_2x_2 - a_1y_1 - a_2y_2)\\
&\quad = (a_1x_1 + a_1y_1, 2a_1x_1, a_1x_1 - a_1y_1) + (a_2x_2 + a_2y_2, 2a_2x_2, a_2x_2 - a_2y_2)\\
&\quad = a_1(x_1 + y_1, 2x_1, x_1 - y_1) + a_2(x_2 + y_2, 2x_2, x_2 - y_2)\\
&\quad = a_1f(\mathbf{X}_1) + a_2f(\mathbf{X}_2)
\end{aligned}$$

LECTOR *Good grief. You can* see *it's linear without all that.*

AUCTOR *I will try to avoid such tedium in the future. But perhaps we ought to be precise about* how *you can see it's linear.*

The work involved here (and in Example 3.1) could be avoided if we knew more about linear maps. Note that this one sends (x,y) into $(x + y, 2x, x - y)$, a vector whose entries are linear combinations of x and y. Eventually (Prob. 3, Sec. 4.1) we'll prove that this is characteristic of linear maps from n-space to m-space. In other words, we can say that

> f: $\mathcal{F}^n \to \mathcal{F}^m$ is linear if and only if it sends $(x_1, \ldots, x_n)$ into a vector $(y_1, \ldots, y_m)$ whose entries are linear combinations of $x_1, \ldots, x_n$.

In notation that is suggestive of things to come in Chap. 4, we can write

$$\begin{aligned}
y_1 &= a_{11}x_1 + a_{12}x_2 + \cdots + a_{1n}x_n\\
y_2 &= a_{21}x_1 + a_{22}x_2 + \cdots + a_{2n}x_n\\
&\cdots\cdots\cdots\cdots\cdots\cdots\cdots\cdots\\
y_m &= a_{m1}x_1 + a_{m2}x_2 + \cdots + a_{mn}x_n
\end{aligned}$$

where the a_{ij} are scalars. Note the matrix lurking in the wings! (See Prob. 28, Sec. 2.4, for the $m \times n$ array we have in mind.) This is the second time in the book that a matrix has begged for attention.

EXAMPLE 3.5

Suppose that S is the vector space of real-valued functions with derivatives of all orders on the interval J. Define D: $S \to S$ by $D(\phi) = \phi'$ for each $\phi \in S$. Then D is a linear operator, a statement that merely summarizes the familiar property of differentiation,

$$D(c_1\phi_1 + c_2\phi_2) = c_1D(\phi_1) + c_2D(\phi_2)$$

EXAMPLE 3.6

Let S be the vector space of 2×2 matrices with entries in $\mathcal{F}$ and define f: $S \to \mathcal{F}^4$ by

$$f\left(\begin{bmatrix} a & b \\ c & d \end{bmatrix}\right) = (a,b,c,d)$$

Then f is linear. For addition of two matrices in S

$$\begin{bmatrix} a & b \\ c & d \end{bmatrix} + \begin{bmatrix} e & f \\ g & h \end{bmatrix} = \begin{bmatrix} a+e & b+f \\ c+g & d+h \end{bmatrix}$$

is duplicated in $\mathfrak{F}^4$ by addition of their images

$$(a,b,c,d) + (e,f,g,h) = (a+e, b+f, c+g, d+h)$$

while scalar multiplication in S

$$k\begin{bmatrix} a & b \\ c & d \end{bmatrix} = \begin{bmatrix} ka & kb \\ kc & kd \end{bmatrix}$$

corresponds to the same scalar multiplication in $\mathfrak{F}^4$

$$k(a,b,c,d) = (ka,kb,kc,kd)$$

That is,

$$f(A+B) = f(A) + f(B) \qquad \text{and} \qquad f(kA) = kf(A)$$

where A and B are 2×2 matrices and $k \in \mathfrak{F}$.

EXAMPLE 3.7

The "transpose" of a 2×2 matrix

$$A = \begin{bmatrix} a & b \\ c & d \end{bmatrix}$$

is the matrix

$$A^T = \begin{bmatrix} a & c \\ b & d \end{bmatrix}$$

obtained from A by writing its rows as columns. (We'll discuss this idea in detail in Sec. 4.3.) If S is the vector space of 2×2 matrices, the map f: $S \to S$ defined by $f(A) = A^T$ is linear, that is,

$$(A+B)^T = A^T + B^T \qquad \text{and} \qquad (kA)^T = kA^T$$

(Confirm this result!)

EXAMPLE 3.8

Define f: $\mathfrak{R}^2 \to \mathfrak{R}^2$ by $f(x,y) = (x',y')$, where (x',y') is the vector obtained from (x,y) by rotating (x,y) through an angle θ. (See Fig. 3.2.) By elementary trigonometry we have

$$x' = x\cos\theta - y\sin\theta$$
$$y' = x\sin\theta + y\cos\theta$$

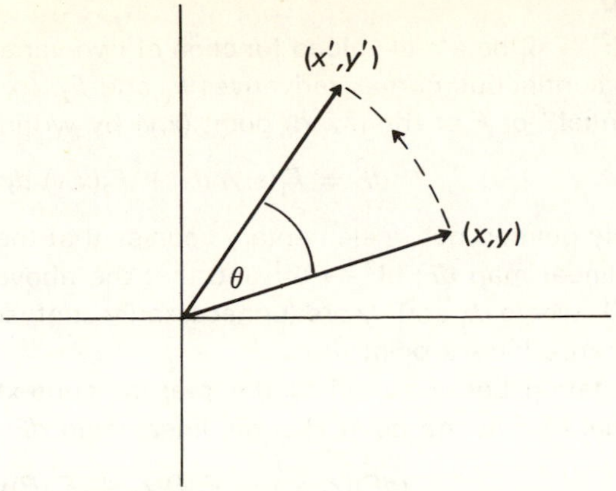

Fig. 3.2 Rotation of the plane

from which it is not hard to show that f is linear. (See Prob. 10.) Thus a "rotation of the plane" is a linear operator.†

EXAMPLE 3.9

The *dot product* of two vectors $\mathbf{X} = (x_1, \ldots, x_n)$ and $\mathbf{Y} = (y_1, \ldots, y_n)$ in $\mathfrak{F}^n$ is defined to be the scalar

$$\mathbf{X} \cdot \mathbf{Y} = x_1 y_1 + \cdots + x_n y_n = \sum_{j=1}^{n} x_j y_j \ddagger$$

It is not hard to show that

$$(a\mathbf{X} + b\mathbf{Y}) \cdot \mathbf{Z} = a(\mathbf{X} \cdot \mathbf{Z}) + b(\mathbf{Y} \cdot \mathbf{Z})$$

for all scalars a and b and vectors **X**, **Y**, and **Z**. We may translate this as linearity of the map $f\colon \mathfrak{F}^n \to \mathfrak{F}^1$ defined by $f(\mathbf{X}) = \mathbf{X} \cdot \mathbf{A}$, where **X** is the "variable" and **A** is "constant." (Confirm this!) In fact, since the formula

$$\mathbf{Z} \cdot (a\mathbf{X} + b\mathbf{Y}) = a(\mathbf{Z} \cdot \mathbf{X}) + b(\mathbf{Z} \cdot \mathbf{Y})$$

is also true (why?), the map $g\colon \mathfrak{F}^n \to \mathfrak{F}^1$ defined by $g(\mathbf{X}) = \mathbf{A} \cdot \mathbf{X}$ is linear, too. For this reason the dot product on $\mathfrak{F}^n$ (which is really a function of two variables) is called "bilinear," that is, linear in each variable.

† Do not confuse this term with "rotation of axes" as studied in analytic geometry, where the points of the plane stay put and the coordinate system is rotated (the basis is changed!). Here the coordinate system is fixed and every point in the plane is mapped into a new point (except for the origin).

‡ No doubt you have encountered this equation before, at least in $\mathfrak{R}^2$ and $\mathfrak{R}^3$. (Also see Prob. 53, Sec. 2.6.) Later on in Chap. 10 we'll use such products to discuss geometric properties of vector spaces.

EXAMPLE 3.10

Let F: $\mathcal{R}^2 \to \mathcal{R}$ be a real-valued function of two variables which is continuous and has continuous partial derivatives F_1 and F_2. In calculus one defines the "differential" of F at the (fixed) point (x,y) by writing

$$dF = F_1(x,y)\,dx + F_2(x,y)\,dy$$

It is rarely pointed out in elementary courses that the differential really represents a linear map dF: $\mathcal{R}^2 \to \mathcal{R}^1$ and that the above formula gives its value at (dx,dy), where dx and dy are (unnecessarily confusing) labels for the coordinates of an arbitrary point in $\mathcal{R}^2$.

In notation better suited to the present context, we may say that the differential of F at the point P is the linear map dF: $\mathcal{R}^2 \to \mathcal{R}^1$ defined by

$$(dF)(x_1,x_2) = F_1(P)x_1 + F_2(P)x_2$$

Sometimes this is written in the form of a dot product,

$$(dF)(\mathbf{X}) = \mathbf{A} \cdot \mathbf{X}$$

where $\mathbf{X} = (x_1,x_2)$ and $\mathbf{A}$ is the "gradient" of F at P, usually denoted by

$$\nabla F(P) = (F_1(P), F_2(P))$$

Since $\mathbf{A} = \nabla F(P)$ is fixed (for a given P), the linearity of dF as a function of $\mathbf{X}$ follows from our remarks in the preceding example.

EXAMPLE 3.11

Let S be $\mathcal{R}^2$ or $\mathcal{R}^3$ and let U be a subspace of S. The "orthogonal complement" of U, denoted by $U^\perp$, is the set of all vectors in S that are perpendicular to every vector in U, that is,

$$U^\perp = \{\mathbf{Y} \in S \,|\, \mathbf{X} \cdot \mathbf{Y} = 0 \text{ for every } \mathbf{X} \in U\}^\dagger$$

In Chap. 11 we shall prove the *projection theorem*, which in the present context says that every vector $\mathbf{X}$ in S can be written (uniquely) in the form

$$\mathbf{X} = \mathbf{X}_1 + \mathbf{X}_2 \qquad \text{where } \mathbf{X}_1 \in U \quad \text{and} \quad \mathbf{X}_2 \in U^\perp$$

For example, if $S = \mathcal{R}^3$ and U is a plane through the origin, then $U^\perp$ is the line through the origin perpendicular to U. The expression of $\mathbf{X} \in \mathcal{R}^3$ as a sum of elements of U and $U^\perp$ may then be visualized as in Fig. 3.3.

Because $\mathbf{X}_1$ and $\mathbf{X}_2$ are perpendicular, they are called the "orthogonal projections" of $\mathbf{X}$ on U and $U^\perp$, respectively. The map f: $S \to S$ defined by $f(\mathbf{X}) = \mathbf{X}_1$ "projects" each $\mathbf{X}$ onto U; it is called the "orthogonal projection"

† See Prob. 53, Sec. 2.6, where we pointed out that $\mathbf{X}$ and $\mathbf{Y}$ are perpendicular ("orthogonal") if and only if $\mathbf{X} \cdot \mathbf{Y} = 0$ (with the convention that the zero vector is perpendicular to every vector). We also asked you to prove that $U^\perp$ is a subspace of S.

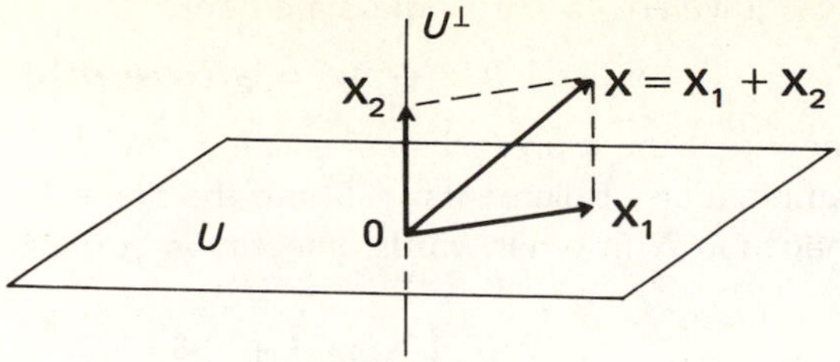

Fig. 3.3 Orthogonal projections

of S on U.† Similarly, $g\colon S \to S$ defined by $g(\mathbf{X}) = \mathbf{X}_2$ is called the orthogonal projection of S on $U^{\perp}$.

To show that f is linear, take any $\mathbf{X}$ and $\mathbf{Y}$ in S and write

$$\mathbf{X} = \mathbf{X}_1 + \mathbf{X}_2 \quad \text{and} \quad \mathbf{Y} = \mathbf{Y}_1 + \mathbf{Y}_2$$

where $\mathbf{X}_1$ and $\mathbf{Y}_1$ are in U and $\mathbf{X}_2$ and $\mathbf{Y}_2$ are in $U^{\perp}$. If a and b are scalars, we have

$$a\mathbf{X} + b\mathbf{Y} = (a\mathbf{X}_1 + b\mathbf{Y}_1) + (a\mathbf{X}_2 + b\mathbf{Y}_2)$$

where $a\mathbf{X}_1 + b\mathbf{Y}_1 \in U$ and $a\mathbf{X}_2 + b\mathbf{Y}_2 \in U^{\perp}$ because U and $U^{\perp}$ are subspaces (hence closed relative to addition and scalar multiplication). This means that the orthogonal projection of $a\mathbf{X} + b\mathbf{Y}$ on U is $a\mathbf{X}_1 + b\mathbf{Y}_1$; that is,

$$f(a\mathbf{X} + b\mathbf{Y}) = a\mathbf{X}_1 + b\mathbf{Y}_1 = af(\mathbf{X}) + bf(\mathbf{Y})$$

The same argument shows that $g(a\mathbf{X} + b\mathbf{Y}) = a\mathbf{X}_2 + b\mathbf{Y}_2 = ag(\mathbf{X}) + bg(\mathbf{Y})$, so g is also linear.

These examples should give you plenty of food for thought. If you haven't absorbed them yet, don't worry about it; we'll return to them repeatedly in the sequel. They are intended to suggest that the concept of a linear map is a broad one, occurring in many parts of mathematics. *One may describe linear algebra as the study of linear maps*. By learning something about such functions we'll be acquiring information with a greater range of applications than might at first appear to be the case.

Problems

1. Explain why the linearity properties

$$f(x + y) = f(x) + f(y) \quad \text{and} \quad f(kx) = kf(x)$$

† Note the two uses of the term "orthogonal projection," one referring to the functional value $f(\mathbf{X}) = \mathbf{X}_1$, the other to the function itself. Such duality is not uncommon; recall, for example, the use of the term *logarithm* to refer either to $f(x) = \log x$ or to f itself.

are equivalent to the single statement

$$f(ax + by) = af(x) + bf(y)$$

★ **2.** Show that the map $f: S \to T$ is linear if and only if it sends every linear combination of elements in S into the same linear combination of their images in T. In other words, linearity is equivalent to the statement that

$$f\left(\sum_{j=1}^{n} a_j u_j\right) = \sum_{j=1}^{n} a_j f(u_j)$$

where n is any positive integer, $a_1, \ldots, a_n$ are any scalars, and $u_1, \ldots, u_n$ are any elements of S.

★ **3.** Let $f: S \to T$ be a linear map.
a. Show that $f(0) = 0$; that is, a linear map always sends the zero element of S into the zero element of T.
b. Prove that $\{x \in S \mid f(x) = 0\}$ is a subspace of S. (This set is called the "kernel" of f; according to part *a* of this problem, it always contains 0.)

4. Show that the map $f: \mathcal{R}^1 \to \mathcal{R}^1$ defined by $f(x) = 2x + 1$ is not linear. (In elementary mathematics the function $f(x) = ax + b$ is usually called linear because its graph is a straight line. But unless $b = 0$ it is not linear in the sense we have used here: it does not preserve sums and scalar products.)

5. Show that the map $f: \mathcal{R}^2 \to \mathcal{R}^1$ defined by $f(x,y) = \sin(x + y)$ is not linear.

6. Which of the following maps are linear?
a. $f: \mathcal{R}^2 \to \mathcal{R}^1$ defined by $f(x,y) = x - 2y$
b. $f: \mathcal{R}^3 \to \mathcal{R}^2$ defined by $f(x,y,z) = (y - x, 2x + y + z)$
c. $f: \mathcal{R}^2 \to \mathcal{R}^2$ defined by $f(x,y) = (x - y, 2y)$
d. $f: \mathcal{R}^1 \to \mathcal{R}^3$ defined by $f(x) = (2x, 0, -x)$

★ **7.** For each complex number $z = a + bi$ define $f(z) = \bar{z}$, where $\bar{z} = a - bi$ is the conjugate of z. (See Prob. 43, Sec. 1.3.)
a. Explain why $f(u + v) = f(u) + f(v)$ for all complex numbers u and v.
b. Show that although $f(ku) = kf(u)$ for all $k \in \mathcal{R}$ and $u \in \mathcal{C}$, in general this is not the case if k is allowed to be any complex number.

The point of this problem is that the reference to the base field $\mathfrak{F}$ in our definition of linear map is not an idle one. If $\mathcal{C}$ is regarded as a vector space over $\mathcal{R}$, the map $f: \mathcal{C} \to \mathcal{C}$ defined by $f(z) = \bar{z}$ is linear, since the scalars are real. But the map $f: \mathcal{C}^1 \to \mathcal{C}^1$ defined by the same formula is not linear, since $\mathcal{C}^1$ specifies $\mathcal{C}$ as a vector space over $\mathcal{C}$. (See Prob. 2*d*, Sec. 2.1.)

8. Confirm that the map in Example 3.7 is linear by deriving the formulas

$$(A + B)^T = A^T + B^T \qquad \text{and} \qquad (kA)^T = kA^T$$

9. The "conjugate transpose" of a 2×2 matrix A is the matrix $A^* = \bar{A}^T$, where $\bar{A}$ (the "conjugate" of A) is A with its entries replaced by their conjugates.

a. If $A = \begin{bmatrix} 1+i & 1 \\ 2 & i \end{bmatrix}$, what is A^*?

b. Explain why $A^* = A^T$ if the base field is $\mathfrak{F} = \mathfrak{R}$.

c. Derive the formulas

$$(A + B)^* = A^* + B^* \qquad \text{and} \qquad (kA)^* = \bar{k}A^*$$

where A and B are 2×2 matrices with entries in $\mathfrak{F}$ and $k \in \mathfrak{F}$.

d. Let S be the vector space of 2×2 matrices with entries in $\mathfrak{F}$ and define $f: S \to S$ by $f(A) = A^*$. Is f linear?

10. Show that the rotation of the plane described in Example 3.8 is a linear operator.

11. In each case compute the dot product of $\mathbf{X}$ and $\mathbf{Y}$. (See Example 3.9.)

a. $\mathbf{X} = (1,-1,3,2)$ and $\mathbf{Y} = (2,-1,0,1)$ in $\mathfrak{R}^4$

b. $\mathbf{X} = (1,3,-1)$ and $\mathbf{Y} = (-1,1,2)$ in $\mathfrak{R}^3$

c. $\mathbf{X} = (1 + i, 1)$ and $\mathbf{Y} = (i, 2 - i)$ in $\mathfrak{C}^2$

12. Confirm that the dot product on $\mathfrak{R}^n$ is "positive-definite"; that is,

$$\mathbf{X} \cdot \mathbf{X} \geq 0 \qquad \text{with equality holding if and only if } \mathbf{X} = \mathbf{0}$$

Does this property hold for the dot product on $\mathfrak{C}^n$?

13. Explain why the dot product on $\mathfrak{F}^n$ is "symmetric"; that is,

$$\mathbf{X} \cdot \mathbf{Y} = \mathbf{Y} \cdot \mathbf{X} \qquad \text{for all } \mathbf{X} \text{ and } \mathbf{Y}$$

14. Confirm that the dot product on $\mathfrak{F}^n$ is "bilinear"; that is,

$$(a\mathbf{X} + b\mathbf{Y}) \cdot \mathbf{Z} = a(\mathbf{X} \cdot \mathbf{Z}) + b(\mathbf{Y} \cdot \mathbf{Z})$$

and

$$\mathbf{Z} \cdot (a\mathbf{X} + b\mathbf{Y}) = a(\mathbf{Z} \cdot \mathbf{X}) + b(\mathbf{Z} \cdot \mathbf{Y})$$

for all scalars a and b and vectors $\mathbf{X}$, $\mathbf{Y}$, and $\mathbf{Z}$.

15. What is the formula for $f(x,y,z)$ if $f: \mathfrak{R}^3 \to \mathfrak{R}^1$ is defined by $f(\mathbf{X}) = \mathbf{A} \cdot \mathbf{X}$, where $\mathbf{A} = (2,-1,3)$? Is f linear?

16. Let S be the vector space of real-valued functions with derivatives of all orders on $(-\infty,\infty)$ and let T be the subset of S consisting of functions whose graphs contain the origin.

a. Why is T a subspace of S?

b. For each $f \in S$, define $I(f) = g$, where $g(x) = \int_0^x f(t)\,dt$. Why is I a map from S to T?

c. Prove that $I: S \to T$ is linear.

17. Let S be $\mathfrak{R}^2$ or $\mathfrak{R}^3$ and let U be a subspace of S. If $f: S \to S$ and $g: S \to S$ are the orthogonal projections of S on U and $U^\perp$, respectively (Example 3.11), explain why $f + g = i$, the identity map on S.

18. In the context of Prob. 17, what are f and g if U is the zero subspace of S? What are they if $U = S$?

19. If U is the straight line $2x + 3y = 0$ in $\mathcal{R}^2$, then $U^\perp$ is the line through the origin perpendicular to U, with equation $3x - 2y = 0$. (See Prob. 53*d*, Sec. 2.6.)

a. By methods to be discussed in Chap. 11 (or by more elementary means which you might want to check out), it can be shown that the orthogonal projection of $\mathcal{R}^2$ on U is the linear operator $f\colon \mathcal{R}^2 \to \mathcal{R}^2$ defined by

$$f(x,y) = \tfrac{1}{13}(9x - 6y, -6x + 4y)$$

Confirm that this map sends each vector $\mathbf{X} = (x,y)$ into U.

b. Let $g\colon \mathcal{R}^2 \to \mathcal{R}^2$ be the orthogonal projection of $\mathcal{R}^2$ on $U^\perp$. Use the fact that $f + g = i$ (Prob. 17) to find $g(x,y)$.

c. Confirm that g sends each vector $\mathbf{X} = (x,y)$ into $U^\perp$.

d. Given that $\mathbf{X} = (x,y)$, the orthogonal projections of $\mathbf{X}$ on U and $U^\perp$ are

$$\mathbf{X}_1 = f(x,y) \in U \quad \text{and} \quad \mathbf{X}_2 = g(x,y) \in U^\perp$$

respectively. The orthogonal projections are supposed to be perpendicular; confirm that they are by showing that $\mathbf{X}_1 \cdot \mathbf{X}_2 = 0$.

20. If U is the plane $x - y - z = 0$ in $\mathcal{R}^3$, then $U^\perp$ is the line through the origin perpendicular to U, with parametric equations $x = t$, $y = -t$, and $z = -t$. Given that the orthogonal projection of $\mathcal{R}^3$ on U is defined by

$$f(x,y,z) = \tfrac{1}{3}(2x + y + z, x + 2y - z, x - y + 2z)$$

repeat Prob. 19.

3.2 The Law of Nullity

The analysis of linear maps is one of the major topics of linear algebra; much of the rest of this book will be devoted to it in one way or another. In this section we derive a theorem about linear maps discovered by James Sylvester (1814–1897), one of the pioneers of linear algebra. It will be of great importance in the sequel, as you will see.

KERNEL

Let $f\colon S \to T$ be a linear map from S to T, where S and T are vector spaces over $\mathfrak{F}$. The *kernel* of f is the set of elements of S which f maps into the zero element of T; that is,

$$\ker f = \{x \in S \mid f(x) = 0\}$$

Sometimes the kernel is also called the *null space* of f.

EXAMPLE 3.12

Suppose that $f\colon \mathcal{R}^1 \to \mathcal{R}^1$ is defined by $f(x) = 2x$. Then f is linear and

$$\ker f = \{x \in \mathcal{R}^1 \mid f(x) = 0\} = \{0\}$$

Thus the only element that maps into 0 is 0.

EXAMPLE 3.13

The other extreme, in which everything maps into 0, is the zero map $0\colon S \to T$ defined by $0(x) = 0$, $x \in S$. In this case, we have $\ker f = S$.

EXAMPLE 3.14

Let f be the orthogonal projection of $\mathcal{R}^3$ on the xy plane, defined by $f(x,y,z) = (x,y,0)$. (See Fig. 3.4. For a definition of orthogonal projection, see Example 3.11.) It is geometrically apparent that the set of points f sends into the origin is the z axis. To confirm this by the definition, we write

$$\begin{aligned}\ker f &= \{(x,y,z) \mid f(x,y,z) = (0,0,0)\} = \{(x,y,z) \mid (x,y,0) = (0,0,0)\} \\ &= \{(x,y,z) \mid x = 0,\ y = 0,\ z \in \mathcal{R}\} = \{(0,0,z) \mid z \in \mathcal{R}\}\end{aligned}$$

Thus the kernel of a linear map $f\colon S \to T$ may consist of zero alone (Example 3.12), or all of S (Example 3.13), or something in between (Example 3.14). Whatever the case, the kernel is always a subspace of S, as we now prove.

THEOREM 3.1

If $f\colon S \to T$ is linear, then $\ker f$ is a subspace of S.

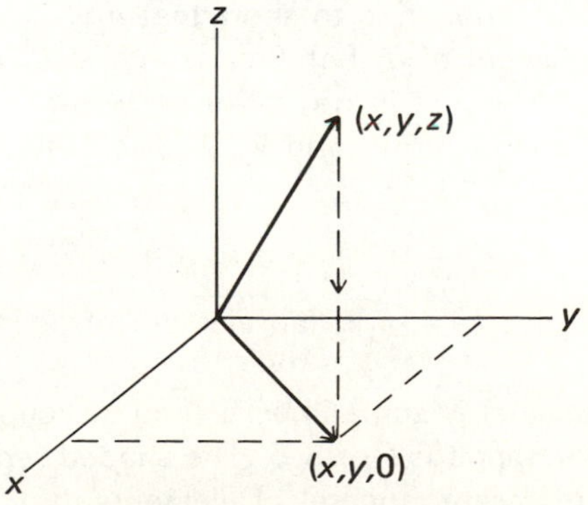

Fig. 3.4 Orthogonal projection on the xy plane

Proof

Since a linear map always sends the zero element of S into the zero element of T (Problem 3, Sec. 3.1), we have $0 \in \ker f$ and hence $\ker f$ is nonempty. It is a subspace of S provided that

$$x \in \ker f \text{ and } y \in \ker f \quad \Rightarrow \quad ax + by \in \ker f \qquad \text{for all scalars } a \text{ and } b$$

(See Corollary 2.14a.) But this follows from

$$f(ax + by) = af(x) + bf(y) = a0 + b0 = 0 \; \blacksquare$$

RANGE

Let $f\colon S \rightarrow T$ be a linear map. The *range* of f is the subset of T defined by

$$f(S) = \operatorname{rng} f = \{y \in T \mid y = f(x),\; x \in S\}$$

In other words, the range of f (as with every function) is the set of functional values $f(x)$ one gets when x takes on all values in the domain. When f is linear, however, its range is not merely a subset of T, but a subspace.

THEOREM 3.2

If $f\colon S \rightarrow T$ is linear, then $\operatorname{rng} f$ is a subspace of T.

Proof

Since $f(0) = 0$, $\operatorname{rng} f$ contains at least the zero element of T, so it is not empty. To prove that it is a subspace, take any y_1 and y_2 in $\operatorname{rng} f$ and any scalars a_1 and a_2. The problem is to show that $a_1y_1 + a_2y_2 \in \operatorname{rng} f$, which involves naming an element of S that f maps into $a_1y_1 + a_2y_2$.

Since y_1 and y_2 are in $\operatorname{rng} f$, we know that there are elements x_1 and x_2 in S which f maps into y_1 and y_2, respectively: $f(x_1) = y_1$ and $f(x_2) = y_2$. It follows that

$$f(a_1x_1 + a_2x_2) = a_1f(x_1) + a_2f(x_2) = a_1y_1 + a_2y_2$$

That is, f maps $a_1x_1 + a_2x_2$ (an element of S) into $a_1y_1 + a_2y_2$. $\blacksquare$

A diagram that is sometimes helpful in keeping straight the locations of the kernel and range is Fig. 3.5. The shaded region in S (which may consist of 0 alone) represents the set of elements that f sends into 0, while in T it represents the set of images of elements of S.

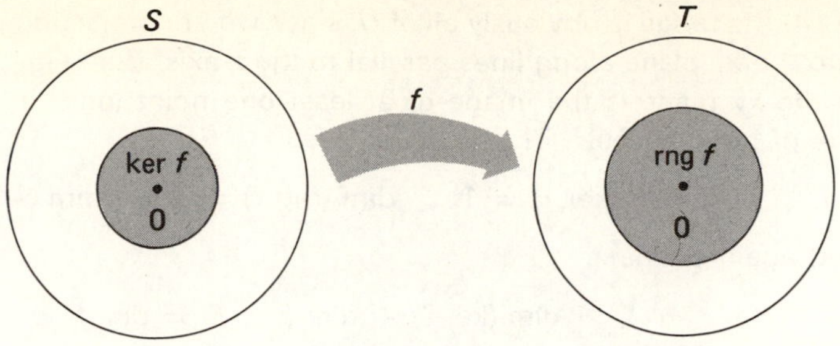

Fig. 3.5 Kernel and range of a linear map

EXAMPLE 3.15

Define $f\colon \mathcal{R}^2 \to \mathcal{R}^3$ by $f(x,y) = (x + y, 2x, x - y)$. (See Example 3.4.) The range of f is the subset of $\mathcal{R}^3$ defined by

$$\text{rng } f = \{(x,y,z) \mid x = s + t,\ y = 2s,\ z = s - t,\ (s,t) \in \mathcal{R}^2\}$$

(Why?) That is, rng f is the set of points (x,y,z) characterized by the parametric equations

$$x = s + t \qquad y = 2s \qquad z = s - t$$

Eliminating the parameters s and t yields the equation $x - y + z = 0$. (Confirm!) This equation represents a plane through the origin, that is, a subspace of $\mathcal{R}^3$. (See Prob. 44, Sec. 2.6.)

The kernel of f is the subspace of $S = \mathcal{R}^2$ defined by

$$\ker f = \{(x,y) \mid (x + y, 2x, x - y) = (0,0,0)\}$$

This is the set of points (x,y) satisfying the system

$$x + y = 0 \qquad 2x = 0 \qquad x - y = 0$$

of which the only solution (as you can see by inspection) is $(x,y) = (0,0)$. Thus ker f is the zero subspace of S. Note that

$$\dim(\ker f) = 0 \qquad \dim(\text{rng } f) = 2 \qquad \dim S = 2$$

Consequently the formula

$$\dim(\ker f) + \dim(\text{rng } f) = \dim S$$

holds in this case. That it holds in general (when S is finite-dimensional) is the ''law of nullity,'' which we are going to prove. But first let's look at another example.

EXAMPLE 3.16

If U is the xy plane in $S = \mathcal{R}^3$, the projection $f\colon S \to S$ defined by $f(x,y,z) = (x,y,0)$ has a kernel consisting of the z axis in $\mathcal{R}^3$, as we have seen (Example

3.14). Its range is obviously all of U, since we are projecting points of 3-space into the xy plane along lines parallel to the z axis. (See Fig. 3.4.) Every point of the xy plane is the image of at least one point (and, in fact, of a whole line of points) in $\mathcal{R}^3$. Thus

$$\dim(\ker f) = 1 \qquad \dim(\operatorname{rng} f) = 2 \qquad \dim S = 3$$

and again we have

$$\dim(\ker f) + \dim(\operatorname{rng} f) = \dim S$$

These examples should help make the next theorem less mysterious.

THEOREM 3.3 Sylvester's Law of Nullity

If S is finite-dimensional and $f: S \to T$ is linear, then $\dim(\ker f) + \dim(\operatorname{rng} f) = \dim S$.

Proof

Let $n = \dim S$, $k = \dim(\ker f)$, and $r = n - k$; the problem is to show that $\dim(\operatorname{rng} f) = r$. This is trivial if S is a zero space or if $\ker f$ is all of S, so assume that n is positive and $k < n$. (According to Theorems 2.16 and 2.17, we have $k \leq n$, with equality holding only if $\ker f = S$.)

The idea of the proof (which virtually generates itself, as you will see) is to produce bases for ker f, rng f, and S with the right number of elements. Start by choosing any basis for ker f, say $\{u_{r+1}, \ldots, u_n\}$, and build it up to a basis $\{u_1, \ldots, u_n\}$ for S.† To visualize where these elements are, look at Fig. 3.5. The elements $u_1, \ldots, u_r$ are outside the shaded part of S, while $u_{r+1}, \ldots, u_n$ (if they occur) are in the kernel.

To produce a basis for rng f (the shaded part of T in Fig. 3.5), we have to move from S to T. The map f provides the means for doing this, and the images of $u_1, \ldots, u_n$ are the only ones available for inspection. There is no point in looking at $f(u_{r+1}), \ldots, f(u_n)$, for they are all 0. (Why?) We are therefore irresistibly drawn to look at

$$v_1 = f(u_1), \quad \ldots, \quad v_r = f(u_r)$$

and to claim that these r images constitute a basis for rng f.

LECTOR *I must admit you've been crafty.*

AUCTOR *Nothing to it. How else could we get this proof off the ground?*

To finish the proof, we need to show two things. First, $v_1, \ldots, v_r$ span rng f, for if y is any element of rng f, there is an element $x \in S$ such that

† We have put $n - r$ elements in the basis for ker f because $\dim(\ker f) = k = n - r$. Labeling them $u_{r+1}, \ldots, u_n$ is purely a matter of convenience. Of course k may be 0, in which case these elements do not appear (and $r = n$). In that event start by choosing any basis $\{u_1, \ldots, u_n\}$ for S (and treat the occurrence of $u_{r+1}, \ldots, u_n$ in the rest of the proof as vacuous).

$f(x) = y$. The only way to get a handle on x is to write it in terms of the basis for S:

$$x = \sum_{j=1}^{n} a_j u_j$$

To get back to y, use f (which is linear):

$$y = f(x) = f\left(\sum_{j=1}^{n} a_j u_j\right) = \sum_{j=1}^{n} a_j f(u_j) \qquad \text{(Prob. 2, Sec. 3.1)}$$

But $u_{r+1}, \ldots, u_n$ are in ker f, so $f(u_j) = 0$, where $j = r + 1, \ldots, n$, and the above sum reduces to

$$y = \sum_{j=1}^{r} a_j f(u_j) = \sum_{j=1}^{r} a_j v_j$$

Thus y is a linear combination of $v_1, \ldots, v_r$, as advertised.

The second thing we must show is that $v_1, \ldots, v_r$ are linearly independent. Suppose

$$\sum_{j=1}^{r} a_j v_j = 0$$

that is,

$$\sum_{j=1}^{r} a_j f(u_j) = 0$$

Since f is linear, this can be written

$$f\left(\sum_{j=1}^{r} a_j u_j\right) = 0$$

which means that the element

$$x = \sum_{j=1}^{r} a_j u_j$$

is in ker f. If ker f is the zero space, then $x = 0$ and the linear independence of $u_1, \ldots, u_r$ implies that each a_j is 0. If it is not the zero space, then x can be written as a linear combination of $u_{r+1}, \ldots, u_n$, say

$$x = \sum_{j=r+1}^{n} b_j u_j$$

and we have

$$\sum_{j=1}^{r} a_j u_j = \sum_{j=r+1}^{n} b_j u_j$$

Writing this in the form

$$a_1 u_1 + \cdots + a_r u_r - b_{r+1} u_{r+1} - \cdots - b_n u_n = 0$$

we conclude from the linear independence of $u_1, \ldots, u_n$ that each coefficient is 0. In particular, each a_j is 0. Hence whatever ker f is, we have shown that

$$\sum_{j=1}^{r} a_j v_j = 0 \quad \Rightarrow \quad a_j = 0 \qquad j = 1, \ldots, r \; \blacksquare$$

The reason this theorem is referred to as the "law of nullity" is that some writers call ker f the "null space" of f and its dimension the "nullity" of f. They also call dim (rng f) the "rank" of f. Then Sylvester's law reads:

Nullity plus rank equals dimension of domain.

Later on (after we have discussed matrices) we'll see why such terminology is appropriate.

EXAMPLE 3.17

Define f: $\mathcal{R}^2 \to \mathcal{R}^1$ by $f(x,y) = y - 2x$. Since

$$\ker f = \{(x,y) \mid y - 2x = 0\}$$

(a straight line through the origin in $\mathcal{R}^2$), we have dim (ker f) = 1. Hence

$$\dim (\text{rng } f) = \dim \mathcal{R}^2 - \dim (\ker f) = 2 - 1 = 1$$

which means that rng $f = \mathcal{R}^1$. (See Theorem 2.17.) Thus we may conclude that f maps $\mathcal{R}^2$ "onto" $\mathcal{R}^1$, not merely "into"; that is, every element of $\mathcal{R}^1$ is the image of some element $(x,y) \in \mathcal{R}^2$. (The shaded part of T in Fig. 3.5 is all of T.) It is certainly not very hard to see this directly! However, Sylvester's law makes it unnecessary to argue the question.

EXAMPLE 3.18

Define f: $\mathcal{R}^2 \to \mathcal{R}^3$ by $f(x,y) = (x + y, 2x, x - y)$. The kernel of f is the zero subspace of $\mathcal{R}^2$, so we have dim (rng f) $= 2 - 0 = 2$. Hence rng f is a plane through the origin in $\mathcal{R}^3$, a fact it took some work to discover before. (See Example 3.15, where we had to eliminate two parameters from a system of three equations for x, y, z to characterize rng f by the equation $x - y + z = 0$.) The law of nullity is a shortcut.

Problems

21. Prove Sylvester's law of nullity in the special cases where S is a zero space or ker f is all of S.

22. Each of the following maps is linear. (See Prob. 6, Sec. 3.1.) In each case, find the kernel, its dimension, and describe it geometrically.
a. $f\colon \mathcal{R}^2 \to \mathcal{R}^1$ defined by $f(x,y) = x - 2y$
b. $f\colon \mathcal{R}^3 \to \mathcal{R}^2$ defined by $f(x,y,z) = (y - x,\ 2x + y + z)$
c. $f\colon \mathcal{R}^2 \to \mathcal{R}^2$ defined by $f(x,y) = (x - y,\ 2y)$
d. $f\colon \mathcal{R}^1 \to \mathcal{R}^3$ defined by $f(x) = (2x, 0, -x)$

23. In each part of Prob. 22, find the range (without using Sylvester's law), name its dimension, and describe it geometrically.

24. In each part of Prob. 22 check Sylvester's law of nullity.

25. Define $f\colon \mathcal{C} \to \mathcal{C}$ by $f(z) = \bar{z}$ (where $\mathcal{C}$ is regarded as a vector space over $\mathcal{R}$). Then f is linear (Prob. 7, Sec. 3.1); find the kernel and range and check the law of nullity.

26. Let S be the vector space of 2×2 matrices with entries in $\mathfrak{F}$. Find the kernel and range of each of the following linear maps and check the law of nullity.
a. $f\colon S \to \mathfrak{F}^4$ defined by

$$f\left(\begin{bmatrix} a & b \\ c & d \end{bmatrix}\right) = (a,b,c,d)$$

(See Example 3.6.)
b. $f\colon S \to S$ defined by $f(A) = A^T$. (See Example 3.7.)

27. Let f be the rotation of the plane described in Example 3.8. Prove that the kernel of f is the zero subspace of $\mathcal{R}^2$ and use the law of nullity to conclude that rng $f = \mathcal{R}^2$. (In other words, you will show that every vector in the plane is the image of some vector which is rotated into it, a fact that is geometrically apparent.)

28. Let D be the linear operator defined in Example 3.5. What is ker D? What does the law of nullity say in this case?

29. Let S be the vector space of real-valued functions with derivatives of all orders on $(-\infty,\infty)$ and let T be the subspace of S consisting of functions whose graphs contain the origin. Find the kernel and range of the linear map $I\colon S \to T$ which sends each $f \in S$ into $g(x) = \int_0^x f(t)\,dt$. (See Prob. 16, Sec. 3.1.)

30. The orthogonal projection of $\mathcal{R}^2$ on the line $U = \{(x,y) \mid 2x + 3y = 0\}$ is defined by $f(x,y) = \frac{1}{13}(9x - 6y,\ -6x + 4y)$. (See Prob. 19, Sec. 3.1.) Find the kernel and range of f and describe them geometrically.

31. The orthogonal projection of $\mathcal{R}^3$ on the plane $U = \{(x,y,z) | x - y - z = 0\}$ is defined by $f(x,y,z) = \frac{1}{3}(2x + y + z, x + 2y - z, x - y + 2z)$. (See Prob. 20, Sec. 3.1.) Find the kernel and range of f and describe them geometrically.

★ **32.** Let U be a subspace of S (where S is $\mathcal{R}^2$ or $\mathcal{R}^3$) and let $f: S \to S$ be the orthogonal projection of S on U (Example 3.11).

a. Prove that $\ker f = U^\perp$.

b. Prove that $\text{rng } f = U$.

c. Use Sylvester's law of nullity to conclude that $\dim U + \dim U^\perp = \dim S$. (This is a special case of the projection theorem, to be proved in Chap. 11. Also see Prob. 53, Sec. 2.6.)

★ **33.** Prove that if S is finite-dimensional and $f: S \to T$ is a linear map whose range is all of T, then T is finite-dimensional and $\dim T \leq \dim S$. (Thus a linear map cannot carry a finite-dimensional space "onto" a higher-dimensional space, but only "into." See the next section for a formal definition of "onto.")

★ **34.** Let $f: S \to T$ be a linear map and suppose that $y \in \text{rng } f$. Then there is an element $x_p \in S$ such that $f(x_p) = y$; show that *every* element that f maps into y can be written as the sum of an element of $\ker f$ and the element x_p. Conversely, show that every such sum maps into y.

The point of this problem is that the solution set of the equation $f(x) = y$ (where y is given) is completely determined if *one* solution is known and if the solution set of the "homogeneous" equation $f(x) = 0$ (the kernel of f) has been found. (See Fig. 3.6.) This fact has many important applications, one of which is outlined in Prob. 35.

35. Let $L: S \to S$ be the second-order differential operator defined by $L(\phi) = \phi'' + \phi$, where S is the vector space of real-valued functions with derivatives of all orders on $(-\infty,\infty)$.

a. Prove that L is linear.

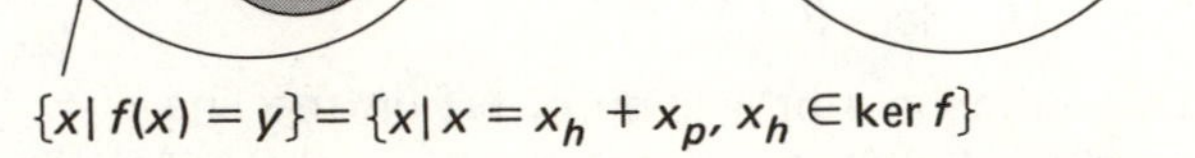

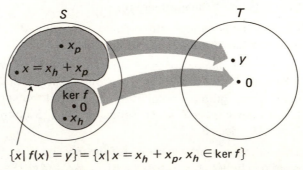

Fig. 3.6 Solution sets of $f(x) = y$ and $f(x) = 0$

b. Explain why ker L is the set of solutions of the differential equation $y'' + y = 0$. (See Example 2.8.)

c. Show that if $\phi_1(x) = \cos x$ and $\phi_2(x) = \sin x$, then $L(c_1\phi_1 + c_2\phi_2) = 0$ for all scalars c_1 and c_2. (Thus every linear combination of cos and sin is a solution of $y'' + y = 0$. Conversely, it can be shown that every solution is of this form, which means that cos and sin constitute a basis for ker L. The solution set of the homogeneous equation $y'' + y = 0$ is therefore known.)

d. Show that $\phi_p(x) = x^2 - 2$ is a solution of the equation $y'' + y = x^2$. Why does this imply that $g(x) = x^2$ is an element of rng L?

e. Use Prob. 34 to explain why every solution of $y'' + y = x^2$ is of the form $\phi(x) = c_1 \cos x + c_2 \sin x + (x^2 - 2)$.

f. Use part *e* to find the unique solution of $y'' + y = x^2$ satisfying the initial conditions $\phi(0) = 1$, $\phi'(0) = -2$.

3.3 Invertible Linear Maps

Let S be the vector space of 2×2 matrices with entries in $\mathfrak{F}$. As we have seen (Example 3.6), the map $f: S \to \mathfrak{F}^4$ defined by

$$f\left(\begin{bmatrix} a & b \\ c & d \end{bmatrix}\right) = (a,b,c,d)$$

is linear; if A and B are matrices in S and $\mathbf{X}$ and $\mathbf{Y}$ are their vector images in $\mathfrak{F}^4$, then

$$f(A + B) = \mathbf{X} + \mathbf{Y} \qquad \text{and} \qquad f(kA) = k\mathbf{X} \qquad (k \in \mathfrak{F})$$

Sums and scalar products are therefore preserved in the process of mapping matrices into vectors; the algebraic structure of S is carried over into $\mathfrak{F}^4$.

When we generalized these remarks to an arbitrary linear map $f: S \to T$, we warned that this duplication of algebraic structure does not necessarily mean that S and T are much alike, for there may be a constriction taking place. An extreme example of this is the zero map, which sends everything in S into the zero element of T. In this case the duplication of the algebraic structure of S is compressed into the zero subspace of T and is largely meaningless.

Another example is the projection $f: \mathcal{R}^3 \to \mathcal{R}^2$ defined by $f(x,y,z) = (x,y)$.† Although the structure of 3-space is imitated in the xy plane into which it is projected, the replica is by no means faultless. Whole lines in $\mathcal{R}^3$ end up as points in $\mathcal{R}^2$. For example, given a point like (2,5) in the range, we cannot

† Since the range is not a subspace of the domain, this is not really a projection, whereas the map $f(x,y,z) = (x,y,0)$ is. But it is convenient to think of it as one; imagine $\mathcal{R}^2$ embedded in $\mathcal{R}^3$ as the xy plane.

even tell where it came from in $\mathcal{R}^3$ because it is the image of infinitely many points in the domain, namely all the points $(2,5,z)$ in $\mathcal{R}^3$.

An example in the other direction is the linear map $f\colon \mathcal{R}^2 \to \mathcal{R}^3$ defined by $f(x,y) = (x + y, 2x, x - y)$. In Example 3.15 we found that the range of f is the plane $x - y + z = 0$ in $\mathcal{R}^3$. The duplication in this plane of the algebraic structure of $\mathcal{R}^2$ is precise, for each of its points is the image of one and only one point in $\mathcal{R}^2$. (Can you prove this?) But the plane is only part of $\mathcal{R}^3$; other points are not even touched by the map. (See Fig. 3.7.)

Returning to the map $f\colon S \to \mathfrak{F}^4$ of 2×2 matrices into 4-tuples, we observe that none of these difficulties occurs. Each vector $(a,b,c,d) \in \mathfrak{F}^4$ is the image of one and only one matrix

$$\begin{bmatrix} a & b \\ c & d \end{bmatrix}$$

in S; there is an exact correspondence between what goes on in S and what goes on in $\mathfrak{F}^4$. We could even turn things around and say that the map $f^{-1}\colon \mathfrak{F}^4 \to S$ defined by

$$f^{-1}(a,b,c,d) = \begin{bmatrix} a & b \\ c & d \end{bmatrix}$$

carries the structure of $\mathfrak{F}^4$ back onto S in a one-to-one fashion. (The linearity of the inverse map should be checked, but in this case it is pretty apparent.)

Such a linear map (we'll give precise definitions shortly) is called an "isomorphism" (Greek for "identical structure"). Two vector spaces which are related by an isomorphism are for all practical purposes the same thing; the only distinction between them is that the notation used for their elements and operations may be different. Thus in the above example we may say that the space S of 2×2 matrices is nothing more than 4-space in disguise. A matrix

$$\begin{bmatrix} a & b \\ c & d \end{bmatrix}$$

doesn't look like a vector (a,b,c,d), but the difference is purely notational.†

ONTO

Let $f\colon S \to T$ be a map (not necessarily linear). If the range of f is all of T, so that every element of T is the image of at least one element of S, we say that f maps S *onto* T, or simply that f is *onto*. (Some writers call f *surjective*.)

† Later on, when matrix multiplication is introduced, this will no longer be the case, for we will not duplicate the definition in $\mathfrak{F}^4$. But then S will have a different algebraic structure, not confined to our present discussion of vector spaces.

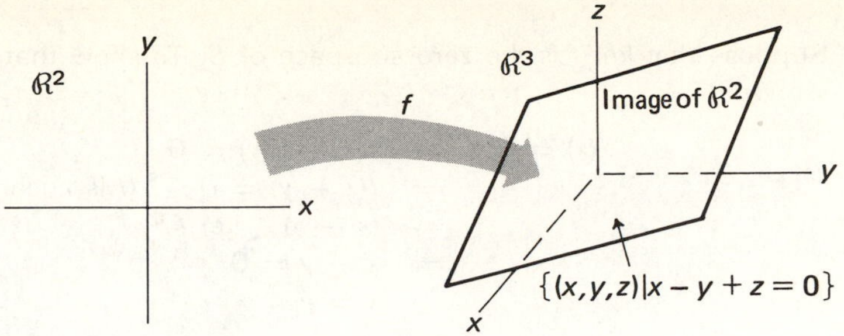

Fig. 3.7 Duplication of $\mathcal{R}^2$ in $\mathcal{R}^3$

The essential property of an "onto" map is that if $y \in T$, there is at least one element $x \in S$ such that $f(x) = y$. Note, however, that this is not enough to guarantee identical structure of S and T (even when f is linear). The map $f: \mathcal{R}^3 \to \mathcal{R}^2$ defined by $f(x,y,z) = (x,y)$ is onto (why?), but there are infinitely many points of $\mathcal{R}^3$ mapping into each point of $\mathcal{R}^2$.

ONE-TO-ONE

Let $f: S \to T$ be a map (not necessarily linear). If distinct elements of S are mapped into distinct elements of T, that is, if

$$x \neq y \quad \Rightarrow \quad f(x) \neq f(y)$$

we call f a *one-to-one* map. (Some writers call f *injective*.) Equivalently, f is one-to-one if

$$f(x) = f(y) \quad \Rightarrow \quad x = y$$

The essential property of a one-to-one map is that no element of T is the image of more than one element of S. Note that this (even with linearity) is not enough to guarantee identical structure either! The map $f: \mathcal{R}^2 \to \mathcal{R}^3$ defined by $f(x,y) = (x + y, 2x, x - y)$ is one-to-one, but some points in $\mathcal{R}^3$ are not images of anything. Thus f is not an onto map.†

A simple test for discovering whether a *linear* map is one-to-one is the following.

THEOREM 3.4

The linear map $f: S \to T$ is one-to-one if and only if ker f is the zero subspace of S.

† However, see Prob. 49. If f is changed to $f: \mathcal{R}^2 \to T$, where $T = \text{rng } f$, it becomes onto.

Proof

Suppose that ker f is the zero subspace of S. To show that f is one-to-one, we write

$$\begin{aligned} f(x) = f(y) &\Rightarrow f(x) - f(y) = 0 \\ &\Rightarrow f(x - y) = 0 \qquad (f \text{ is linear}) \\ &\Rightarrow (x - y) \in \ker f \\ &\Rightarrow x - y = 0 \\ &\Rightarrow x = y \end{aligned}$$

The "only if" part of the theorem is left for the problems. ∎

To see how this theorem simplifies things, look again at the map $f\colon \mathcal{R}^2 \to \mathcal{R}^3$ defined by $f(x,y) = (x + y, 2x, x - y)$. We asked you earlier whether you could prove that this map is one-to-one; with only the definition at hand, it is necessary to do something like the following. Suppose that $f(x_1,y_1) = f(x_2,y_2)$. Then

$$\begin{aligned} x_1 + y_1 &= x_2 + y_2 \\ 2x_1 &= 2x_2 \\ x_1 - y_1 &= x_2 - y_2 \end{aligned}$$

which implies that $x_1 = x_2$ and $y_1 = y_2$. That is,

$$f(\mathbf{X}_1) = f(\mathbf{X}_2) \quad \Rightarrow \quad \mathbf{X}_1 = \mathbf{X}_2$$

However, this is the hard way. One can see by inspection that if

$$f(x,y) = (x + y, 2x, x - y) = (0,0,0)$$

then

$$(x,y) = (0,0)$$

The only element that f maps into zero is zero. Thus ker f is the zero subspace of $\mathcal{R}^2$ and by Theorem 3.4 f is one-to-one.

This state of affairs is remarkable when you think about it! What the theorem says is that if the zero element of T is the image of only one element of S, then *no* element of T can be the image of more than one element of S. We lift ourselves by our bootstraps from what seems like an insignificant scrap of information to a general statement about the map. Of course the reason it works is that f is linear, as you can see in the proof. This property gives the map a coherence that spreads the uniqueness of the "preimage" of zero throughout the whole domain.

Maps which are both one-to-one and onto are extremely important in mathematics, *for they can be reversed*. To see what we mean by this, let $f\colon S \to T$ be such a map and take any $y \in T$. Since f is onto, there exists at least one $x \in S$ such that $f(x) = y$; moreover, there is only one such x because f is one-to-one. It therefore makes sense to say that each element $y \in T$ can be mapped into a definite element $x \in S$. More precisely, define $g: T \to S$ by $g(y) = x$, where $y \in T$ and x is the unique element of S satisfying $f(x) = y$. Then g is the "inverse" of f, formally defined as follows.

INVERSE

Let $f: S \to T$ be a map (not necessarily linear) which is both one-to-one and onto. The map $g: T \to S$ defined by $g(y) = x$, where $y \in T$ and x is the unique element of S satisfying $f(x) = y$, is called the *inverse* of f and is denoted by f^{-1}. In these circumstances f is said to be *invertible* (also called a *one-to-one correspondence* between S and T).

ISOMORPHISM

An invertible linear map from S to T (where S and T are vector spaces over $\mathfrak{F}$) is called an *isomorphism*. If such a map exists, we say that S is *isomorphic* to T.

To see how the inverse of an invertible map may be found explicitly, consider the map $f: \mathcal{R}^2 \to \mathcal{R}^2$ defined by $f(x,y) = (x',y')$, where (for a given θ)

$$\begin{aligned} x' &= x\cos\theta - y\sin\theta \\ y' &= x\sin\theta + y\cos\theta \end{aligned}$$

This is a rotation of the plane through the angle θ; you have already proved that it is linear. (See Example 3.8 and Prob. 10, Sec. 3.1.) Note that ker f is the set of points (x,y) satisfying the system

$$\begin{aligned} x\cos\theta - y\sin\theta &= 0 \\ x\sin\theta + y\cos\theta &= 0 \end{aligned}$$

for which the only solution is $(x,y) = (0,0)$. (See Prob. 27, Sec. 3.2.) Hence ker f is the zero subspace of $\mathcal{R}^2$ and f is one-to-one. To prove that it is onto, use Sylvester's law of nullity to write

$$\dim(\text{rng } f) = \dim \mathcal{R}^2 - \dim(\ker f) = 2 - 0 = 2$$

from which it follows that rng $f = \mathcal{R}^2$. (This implication is automatic in the case of a linear map $f: S \to T$ when S and T have the same dimension. See Prob. 46.)

Thus f is invertible; its inverse is the map $f^{-1}: \mathcal{R}^2 \to \mathcal{R}^2$ defined by $f^{-1}(x',y') = (x,y)$, where (x,y) is the unique element of $\mathcal{R}^2$ satisfying $f(x,y) =$

(x',y'). However, this is not explicit. What we want is a formula for $f^{-1}(x',y')$ similar to the formula

$$f(x,y) = (x \cos \theta - y \sin \theta,\ x \sin \theta + y \cos \theta)$$

which defines f. Given that $(x',y') \in \mathbb{R}^2$, we therefore solve the system

$$\begin{aligned} x \cos \theta - y \sin \theta &= x' \\ x \sin \theta + y \cos \theta &= y' \end{aligned}$$

for x and y in terms of x' and y'. The result, which you can check by Cramer's rule† or by solving directly, is

$$\begin{aligned} x &= x' \cos \theta + y' \sin \theta \\ y &= -x' \sin \theta + y' \cos \theta \end{aligned}$$

Thus the desired formula for f^{-1} is

$$f^{-1}(x',y') = (x' \cos \theta + y' \sin \theta,\ -x' \sin \theta + y' \cos \theta)^{\ddagger}$$

It is no doubt apparent in this example that the inverse is itself an invertible linear map and that $(f^{-1})^{-1} = f$. We prove this in general as follows.

THEOREM 3.5

If $f: S \rightarrow T$ is an invertible linear map, so is $f^{-1}: T \rightarrow S$.

Proof

First we show that f^{-1} is linear. Take any elements u and v in T and let $x = f^{-1}(u)$ and $y = f^{-1}(v)$. Then if a and b are scalars, we have

$$f(ax + by) = af(x) + bf(y) = au + bv$$

from which

$$f^{-1}(au + bv) = ax + by = af^{-1}(u) + bf^{-1}(v)$$

To show that f^{-1} is one-to-one, suppose that u and v are elements of T such that $f^{-1}(u) = f^{-1}(v)$. Then if $x = f^{-1}(u)$ and $y = f^{-1}(v)$, we have $x = y$ and hence $f(x) = f(y)$; that is, $u = v$.¶

The proof that f^{-1} is onto is left for the problems. ∎

COROLLARY 3.5a

If $f: S \rightarrow T$ is an invertible linear map, then $(f^{-1})^{-1} = f$.

† You may have encountered this in an earlier course. For its statement in the 2×2 case, see Sec. 7.1; a more general version is derived in Sec. 8.5.

‡ We have used an elephant gun to shoot a mouse. For an easier way, see Prob. 40.

¶ If f is *any* function, then $x = y \Rightarrow f(x) = f(y)$. (Why?)

Proof

Let $g = f^{-1}$. Then g is an invertible linear map by Theorem 3.5; let $h = g^{-1}$. The problem is to prove that h and f are the same maps, that is, to confirm that $h(x) = f(x)$ for every $x \in S$. Take any $x \in S$ and let $y = f(x)$. Then, by definition of inverse, we have $g(y) = x$ and hence $h(x) = y$. Therefore $h(x) = f(x)$. ∎

Problems

36. Define $f\colon \mathcal{R} \to \mathcal{R}$ by $f(x) = 3x - 2$. Show that f is invertible but not linear. What is the formula for $f^{-1}(x)$?†

37. Show that the map $f\colon \mathcal{R} \to \mathcal{R}$ defined by $f(x) = x^2$ is neither linear, onto, nor one-to-one.

38. Which of the following linear maps are one-to-one? Which are onto? Which are invertible? (See Probs. 22 and 23, Sec. 3.2.)
   a. $f\colon \mathcal{R}^2 \to \mathcal{R}^1$ defined by $f(x,y) = x - 2y$
   b. $f\colon \mathcal{R}^3 \to \mathcal{R}^2$ defined by $f(x,y,z) = (y - x, 2x + y + z)$
   c. $f\colon \mathcal{R}^2 \to \mathcal{R}^2$ defined by $f(x,y) = (x - y, 2y)$
   d. $f\colon \mathcal{R}^1 \to \mathcal{R}^3$ defined by $f(x) = (2x, 0, -x)$

39. Find the inverse of each of the following invertible linear maps.
   a. $f\colon \mathcal{R}^2 \to \mathcal{R}^2$ defined by $f(x,y) = (x + y, 2x + 3y)$
   b. $f\colon \mathcal{C} \to \mathcal{C}$ defined by $f(z) = \bar{z}$ (where $\mathcal{C}$ is regarded as a vector space over $\mathcal{R}$)
   c. $f\colon S \to S$ defined by $f(A) = A^T$, where S is the vector space of 2×2 matrices with entries in $\mathfrak{F}$.

40. Let $f\colon \mathcal{R}^2 \to \mathcal{R}^2$ be the rotation

$$f(x,y) = (x \cos\theta - y \sin\theta,\ x \sin\theta + y \cos\theta)$$

Since f^{-1} reverses f, it must be a rotation through $-\theta$. Hence the formula for $f^{-1}(x,y)$ should drop out if we use the formula for $f(x,y)$ with θ replaced by $-\theta$. Confirm that this works. (You may also want to replace (x,y) by (x',y') to get the formula in the text. However, the letters used are immaterial.)

41. Is the differential operator D in Example 3.5 an isomorphism?

42. Let S be the vector space of real-valued functions with derivatives of all orders on $(-\infty,\infty)$ and let T be the subspace of S consisting of functions whose graphs contain the origin. Explain why the linear map

† You may feel that we should be asking for $f^{-1}(y)$ since $y = f(x) \Leftrightarrow x = f^{-1}(y)$. Once $f^{-1}(y)$ is found, however, we can just as well use x for the independent variable. The letter used is unimportant.

$I: S \to T$ which sends each $f \in S$ into $g(x) = \int_0^x f(t)\, dt$ is invertible and find its inverse. (See Prob. 16, Sec. 3.1, and Prob. 29, Sec. 3.2.)

The result is shocking, for it says that a vector space can be isomorphic to a proper subspace of itself. This situation is impossible when the space is finite-dimensional (see Prob. 48), which may be some consolation.

43. Prove that if $f: S \to T$ is a one-to-one linear map, then ker f is the zero subspace of S (the "only if" part of Theorem 3.4).

44. Finish the proof of Theorem 3.5 by showing that the inverse of an invertible linear map is onto. Also confirm that Theorem 3.5 and Corollary 3.5a are true if "linear" is left out of their statements; that is, the inverse of an invertible map f (not necessarily linear) is invertible, and $(f^{-1})^{-1} = f$.

★ **45.** Prove that if $f: S \to T$ is a linear map and dim S = dim T, then f is one-to-one if and only if it is onto. (Thus the notions of one-to-one and onto are equivalent in this case, and f is invertible if it has either one of these properties.)

46. Let $f: S \to T$ be a linear map. Prove the following.

a. If $u_1, \ldots, u_n$ span S and f is onto, then $v_1, \ldots, v_n$ span T, where $v_j = f(u_j)$ for $j = 1, \ldots, n$.

b. If $u_1, \ldots, u_n$ are linearly independent in S and f is one-to-one, then $v_1, \ldots, v_n$ are linearly independent in T.

Thus an onto linear map "preserves spanning," while a one-to-one linear map "preserves linear independence."

★ **47.** Suppose that $f: S \to T$ is an invertible linear map. Prove the following.

a. If $\{u_1, \ldots, u_n\}$ is a basis for S, then $\{v_1, \ldots, v_n\}$ is a basis for T, where $v_j = f(u_j)$ for $j = 1, \ldots, n$. (Thus an isomorphism "preserves bases.")

b. If S is finite-dimensional, so is T, and dim S = dim T. (Thus an isomorphism "preserves dimension.")

c. If T is finite-dimensional, so is S, and dim S = dim T. *Hint:* Use Theorem 3.5.

48. Suppose that S is isomorphic to a subspace T. If either space is finite-dimensional, why must $T = S$?

49. As we saw in the text, the linear map $f(x,y) = (x + y, 2x, x - y)$ is one-to-one and its range is the plane $x - y + z = 0$ in $\mathcal{R}^3$. (See Fig. 3.7.) Thus as a map from $\mathcal{R}^2$ to $\mathcal{R}^3$ it is not onto. Let $S = \mathcal{R}^2$ and T = rng f. Why is $f: S \to T$ an isomorphism?

The point of this problem is that a one-to-one linear map which is not onto may nevertheless be regarded as an isomorphism from its domain *to its range*. Of course this involves a change in the map; $f: \mathcal{R}^2 \to \mathcal{R}^3$ and $f: S \to T$ are different functions.

3.4 Concrete Spaces Revisited

In Chap. 1 we confined our attention to vector spaces whose elements are n-tuples of numbers, referring to such spaces as "concrete" because they are obvious generalizations of the cartesian plane. Then we let the algebraic properties shared by these spaces serve as a set of axioms for vector spaces in general; in Chap. 2 we asked you to forget about n-tuples and concentrate on the abstract meaning of "vector space" as though nothing were known about its elements beyond what is contained in the definition. We even showed you an example of an "infinite-dimensional" space to convince you that not all vector spaces are like the ones studied in Chap. 1. (See Example 2.3.)

However, this has not been altogether honest, for the fact is that nonzero finite-dimensional spaces really *are* like n-tuple spaces, so much so that the distinction is entirely a matter of notation. We have already pointed this out in connection with the space S of 2×2 matrices with entries in $\mathfrak{F}$, observing that the map $f: S \to \mathfrak{F}^4$ defined by

$$f\left(\begin{bmatrix} a & b \\ c & d \end{bmatrix}\right) = (a,b,c,d)$$

is an isomorphism. Thus f not only preserves the algebraic structure of S (like all linear maps), but also sends S *onto* $\mathfrak{F}^4$ in a one-to-one fashion. Moreover (see Prob. 47 in Sec. 3.3) any basis for S is sent into a basis for $\mathfrak{F}^4$. To illustrate this fact, recall that a basis for S is $\{E_{11},E_{12},E_{21},E_{22}\}$, where

$$E_{11} = \begin{bmatrix} 1 & 0 \\ 0 & 0 \end{bmatrix} \quad E_{12} = \begin{bmatrix} 0 & 1 \\ 0 & 0 \end{bmatrix} \quad E_{21} = \begin{bmatrix} 0 & 0 \\ 1 & 0 \end{bmatrix} \quad E_{22} = \begin{bmatrix} 0 & 0 \\ 0 & 1 \end{bmatrix}$$

(See Prob. 4, Sec. 2.1.) Then

$$f(E_{11}) = (1,0,0,0) \qquad f(E_{12}) = (0,1,0,0)$$
$$f(E_{21}) = (0,0,1,0) \qquad f(E_{22}) = (0,0,0,1)$$

the standard basis elements of $\mathfrak{F}^4$.

It seems reasonable to expect that we ought to be able to do something like this with any finite-dimensional space. We can leave zero spaces out of account, since their structure is trivial; all other finite-dimensional spaces come equipped with a basis. To be explicit, suppose that S is a vector space over $\mathfrak{F}$ with basis $\{u_1, \ldots, u_n\}$. What we want to do is strip away its disguise and expose its algebraic structure as none other than $\mathfrak{F}^n$.

LECTOR *This is beginning to sound like* Trent's Last Case.

AUCTOR *More like* The Man Who Was Thursday.

LECTOR *Why?*

AUCTOR *The disguise hides something benign.*

The problem is how to set up a correspondence between the elements of S (of which we know next to nothing) and the n-tuples of $\mathfrak{F}^n$. We do know that every $x \in S$ can be written (uniquely) in the form $x = x_1u_1 + \cdots + x_nu_n$; the clue is in the coefficients, the scalars $x_1, \ldots, x_n$ in $\mathfrak{F}$. Why not let $\mathbf{X} = (x_1, \ldots, x_n)$ and call $\mathbf{X}$ the image of x? More precisely, we introduce the following terminology.

COORDINATE VECTOR

Let $\{u_1, \ldots, u_n\}$ be a basis for the vector space S and suppose that

$$x = \sum_{j=1}^{n} x_j u_j$$

is any element of S. The n-tuple of coefficients in this expression, namely $\mathbf{X} = (x_1, \ldots, x_n)$, is called the *coordinate vector* of x relative to the given basis.†

EXAMPLE 3.19

The functions $\phi_1(x) = \cos x$ and $\phi_2(x) = \sin x$ constitute a basis for the vector space of solutions of the differential equation $y'' + y = 0$. (See Example 2.8.) The function $\phi(x) = 3 \cos x - \sin x$ (that is, $\phi = 3\phi_1 - \phi_2$) is an element of this space; its coordinate vector relative to $\{\phi_1, \phi_2\}$ is $(3, -1)$ in $\mathfrak{R}^2$.

This example raises a question. What if we wanted the coordinate vector of ϕ relative to $\{\phi_2, \phi_1\}$? Obviously the order in which the elements of a basis are listed makes a difference! Strictly speaking, this means we should not use braces, since $\{\phi_1, \phi_2\}$ and $\{\phi_2, \phi_1\}$ are the same sets. The bases we are talking about are really ordered pairs of elements of S, namely (ϕ_1, ϕ_2) and (ϕ_2, ϕ_1).

More generally, the above definition of coordinate vector is unambiguous only if we agree that the given basis is an ordered n-tuple of vectors, $(u_1, \ldots, u_n)$. Since this is easily confused with an ordered n-tuple of *scalars,* we'll continue to use braces, asking you to remember that for certain purposes in this book set notation is understood to imply order as well as content.

EXAMPLE 3.20

The coordinate vector of

$$\begin{bmatrix} 2 & -1 \\ 0 & 1 \end{bmatrix}$$

† For this definition to be unambiguous, the basis must be "ordered." See the remarks following Example 3.19.

relative to the basis $\{E_{11}, E_{12}, E_{21}, E_{22}\}$ for the vector space of 2×2 matrices is $(2, -1, 0, 1)$ in $\mathfrak{F}^4$.

This example indicates that the map

$$f\left(\begin{bmatrix} a & b \\ c & d \end{bmatrix}\right) = (a,b,c,d)$$

is nothing more than a transformation of 2×2 matrices into their coordinate vectors relative to $\{E_{11}, E_{12}, E_{21}, E_{22}\}$. The same idea in an n-dimensional space S with basis $\{u_1, \ldots, u_n\}$ is to send each element $x = x_1u_1 + \cdots + x_nu_n$ into its coordinate vector $\mathbf{X} = (x_1, \ldots, x_n)$. That is, define the map $f\colon S \to \mathfrak{F}^n$ by $f(x) = \mathbf{X}$. (See Fig. 3.8.)

THEOREM 3.6

Let S be an n-dimensional vector space over $\mathfrak{F}$ with (ordered) basis $\{u_1, \ldots, u_n\}$. The coordinate vector map $f\colon S \to \mathfrak{F}^n$ which sends each element $x = x_1u_1 + \cdots + x_nu_n$ into $\mathbf{X} = (x_1, \ldots, x_n)$ is an isomorphism.

Proof

Let $x = \sum_{j=1}^n x_ju_j$ and $y = \sum_{j=1}^n y_ju_j$ be any elements of S and let $\mathbf{X} = (x_1, \ldots, x_n)$ and $\mathbf{Y} = (y_1, \ldots, y_n)$ be their coordinate vectors relative to the given basis. Since

$$x + y = \sum_{j=1}^{n}(x_j + y_j)u_j$$

the definition of coordinate vector yields

$$\begin{aligned} f(x + y) = (x_1 + y_1, \ldots, x_n + y_n) &= (x_1, \ldots, x_n) + (y_1, \ldots, y_n) \\ &= \mathbf{X} + \mathbf{Y} = f(x) + f(y) \end{aligned}$$

Similarly, $f(kx) = kf(x)$ for each $k \in \mathfrak{F}$ and $x \in S$. (See the problems.) Hence f is linear.

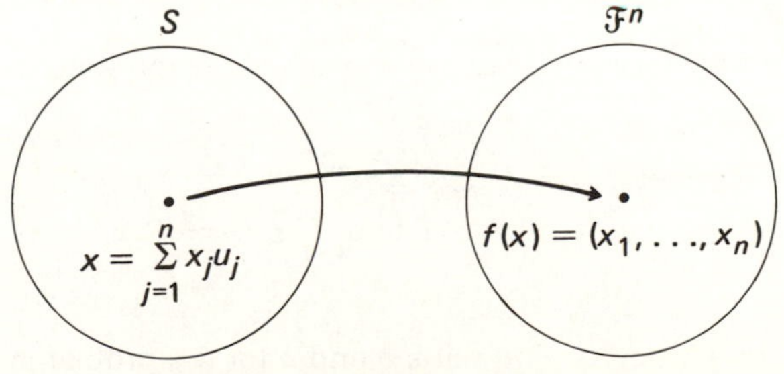

Fig. 3.8 Coordinate vector map

To prove that f is invertible, we need only show that it is onto, because S and $\mathfrak{F}^n$ have the same dimension. (See Prob. 45, Sec. 3.3.) This is obvious, for if $\mathbf{X} = (x_1, \ldots, x_n)$ is any vector in $\mathfrak{F}^n$, the element $x = \sum_{j=1}^{n} x_j u_j$ in S is sent into $\mathbf{X}$ by f. ∎

COROLLARY 3.6a

Every n-dimensional vector space over $\mathfrak{F}$ $(n > 0)$ is isomorphic to $\mathfrak{F}^n$.

In plain English, the only distinction between S and $\mathfrak{F}^n$ is notational. The elements of S may not look like n-tuples (they may be solutions of a differential equation, for example), but they act like n-tuples.

This should do two things for your understanding of vector spaces. First, it should make the abstract discussions of Chap. 2 seem less abstract (provided that the space in question is finite-dimensional); for it is clear that we were talking about n-tuple space all the time, despite appearances. Second, it should magnify the importance of the concrete spaces we discussed in Chap. 1. They are not such special cases as our language in Chap. 2 made them out to be, but are prototypes of finite-dimensional vector spaces in general.

Corollary 3.6a suggests that any two n-dimensional spaces over $\mathfrak{F}$ are isomorphic (since each is isomorphic to $\mathfrak{F}^n$). Before we can be precise about this, however, we need to sharpen our terminology.

ISOMORPHIC

Let S and T be vector spaces over $\mathfrak{F}$. If there exists an invertible linear map from S to T, we say that S is *isomorphic* to T, and write $S \approx T$.

This definition is a repetition of the last sentence of our definition of "isomorphism" in Sec. 3.3. However now we are concentrating not so much on the map itself as on the relation between vector spaces that the existence of such a map implies. The next theorem shows the similarity of this relation to the equals relation in ordinary algebra.

THEOREM 3.7

The isomorphism relation on the set of vector spaces over $\mathfrak{F}$ is

a. Reflexive: $S \approx S$

b. Symmetric: $S \approx T \Rightarrow T \approx S$

c. Transitive: $S \approx T$ and $T \approx U \Rightarrow S \approx U$

Proof

We'll do part *c*, leaving parts *a* and *b* for the problems. If $S \approx T$ and $T \approx U$, there exist invertible linear maps $f\colon T \rightarrow U$ and $g\colon S \rightarrow T$. The composite map

$h: S \to U$ defined by

$$h(x) = f[g(x)] \qquad x \in S$$

is linear because

$$\begin{aligned} h(ax + by) = f[g(ax + by)] &= f[ag(x) + bg(y)] \\ &= af[g(x)] + bf[g(y)] = ah(x) + bh(y) \end{aligned}$$

It is one-to-one because

$$\begin{aligned} h(x) = h(y) \quad &\Rightarrow \quad f[g(x)] = f[g(y)] \\ &\Rightarrow \quad g(x) = g(y) \qquad (f \text{ is one-to-one}) \\ &\Rightarrow \quad x = y \qquad (g \text{ is one-to-one}) \end{aligned}$$

It is onto because

$$\begin{aligned} z \in U \quad &\Rightarrow \quad \text{There exists } y \in T \text{ such that } f(y) = z \qquad (f \text{ is onto}) \\ &\Rightarrow \quad \text{There exists } x \in S \text{ such that } g(x) = y \qquad (g \text{ is onto}) \\ &\Rightarrow \quad h(x) = f[g(x)] = f(y) = z \end{aligned}$$

Hence h is an invertible linear map from S to U, which proves that $S \approx U$. ▮

These properties of the isomorphism relation allow us to be careless in our usage. A few lines back, for instance, we suggested that "any two n-dimensional spaces over $\mathfrak{F}$ are isomorphic." Strictly speaking, this statement was premature, since it is not clear whether we mean that S is isomorphic to T or that T is isomorphic to S. However, since the isomorphism relation is symmetric, it makes no difference!

EQUIVALENCE RELATION

Any relation (on a given set) with the reflexive, symmetric, and transitive properties is called an *equivalence relation*. It partitions the set into *equivalence classes*, each of which consists of elements that are *equivalent*, while elements of different classes are not equivalent.†

EXAMPLE 3.21

The "equals relation" on a given set partitions the set into equivalence classes of one element each; in this case, the partitioning is trivial.

† The language in Theorem 3.7, and in this definition, is not all it should be. We have not defined what a "relation on a set" is, nor what a "partition" of a set is. The vagueness is deliberate; we are not concerned in this book with a return to the first principles of set theory, but are muddling along in the hope that the essential ideas are clear.

EXAMPLE 3.22

The "less than" relation on $\mathcal{R}$ is not an equivalence relation, for although it is transitive ($a < b$ and $b < c \Rightarrow a < c$), it is neither reflexive nor symmetric.

THEOREM 3.8

Two finite-dimensional vector spaces over $\mathfrak{F}$ are isomorphic if and only if they have the same dimension.

Proof

Let S and T be the spaces, and suppose that they have the same dimension n. If $n = 0$, they are obviously isomorphic (being zero spaces); so we'll assume that $n > 0$. Then $S \approx \mathfrak{F}^n$ and $T \approx \mathfrak{F}^n$ (Corollary 3.6a). Since the isomorphism relation is symmetric, $\mathfrak{F}^n \approx T$; since it is transitive, $S \approx T$.

Conversely, suppose that $S \approx T$. Then $\dim S = \dim T$ by Prob. 47, Sec. 3.3. ▮

Thus the isomorphism relation on the set of finite-dimensional vector spaces over $\mathfrak{F}$ partitions the set into equivalence classes consisting of zero spaces, 1-dimensional spaces, 2-dimensional spaces, and so on. Each class has infinitely many members (all "equivalent" in the sense of isomorphism). For example, $\mathfrak{F}^4$ and the space of 2×2 matrices with entries in $\mathfrak{F}$ are in the equivalence class of 4-dimensional spaces over $\mathfrak{F}$.

The simplest "representative" of the class of n-dimensional spaces ($n > 0$) is $\mathfrak{F}^n$. Every other space in this class is isomorphic to $\mathfrak{F}^n$ (and therefore differs from it only in the notation used for its elements and operations); no space in a different class is isomorphic to it. Such a classification of abstract vector spaces by dimension (all spaces of a given dimension being essentially the same) is a major simplification of the theory.

Problems

50. What is the coordinate vector of $(-1,5,0)$ relative to the standard basis for $\mathcal{R}^3$? If $S = \mathfrak{F}^n$ in Theorem 3.6 and we use the standard basis, what is the coordinate vector map f?

51. What is the coordinate vector of $(-1,5,0)$ relative to the basis $\{(0,-1,1), (-2,3,3), (3,1,1)\}$ for $\mathcal{R}^3$?

52. Let S be the vector space of linear combinations of the functions $\phi_1(x) = e^x$ and $\phi_2(x) = e^{-x}$. What is the coordinate vector of $\sinh x = \frac{1}{2}(e^x - e^{-x})$ relative to the basis $\{\phi_1, \phi_2\}$ for S? Relative to $\{\phi_2, \phi_1\}$?

53. Let $f: S \to \mathfrak{F}^n$ be the coordinate vector map of Theorem 3.6. Finish the proof of linearity by showing that $f(kx) = kf(x)$ for each $k \in \mathfrak{F}$ and $x \in S$.

54. In the proof of Theorem 3.6 we did not show that f is one-to-one because it is enough to show that it is onto. However, prove it directly.

55. Finish the proof of Theorem 3.7 by showing that the isomorphism relation on the set of vector spaces over $\mathfrak{F}$ is reflexive and symmetric.

56. In our proof of transitivity of the isomorphism relation (Theorem 3.7) we showed that the composite map h is both one-to-one and onto. Why didn't we stop with one of these properties and not bother with the other?

57. Why isn't the "less than" relation on $\mathfrak{R}$ reflexive and symmetric?

58. What properties of an equivalence relation does "$\leq$" possess?

59. In our proof of Theorem 3.8 we said that zero spaces are "obviously isomorphic." Explain why.

LECTOR *You mean explain why you said it?*

AUCTOR *Explain why it's true.*

LECTOR *I would appreciate your not calling anything obvious if you're going to ask me to prove it.*

AUCTOR *I'll try.*

3.5 The Vector Space of Linear Maps

So far in this chapter we have been considering linear maps individually. Now we are going to look at the class of linear maps from S to T, where S and T are given vector spaces over $\mathfrak{F}$. Like all sets encountered in mathematics, this class has no inherent algebraic structure; it is up to us to impose whatever operations we find natural and convenient. In this section we'll define addition and multiplication by a scalar with the object of making a vector space of the class of linear maps.

ADDITION AND SCALAR MULTIPLICATION OF MAPS

Let S and T be vector spaces over $\mathfrak{F}$. We denote the class of linear maps from S to T by $L(S,T)$, or simply $L(S)$ when $T = S$.† Addition and scalar multiplication in L are defined by

$$f \in L,\ g \in L \Rightarrow f + g \in L \qquad \text{where } (f+g)(x) = f(x) + g(x),\ x \in S$$

$$c \in \mathfrak{F},\ f \in L \Rightarrow cf \in L \qquad \text{where } (cf)(x) = cf(x),\ x \in S$$

---

† Recall from Sec. 3.1 that a linear map from S into itself is called a "linear operator." Hence $L(S)$ is the class of all linear operators defined on S.

EXAMPLE 3.23

The sum of the maps $f\colon \mathcal{R}^3 \to \mathcal{R}^2$ and $g\colon \mathcal{R}^3 \to \mathcal{R}^2$ defined by

$$f(x,y,z) = (x - y + z,\ 3y) \qquad \text{and} \qquad g(x,y,z) = (y + 2z,\ x - y)$$

is the map $f + g\colon \mathcal{R}^3 \to \mathcal{R}^2$ defined by

$$\begin{aligned}(f + g)(x,y,z) &= f(x,y,z) + g(x,y,z)\\ &= (x - y + z,\ 3y) + (y + 2z,\ x - y) = (x + 3z,\ x + 2y)\end{aligned}$$

The product of f and the scalar 2 is $2f\colon \mathcal{R}^3 \to \mathcal{R}^2$ defined by

$$(2f)(x,y,z) = 2f(x,y,z) = 2(x - y + z,\ 3y) = (2x - 2y + 2z,\ 6y)$$

As you can see, both these maps are linear, which suggests that in general $L(S,T)$ is closed relative to addition and scalar multiplication.† Indeed, it is a vector space, as we now prove.

THEOREM 3.9

Let S and T be vector spaces oves $\mathfrak{F}$. Then $L(S,T)$ is a vector space over $\mathfrak{F}$.

Proof

We must verify the laws of a vector space given in Sec. 2.1 (with L in place of S). This verification is tedious, so we won't do it all, but just enough to indicate what is involved.

Law 1: L is closed relative to addition. To show this, we must prove that the sum of linear maps is itself a linear map. (See the problems.)

Law 2: If f, g, h are linear maps from S to T, then $(f + g) + h = f + (g + h)$; that is,

$$(f + g)(x) + h(x) = f(x) + (g + h)(x) \qquad \text{for every } x \in S$$

Since the definition of addition in L tells us that

$$(f + g)(x) = f(x) + g(x) \qquad \text{and} \qquad (g + h)(x) = g(x) + h(x)$$

we are asserting that

$$[f(x) + g(x)] + h(x) = f(x) + [g(x) + h(x)] \qquad \text{for each } x \in S$$

This is correct because $f(x)$, $g(x)$, $h(x)$ are in T and T is a vector space.

Law 3: L contains a zero element, for the zero map $0\colon S \to T$, defined by $0(x) = x$, has the property that $f + 0 = 0 + f = f$ for each $f \in L$.

† We have assumed this fact prematurely in the above definition, where we wrote $f \in L$, $g \in L \Rightarrow f + g \in L$ and $c \in \mathfrak{F}$, $f \in L \Rightarrow cf \in L$.

Law 4: Corresponding to each linear map f there is a linear map g satisfying $f + g = g + f = 0$. To prove this, note that $-1f$ is linear (see Law 6 below) and if $g = -1f$, we have (for each $x \in S$)

$$\begin{aligned} f(x) + g(x) &= f(x) + (-1)f(x) && \text{(scalar multiplication in } L) \\ &= 0 && \text{(Theorem 2.3 applied in } T) \end{aligned}$$

Hence $f + g$ is the zero map. A similar argument shows that $g + f$ is also the zero map.

Law 5: If f and g are elements of L, then $f + g = g + f$. (See the problems.)

Law 6: L is closed relative to scalar multiplication; that is, if $c \in \mathfrak{F}$ and $f \in L$, then $cf \in L$. To prove this, let a and b be any scalars and suppose that x and y are elements of S. Then

$$\begin{aligned} (cf)(ax + by) &= cf(ax + by) && \text{(definition of } cf) \\ &= c[af(x) + bf(y)] && (f \text{ is linear}) \\ &= c[af(x)] + c[bf(y)] && \text{(Law 8 in } T) \\ &= (ca)f(x) + (cb)f(y) && \text{(Law 7 in } T) \\ &= (ac)f(x) + (bc)f(y) && \text{(multiplication is commutative in } \mathfrak{F}) \\ &= a[cf(x)] + b[cf(y)] && \text{(Law 7 in } T) \\ &= a(cf)(x) + b(cf)(y) && \text{(definition of } cf)^{\dagger} \end{aligned}$$

Law 7: If a and b are scalars and $f \in L$, then $(ab)f = a(bf)$.

Law 8: If a and b are scalars and f and g are linear maps, then

$$(a + b)f = af + bf \qquad \text{and} \qquad a(f + g) = af + ag$$

Law 9: If $f \in L$, then $1f = f$. We leave the proof of Laws 7 through 9 as "trivial" problems. ∎

THEOREM 3.10

If S and T are finite-dimensional vector spaces over $\mathfrak{F}$, then $\dim L(S,T) = (\dim S)(\dim T)$.

Proof

If either S or T is a zero space, the only linear map from S to T is the zero map; the theorem is trivial in this case. Assuming that $\dim S = n > 0$ and $\dim T = m > 0$, the problem is to name mn linearly independent maps which span $L(S,T)$. This is not hard if you know ahead of time what to try. But the definition of such maps might be mysterious now; so we are going to defer

† This is a good example of the kind of thing textbook writers mean when they say "obvious" or "trivial." There is a temptation to dismiss such arguments and, on the whole, we yield to it whenever it is decent to do so. On the other hand, nobody should call a proof trivial unless he can do it himself. (Perhaps a good analogy is the "gimme" putt in golf. Under what conditions do we insist on sinking it instead of picking it up?)

the proof of this theorem until we have discussed the matrix representation of a linear map. (However, see the following example.)

EXAMPLE 3.24

If $S = \mathcal{R}^3$ and $T = \mathcal{R}^2$, Theorem 3.10 says that

$$\dim L(S,T) = (\dim S)(\dim T) = 6$$

To see why this is true, observe that a linear map from $\mathcal{R}^3$ to $\mathcal{R}^2$ is completely determined if we know its effect on $\mathbf{E}_1$, $\mathbf{E}_2$, $\mathbf{E}_3$, the standard basis elements of $\mathcal{R}^3$. For example, let us suppose that $f\colon \mathcal{R}^3 \to \mathcal{R}^2$ is linear and that

$$f(\mathbf{E}_1) = f(1,0,0) = (2,-1)$$

$$f(\mathbf{E}_2) = f(0,1,0) = (0,3)$$

$$f(\mathbf{E}_3) = f(0,0,1) = (5,1)$$

Then if (x,y,z) is any vector in $\mathcal{R}^3$, we have

$$(x,y,z) = x(1,0,0) + y(0,1,0) + z(0,0,1) = x\mathbf{E}_1 + y\mathbf{E}_2 + z\mathbf{E}_3$$

and hence

$$\begin{aligned} f(x,y,z) &= f(x\mathbf{E}_1 + y\mathbf{E}_2 + z\mathbf{E}_3) \\ &= xf(\mathbf{E}_1) + yf(\mathbf{E}_2) + zf(\mathbf{E}_3) \\ &= x(2,-1) + y(0,3) + z(5,1) \\ &= (2x + 5z, -x + 3y + z) \end{aligned}$$

Thus the three functional "values" $f(\mathbf{E}_1)$, $f(\mathbf{E}_2)$, $f(\mathbf{E}_3)$ (together with the assumption of linearity) are sufficient to determine the formula defining f!

Moreover, this formula can be written as

$$f(x,y,z) = 2(x,0) + 0(y,0) + 5(z,0) + (-1)(0,x) + 3(0,y) + 1(0,z)$$

which says that f is a linear combination of the maps

$$f_{ij}\colon \mathcal{R}^3 \to \mathcal{R}^2 \qquad i = 1, 2;\ j = 1, 2, 3$$

defined by

$$\begin{array}{lll} f_{11}(x,y,z) = (x,0) & f_{12}(x,y,z) = (y,0) & f_{13}(x,y,z) = (z,0) \\ f_{21}(x,y,z) = (0,x) & f_{22}(x,y,z) = (0,y) & f_{23}(x,y,z) = (0,z) \end{array}$$

More generally, suppose that $f\colon \mathcal{R}^3 \to \mathcal{R}^2$ is any element of $L(S,T)$. Denote the images of $\mathbf{E}_1$, $\mathbf{E}_2$, $\mathbf{E}_3$ by

$$\mathbf{W}_1 = f(\mathbf{E}_1) = (a_{11},a_{21}) \qquad \mathbf{W}_2 = f(\mathbf{E}_2) = (a_{12},a_{22}) \qquad \mathbf{W}_3 = f(\mathbf{E}_3) = (a_{13},a_{23})$$

Then (as before)

$$\begin{aligned} f(x,y,z) &= xf(\mathbf{E}_1) + yf(\mathbf{E}_2) + zf(\mathbf{E}_3) \\ &= x\mathbf{W}_1 + y\mathbf{W}_2 + z\mathbf{W}_3 \\ &= (a_{11}x + a_{12}y + a_{13}z, a_{21}x + a_{22}y + a_{23}z) \end{aligned}$$

This proves that every linear map from $\mathcal{R}^3$ to $\mathcal{R}^2$ has this form (as predicted in Sec. 3.1 following Example 3.4). Since

$$f = a_{11}f_{11} + a_{12}f_{12} + a_{13}f_{13} + a_{21}f_{21} + a_{22}f_{22} + a_{23}f_{23}$$

we conclude that the maps

$$f_{11}, \quad f_{12}, \quad f_{13}, \quad f_{21}, \quad f_{22}, \quad f_{23}$$

span $L(S,T)$. The maps are also linearly independent, because the only linear combination of them which is the zero map is the one having zero coefficients. (Confirm this!) Hence the maps constitute a basis for $L(S,T)$ and dim $L(S,T) = 6$, as advertised.

The above remarks can be put another way. Given the vectors $\mathbf{W}_1$, $\mathbf{W}_2$, $\mathbf{W}_3$ in $\mathcal{R}^2$, there is *at most one* linear map f: $\mathcal{R}^3 \to \mathcal{R}^2$ having the prescribed values

$$f(\mathbf{E}_1) = \mathbf{W}_1 \qquad f(\mathbf{E}_2) = \mathbf{W}_2 \qquad f(\mathbf{E}_3) = \mathbf{W}_3$$

because these values (as we have seen) determine the formula defining f. This is a uniqueness statement. The corresponding existence statement is that *at least one* such map exists; it is the map

$$f(x,y,z) = x\mathbf{W}_1 + y\mathbf{W}_2 + z\mathbf{W}_3$$

These remarks are true in general, as we now show.

THEOREM 3.11

Let S be a vector space over $\mathcal{F}$ with basis $\{u_1, \ldots, u_n\}$ and let T be any vector space over $\mathcal{F}$. Given the elements $w_1, \ldots, w_n$ of T (not necessarily distinct), there exists one and only one linear map f: $S \to T$ such that $f(u_j) = w_j$, where $j = 1, \ldots, n$.

Proof

If such a map f exists, it must be unique. To see why, take any $x = \sum_{i=1}^n x_iu_i$ in S. Since f is linear, we have

$$f(x) = f\left(\sum_{i=1}^n x_iu_i\right) = \sum_{i=1}^n x_if(u_i) = \sum_{i=1}^n x_iw_i$$

The values $w_1, \ldots, w_n$ are prescribed; that is, we know what they are. Hence the effect of f on each $x \in S$ is determined. To put it differently, there is at most one linear map f that can fit the prescription $f(u_j) = w_j$, where $j = 1, \ldots, n$.

To prove that such a map exists, we must define $f(x)$ for each $x \in S$ and show that the resulting function is linear and has the prescribed values $f(u_j) = w_j$, where $j = 1, \ldots, n$. We do this by a process known as "linear extension," using as a clue the formula

$$f(x) = \sum_{i=1}^{n} x_i w_i$$

which we have just shown must hold if f exists:

$$\text{For each } x = \sum_{i=1}^{n} x_i u_i \text{ in } S, \textit{ define } f(x) = \sum_{i=1}^{n} x_i w_i \text{ in } T$$

This certainly describes a map from S to T; the problem is to show that it has the required properties.

LECTOR *You're being devious.*
AUCTOR *The problem is to name a linear map such that $f(u_j) = w_j$, right?*
LECTOR *Right.*
AUCTOR *I've just named it, by telling you what $f(x)$ is for each x.*
LECTOR *It'll never sell.*
AUCTOR *I haven't tried to sell it yet.*
LECTOR *But you're about to.*
AUCTOR *Well, yes, but you needn't buy it unless you're convinced it works.*
LECTOR *Who would buy a used map from a mathematician?*

Observe first that f does in fact take on the prescribed values w_j at the basis elements u_j. This is because the expression of each u_j in terms of $u_1, \ldots, u_n$ is

$$u_j = \sum_{i=1}^{n} \delta_{ij} u_i \qquad \text{where } \delta_{ij} \text{ is Kronecker's delta}$$

By definition of f it follows that

$$f(u_j) = \sum_{i=1}^{n} \delta_{ij} w_i = w_j \qquad j = 1, \ldots, n$$

Moreover, f is linear, for if a and b are any scalars and

$$x = \sum_{i=1}^{n} x_i u_i \qquad \text{and} \qquad y = \sum_{i=1}^{n} y_i u_i$$

are any elements of S, we have

$$ax + by = \sum_{i=1}^{n} (ax_i + by_i) u_i$$

and our definition of f yields

$$f(ax + by) = \sum_{i=1}^{n}(ax_i + by_i)w_i$$

$$= a\sum_{i=1}^{n} x_i w_i + b\sum_{i=1}^{n} y_i w_i = af(x) + bf(y) \blacksquare$$

COROLLARY 3.11a

The linear maps $f\colon S \to T$ and $g\colon S \to T$ are identical if and only if $f(u_j) = g(u_j)$ for $j = 1, \ldots, n$, where $\{u_1, \ldots, u_n\}$ is a basis for S.

Proof

This is left for the problems.

Problems

60. Finish the proof of Theorem 3.9 by showing the following.

a. The sum of two linear maps from S to T is itself a linear map from S to T.

b. If f and g are elements of $L(S,T)$, then $f + g = g + f$.

c. If a and b are scalars and $f \in L$, then $(ab)f = a(bf)$.

d. If a and b are scalars and f and g are elements of L, then

$$(a + b)f = af + bf \qquad \text{and} \qquad a(f + g) = af + ag$$

e. If $f \in L$, then $1f = f$.

61. Let $f\colon \mathcal{R}^2 \to \mathcal{R}^3$ be the linear map whose values at the standard basis elements of $\mathcal{R}^2$ are $f(\mathbf{E}_1) = (1,0,-1)$ and $f(\mathbf{E}_2) = (2,1,5)$. What is the formula for $f(x,y)$, where (x,y) is any element of $\mathcal{R}^2$?

62. If $\phi_1(x) = 1$, $\phi_2(x) = x$, $\phi_3(x) = x^2$, then $\{\phi_1,\phi_2,\phi_3\}$ is a basis for the vector space of polynomials of the form $\phi(x) = ax^2 + bx + c$.

a. What linear operator on this space sends ϕ_1, ϕ_2, ϕ_3 into $-\phi_1$, $-\phi_2$, $2\phi_1 - \phi_3$, respectively?

b. If $\phi(x) = x^2 + 1$, what polynomial is ϕ sent into by this map?

c. Is this map invertible?

63. Let f_{11}, f_{12}, f_{13}, f_{21}, f_{22}, f_{23} be the elements of $L(\mathcal{R}^3,\mathcal{R}^2)$ defined in Example 3.24. Prove that they are linearly independent.

64. Show that every linear operator on $\mathcal{R}^2$ has the form

$$f(x,y) = (a_{11}x + a_{12}y,\ a_{21}x + a_{22}y)$$

where the a_{ij} are scalars.

65. Name four linearly independent maps which span $L(\mathcal{R}^2)$, thus proving that $\dim L(\mathcal{R}^2) = 4$.

66. Prove Corollary 3.11a.

Review Quiz

True or false?

1. If $\dim S = n$ and $\dim T = m$, the set of linear maps from S to T is a vector space of dimension mn.
2. The only linear map $f: S \to T$ whose kernel is S is the zero map.
3. A linear map from S to T sends a basis for S into a basis for T.
4. Every linear map from $\mathcal{R}^2$ to $\mathcal{R}^1$ has the form $f(x,y) = ax + by$.
5. James Sylvester fought in the War of 1812.
6. A linear map cannot send a finite-dimensional space onto a higher-dimensional space.
7. If $f: S \to T$ sends every linear combination of arbitrary x and y in S into the same linear combination of their images in T, then f is linear.
8. If m and b are real, the map $f: \mathcal{R} \to \mathcal{R}$ defined by $f(x) = mx + b$ is linear.
9. If $f(x) \neq f(y) \Rightarrow x \neq y$, then f is a one-to-one map.
10. There is only one linear map $f: \mathcal{R}^1 \to \mathcal{R}^2$ with the property that $f(1) = (1,1)$.
11. The kernel of the orthogonal projection of $\mathcal{R}^2$ on the line $y = x$ is the line $y = -x$.
12. If S is finite-dimensional and f is a linear operator on S which is either one-to-one or onto, then f^{-1} exists.
13. If S is the vector space of linear operators on $\mathcal{R}^2$, then $\dim S = 4$.
14. The map $f: \mathcal{C}^1 \to \mathcal{C}^1$ defined by $f(z) = \bar{z}$ is linear.
15. If the kernel of a linear operator on S is the zero subspace of S, the operator maps S onto itself.
16. The map $f: \mathcal{R}^3 \to \mathcal{R}^2$ defined by $f(x,y,z) = (x + y - z, x + 1)$ is linear.
17. If the kernel of the linear map $f: S \to T$ is not the zero subspace of S, then for every $y \in \operatorname{rng} f$ there are at least two elements of S which f maps into y.
18. Every linear operator on $\mathcal{R}^2$ has the form $f(x,y) = (ax + by, cx + dy)$.
19. If S is the vector space of real-valued functions with derivatives of all orders on $(-\infty,\infty)$, the operator $D: S \to S$ defined by $D(\phi) = \phi'$ is invertible.

20. The kernel of $f\colon \mathcal{R}^3 \to \mathcal{R}^1$ defined by $f(\mathbf{X}) = \mathbf{A}\cdot\mathbf{X}$, where $\mathbf{A}$ is nonzero, is a plane.

21. If S is the vector space of real-valued functions continuous on $[a,b]$, then the map $I\colon S \to R$ defined by $I(f) = \int_a^b f(x)\,dx$ is linear.

22. If $f\colon S \to T$ is linear and x and y are sent into the same element of T, then $x = y + z$, where $z \in \ker f$.

23. If f and g are linear operators on $\mathcal{R}^2$ such that $f(1,-1) = g(1,-1)$ and $f(-1,1) = g(-1,1)$, then $f(x,y) = g(x,y)$ for all $(x,y) \in \mathcal{R}^2$.

24. The inverse of an invertible linear map is linear.

25. If $f\colon S \to T$ is linear and one-to-one, and $u_1, \ldots, u_n$ span S, then $f(u_1), \ldots, f(u_n)$ span T.

26. If S is the vector space of 2×2 matrices, the map $f\colon S \to S$ defined by $f(A) = A^T$ is an isomorphism.

27. If U is the zero subspace of $\mathcal{R}^3$, then $U^\perp = \mathcal{R}^3$.

28. The coordinate vector of $(2,-5)$ relative to the basis $\{(-1,0), (0,-1)\}$ for $\mathcal{R}^2$ is $(-2,5)$.

29. If S is isomorphic to T, then T is isomorphic to S.

30. The dot product of $(1,i)$ and $(i,-i)$ in $\mathcal{C}^2$ is $1 + i$.

31. If f is a linear map from $\mathcal{R}^2$ to $\mathcal{R}^3$, the nullity plus the rank of f is 3.

32. The coordinate vector of an element of n-space relative to the standard basis is the element itself.

33. The vector space of 2×2 matrices of the form

$$\begin{bmatrix} a & b \\ -b & a \end{bmatrix}$$

where a and b are real, is isomorphic to $\mathcal{R}^2$.

34. If $f\colon S \to T$ is a map and $f(0) \neq 0$, then f is not linear.

35. The orthogonal complement of the xy plane in $\mathcal{R}^3$ is the z axis.

36. The range of $f\colon \mathcal{R}^1 \to \mathcal{R}^2$ defined by $f(t) = (t^2,t^3)$ is a subspace of $\mathcal{R}^2$.

37. In the vector space of polynomials with real coefficients, the map that sends $p(x)$ into $xp(x)$ is one-to-one.

CHAPTER 4

Matrices

Given a linear map from one finite-dimensional vector space to another, it is easy to associate a matrix with the map. In this chapter we describe how this association is made and examine the relationship between linear maps and matrices. It turns out, as we shall see, that the two are essentially the same thing (in the sense of isomorphism). Hence if linear algebra can be described as the study of linear maps (as we did in Sec. 3.1), it can equally well be regarded as the theory of matrices, at least in the context of finite-dimensional vector spaces.

4.1 The Matrix Representation of a Linear Map

Let $f\colon \mathcal{R}^2 \to \mathcal{R}^3$ be the linear map defined by $f(x,y) = (x + y, 2x, x - y)$. If $\{\mathbf{E}_1,\mathbf{E}_2\}$ is the standard basis for $\mathcal{R}^2$, we have

$$f(\mathbf{E}_1) = f(1,0) = (1,2,1) \qquad \text{and} \qquad f(\mathbf{E}_2) = f(0,1) = (1,0,-1)$$

These 3-tuples can be used to construct a 3×2 matrix, namely

$$A = \begin{bmatrix} 1 & 1 \\ 2 & 0 \\ 1 & -1 \end{bmatrix}$$

an object which it is natural to regard as a concrete representation of f.

LECTOR *Why didn't you write*

$$A = \begin{bmatrix} 1 & 2 & 1 \\ 1 & 0 & -1 \end{bmatrix}$$

AUCTOR *I'd rather not explain it just now.*

LECTOR *May I do it either way?*

AUCTOR *No.*

LECTOR *You're an arbitrary sort of fellow at times.*

AUCTOR *I have my reasons.*†

---

† Some writers *do* use the 3-tuples as rows rather than columns. You must pay attention to which convention is adopted, since the statements of many later results depend on it.

Before we can call A the "matrix of f," we must dispose of certain questions. First, given a linear map $f: S \to T$ (where S and T are nonzero finite-dimensional vector spaces over $\mathfrak{F}$), does the above procedure lead to a definite matrix representation of f? Obviously not! Unless S is an n-tuple space (like $\mathfrak{R}^2$ in the above example), we can't even talk about its "standard basis"; that concept is restricted to vector spaces of the form $S = \mathfrak{F}^n$. Moreover, the "values" of f in T are not m-tuples ($m = 3$ in the above example) unless T is a vector space of the form $\mathfrak{F}^m$.

Nevertheless, there is a way around this; for if S is a nonzero n-dimensional space, we know that it has a basis, say

$$\alpha = \{u_1, \ldots, u_n\}^\dagger$$

Similarly, if T is m-dimensional, where $m > 0$, it has a basis

$$\beta = \{v_1, \ldots, v_m\}$$

Each functional value $f(u_j)$, where $j = 1, \ldots, n$, is an element of T; hence it may be uniquely expressed as a linear combination of the v_i, where $i = 1, \ldots, m$. To be explicit, we write

$$f(u_1) = a_{11}v_1 + a_{21}v_2 + \cdots + a_{m1}v_m = \sum_{i=1}^{m} a_{i1}v_i$$

$$f(u_2) = a_{12}v_1 + a_{22}v_2 + \cdots + a_{m2}v_m = \sum_{i=1}^{m} a_{i2}v_i$$

$$\cdots\cdots\cdots\cdots\cdots\cdots\cdots\cdots\cdots\cdots\cdots\cdots$$

$$f(u_n) = a_{1n}v_1 + a_{2n}v_2 + \cdots + a_{mn}v_m = \sum_{i=1}^{m} a_{in}v_i$$

(We use double subscripts to avoid running out of letters for the coefficients.) In other words, for each $j = 1, \ldots, n$, we have

$$f(u_j) = a_{1j}v_1 + a_{2j}v_2 + \cdots + a_{mj}v_m = \sum_{i=1}^{m} a_{ij}v_i$$

The coefficients $a_{1j}, a_{2j}, \ldots, a_{mj}$ appearing in this expression are definite (assuming that the map $f: S \to T$ and the ordered bases α and β are given). Hence the $m \times n$ matrix

† This is the first time in the book that we have designated a basis by a single letter. Recall from Sec. 3.4 that we are not merely talking about a set, but an ordered n-tuple of vectors $\alpha = (u_1, \ldots, u_n)$. We agreed then that because this notation can lead to some confusion with an n-tuple of scalars, we would continue to use braces, intending the set notation to imply order as well as content.

$$A = \begin{bmatrix} a_{11} & a_{12} & \cdots & a_{1n} \\ a_{21} & a_{22} & \cdots & a_{2n} \\ \cdots & \cdots & \cdots & \cdots \\ a_{m1} & a_{m2} & \cdots & a_{mn} \end{bmatrix}$$

whose jth *column,* where $j = 1, \ldots, n$, is the m-tuple $(a_{1j}, a_{2j}, \ldots, a_{mj})$ is a unique representation of the map (relative to the given bases). Note that $(a_{1j}, a_{2j}, \ldots, a_{mj})$ is the coordinate vector of $f(u_j)$ relative to the basis β. (See Sec. 3.4.) Thus we look at the *image* of the jth element of the basis α; its coordinate vector relative to β becomes the jth column of A.

MATRIX REPRESENTATION

Let $f: S \to T$ be a linear map, where S and T have (ordered) bases $\alpha = \{u_1, \ldots, u_n\}$ and $\beta = \{v_1, \ldots, v_m\}$, respectively. The *matrix of f relative to α and β* is the $m \times n$ array

$$A = \begin{bmatrix} a_{11} & a_{12} & \cdots & a_{1n} \\ a_{21} & a_{22} & \cdots & a_{2n} \\ \cdots & \cdots & \cdots & \cdots \\ a_{m1} & a_{m2} & \cdots & a_{mn} \end{bmatrix}$$

whose jth column, where $j = 1, \ldots, n$, is the coordinate vector of $f(u_j)$ relative to β. We denote it by $[f]_{\alpha\beta}$, or simply $[f]$ when the bases are understood. When $T = S$ and $\beta = \alpha$ (that is, when f is a linear operator on S and only one basis is used), we call A the *matrix of f relative to* α, and write $[f]_\alpha$.†

EXAMPLE 4.1

Define $f: \mathcal{R}^2 \to \mathcal{R}^3$ by $f(x,y) = (x + y, 2x, x - y)$. Then, as we have seen,

$$f(\mathbf{E}_1) = (1,2,1) \qquad \text{and} \qquad f(\mathbf{E}_2) = (1,0,-1)$$

where $\alpha = \{\mathbf{E}_1, \mathbf{E}_2\}$ is the standard basis for $\mathcal{R}^2$. In writing

$$A = \begin{bmatrix} 1 & 1 \\ 2 & 0 \\ 1 & -1 \end{bmatrix}$$

† Otherwise we are stuck with the ponderous "matrix of f relative to α and α," and the redundant notation $[f]_{\alpha\alpha}$.

we made no reference to a basis β for $\mathbb{R}^3$. But we had in mind the standard basis

$$\beta = \{(1,0,0), (0,1,0), (0,0,1)\}$$

since

$$f(\mathbf{E}_1) = 1(1,0,0) + 2(0,1,0) + 1(0,0,1)$$
$$f(\mathbf{E}_2) = 1(1,0,0) + 0(0,1,0) + (-1)(0,0,1)$$

The coordinate vectors of $f(\mathbf{E}_1)$ and $f(\mathbf{E}_2)$ relative to β are

$$(a_{11},a_{21},a_{31}) = (1,2,1) \qquad \text{and} \qquad (a_{12},a_{22},a_{32}) = (1,0,-1)$$

respectively; these are the columns of A.† The reason we didn't bother with β before is that the 3-tuples (1,2,1) and (1,0,−1) were staring us in the face. It is superfluous to find their coordinate vectors relative to β, since these are the same 3-tuples over again.

In general, however, it is necessary to work with both bases explicitly, as in the next example.

EXAMPLE 4.2

Let f be the same map as in Example 4.1, but this time use the bases

$$\alpha = \{\mathbf{U}_1,\mathbf{U}_2\} \qquad \text{and} \qquad \beta = \{\mathbf{V}_1,\mathbf{V}_2,\mathbf{V}_3\}$$

where

$$\mathbf{U}_1 = (1,1) \qquad \mathbf{U}_2 = (0,-1)$$

and

$$\mathbf{V}_1 = (1,-1,0) \qquad \mathbf{V}_2 = (0,2,1) \qquad \mathbf{V}_3 = (2,1,1)$$

Then

$$f(\mathbf{U}_1) = f(1,1) = (2,2,0) = -6\mathbf{V}_1 - 4\mathbf{V}_2 + 4\mathbf{V}_3$$
$$f(\mathbf{U}_2) = f(0,-1) = (-1,0,1) = 5\mathbf{V}_1 + 4\mathbf{V}_2 - 3\mathbf{V}_3$$

(Confirm this result!) Hence the matrix associated with f is now

$$[f]_{\alpha^\beta} = \begin{bmatrix} -6 & 5 \\ -4 & 4 \\ 4 & -3 \end{bmatrix}$$

† Because these become columns, many writers introduce the concept of "column vector" at this point, writing (a_{11},a_{21},a_{31}) and (a_{12},a_{22},a_{32}) as

$$\begin{pmatrix} a_{11} \\ a_{21} \\ a_{31} \end{pmatrix} \quad \text{and} \quad \begin{pmatrix} a_{12} \\ a_{22} \\ a_{32} \end{pmatrix}$$

(The ordinary notation is then called "row vector.") This agreement lends additional precision to some discussions, as we shall see. However, it is cumbersome to write (and print) column vectors. We'll use them sparingly.

Thus there are many different matrix representations of a given linear map. But once the bases have been chosen, there is only one. This disposes of the first question we raised: Does a linear map $f: S \to T$ lead to a definite matrix? The answer is yes, provided that S and T are nonzero finite-dimensional spaces whose (ordered) bases have been specified.†

THEOREM 4.1

Suppose that S and T are vector spaces over $\mathfrak{F}$ with (ordered) bases $\alpha = \{u_1, \ldots, u_n\}$ and $\beta = \{v_1, \ldots, v_m\}$, respectively. Given a linear map $f: S \to T$, there is a unique $m \times n$ matrix (with entries in $\mathfrak{F}$) representing f relative to α and β. Its jth column, where $j = 1, \ldots, n$, is the coordinate vector of $f(u_j)$ relative to β.

Proof

Refer to the above discussion.

A second question arises naturally at this point. How about going in reverse? Assuming that S and T with their (ordered) bases $\alpha = \{u_1, \ldots, u_n\}$ and $\beta = \{v_1, \ldots, v_m\}$ are given, suppose that an $m \times n$ matrix A is named. Is there a linear map $f: S \to T$ whose matrix relative to α and β is A? If so, is it unique, or are there many maps with this same matrix representation?

To get an idea about this, look at the matrix

$$A = \begin{bmatrix} 1 & 1 \\ 2 & 0 \\ 1 & -1 \end{bmatrix}$$

in connection with the spaces $S = \mathcal{R}^2$ and $T = \mathcal{R}^3$ and their standard bases. Of course we know that there is a linear map $f: \mathcal{R}^2 \to \mathcal{R}^3$ having A as its matrix; it is the map $f(x,y) = (x + y, 2x, x - y)$ in Example 4.1. But suppose that we didn't know this. It is possible to construct the map by working with the matrix. What we want is a linear map $f: \mathcal{R}^2 \to \mathcal{R}^3$ such that

$$f(\mathbf{E}_1) = (1,2,1) \qquad \text{and} \qquad f(\mathbf{E}_2) = (1,0,-1)$$

As we have seen before (Sec. 3.5), these values of f are sufficient to determine it completely. If $(x,y) = x\mathbf{E}_1 + y\mathbf{E}_2$ is any element of $\mathcal{R}^2$, the linearity of f implies that

$$\begin{aligned} f(x,y) &= f(x\mathbf{E}_1 + y\mathbf{E}_2) = xf(\mathbf{E}_1) + yf(\mathbf{E}_2) \\ &= x(1,2,1) + y(1,0,-1) = (x + y, 2x, x - y) \end{aligned}$$

† If S or T is a zero space, the only possible linear map from S to T is the zero map; no matrix representation is defined in this case. On the other hand, if $n = \dim S$ and $m = \dim T$ are positive, the matrix representation of the zero map (relative to any bases) is the $m \times n$ array whose entries are all 0. (Why?)

Thus the formula $f(x,y) = (x + y, 2x, x - y)$ which defines the map can be recovered from the matrix.

We claim this can always be done, for suppose that S and T, with (ordered) bases $\alpha = \{u_1, \ldots, u_n\}$ and $\beta = \{v_1, \ldots, v_m\}$, are specified ahead of time, and an $m \times n$ matrix A is given. Denote the entry in the ith row and jth column of A by a_{ij}, where $i = 1, \ldots, m$ and $j = 1, \ldots, n$, and let

$$w_j = \sum_{i=1}^{m} a_{ij} v_i \qquad j = 1, \ldots, n$$

Then Theorem 3.11 guarantees the existence of a unique linear map $f: S \to T$ satisfying

$$f(u_j) = w_j \qquad j = 1, \ldots, n$$

Moreover, this is precisely the map which we agreed earlier has A as its matrix relative to α and β.

THEOREM 4.2

Suppose that S and T are vector spaces over $\mathfrak{F}$ with (ordered) bases $\alpha = \{u_1, \ldots, u_n\}$ and $\beta = \{v_1, \ldots, v_m\}$, respectively. Given an $m \times n$ matrix A (with entries in $\mathfrak{F}$), there is a unique linear map $f: S \to T$ whose matrix relative to α and β is A. Its value at u_j, for $j = 1, \ldots, n$, is

$$f(u_j) = \sum_{i=1}^{m} a_{ij} v_i$$

where a_{ij} is the entry in the ith row and jth column of A.

Proof

Refer to the above discussion.

One need only read Theorems 4.1 and 4.2 together to see the intimate relation between linear maps and matrices! In fact these theorems establish a one-to-one correspondence between the set of linear maps from S to T and the set of $m \times n$ matrices with entries in $\mathfrak{F}$ (provided that the bases for S and T are specified). To be precise, denote these sets by $L(S, T)$ and $M_{m,n}(\mathfrak{F})$, respectively (or simply L and M when the context is understood). For each map $f \in L$, define $\phi(f) = A$, where $A = [f]_{\alpha^\beta}$ is the matrix of f relative to (fixed) bases α and β for S and T. Then $\phi: L \to M$ is a one-to-one mapping of L onto M. Its inverse is the function $\phi^{-1}: M \to L$ defined by $\phi^{-1}(A) = f$, where f is the unique linear map from S to T whose matrix (relative to α and β) is A. It would be natural at this point to call ϕ an isomorphism, tying L and M

together as vector spaces with identical structure. But when we come to think of it, M has no structure! Unlike L, on which we have imposed definitions of addition and scalar multiplication which make it a vector space, M is just a set. The obvious remedy is to define addition of matrices and multiplication of a matrix by a scalar, and to do this in such a way that the expected isomorphism will materialize. See the next section.

Problems

1. Using the standard basis for each space, find the matrix representation of each of the following linear maps.
   a. $f\colon \mathcal{R}^3 \to \mathcal{R}^2$ defined by $f(x,y,z) = (x - 2y + z, 2y)$
   b. $f\colon \mathcal{R}^2 \to \mathcal{R}^2$ defined by $f(x,y) = (x - y, 2x + y)$
   c. $f\colon \mathcal{R}^3 \to \mathcal{R}^1$ defined by $f(\mathbf{X}) = \mathbf{A} \cdot \mathbf{X}$, where $\mathbf{A} = (2,1,-1)$
   d. $f\colon \mathcal{R}^1 \to \mathcal{R}^3$ defined by $f(x) = (2x,-x,x)$
   e. $f\colon \mathcal{R}^3 \to \mathcal{R}^3$ defined by $f(x,y,z) = (x,y,z)$
   f. $f\colon \mathcal{R}^3 \to \mathcal{R}^2$ defined by $f(x,y,z) = (0,0)$
   g. $f\colon \mathcal{R}^1 \to \mathcal{R}^1$ defined by $f(x) = 3x$. *Note:* If we agree to identify 1×1 matrices with numbers, the matrix of this function is just the slope of its graph.
2. The map in part *c* of Prob. 1 is determined by the vector **A**; that is, when **A** is given, f is known. Comment on the relationship between this and the matrix representation of f. [Note that it wouldn't be as simple if we had adopted the convention that the coordinate vectors of $f(u_1)$, $f(u_2)$, . . . are the *rows* of $[f]$ instead of the columns.]

★ 3. Prove that the map $f\colon \mathfrak{F}^n \to \mathfrak{F}^m$ is linear if and only if it sends $(x_1, \ldots, x_n)$ into a vector $(y_1, \ldots, y_m)$ whose entries are linear combinations of $x_1, \ldots, x_n$, that is,

$$\begin{aligned} y_1 &= a_{11}x_1 + a_{12}x_2 + \cdots + a_{1n}x_n \\ y_2 &= a_{21}x_1 + a_{22}x_2 + \cdots + a_{2n}x_n \\ &\cdots\cdots\cdots\cdots\cdots\cdots \\ y_m &= a_{m1}x_1 + a_{m2}x_2 + \cdots + a_{mn}x_n \end{aligned}$$

where the a_{ij} are scalars. (This justifies the remark following Example 3.4.)

★ 4. Suppose that $f\colon \mathfrak{F}^n \to \mathfrak{F}^m$ is defined by $f(x_1, \ldots, x_n) = (y_1, \ldots, y_m)$, where the y_i are given in terms of the x_j by the formulas in Prob. 3, that is,

$$y_i = a_{i1}x_1 + a_{i2}x_2 + \cdots + a_{in}x_n \qquad i = 1, \ldots, m$$

Explain why the coefficients in this expression are the entries of the ith *row* of the matrix of f relative to the standard bases.

This problem can cause confusion if it is not understood! But it is also helpful, since it enables us to read off the matrix of a map from n-space to m-space without finding $f(\mathbf{E}_j)$, where $j = 1, \ldots, n$. For example, the matrix of

$$f(x,y) = (x + y, 2x, x - y)$$

(Example 4.1) can be found by inspection of the coefficients in the expressions

$$x + y = 1x + 1y \qquad 2x = 2x + 0y \qquad x - y = 1x + (-1)y$$

The *rows* of $[f]$ are $(1,1)$, $(2,0)$, $(1,-1)$, that is,

$$[f] = \begin{bmatrix} 1 & 1 \\ 2 & 0 \\ 1 & -1 \end{bmatrix}$$

Note, however, that this works only when the map is from n-space to m-space and the standard bases are used.

5. Using the standard basis for $\mathcal{R}^2$, find the matrix of the rotation of the plane described in Example 3.8. What is the matrix of the inverse map? (See Sec. 3.3.)

6. Let S be the vector space of solutions of the differential equation $y'' + y = 0$. A basis for S is $\alpha = \{\phi_1, \phi_2\}$, where $\phi_1(x) = \cos x$ and $\phi_2(x) = \sin x$. (See Example 2.8.) Find the matrix (relative to α) of each of the following linear operators.
a. $D: S \to S$ defined by $D(\phi) = \phi'$
b. $L: S \to S$ defined by $L(\phi) = \phi'' + \phi$

7. The matrix of $f(x,y) = (x + y, 2x, x - y)$ relative to the standard bases is

$$A = \begin{bmatrix} 1 & 1 \\ 2 & 0 \\ 1 & -1 \end{bmatrix}$$

What does this become if the standard basis for $\mathcal{R}^2$ is replaced by $\alpha = \{\mathbf{E}_2, \mathbf{E}_1\}$? (The point of this problem is that *order* is important in a basis. See Sec. 3.4, where we made the same remark in connection with coordinate vectors.)

8. Define $f: \mathcal{R}^3 \to \mathcal{R}^1$ by $f(x,y,z) = x - 2y + z$. If

$$\alpha = \{(0,2,1), (1,0,1), (0,0,3)\}$$

what is the matrix of f relative to α and the standard basis for $\mathcal{R}^1$?

9. Let $f\colon \mathcal{C} \to \mathcal{C}$ be the linear map defined by $f(z) = \bar{z}$ (where $\mathcal{C}$ is regarded as a vector space over $\mathcal{R}$, as in Prob. 7, Sec. 3.1).

a. What is the matrix of f relative to the basis $\{1,i\}$ for $\mathcal{C}$?

b. The result in part *a* is

$$\begin{bmatrix} 1 & 0 \\ 0 & -1 \end{bmatrix}$$

which may also be regarded as the matrix of the linear operator on $\mathcal{R}^2$ sending (x,y) into $(x,-y)$; this is a reflection of the plane in the x axis. Since $\mathcal{C}$ and $\mathcal{R}^2$ are isomorphic as vector spaces over $\mathcal{R}$ (why?), we should be able to interpret f as a reflection of the ''complex plane'' in the ''real axis.'' Confirm that this is the case. (Draw a picture!)

10. Let S be the vector space of 2×2 matrices with entries in $\mathcal{F}$ and let $\alpha = \{E_{11},E_{12},E_{21},E_{22}\}$, where E_{ij} is the 2×2 matrix with 1 in the i,j position and 0 elsewhere.

a. If $f\colon S \to \mathcal{F}^4$ is the coordinate vector map defined by

$$f\left(\begin{bmatrix} a & b \\ c & d \end{bmatrix}\right) = (a,b,c,d)$$

what is the matrix of f relative to α and the standard basis for $\mathcal{F}^4$?

b. If $f\colon S \to S$ is the linear operator defined by $f(A) = A^T$, what is the matrix of f relative to α?

11. Let $f\colon S \to \mathcal{F}^n$ be the coordinate vector map defined by $f(x) = \mathbf{X}$, where S is an n-dimensional vector space over $\mathcal{F}$ with basis $\alpha = \{u_1, \ldots, u_n\}$ and $\mathbf{X} = (x_1, \ldots, x_n)$ is the coordinate vector of $x = x_1u_1 + \cdots + x_nu_n$ relative to α. (See Sec. 3.4.) Explain why the matrix of f relative to α and the standard basis for $\mathcal{F}^n$ is the $n \times n$ matrix whose columns are $\mathbf{E}_1, \ldots, \mathbf{E}_n$.

12. For each $m \times n$ matrix A, define a linear map $f\colon \mathcal{F}^n \to \mathcal{F}^m$ whose matrix relative to the standard bases is A.

a. $A = \begin{bmatrix} 2 & -3 & 1 \\ 1 & 5 & 0 \end{bmatrix}$ *b.* $A = \begin{bmatrix} i & 0 \\ 1 & i \end{bmatrix}$

c. $A = [1 \quad 1 \quad 2]$ *d.* $A = [2]$

★ **13.** If A is an $m \times n$ matrix with entries in $\mathcal{F}$, we know there is a unique linear map $f\colon \mathcal{F}^n \to \mathcal{F}^m$ whose matrix relative to the standard bases is A. Explain why $f(\mathbf{E}_j)$ is the jth column of A (given that $\mathbf{E}_j$ is the jth element of the standard basis for $\mathcal{F}^n$, where $j = 1, \ldots, n$).

★ **14.** Let S be a vector space over $\mathcal{F}$ with basis $\alpha = \{u_1, \ldots, u_n\}$ and let $i\colon S \to S$ be the identity map on S. Prove that the matrix of i relative to α is the $n \times n$ matrix whose columns are the standard basis vectors in $\mathcal{F}^n$.

Thus $[i]$ is the matrix with δ_{ij} as its i,j entry; it has 1s down the "main diagonal" and 0s elsewhere. We call it the $n \times n$ *identity matrix* and denote it by I_n, or simply I when n is understood. (See Prob. 1*e* for the 3×3 case, Prob. 10*a* for the 4×4 case, and Prob. 11 for another context in which the $n \times n$ case arises.)

15. Let i be the identity map on $\mathcal{R}^2$. Find $[i]_{\alpha^\beta}$, where $\alpha = \{\mathbf{E}_1,\mathbf{E}_2\}$ and $\beta = \{\mathbf{E}_2,\mathbf{E}_1\}$.

The point of this problem is that the identity map is represented by the identity matrix only if one basis for the space is used. It is correct to say that $[i]_{\alpha^\alpha} = I$, but here we found $[i]_{\alpha^\beta}$, where $\beta \neq \alpha$.

16. Let U be the plane $x - y - z = 0$ in $\mathcal{R}^3$ and define $f\colon \mathcal{R}^3 \to U$ by

$$f(x,y,z) = \tfrac{1}{3}(2x + y + z, x + 2y - z, x - y + 2z)$$

the formula for the orthogonal projection of $\mathcal{R}^3$ on U (Prob. 20, Sec. 3.1).

a. Find a matrix representation of f. *Hint:* It is natural to use the standard basis for $\mathcal{R}^3$, but U doesn't have one. Choose your own basis.

b. The procedure in part *a* is awkward. Regard f as a linear operator on $\mathcal{R}^3$ (rather than a map from $\mathcal{R}^3$ to U) and use the standard basis for $\mathcal{R}^3$ to find $[f]$.

This raises a question. How can the same map have a 2×3 and a 3×3 matrix representation? The answer is that we are not dealing with the same maps in parts *a* and *b* of the problem. The definition of a function $f\colon A \to B$ involves not only a rule, or formula, but also the sets A and B; the maps $f\colon\mathcal{R}^3 \to U$ and $f\colon \mathcal{R}^3 \to \mathcal{R}^3$ are different. (The first, for example, is onto; the second isn't.) Note that the second is the one we call the orthogonal projection of $\mathcal{R}^3$ on U. (See Example 3.11.)

17. Define $f\colon \mathcal{R}^3 \to \mathcal{R}^2$ and $g\colon \mathcal{R}^3 \to \mathcal{R}^2$ by

$$f(x,y,z) = (x - y + z, 3y) \qquad \text{and} \qquad g(x,y,z) = (y + 2z, x - y)$$

and let A and B be their matrices relative to the standard bases.

a. Find A and B.

b. Defining addition and scalar multiplication of matrices in the natural way (Example 2.2), find $A + B$ and $2A$.

c. Find $f + g$ and $2f$. (See Example 3.23.)

d. Confirm that the matrix of $f + g$ is $A + B$ and the matrix of $2f$ is $2A$, that is,

$$[f + g] = [f] + [g] \qquad \text{and} \qquad [2f] = 2[f]$$

The sets L and M discussed at the end of this section are in this case $L(\mathcal{R}^3,\mathcal{R}^2)$ and $M_{2,3}(\mathcal{R})$. The first is already a vector space (Sec. 3.5); by imposing the "natural" definitions of addition and scalar multiplication on M, we have made it a vector space, too. (See the next section.) Hence it makes sense to inquire whether the map $\phi: L \to M$ defined by $\phi(f) = [f]$ is linear. In view of the results in part *d*, how do things look?

18. Define $f: \mathcal{R}^3 \to \mathcal{R}^4$ and $g: \mathcal{R}^2 \to \mathcal{R}^3$ by

$$f(x,y,z) = (3x + z, z - x, 2y, y + z)$$

and

$$g(x,y) = (x, x - 2y, x + y)$$

a. Find $A = [f]$ and $B = [g]$.

b. Let $h: \mathcal{R}^2 \to \mathcal{R}^4$ be the composite map defined by $h(x,y) = f[g(x,y)]$. Find $C = [h]$. (Later on matrix multiplication will be defined in terms of composition of linear maps and we'll have $C = AB$. If you already know how to multiply matrices, you might check this out.)

4.2 The Isomorphism between Maps and Matrices

Let S and T be vector spaces over $\mathfrak{F}$. In Sec. 3.5 we proved that $L(S,T)$, the set of linear maps from S to T, is a vector space over $\mathfrak{F}$. We also stated (without proof) that if S and T are finite-dimensional, then

$$\dim L(S,T) = (\dim S)(\dim T)$$

One of our objectives in this section is to supply a proof of this fact by using matrix representations of the maps in L. But our main purpose is to show that $L(S,T)$, the set of linear maps from S to T, and $M_{m,n}(\mathfrak{F})$, the set of $m \times n$ matrices with entries in $\mathfrak{F}$, are isomorphic (where $n = \dim S$ and $m = \dim T$). If this is the case, then each set can be used to answer questions about the other—an idea that turns out to be useful indeed.

As we pointed out at the end of the last section, our first order of business is to define addition and scalar multiplication of matrices, since otherwise $M_{m,n}(\mathfrak{F})$ has no algebraic structure to compare with that of $L(S,T)$.

MATRIX NOTATION

Let A be an $m \times n$ matrix with entries in $\mathfrak{F}$. If the entry in the ith row and jth column is designated by a_{ij}, where $i = 1, \ldots, m$ and $j = 1, \ldots, n$, we write $A = [a_{ij}]$.†

† When A is the matrix of a linear map f, we have agreed to write $A = [f]$. This should be read "matrix of f," whereas $[a_{ij}]$ means "matrix whose i,j entry is a_{ij}." There should not be any occasion for confusion between the two.

This notation is shorthand for the gruesome business of writing out A in full:

$$A = \begin{bmatrix} a_{11} & a_{12} & \cdots & a_{1n} \\ a_{21} & a_{22} & \cdots & a_{2n} \\ \cdots & \cdots & \cdots & \cdots \\ a_{m1} & a_{m2} & \cdots & a_{mn} \end{bmatrix}$$

Whenever we put down $A = [a_{ij}]$, we should have in our mind's eye the whole array of m rows and n columns.

ADDITION AND SCALAR MULTIPLICATION OF MATRICES

Let m and n be positive integers. We denote the class of $m \times n$ matrices with entries in $\mathfrak{F}$ by $M_{m,n}(\mathfrak{F})$, or simply $M_n(\mathfrak{F})$ when $m = n$. Addition and scalar multiplication in M are defined by

$$[a_{ij}] + [b_{ij}] = [a_{ij} + b_{ij}] \quad \text{and} \quad c[a_{ij}] = [ca_{ij}]$$

EXAMPLE 4.3

If

$$A = \begin{bmatrix} 1 & -1 & 1 \\ 0 & 3 & 0 \end{bmatrix} \quad \text{and} \quad B = \begin{bmatrix} 0 & 1 & 2 \\ 1 & -1 & 0 \end{bmatrix}$$

are the 2×3 matrices of Prob. 17, Sec. 4.1, then

$$A + B = \begin{bmatrix} 1 & 0 & 3 \\ 1 & 2 & 0 \end{bmatrix} \quad \text{and} \quad 2A = \begin{bmatrix} 2 & -2 & 2 \\ 0 & 6 & 0 \end{bmatrix}$$

THEOREM 4.3

$M_{m,n}(\mathfrak{F})$ is a vector space over $\mathfrak{F}$ of dimension mn.

Proof

It is tiresome to check that M is a vector space, so we'll skip it. (See Theorem 3.9 for the sort of tedium we are omitting.) We merely point out that the zero element of M is the $m \times n$ matrix whose entries are all 0, and the additive inverse of $A = [a_{ij}] \in M$ is the matrix

$$-A = -1A = [-a_{ij}] \in M$$

To prove that the dimension of M is mn, we name a basis consisting of mn elements. For each $i = 1, \ldots, m$ and $j = 1, \ldots, n$, let E_{ij} be the $m \times n$ matrix with 1 in the i,j position and 0 elsewhere. There are mn such matrices (why?); we leave it to you to show that they span M and are linearly independent. ∎

EXAMPLE 4.4

The six matrices

$$E_{11} = \begin{bmatrix} 1 & 0 & 0 \\ 0 & 0 & 0 \end{bmatrix} \quad E_{12} = \begin{bmatrix} 0 & 1 & 0 \\ 0 & 0 & 0 \end{bmatrix} \quad E_{13} = \begin{bmatrix} 0 & 0 & 1 \\ 0 & 0 & 0 \end{bmatrix}$$

$$E_{21} = \begin{bmatrix} 0 & 0 & 0 \\ 1 & 0 & 0 \end{bmatrix} \quad E_{22} = \begin{bmatrix} 0 & 0 & 0 \\ 0 & 1 & 0 \end{bmatrix} \quad E_{23} = \begin{bmatrix} 0 & 0 & 0 \\ 0 & 0 & 1 \end{bmatrix}$$

constitute a basis for $M_{2,3}(\mathfrak{F})$.

THEOREM 4.4

Let S and T be vector spaces over $\mathfrak{F}$ with (ordered) bases $\alpha = \{u_1, \ldots, u_n\}$ and $\beta = \{v_1, \ldots, v_m\}$, respectively. Then $L(S,T)$ and $M_{m,n}(\mathfrak{F})$ are isomorphic by virtue of the map which sends each $f \in L$ into its matrix relative to α and β.

Proof

The map in question is $\phi: L \to M$ defined by $\phi(f) = [f]$. (We are suppressing reference to the bases, as agreed in Sec. 4.1.) We have already proved in Sec. 4.1 (Theorems 4.1 and 4.2) that ϕ is one-to-one and onto, so the only thing that needs verifying is whether it is linear.†

Let f and g be any elements of L (linear maps from S to T) and suppose that c and d are scalars. Let

$$A = \phi(f) = [f] = [a_{ij}] \qquad \text{and} \qquad B = \phi(g) = [g] = [b_{ij}]$$

To prove that ϕ is linear, we have to show that

$$\phi(cf + dg) = c\phi(f) + d\phi(g) = cA + dB$$

But

$$\phi(cf + dg) = [cf + dg]$$

the matrix of the map $cf + dg$ relative to α and β. We propose to prove that the jth column ($j = 1, \ldots, n$) of this matrix is the same as the jth column of $cA + dB$.

To do this, we use the isomorphism $\beta: T \to \mathfrak{F}^m$ which sends each element of T into its coordinate vector relative to β.‡ The jth column of $[cf + dg]$ is the coordinate vector of

$$(cf + dg)(u_j) = cf(u_j) + dg(u_j)$$

† Look back to Prob. 17, Sec. 4.1, and note that in part d you proved that $\phi(f + g) = \phi(f) + \phi(g)$ and $\phi(2f) = 2\phi(f)$. What we want to do now is establish linearity in the general case.

‡ This is the linear map $\beta(x) = \mathbf{X}$, where $x = x_1v_1 + \cdots + x_mv_m$ and $\mathbf{X} = (x_1, \ldots, x_m)$. (See Theorem 3.6.) There shouldn't be any confusion between the basis β and the map β. We avoid excessive notation by using the same letter to represent both.

namely

$$\beta[cf(u_j) + dg(u_j)] = c\beta[f(u_j)] + d\beta[g(u_j)]$$

But $\beta[f(u_j)]$ and $\beta[g(u_j)]$ are the jth columns of A and B, so

$$\begin{aligned} c\beta[f(u_j)] + d\beta[g(u_j)] &= c(a_{1j}, \ldots, a_{mj}) + d(b_{1j}, \ldots, b_{mj}) \\ &= (ca_{1j} + db_{1j}, \ldots, ca_{mj} + db_{mj}) \end{aligned}$$

This is the jth column of $cA + dB$. ∎

COROLLARY 4.4a

If $\dim S = n$ and $\dim T = m$, then $\dim L(S,T) = mn$.

Proof

Isomorphic vector spaces have the same dimension (Theorem 3.8.) ∎

This corollary is a restatement of Theorem 3.10, the proof of which is now complete. As we noted then, the problem in proving that $\dim L(S,T) = mn$ is to name mn linearly independent maps which span $L(S,T)$. We have avoided the explicit construction of such a basis by showing instead that $L(S,T)$ is isomorphic to $M_{m,n}(\mathfrak{F})$, which we know has dimension mn. (Note how easy it was in Theorem 4.3 to name a basis for the vector space of $m \times n$ matrices.)

Now, however, it is easy to construct a basis for $L(S,T)$; simply define $f_{ij}\colon S \to T$ to be the map whose matrix relative to (fixed) bases for S and T is E_{ij}!

EXAMPLE 4.5

A basis for $L(\mathfrak{R}^3,\mathfrak{R}^2)$ is $\{f_{11},f_{12},f_{13},f_{21},f_{22},f_{23}\}$, where

$$\begin{array}{lll} f_{11}(x,y,z) = (x,0) & f_{12}(x,y,z) = (y,0) & f_{13}(x,y,z) = (z,0) \\ f_{21}(x,y,z) = (0,x) & f_{22}(x,y,z) = (0,y) & f_{23}(x,y,z) = (0,z) \end{array}$$

See Example 3.24, where we came up with this result directly, without the help of matrices.

This example provides a good illustration of what isomorphisms can do in algebra. In the present case, we answered a hard question about maps by dealing with an easy question about matrices. You'll find as we proceed that it works both ways; we move from one realm to the other depending on which seems more transparent at the moment.

Problems

19. Convince yourself that $M_{m,n}(\mathfrak{F})$ is a vector space over $\mathfrak{F}$.

20. Show that the matrices E_{ij} defined in the proof of Theorem 4.3 constitute a basis for $M_{m,n}(\mathfrak{F})$.

21. Give an alternate proof of the linearity of ϕ in Theorem 4.4 by comparing the i,j entries of $\phi(cf + dg)$ and $c\phi(f) + d\phi(g)$. (We compared *columns*.) *Hint:* If $f: S \to T$ is a linear map, the i,j entry of its matrix relative to α and β is the ith coefficient in the expression $f(u_j) = \Sigma_{i=1}^{m} a_{ij}v_i$. (See Sec. 4.1.)

22. Since the function $\phi: L(\mathcal{R}^3,\mathcal{R}^2) \to M_{2,3}(\mathcal{R})$ defined by $\phi(f) = [f]$ is itself a linear map from one finite-dimensional space to another, it can be represented by a matrix. Using the bases

$$\{f_{11},f_{12},f_{13},f_{21},f_{22},f_{23}\} \qquad \text{and} \qquad \{E_{11},E_{12},E_{13},E_{21},E_{22},E_{23}\}$$

which we exhibited in Examples 4.4 and 4.5, find $[\phi]$.

★ **23.** If $f: S \to T$ and $g: S \to T$ are linear maps and c is a scalar (where S and T are vector spaces with specified bases), why is the following true?

$$[f + g] = [f] + [g] \qquad \text{and} \qquad [cf] = c[f]$$

24. To test your understanding of matrix notation, here is a definition from the next chapter. Let $A = [a_{ij}]$ and $B = [b_{ij}]$ be $p \times q$ and $q \times r$, respectively. Their *product* (in the order named) is the $p \times r$ matrix $C = [c_{ij}]$, where

$$c_{ij} = \sum_{k=1}^{q} a_{ik}b_{kj} \qquad i = 1, \ldots, p;\ j = 1, \ldots, r$$

a. Explain why this definition amounts to saying that the i,j entry of AB is the dot product of the ith row of A and the jth column of B (each regarded as a vector in q-space).

b. Use the definition to find AB if

$$A = \begin{bmatrix} 3 & -1 & 1 \\ 1 & 5 & 0 \end{bmatrix} \qquad \text{and} \qquad B = \begin{bmatrix} 1 & -2 & 4 \\ 0 & 1 & -1 \\ 1 & 2 & 0 \end{bmatrix}$$

c. Prove that $(kA)B = k(AB) = A(kB)$ for every scalar k, where A is $p \times q$ and B is $q \times r$.

d. Prove the distributive law $A(B + C) = AB + AC$, where A is $p \times q$ and B and C are $q \times r$.

25. Let

$$A = \begin{bmatrix} 2 & 0 & -1 \\ 3 & 5 & 1 \end{bmatrix}$$

Regarding the elements of $\mathcal{R}^3$ and $\mathcal{R}^2$ as "column vectors" (more properly, 3×1 and 2×1 matrices), define $f: \mathcal{R}^3 \to \mathcal{R}^2$ by $f(\mathbf{X}) = A\mathbf{X}$ for each $\mathbf{X} \in \mathcal{R}^3$ (where $A\mathbf{X}$ means matrix multiplication as defined in Problem 24).

a. What is the explicit formula for f? In other words, find $f(x,y,z)$.†
b. Use the definition $f(\mathbf{X}) = A\mathbf{X}$ (not the explicit formula) to prove that f is linear. *Hint:* See parts *c*, *d* of Prob. 24.
c. Confirm that $f(\mathbf{E}_1)$, $f(\mathbf{E}_2)$, $f(\mathbf{E}_3)$ are the columns of A.
d. Conclude that $[f] = A$ (relative to the standard bases).

LECTOR *Well, of course.*
AUCTOR *Why "of course"?*
LECTOR *If* $[f] \neq A$, *I'd want a refund.*
AUCTOR *See our conversation at the beginning of Sec. 4.1. We had to plan ahead.*

---

26. Generalize Prob. 25 to the map $f\colon \mathfrak{F}^n \to \mathfrak{F}^m$ defined by $f(\mathbf{X}) = A\mathbf{X}$, where A is an $m \times n$ matrix with entries in $\mathfrak{F}$ and $\mathbf{X} \in \mathfrak{F}^n$. What is required is to prove that f is linear, that $f(\mathbf{E}_j)$ is the jth column of A, and hence that $[f] = A$.

4.3 The Transpose of a Matrix

In Example 3.7 we defined the transpose of a 2×2 matrix

$$A = \begin{bmatrix} a & b \\ c & d \end{bmatrix}$$

to be the matrix

$$A^T = \begin{bmatrix} a & c \\ b & d \end{bmatrix}$$

obtained from A by writing its rows as columns. More generally, we make the following definition.

TRANSPOSE

Let A be an $m \times n$ matrix with entries in $\mathfrak{F}$. The *transpose* of A is the $n \times m$ matrix A^T whose i,j entry is the j,i entry of A; that is, if $A = [a_{ij}]$, then $A^T = [b_{ij}]$, where $b_{ij} = a_{ji}$, for $i = 1, \ldots, n$ and $j = 1, \ldots, m$.

---

† Strictly speaking, this should read

$$f\left(\begin{bmatrix} x \\ y \\ z \end{bmatrix}\right)$$

since **X** is being regarded as a 3×1 matrix for purposes of matrix multiplication. But we shall not insist on this; the elements of $\mathcal{R}^3$ and $\mathcal{R}^2$ may still be written as ordinary vectors.

EXAMPLE 4.6

If

$$A = \begin{bmatrix} a_{11} & a_{12} & a_{13} \\ a_{21} & a_{22} & a_{23} \end{bmatrix} = \begin{bmatrix} 2 & 0 & 5 \\ 1 & -2 & -2 \end{bmatrix}$$

then

$$A^T = \begin{bmatrix} b_{11} & b_{12} \\ b_{21} & b_{22} \\ b_{31} & b_{32} \end{bmatrix} = \begin{bmatrix} a_{11} & a_{21} \\ a_{12} & a_{22} \\ a_{13} & a_{23} \end{bmatrix} = \begin{bmatrix} 2 & 1 \\ 0 & -2 \\ 5 & -2 \end{bmatrix}$$

As this example suggests, A^T is simply A with its rows and columns interchanged; that is, the ith row of A becomes the ith column of A^T, where $i = 1, \ldots, m$, and the jth column of A becomes the jth row of A^T, where $j = 1, \ldots, n$. We leave it to you to explain why this is true in general, and to conclude that

$$(A^T)^T = A$$

THEOREM 4.5

If A and B are $m \times n$ matrices and k is a scalar, then

$$(A + B)^T = A^T + B^T \qquad \text{and} \qquad (kA)^T = kA^T$$

Proof

Let $A = [a_{ij}]$ and $B = [b_{ij}]$. The i,j entry of $(A + B)^T$ is the j,i entry of $A + B$, namely $a_{ji} + b_{ji}$. The i,j entry of $A^T + B^T$ is the sum of the i,j entries of A^T and B^T, that is, the sum of the j,i entries of A and B. Hence it is also $a_{ji} + b_{ji}$. Thus $(A + B)^T = A^T + B^T$. The proof of $(kA)^T = kA^T$ is similar. ∎

EXAMPLE 4.7

Let

$$A = \begin{bmatrix} 2 & 0 & 5 \\ 1 & -2 & -2 \end{bmatrix} \qquad \text{and} \qquad B = \begin{bmatrix} 1 & -1 & -3 \\ 0 & 2 & 1 \end{bmatrix}$$

Then

$$A + B = \begin{bmatrix} 3 & -1 & 2 \\ 1 & 0 & -1 \end{bmatrix}$$

and hence

$$(A + B)^T = \begin{bmatrix} 3 & 1 \\ -1 & 0 \\ 2 & -1 \end{bmatrix}$$

On the other hand,

$$A^T = \begin{bmatrix} 2 & 1 \\ 0 & -2 \\ 5 & -2 \end{bmatrix} \quad \text{and} \quad B^T = \begin{bmatrix} 1 & 0 \\ -1 & 2 \\ -3 & 1 \end{bmatrix}$$

from which

$$A^T + B^T = \begin{bmatrix} 3 & 1 \\ -1 & 0 \\ 2 & -1 \end{bmatrix}$$

Hence $(A + B)^T = A^T + B^T$, as advertised in Theorem 4.5.

SYMMETRIC MATRIX

Let A be an $n \times n$ matrix with real entries. If $A^T = A$, then A is said to be *symmetric*. In other words, a symmetric matrix is a square matrix with real entries which is unaffected when its rows and columns are interchanged.†

EXAMPLE 4.8

Let $\{\mathbf{U}_1, \ldots, \mathbf{U}_n\}$ be a basis for $\mathcal{R}^n$ and define the matrix $A = [a_{ij}]$ by

$$a_{ij} = \mathbf{U}_i \cdot \mathbf{U}_j \qquad \text{for all } i \text{ and } j$$

Then $A^T = A$. (Why?) For example, if

$$\mathbf{U}_1 = (1,2,-1) \qquad \mathbf{U}_2 = (3,5,0) \qquad \mathbf{U}_3 = (-1,2,2)$$

we find that

$$A = \begin{bmatrix} 6 & 13 & 1 \\ 13 & 34 & 7 \\ 1 & 7 & 9 \end{bmatrix} = A^T$$

You can see at a glance why such a matrix is called symmetric.

Later on we are going to investigate the "standard inner product" on $\mathfrak{F}^n$, defined by

$$\mathbf{X} * \mathbf{Y} = x_1\bar{y}_1 + \cdots + x_n\bar{y}_n$$

where $\mathbf{X} = (x_1, \ldots, x_n)$ and $\mathbf{Y} = (y_1, \ldots, y_n)$. (Note that this reduces to the dot product when $\mathfrak{F} = \mathcal{R}$. Why?) If $\{\mathbf{U}_1, \ldots, \mathbf{U}_n\}$ is a basis for $\mathfrak{F}^n$ and

† There is no apparent reason why we shouldn't use this terminology in connection with matrices having complex entries; some writers do. The reason we are not allowing it will not appear until near the end of the book.

$A = [a_{ij}]$ is defined by $a_{ij} = \mathbf{U}_i * \mathbf{U}_j$ (as in Example 4.8), it is no longer true (unless $\mathfrak{F} = \mathfrak{R}$) that $A^T = A$. Instead we have $A^T = \bar{A}$, where $\bar{A}$, (the "conjugate" of A) is A with its entries replaced by their conjugates. This may help to motivate our next definition (which will be of no immediate use—we include it here because the idea is similar to the ordinary transpose).

CONJUGATE TRANSPOSE

Let A be an $m \times n$ matrix with entries in $\mathfrak{F}$. The *conjugate transpose* of A is the $n \times m$ matrix A^* whose i,j entry is the conjugate of the j,i entry of A. That is, if $A = [a_{ij}]$, then $A^* = [b_{ij}]$, where

$$b_{ij} = \bar{a}_{ji} \qquad i = 1, \ldots, n;\ j = 1, \ldots, m$$

EXAMPLE 4.9

If

$$A = \begin{bmatrix} a_{11} & a_{12} & a_{13} \\ a_{21} & a_{22} & a_{23} \end{bmatrix} = \begin{bmatrix} 1 & 1+i & -i \\ 0 & 3-i & 5 \end{bmatrix}$$

then

$$A^* = \begin{bmatrix} b_{11} & b_{12} \\ b_{21} & b_{22} \\ b_{31} & b_{32} \end{bmatrix} = \begin{bmatrix} \bar{a}_{11} & \bar{a}_{21} \\ \bar{a}_{12} & \bar{a}_{22} \\ \bar{a}_{13} & \bar{a}_{23} \end{bmatrix} = \begin{bmatrix} 1 & 0 \\ 1-i & 3+i \\ i & 5 \end{bmatrix}$$

This suggests that

$$A^* = \bar{A}^T$$

a formula we ask you to prove in the problems. If $\mathfrak{F} = \mathfrak{R}$, we have $A^* = A^T$; in that case, the conjugate transpose and transpose are identical. Also note that

$$(A^*)^* = A$$

THEOREM 4.6

If A and B are $m \times n$ matrices and k is a scalar, then

$$(A + B)^* = A^* + B^* \qquad \text{and} \qquad (kA)^* = \bar{k}A^*$$

Proof

We leave the first formula for the problems. To prove the second, let $A = [a_{ij}]$ and observe that the i,j entry of $(kA)^*$ is the conjugate of the j,i entry of kA:

$$\overline{ka_{ji}} = \bar{k}\bar{a}_{ji}$$

(See Prob. 43*e*, Sec. 1.3.) Since this is the i,j entry of $\bar{k}A^*$, we have $(kA)^* = \bar{k}A^*$. ∎

EXAMPLE 4.10

If

$$A = \begin{bmatrix} 1 & 1+i & -i \\ 0 & 3-i & 5 \end{bmatrix}$$

and $k = i$, then

$$kA = \begin{bmatrix} i & -1+i & 1 \\ 0 & 1+3i & 5i \end{bmatrix}$$

and hence

$$(kA)^* = \begin{bmatrix} -i & 0 \\ -1-i & 1-3i \\ 1 & -5i \end{bmatrix}$$

On the other hand,

$$A^* = \begin{bmatrix} 1 & 0 \\ 1-i & 3+i \\ i & 5 \end{bmatrix}$$

from which

$$\bar{k}A^* = -iA^* = \begin{bmatrix} -i & 0 \\ -1-i & 1-3i \\ 1 & -5i \end{bmatrix}$$

Hence $(kA)^* = \bar{k}A^*$.

HERMITIAN MATRIX

Let A be a square matrix with entries in $\mathfrak{F}$. If $A^* = A$, then A is said to be *Hermitian* (in honor of Charles Hermite, 1822–1901).

EXAMPLE 4.11

Let $\{\mathbf{U}_1, \ldots, \mathbf{U}_n\}$ be a basis for $\mathfrak{F}^n$. If $A = [a_{ij}]$, where

$$a_{ij} = \mathbf{U}_i * \mathbf{U}_j \qquad \text{for all } i \text{ and } j$$

then (as we observed following Example 4.8) $A^T = \bar{A}$. Hence

$$A^* = \bar{A}^T = (A^T)^T = A$$

which means that A is Hermitian. For example, if the basis elements in $\mathcal{C}^2$ are

$$\mathbf{U}_1 = (1 - i, 1) \qquad \mathbf{U}_2 = (i,i)$$

we find that

$$A = \begin{bmatrix} 3 & -1 - 2i \\ -1 + 2i & 2 \end{bmatrix} = A^*$$

(Confirm this!) Hence A is Hermitian.

It is worth noting that a matrix with real entries is Hermitian if and only if it is symmetric, since the formulas $A^T = A$ and $A^* = A$ are equivalent when the field of scalars is $\mathcal{R}$. But a complex matrix satisfying $A^T = A$ is not necessarily Hermitian, as shown by the example

$$A = \begin{bmatrix} 1 & i \\ i & 2 \end{bmatrix}$$

Here we have $A^T = A$, but

$$A^* = \begin{bmatrix} 1 & -i \\ -i & 2 \end{bmatrix} \neq A$$

This is one of the reasons we restrict the definition of "symmetric" to real matrices; we want symmetric matrices to be a special case of Hermitian matrices. Both these classes of matrices are extremely important in linear algebra and its applications; we'll develop their properties in Chaps. 12 and 13.

Problems

★ **27.** Let A be an $m \times n$ matrix. Prove the following.

a. The ith row of A is the ith column of A^T and the jth column of A is the jth row of A^T.

b. $(A^T)^T = A$ *c.* $A^* = \bar{A}^T$ *d.* $(A^*)^* = A$ *e.* $(A^T)^* = (A^*)^T$

28. Complete the proof of Theorem 4.5 by showing that $(kA)^T = kA^T$.

29. Prove that the map f: $M_{m,n}(\mathfrak{F}) \to M_{n,m}(\mathfrak{F})$ defined by $f(A) = A^T$ is an isomorphism.

30. Finish the proof of Theorem 4.6 by showing that $(A + B)^* = A^* + B^*$.

31. Prove that the map f: $M_{m,n}(\mathfrak{F}) \to M_{n,m}(\mathfrak{F})$ defined by $f(A) = A^*$ is invertible. Is it an isomorphism?

32. Let

$$A = \begin{bmatrix} 2+i & 0 & 3i \\ 1 & -i & 1-i \end{bmatrix} \quad \text{and} \quad B = \begin{bmatrix} -i & 2 & 1 \\ i & 1+i & -1 \end{bmatrix}$$

a. Find A^T, B^T, and $(A+B)^T$, and confirm the formula $(A+B)^T = A^T + B^T$.
b. Find A^*, B^*, and $(A+B)^*$, and confirm the formula $(A+B)^* = A^* + B^*$.

33. Let $\{\mathbf{U}_1, \ldots, \mathbf{U}_n\}$ be a basis for $\mathfrak{F}^n$. If $A = [a_{ij}]$, where $a_{ij} = \mathbf{U}_i * \mathbf{U}_j$ for all i and j, explain why $A^T = \bar{A}$.

34. Show that an $n \times n$ matrix A is Hermitian if and only if $A^T = \bar{A}$.

35. Give an example of a Hermitian matrix A for which $A^T \neq A$.

★ **36.** Show that the diagonal entries of a Hermitian matrix are real. (If $A = [a_{ij}]$ is $n \times n$, its "diagonal entries" are $a_{11}, a_{22}, \ldots, a_{nn}$.)

37. Prove that the transpose of a Hermitian matrix is Hermitian. *Hint:* Use Prob. 27*e*. Is the conjugate transpose of a Hermitian matrix Hermitian?

38. Explain why the set of $n \times n$ Hermitian matrices with entries in $\mathfrak{F}$ is an abelian group relative to addition. (See Sec. 2.1 for the definition of "abelian group.")

39. The "adjoint" of a linear operator $f: \mathfrak{F}^n \to \mathfrak{F}^n$ is the linear operator $f^*: \mathfrak{F}^n \to \mathfrak{F}^n$ whose matrix is the conjugate transpose of the matrix of f; that is, $[f^*] = [f]^*$. (As usual, the standard basis is understood.)
a. Find the adjoint of $f: \mathbb{R}^2 \to \mathbb{R}^2$ defined by $f(x,y) = (x-y, 3x+y)$.
b. Confirm that $f(\mathbf{X}) \cdot \mathbf{Y} = \mathbf{X} \cdot f^*(\mathbf{Y})$ for all $\mathbf{X}$ and $\mathbf{Y}$ in $\mathbb{R}^2$.
c. Find the adjoint of $g: \mathbb{C}^2 \to \mathbb{C}^2$ defined by $g(x,y) = (x,iy)$.
d. Confirm that $g(\mathbf{X}) * \mathbf{Y} = \mathbf{X} * g^*(\mathbf{Y})$ for all $\mathbf{X}$ and $\mathbf{Y}$ in $\mathbb{C}^2$.

It is messy (but not hard in principle) to show that the formula in part *d*, of which part *b* is a special case, holds for any linear operator on $\mathfrak{F}^n$. The idea, which turns out to have profound implications, is that when a linear operator is moved "across the star" ("across the dot" in the case of real spaces) it is changed to its adjoint. See Sec. 12.1.

Review Quiz

True or false?

1. If A is a square matrix with real entries, then $(A^T)^* = A$.

2. The matrix of an invertible linear map must be square.

3. If $f: \mathbb{R}^3 \to \mathbb{R}^2$ is a linear map such that $f(\mathbf{E}_j) = (a_{1j}, a_{2j})$, where $j = 1, 2, 3$, then the matrix of f relative to the standard bases is

$$\begin{bmatrix} a_{11} & a_{12} & a_{13} \\ a_{21} & a_{22} & a_{23} \end{bmatrix}$$

4. Every symmetric matrix is Hermitian.

5. If f: $\mathfrak{R}^2 \rightarrow \mathfrak{R}^2$ is the linear operator whose matrix is

$$\begin{bmatrix} 0 & 1 \\ 1 & 0 \end{bmatrix}$$

then the composite map $f^2(\mathbf{X}) = f[f(\mathbf{X})]$ is the identity map on $\mathfrak{R}^2$.

6. If α and β are bases for the n-dimensional space S and i is the identity map on S, then $[i]_{\alpha^\beta} = I_n$ (the $n \times n$ matrix with i,j entry δ_{ij}).

7. The map f: $M_{2,3}(\mathbb{C}) \rightarrow M_{3,2}(\mathbb{C})$ defined by $f(A) = A^*$ is an isomorphism.

8. The map f: $\mathfrak{R}^2 \rightarrow \mathfrak{R}^2$ whose matrix relative to the standard basis is

$$\begin{bmatrix} 1 & 0 \\ 1 & -1 \end{bmatrix}$$

sends (x,y) into $(x + y, -y)$.

9. The set of 3×3 matrices with real entries is a vector space of dimension 9.

10. Let f: $S \rightarrow T$ be a linear map with matrix A relative to given bases for S and T. If the elements of the basis for S are listed in a different order, the effect on A is to rearrange its columns.

11. If A is a matrix with complex entries, then $(kA)^T = kA^T$ for every scalar k.

12. The inverse of

$$\begin{bmatrix} 1 & -1 \\ 2 & 0 \end{bmatrix}$$

relative to addition is

$$\begin{bmatrix} -1 & 1 \\ -2 & 0 \end{bmatrix}$$

13. The matrix

$$\begin{bmatrix} i & 1 \\ 1 & 0 \end{bmatrix}$$

is Hermitian.

14. There is only one linear map from $\mathfrak{R}^2$ to $\mathfrak{R}^3$ with the property that $f(1,0) = f(0,1) = (1,1,1)$.

15. Every linear map from $\mathfrak{R}^3$ to $\mathfrak{R}^2$ has the form $g(x,y,z) = (ax + by + cz, dx + ey + fz)$, where a, b, c, d, e, f are scalars.

CHAPTER 5

Multiplication of Maps and Matrices

The last two chapters have established a close relation between linear maps and matrices. In fact, the vector space of linear maps from S to T and the space of $m \times n$ matrices (where $n = \dim S$ and $m = \dim T$) are abstractly identical—in the sense of isomorphism. Now we are going to push the analogy further by defining "multiplication" of maps and using its properties to motivate a definition of multiplication of matrices. The results are far-reaching in linear algebra and its applications.

5.1 Composition of Linear Maps

Composition of functions is an idea which you have probably encountered in calculus. The subject has also arisen in this book, although we did not make much of it at the time. (See the proof of Theorem 3.7.) Now we are ready to discuss it in some detail.

---

†COMPOSITION OF LINEAR MAPS

Let $f\colon T \to U$ and $g\colon S \to T$ be linear maps, where S, T, and U are vector spaces over $\mathfrak{F}$. (Note that the range of g is a subset of the domain of f.) The map $h\colon S \to U$ defined by $h(x) = f[g(x)]$ for each $x \in S$ is called the *composition* (or *product*) of f and g (in that order), and is denoted by $h = f \circ g$.

A schematic diagram for composition of maps is shown in Fig. 5.1. Here we are thinking of f and g as machines, or computers. When an element within the computer's domain is fed into it, lights flash, wheels turn, and various

† We need not assume that the maps are linear; the idea of composition applies in general. However, we confine our attention to linear maps in this book.

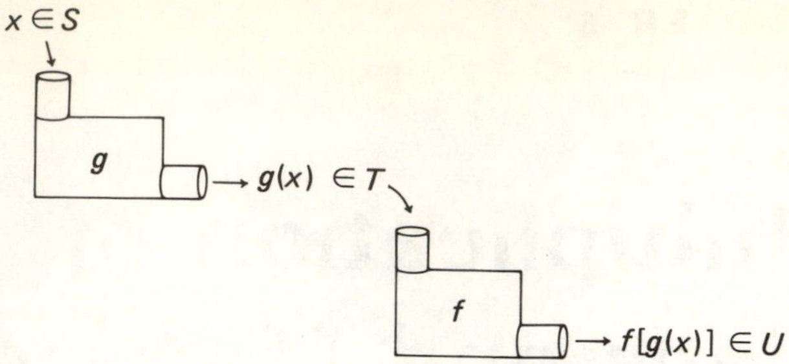

Fig. 5.1 **Composition of maps**

other important things happen; presently the computer produces a corresponding element in its range.

The composite map is a "super computer" consisting of f and g in tandem, as shown in Fig. 5.2.

EXAMPLE 5.1

Define $f\colon \mathcal{R}^3 \to \mathcal{R}^2$ and $g\colon \mathcal{R}^2 \to \mathcal{R}^3$ by

$$f(x,y,z) = (x - y + 2z, 3y) \qquad \text{and} \qquad g(x,y) = (x - y, 2x, x + 3y)$$

The composite map $f \circ g$ is $h\colon \mathcal{R}^2 \to \mathcal{R}^2$ defined by

$$\begin{aligned} h(x,y) &= (f \circ g)(x,y) = f[g(x,y)] \\ &= f(x - y, 2x, x + 3y) = (x + 5y, 6x) \end{aligned}$$

(Confirm this result!)

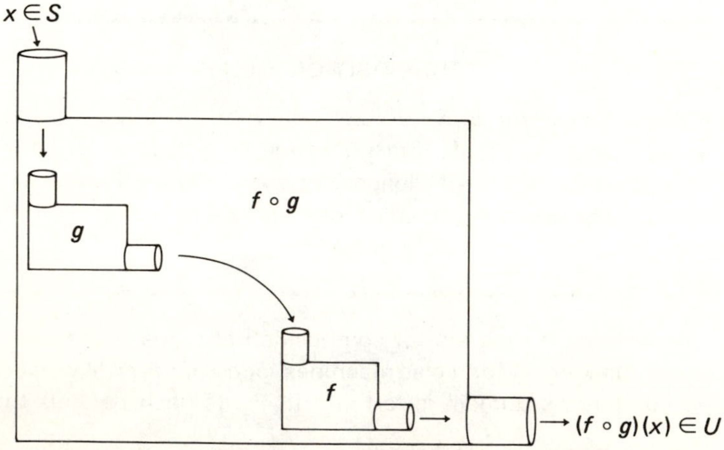

Fig. 5.2 **Composite map**

EXAMPLE 5.2

Let S be the vector space of real-valued functions with derivatives of all orders on $(-\infty, \infty)$. If $D: S \to S$ is the linear operator defined by $D(\phi) = \phi'$, then $D \circ D: S \to S$ is defined by

$$(D \circ D)(\phi) = D[D(\phi)] = D(\phi') = \phi''$$

This map is usually abbreviated as $D \circ D = D^2$, and we write $D^2(\phi) = \phi''$. Thus the notation for second derivatives in calculus is not as simple-minded as it looks!

Composition of linear maps is an operation with algebraic properties that are reminiscent of multiplication in ordinary algebra. The following theorems will indicate what we mean.

THEOREM 5.1

If $f: T \to U$ and $g: S \to T$ are linear maps, so is $f \circ g: S \to U$.

Proof

This has already been done in the proof of Theorem 3.7.

THEOREM 5.2

If $f: U \to V$, $g: T \to U$, and $h: S \to T$ are linear maps, then

$$(f \circ g) \circ h = f \circ (g \circ h)$$

Proof

Let $F = f \circ g$ and $G = g \circ h$; we have to show that $F \circ h = f \circ G$. Take any $x \in S$. Then

$$(F \circ h)(x) = F[h(x)] = (f \circ g)[h(x)] = f[g(h(x))]$$

and

$$(f \circ G)(x) = f[G(x)] = f[(g \circ h)(x)] = f[g(h(x))]$$

Thus since $F \circ h$ and $f \circ G$ have the same effect on each element of S, they must be the same maps. ∎[†]

THEOREM 5.3

If $f: T \to U$, $g: S \to T$, and $h: S \to T$ are linear maps, then

$$f \circ (g + h) = (f \circ g) + (f \circ h) \qquad \text{(the ``left-distributive'' law)}$$

[†] Note that linearity is not used in this argument. The theorem applies to the composition of any three (compatible) maps.

Similarly, if $f\colon T \to U$, $g\colon T \to U$, and $h\colon S \to T$ are linear maps, then

$$(f + g) \circ h = (f \circ h) + (g \circ h) \qquad \text{(the ''right-distributive'' law)}$$

Proof

See the problems.

Taking $U = T = S$ in Theorem 5.1, $V = U = T = S$ in Theorem 5.2, and $U = T = S$ in Theorem 5.3, we obtain statements about composition in $L(S)$, the class of linear operators on S.

COROLLARY 5.3a

$L(S)$ is closed relative to composition.

COROLLARY 5.3b

In $L(S)$ composition is associative.

COROLLARY 5.3c

In $L(S)$ composition is both left-distributive and right-distributive with respect to addition.

These properties, together with the properties of addition already established in $L(S)$ as a vector space (Theorem 3.9 with $T = S$), are characteristic of so many mathematical systems that they have been given a special name.

RING

Let K be an abelian group relative to addition.† If K has a second operation called ''multiplication'' which satisfies the closure, associative, and both distributive laws, it is called a *ring*.

Thus we may summarize the additive and multiplicative properties of $L(S)$ by saying that it is a ring, the so-called ''ring of linear operators.'' The ring with which you are probably most familiar is the set of integers $\{0, \pm 1, \pm 2, \ldots\}$. We won't have much occasion in this book to use the terminology of rings; our only purpose in bringing it up here is to show you how the algebraic properties of linear operators relative to addition and composition can be listed under a single rubric. Note that when we do this we are ignoring multiplication by scalars (which come from outside L). The operations on a ring are purely internal. When all three operations are considered together, we have a mathematical system called a ''linear algebra.'' (See Prob. 10.)

† Recall from Sec. 2.1 that the properties of a group are the closure, associative, identity, and inverse laws. An abelian group also satisfies the commutative law.

Problems

1. If $f: \mathcal{R}^2 \to \mathcal{R}^3$ and $g: \mathcal{R}^2 \to \mathcal{R}^2$ are defined by

$$f(x,y) = (x + y, 2y, x - y) \qquad \text{and} \qquad g(x,y) = (-x, 3x + y)$$

what is the formula defining $f \circ g$? What can be said about $g \circ f$?

2. If f is a rotation of the plane through the angle θ (Example 3.8) what is the formula defining f^2?

3. Let U be a subspace of S (where S is $\mathcal{R}^2$ or $\mathcal{R}^3$) and let $f: S \to S$ and $g: S \to S$ be the orthogonal projections of S on U and $U^\perp$, respectively. (See Example 3.11.)
   a. What are $f \circ g$ and $g \circ f$?
   b. If n is a positive integer, what are f^n and g^n?

4. Use mathematical induction to prove that if $f: S \to S$ is a linear operator and n is any positive integer, then f^n is linear.

5. Show that if $f: T \to U$ and $g: S \to T$ are linear maps, then ker g is a subspace of ker $(f \circ g)$ and rng $(f \circ g)$ is a subspace of rng f.

6. Prove the distributive laws stated in Theorem 5.3.
   a. If $f: T \to U$, $g: S \to T$, and $h: S \to T$ are linear maps, then

$$f \circ (g + h) = (f \circ g) + (f \circ h)$$

   b. If $f: T \to U$, $g: T \to U$, and $h: S \to T$ are linear maps, then

$$(f + g) \circ h = (f \circ h) + (g \circ h)$$

In which of these arguments did you have to use linearity? Is the set of maps from S to S (not necessarily linear) a ring?

7. Give an example of linear operators $f: S \to S$ and $g: S \to S$ which don't commute, that is, $f \circ g \neq g \circ f$. (Thus the ring of linear operators is not commutative relative to multiplication.)

8. A problem that comes up often in mathematics is to solve an equation of the form $f(x) = y$, where $f: S \to T$ is a linear map and y is an element of rng f. Show that if $g: T \to T$ is an ''annihilator'' of y (that is, a linear operator which sends y into 0), then every solution of $f(x) = y$ is also a solution of $(g \circ f)(x) = 0$.

   When y is nonzero, the equation $f(x) = y$ is called ''nonhomogeneous''; the point is that its solutions are to be found among the solutions of the ''homogeneous'' equation $(g \circ f)(x) = 0$. If we can find an annihilator of y, the search for solutions of $f(x) = y$ is confined to a subspace of S—namely, ker $(g \circ f)$. Hopefully this space has a basis, say $\{u_1, \ldots, u_n\}$. Solutions of $f(x) = y$ are then of the form $x = \sum_{j=1}^{n} x_j u_j$, where the x_j are ''undetermined coefficients'' to be found by substitution in $f(x) = y$. (See the next problem for an important application.)

9. To solve the linear nonhomogeneous differential equation $y'' + y = x^2$, we start with the linear operator $L: S \to S$ defined by $L(\phi) = \phi'' + \phi$, where S is the vector space of real-valued functions with derivatives of all orders on $(-\infty, \infty)$. Then the equation has the form $L(y) = g$, where $g(x) = x^2$.

a. Confirm that the linear operator $M: S \to S$ defined by $M(\phi) = \phi'''$ is an annihilator of g.

b. It follows from Prob. 8 that any solution ϕ_p of $L(y) = g$ is also a solution of $(M \circ L)(y) = 0$. Show that this latter equation is $y^{(5)} + y''' = 0$.

c. One learns in differential equations that the general solution of $y^{(5)} + y''' = 0$ is $c_1 + c_2x + c_3x^2 + c_4 \cos x + c_5 \sin x$ (a linear combination of the functions $1, x, x^2, \cos x$, and $\sin x$ which constitute a basis for the space of solutions). If this is to be a solution of $L(y) = g$, why may we assume that it reduces to the form $\phi_p(x) = c_1 + c_2x + c_3x^2$?

d. By substituting $y = c_1 + c_2x + c_3x^2$ in $y'' + y = x^2$ and making use of the linear independence of $1, x, x^2$ in S, determine the unknown coefficients c_1, c_2, c_3. [The result is $\phi_p(x) = x^2 - 2$. See Prob. 35, Sec. 3.2, where this was given as a solution of $y'' + y = x^2$. It was found by the "method of annihilators" indicated here. Some people call it the "method of undetermined coefficients" or the "method of judicious guessing."]

10. Suppose that addition, scalar multiplication, and composition are all considered together in $L(S)$, the class of linear operators on S.

a. Show that if f and g are any elements of L and if k is a scalar, then

$$(kf) \circ g = k(f \circ g) = f \circ (kg)$$

b. Prove that composition may be regarded as a bilinear operation on L, that is,

$$(af + bg) \circ h = a(f \circ h) + b(g \circ h)$$

and

$$h \circ (af + bg) = a(h \circ f) + b(h \circ g)$$

for all f, g, h in L and all scalars a and b.

Thus L is a vector space on which an internal "multiplication" is defined that is associative and bilinear. Such a mathematical system is called a "linear algebra."

5.2 The Inverse of a Linear Map

In Sec. 3.3 we said that the inverse of a map $f: S \to T$ is the map $f^{-1}: T \to S$ defined by $f^{-1}(y) = x$, where $y \in T$ and x is the unique element of S such that $f(x) = y$. The existence of at least one such x for each $y \in T$ is guaranteed if f is onto; its uniqueness is assured if f is one-to-one. We called a map with these properties "invertible," and proved that if f is linear, then f^{-1} is too. Now we are going to look at the algebraic properties of inverse maps in terms of the operation of composition.

As a preliminary to this, recall from Sec. 3.1 (Example 3.3) that the "identity map" on a vector space S is the linear operator $i: S \to S$ defined by $i(x) = x$ for each $x \in S$. When we want to distinguish this from the identity map on another vector space, we use the symbol i_S. Our first theorem (or more properly, its corollary) shows that this map behaves like the unit element 1 in ordinary multiplication.

THEOREM 5.4

If $f: S \to T$ is a linear map, then $f \circ i_S = f$ and $i_T \circ f = f$.

Proof

This is left for the problems.

COROLLARY 5.4a

If S is not a zero space, the identity map on S is a unit element in $L(S)$, that is, i is nonzero and $f \circ i = i \circ f = f$ for all $f \in L(S)$.

The reason S is restricted to be nonzero in this statement is that the only linear operator on a zero space is the zero map. In this case $L(S)$ contains only one element and $i = 0$. Mathematicians do not like to say that a ring contains a "unit element" unless that element is distinct from the zero element. Except for this trivial special case, however, it is correct to say that $L(S)$ is a "ring with unit element."†

THEOREM 5.5

If $f: S \to T$ is an invertible linear map, then

$$f^{-1} \circ f = i_S \qquad \text{and} \qquad f \circ f^{-1} = i_T$$

Proof

Take any $x \in S$ and let $y = f(x)$. Then, by definition of f^{-1}, we have

$$(f^{-1} \circ f)(x) = f^{-1}[f(x)] = f^{-1}(y) = x = i_S(x)$$

Hence $f^{-1} \circ f = i_S$. Similarly $f \circ f^{-1} = i_T$. ∎

COROLLARY 5.5a

If f is an invertible element of $L(S)$, then $f \circ f^{-1} = f^{-1} \circ f = i$.

† In the theory of rings, this phrase has the technical meaning just described. Most rings you will encounter in practice do have a unit element in this sense.

This property is like the equation $x \cdot x^{-1} = x^{-1} \cdot x = 1$ in $\mathfrak{F}$ (which holds for all nonzero x). But note that not every nonzero element of $L(S)$ is invertible; the ring of linear operators differs in this respect from ordinary algebra.

EXAMPLE 5.3

The nonzero map f: $\mathcal{R}^2 \to \mathcal{R}^2$ defined by $f(x,y) = (x - y, y - x)$ is not invertible, for

$$\ker f = \{(x,y) \mid y = x\}$$

which is not the zero subspace of $\mathcal{R}^2$. Hence f is not one-to-one.

EXAMPLE 5.4

To illustrate Theorem 5.5, recall the isomorphism f: $S \to \mathfrak{F}^4$ defined by

$$f\left(\begin{bmatrix} a & b \\ c & d \end{bmatrix}\right) = (a,b,c,d)$$

where S is the vector space of 2×2 matrices with entries in $\mathfrak{F}$. The inverse of f is f^{-1}: $\mathfrak{F}^4 \to S$ defined by

$$f^{-1}(a,b,c,d) = \begin{bmatrix} a & b \\ c & d \end{bmatrix}$$

You can see that for each

$$A = \begin{bmatrix} a & b \\ c & d \end{bmatrix}$$

in S we have

$$(f^{-1} \circ f)(A) = f^{-1}[f(A)] = f^{-1}(a,b,c,d) = A$$

so $f^{-1} \circ f$ is the identity map on S. Similarly, for each $\mathbf{X} = (a,b,c,d) \in \mathfrak{F}^4$, we have

$$(f \circ f^{-1})(\mathbf{X}) = f[f^{-1}(\mathbf{X})] = f(A) = \mathbf{X}$$

and hence $f \circ f^{-1}$ is the identity map on $\mathfrak{F}^4$.

It is worth noting in connection with Theorem 5.5 that f^{-1} is the only linear map with the property $f^{-1} \circ f = i_S$ and $f \circ f^{-1} = i_T$. More precisely, we have the following theorem.

THEOREM 5.6

Let f: $S \to T$ be a linear map and suppose that a linear map g: $T \to S$ exists such that $g \circ f = i_S$ and $f \circ g = i_T$. Then f is invertible and $g = f^{-1}$.

Proof

Observe first that ker f is the zero subspace of S:

$$x \in \ker f \quad \Rightarrow \quad x = i_S(x) = (g \circ f)(x) = g[f(x)] = g(0) = 0$$

By Theorem 3.4, this means that f is one-to-one. It is onto because if $y \in T$, then the element $x = g(y)$ in S has the property that

$$f(x) = f[g(y)] = (f \circ g)(y) = i_T(y) = y$$

Hence f is invertible. To prove that $g = f^{-1}$, take any $y \in T$ and let $x = f^{-1}(y)$. Then $y = f(x)$ and hence

$$g(y) = g[f(x)] = (g \circ f)(x) = i_S(x) = x$$

Thus $g(y) = f^{-1}(y)$ for all $y \in T$, which means that $g = f^{-1}$. ∎

COROLLARY 5.6a

If f and g are elements of $L(S)$ such that $f \circ g = g \circ f = i$, then each is invertible and the other is its inverse.†

THEOREM 5.7

If $f\colon T \to U$ and $g\colon S \to T$ are invertible linear maps, so is $f \circ g\colon S \to U$. Moreover, $(f \circ g)^{-1} = g^{-1} \circ f^{-1}$.

Proof

The map $F = f \circ g$ from S to U is linear because f and g are linear (Theorem 5.1). Since $f^{-1}\colon U \to T$ and $g^{-1}\colon T \to S$ are linear (Theorem 3.5), so is the map $G = g^{-1} \circ f^{-1}$ from U to S. We need show only that $G \circ F = i_S$ and $F \circ G = i_U$ and then conclude from Theorem 5.6 that F is invertible and G is its inverse:

$$\begin{aligned} G \circ F &= (g^{-1} \circ f^{-1}) \circ F = g^{-1} \circ (f^{-1} \circ F) \\ &= g^{-1} \circ [f^{-1} \circ (f \circ g)] = g^{-1} \circ [(f^{-1} \circ f) \circ g] \\ &= g^{-1} \circ (i_T \circ g) = g^{-1} \circ g = i_S \end{aligned}$$

We show that $F \circ G = i_U$ in a similar fashion. ∎

EXAMPLE 5.5

Define $f\colon \mathcal{R}^2 \to \mathcal{R}^2$ and $g\colon \mathcal{R}^2 \to \mathcal{R}^2$ by

$$f(x,y) = (2x + y, x + y) \qquad \text{and} \qquad g(x,y) = (x, x + y)$$

† In general, both $f \circ g = i$ and $g \circ f = i$ must hold before we can say that f and g are inverses. But when S is finite-dimensional, each condition implies the other. (See Prob. 51, Sec. 5.4.) Hence in that case we need check only one.

Then f and g are invertible linear maps; their inverses are defined by

$$f^{-1}(x,y) = (x - y, 2y - x) \qquad \text{and} \qquad g^{-1}(x,y) = (x, y - x)$$

(Confirm this!) Moreover, $f \circ g$: $\mathcal{R}^2 \to \mathcal{R}^2$ is defined by

$$(f \circ g)(x,y) = (3x + y, 2x + y)$$

According to Theorem 5.7, this map should be invertible and $(f \circ g)^{-1}$ should be the same map as $g^{-1} \circ f^{-1}$. You can check that the formula defining $g^{-1} \circ f^{-1}$ is

$$(g^{-1} \circ f^{-1})(x,y) = (x - y, 3y - 2x)$$

What remains is to confirm that $(f \circ g)^{-1}$ is defined by the same formula. (See the problems.)

We end this section by mentioning some additional terminology in connection with rings. We won't make much use of this, but it is worthwhile to have heard of it.

COMMUTATIVE RING; FIELD

A *commutative* ring is a ring in which multiplication is commutative. (Addition always is, by definition.) A commutative ring with unit element which also contains a multiplicative inverse for each of its nonzero elements is called a *field*.

For example, both $\mathcal{R}$ and $\mathcal{C}$ are fields. But $L(S)$ is not. (Why?)

LECTOR *Speaking of fields, yours is rather cluttered with jargon.*

AUCTOR *I am not responsible for it.*

LECTOR *When I get married, I intend to exchange commutative rings with unit elements.*

AUCTOR *If you expect to stay married, you'll have to get serious.*

LECTOR *Increasing and multiplying in accordance with the distributive law, we may have a little group.*

AUCTOR *Please.*

LECTOR *Which we shall proceed to map in a one-to-one fashion into (but not onto) the Peace Corps.*

AUCTOR *Are you finished?*

LECTOR *A singular function, don't you agree?*

AUCTOR *I haven't defined* singular.

LECTOR *No, but you will.*

Problems

11. Prove Theorem 5.4: If $f\colon S \to T$ is a map, then $f \circ i_S = f$ and $i_T \circ f = f$.

12. Prove that the unit element in $L(S)$ is unique; that is, if a linear operator $u\colon S \to S$ exists satisfying $f \circ u = u \circ f = f$ for every $f \in L$, then $u = i$.

13. Finish the proof of Theorem 5.5 by showing that if $f\colon S \to T$ is an invertible linear map, then $f \circ f^{-1} = i_T$.

14. Define $f\colon \mathcal{C} \to \mathcal{C}$ by $f(z) = \bar{z}$ (where $\mathcal{C}$ is regarded as a vector space over $\mathcal{R}$). What is f^2? What is f^{-1}?

15. Let $S = M_{m,n}(\mathfrak{F})$ and $T = M_{n,m}(\mathfrak{F})$ and suppose that $f\colon S \to T$ and $g\colon T \to S$ are the linear maps which send each matrix into its transpose. What are $f \circ g$ and $g \circ f$? What are f^{-1} and g^{-1}?

16. Let S be the vector space of real-valued functions with derivatives of all orders on $(-\infty, \infty)$ and let T be the subspace consisting of functions whose graphs contain the origin. Then the map $I\colon S \to T$ defined by

$$I(f) = g \qquad \text{where } g(x) = \int_0^x f(t)\,dt$$

is linear. (See Prob. 16, Sec. 3.1.)

a. If $D\colon T \to S$ is defined by $D(g) = g'$, what are $D \circ I$ and $I \circ D$? What are I^{-1} and D^{-1}?

b. If T is replaced by S in the definitions of I and D, what are $D \circ I$ and $I \circ D$? Explain why neither $I\colon S \to S$ nor $D\colon S \to S$ is invertible. (This illustrates the fact that in Corollary 5.6a both $f \circ g = i$ and $g \circ f = i$ are essential hypotheses. Also see the footnote following Corollary 5.6a.)

17. Given an alternate proof of Theorem 5.7 as follows.

a. Let $h = f \circ g$. Why is h linear?

b. Show that h is one-to-one by arguing that ker h is the zero subspace of S.

c. Explain why h is onto.

d. Since parts *b* and *c* show that h is invertible, h^{-1} exists. Prove that $h^{-1} = g^{-1} \circ f^{-1}$ by showing that $h^{-1}(z) = (g^{-1} \circ f^{-1})(z)$ for all $z \in U$. *Hint:* Let $y = f^{-1}(z)$ and $x = g^{-1}(y)$; show that $h^{-1}(z) = x$ and $(g^{-1} \circ f^{-1})(z) = x$.

18. Let $f\colon \mathcal{R}^2 \to \mathcal{R}^2$ and $g\colon \mathcal{R}^2 \to \mathcal{R}^2$ be the invertible linear maps defined by

$$f(x,y) = (2x + y,\ x + y) \qquad \text{and} \qquad g(x,y) = (x,\ x + y)$$

a. Confirm the formulas for f^{-1} and g^{-1} given in Example 5.5, namely

$$f^{-1}(x,y) = (x - y,\ 2y - x) \qquad \text{and} \qquad g^{-1}(x,y) = (x,\ y - x)$$

b. Confirm that $f \circ f^{-1} = f^{-1} \circ f = i$.
c. Confirm that $g \circ g^{-1} = g^{-1} \circ g = i$.
d. Confirm the formula for $f \circ g$, namely $(f \circ g)(x,y) = (3x + y, 2x + y)$.
e. Find the formula for $(f \circ g)^{-1}$.
f. Confirm the formula for $g^{-1} \circ f^{-1}$, namely $(g^{-1} \circ f^{-1})(x,y) = (x - y, 3y - 2x)$, and note that it is the same as the formula for $(f \circ g)^{-1}$.

19. Define f: $\mathcal{R}^3 \to \mathcal{R}^2$ and g: $\mathcal{R}^2 \to \mathcal{R}^3$ by

$$f(x,y,z) = (2x - y, x + z) \qquad \text{and} \qquad g(x,y) = (x + 2y, 3y, 2x - y)$$

a. Find the formula defining $f \circ g$.
b. Show that $f \circ g$ is invertible and find the formula defining its inverse.
c. Does $(f \circ g)^{-1} = g^{-1} \circ f^{-1}$? Explain.

20. Convince yourself that the integers 0, ± 1, ± 2, . . . constitute a commutative ring with unit element. Why isn't this a field?

21. Convince yourself that the rational numbers constitute a field. (Note that this is a "subfield" of $\mathcal{R}$.)

22. What properties of a field does $L(S)$ fail to have?

5.3 Multiplication of Matrices

Now we are going to define an operation on matrices that is suggested by composition of linear maps. An example of what we have in mind is given by the maps f: $\mathcal{R}^3 \to \mathcal{R}^4$ and g: $\mathcal{R}^2 \to \mathcal{R}^3$ defined by

$$f(x,y,z) = (3x + z, z - x, 2y, y + x) \quad \text{and} \quad g(x,y) = (x, x - 2y, x + y)$$

The product (composition) of f and g is the map h: $\mathcal{R}^2 \to \mathcal{R}^4$ defined by

$$h(x,y) = f[g(x,y)] = f(x, x - 2y, x + y) = (4x + y, y, 2x - 4y, 2x - y)$$

(Confirm this!) The matrix representations of f, g, and h relative to the standard bases are

$$A = [f] = \begin{bmatrix} 3 & 0 & 1 \\ -1 & 0 & 1 \\ 0 & 2 & 0 \\ 0 & 1 & 1 \end{bmatrix} \qquad B = [g] = \begin{bmatrix} 1 & 0 \\ 1 & -2 \\ 1 & 1 \end{bmatrix}$$

$$C = [h] = [f \circ g] = \begin{bmatrix} 4 & 1 \\ 0 & 1 \\ 2 & -4 \\ 2 & -1 \end{bmatrix}$$

Since C is the matrix of $f \circ g$, it seems natural to call it the "product" of A and B (in that order), that is, $C = AB$.

Thus matrix multiplication corresponds to composition of linear maps. You can see how we constructed the product $C = AB$ in the above example, but it is not obvious how AB could be computed directly from A and B (without any reference to linear maps).† The pattern becomes clear, however, when the general case is considered.

Let S, T, and U be vector spaces over $\mathfrak{F}$ with bases

$$\alpha = \{u_1, \ldots, u_r\} \qquad \beta = \{v_1, \ldots, v_q\} \qquad \gamma = \{w_1, \ldots, w_p\}$$

respectively. Let $f\colon T \to U$ and $g\colon S \to T$ be linear maps and suppose that

$$A = [a_{ij}] = [f]_{\beta^\gamma} \qquad \text{and} \qquad B = [b_{ij}] = [g]_{\alpha^\beta}$$

are their matrix representations relative to the given bases. (Note that these are $p \times q$ and $q \times r$, respectively.) To find

$$C = [c_{ij}] = [f \circ g]_{\alpha^\gamma}$$

which is the $p \times r$ matrix representation of $f \circ g$, we work out the expression for $(f \circ g)(u_j)$, where $j = 1, \ldots, r$, in terms of $w_1, \ldots, w_p$:

$$\begin{aligned}
(f \circ g)(u_j) &= f[g(u_j)] \\
&= f\left(\sum_{k=1}^{q} b_{kj} v_k\right) && \text{(definition of } B = [g]_{\alpha^\beta}) \\
&= \sum_{k=1}^{q} b_{kj} f(v_k) && (f \text{ is linear}) \\
&= \sum_{k=1}^{q} b_{kj}\left(\sum_{i=1}^{p} a_{ik} w_i\right) && \text{(definition of } A = [f]_{\beta^\gamma}) \\
&= \sum_{i=1}^{p}\left(\sum_{k=1}^{q} a_{ik} b_{kj}\right) w_i && j = 1, \ldots, r^{\ddagger}
\end{aligned}$$

Now by definition of $C = [f \circ g]_{\alpha^\gamma}$, we know that

$$(f \circ g)(u_j) = \sum_{i=1}^{p} c_{ij} w_i \qquad j = 1, \ldots, r$$

†We have already stated the definition in Prob. 24, Sec. 4.2. But without any preview there might be some doubt about how to proceed.

‡To see that the order of the "iterated sums" is immaterial, write them out and do some rearranging. A formal argument using mathematical induction can be given, but since the result is essentially a matter of elementary algebra, we prefer not to digress.

Since the matrix representation of $f \circ g$ relative to α and γ is unique, we conclude that the i,j entry of C is

$$c_{ij} = \sum_{k=1}^{q} a_{ik}b_{kj} \qquad i = 1, \ldots, p;\ j = 1, \ldots, r$$

This result may not be very helpful unless it is recognized as the

dot product of the ith *row of A and the* jth *column of B*

each considered as a vector in q-space.

With this idea in mind, return to the example

$$\begin{bmatrix} 3 & 0 & 1 \\ -1 & 0 & 1 \\ 0 & 2 & 0 \\ 0 & 1 & 1 \end{bmatrix} \begin{bmatrix} 1 & 0 \\ 1 & -2 \\ 1 & 1 \end{bmatrix} = \begin{bmatrix} 4 & 1 \\ 0 & 1 \\ 2 & -4 \\ 2 & -1 \end{bmatrix}$$

Each entry in the product can be computed directly (without using maps) as the dot product of the appropriate row of the left-hand factor and column of the right-hand factor. For example, the entry in the first row and second column is

$$c_{12} = \sum_{k=1}^{3} a_{1k}b_{k2} = (3)(0) + (0)(-2) + (1)(1) = 1$$

the dot product of the "row vector" (3,0,1) and the "column vector" $(0,-2,1)$ in $\mathcal{R}^3$.

We remark in passing that both (3,0,1) and $(0,-2,1)$ are being regarded here as ordinary vectors in $\mathcal{R}^3$. We use the terms "row vector" and "column vector" for no other purpose than to identify their source (a row from A and a column from B). Some writers insist that they should be written

$$(3,0,1) \qquad \text{and} \qquad \begin{pmatrix} 0 \\ -2 \\ 1 \end{pmatrix}$$

or even that they are really 1×3 and 3×1 matrices, respectively:

$$\begin{bmatrix} 3 & 0 & 1 \end{bmatrix} \qquad \text{and} \qquad \begin{bmatrix} 0 \\ -2 \\ 1 \end{bmatrix}$$

While there is something to be said for this, we prefer not to say it. It is simpler (whenever confusion is unlikely) to stick with the notation already available.

(See Prob. 33 and the accompanying footnote for a case where we are compelled to alter the notation.)

In any case we now have the clue we need for a definition of matrix multiplication.

MULTIPLICATION OF MATRICES

Let $A = [a_{ij}]$ and $B = [b_{ij}]$ be $p \times q$ and $q \times r$, respectively (with entries in $\mathfrak{F}$). Their *product* AB is the $p \times r$ matrix $C = [c_{ij}]$ whose i,j entry is the dot product of the ith row of A and jth column of B (each considered as a vector in q-space), namely

$$c_{ij} = (a_{i1}, \ldots, a_{iq}) \cdot (b_{1j}, \ldots, b_{qj}) = \sum_{k=1}^{q} a_{ik} b_{kj}$$

Note that in order for this to make sense, the matrices must be *compatible,* that is, A must have the same number of columns as B has rows.

Although the definition of matrix multiplication is independent of linear maps, it is of course motivated by them. Its formulation leads immediately to the important result

$$[f \circ g]_{\alpha^\gamma} = [f]_{\beta^\gamma}[g]_{\alpha^\beta} \qquad \text{or simply} \qquad [f \circ g] = [f][g]$$

when the bases are understood. This says that the matrix of the product of f and g is the product of the matrices of f and g (in that order). Recall from Sec. 4.2 (Prob. 23) that similar formulas hold in connection with addition and scalar multiplication of maps and matrices:

$$[f + g] = [f] + [g] \qquad \text{and} \qquad [cf] = c[f]$$

where $f\colon S \to T$ and $g\colon S \to T$ are linear maps and c is a scalar. In other words, the matrix of the sum of two maps is the sum of their matrices, and the matrix of a scalar multiple of a map is the same scalar multiple of its matrix.

Now we are in a position to develop the algebraic properties of matrix multiplication with a minimum of fuss.

THEOREM 5.8

$M_n(\mathfrak{F})$ is closed relative to matrix multiplication.

Proof

Take $p = q = r = n$ in the definition of matrix multiplication.

THEOREM 5.9

Matrix multiplication is associative.

Proof

Let A, B, C be $p \times q$, $q \times r$, $r \times s$, respectively. If

$$f: \mathfrak{F}^q \rightarrow \mathfrak{F}^p \qquad g: \mathfrak{F}^r \rightarrow \mathfrak{F}^q \qquad h: \mathfrak{F}^s \rightarrow \mathfrak{F}^r$$

are the linear maps whose matrices (relative to the standard bases) are A, B, C, respectively, we have

$$\begin{aligned}(AB)C &= ([f][g])[h] \\ &= [f \circ g][h] \\ &= [(f \circ g) \circ h] \\ &= [f \circ (g \circ h)] \\ &= [f][g \circ h] \\ &= [f]([g][h]) \\ &= A(BC) \ \blacksquare\end{aligned}$$

The beauty of this proof is that the work is done for us; all we have to do is turn the crank. The associativity of matrix multiplication is a trivial consequence of the associativity of composition of linear maps. Compare the above argument with the direct approach, in which we use the definition of matrix multiplication without appealing to linear maps:

Let $A = [a_{ij}]$, $B = [b_{ij}]$, and $C = [c_{ij}]$ be $p \times q$, $q \times r$, and $r \times s$, respectively. To prove that $(AB)C = A(BC)$, we compare their i,j entries; for this purpose, let

$$D = AB = [d_{ij}] \qquad \text{and} \qquad E = (AB)C = DC = [e_{ij}]$$

(Note that all the products are compatible.) The i,j entry of $(AB)C$ is

$$e_{ij} = \sum_{n=1}^{r} d_{in}c_{nj} = \sum_{n=1}^{r}\left(\sum_{m=1}^{q} a_{im}b_{mn}\right)c_{nj}$$

Similarly, let

$$F = BC = [f_{ij}] \qquad \text{and} \qquad G = A(BC) = AF = [g_{ij}]$$

The i,j entry of $A(BC)$ is

$$g_{ij} = \sum_{m=1}^{q} a_{im}f_{mj} = \sum_{m=1}^{q} a_{im}\left(\sum_{n=1}^{r} b_{mn}c_{nj}\right)$$

Since these iterated sums are the same (except for order), we have

$$e_{ij} = g_{ij} \qquad i = 1, \ldots, p;\ j = 1, \ldots, s \ \blacksquare$$

While this proof is not difficult (provided that the notation is judicious), it is messy.

THEOREM 5.10

Matrix multiplication is both left-distributive and right-distributive with respect to addition.

Proof

Suppose that A is $p \times q$ and B and C are $q \times r$. One may prove directly that $A(B + C) = AB + AC$ by comparing the i,j entries of $A(B + C)$ and $AB + AC$. Or one may introduce linear maps

$$f\colon \mathfrak{F}^q \to \mathfrak{F}^p \qquad g\colon \mathfrak{F}^r \to \mathfrak{F}^q \qquad h\colon \mathfrak{F}^r \to \mathfrak{F}^q$$

such that $[f] = A$, $[g] = B$, and $[h] = C$, and then use the distributivity of composition of maps with respect to addition. We leave the details of both approaches for the problems. The right-distributive law $(B + C)A = BA + CA$ may also be proved in either of these ways. ▮

As in Sec. 5.1 we may summarize what we know about addition and multiplication (of square matrices) by stating that $M_n(\mathfrak{F})$ is a ring. That is, it is an abelian group relative to addition and its multiplicative operation is associative and distributive. Moreover, it is a ring with unit element; that is, there is an $n \times n$ matrix I (distinct from the zero matrix) which satisfies

$$AI = IA = A \qquad \text{for every } A \in M_n(\mathfrak{F})$$

This last statement is not obvious unless you already know what I is. (You probably do, from earlier remarks, but let's act ignorant and see what is involved in finding it.)

One approach is to use the identity *map* $i\colon S \to S$ defined by $i(x) = x$ (where S is a vector space over $\mathfrak{F}$ of dimension n). We know that

$$f \circ i = i \circ f = f \qquad \text{for each } f \in L(S)$$

If we compute matrices relative to a basis $\alpha = \{u_1, \ldots, u_n\}$ for S, we have

$$[f][i] = [i][f] = [f]$$

This result suggests that the matrix we want is $I = [i]$, so let's find it. For each $j = 1, \ldots, n$, we have

$$i(u_j) = u_j = \sum_{i=1}^{n} \delta_{ij} u_i$$

Hence, by definition of $[i]$ (relative to α), it follows that

$$I = [\delta_{ij}]$$

which is the $n \times n$ matrix with 1s down the "main diagonal" and 0s elsewhere; for example, the 3×3 identity matrix is

$$I = \begin{bmatrix} 1 & 0 & 0 \\ 0 & 1 & 0 \\ 0 & 0 & 1 \end{bmatrix}$$

(See Prob. 14, Sec. 4.1.)

IDENTITY MATRIX

The $n \times n$ matrix $I_n = [\delta_{ij}]$ is called the *identity matrix* (denoted by I if n is understood).

THEOREM 5.11

If A is an $m \times n$ matrix, then $AI_n = A$ and $I_m A = A$.

Proof

The i,j entry of AI_n is

$$\sum_{k=1}^{n} a_{ik}\delta_{kj} = a_{ij}$$

so $AI_n = A$. Similarly, $I_m A = A$. ∎

COROLLARY 5.11a

The $n \times n$ identity matrix is a unit element in $M_n(\mathfrak{F})$, that is,

$$AI = IA = A \qquad \text{for all } A \in M_n(\mathfrak{F})$$

Thus $M_n(\mathfrak{F})$, like the ring of linear operators, is a ring with unit element. Also like $L(S)$, it is not a commutative ring; that is, AB and BA are not (in general) the same.† If S is a vector space over $\mathfrak{F}$ of dimension n, it seems reasonable to expect that $M_n(\mathfrak{F})$ and $L(S)$ are essentially alike as rings. Of course we have already observed that they are alike as vector spaces; the map $\phi: L \to M$ defined by $\phi(f) = [f]$ is an isomorphism (where $[f]$ is computed relative to a fixed basis for S). However, we have not yet defined what is meant by an isomorphism between rings.

Scalar multiplication is not involved in this question, since the operations on a ring are internal. What we need is a way of describing how the additive and multiplicative structure of L is transferred to M by the map ϕ. The key

† Square matrices which do satisfy $AB = BA$ are said to "commute."

to this is the formula

$$[f \circ g] = [f][g]$$

which in terms of ϕ reads

$$\phi(f \circ g) = \phi(f)\phi(g)$$

In other words, ϕ maps products (compositions) in L into the corresponding products in M, just as the formula

$$\phi(f + g) = \phi(f) + \phi(g)$$

says that sums are mapped into sums. This observation motivates the following definition.

ISOMORPHISM BETWEEN RINGS

Let P and Q be rings. An invertible map $\phi: P \to Q$ with the properties

$$\phi(x + y) = \phi(x) + \phi(y) \qquad \text{and} \qquad \phi(xy) = \phi(x)\phi(y)$$

for all x and y in P is called an *isomorphism* (between rings).

We can go further than this. Why not consider addition, scalar multiplication, and internal multiplication all together? In Prob. 10, Sec. 5.1, we asked you to show that if f and g are any elements of $L(S)$ and k is a scalar, then

$$(kf) \circ g = k(f \circ g) = f \circ (kg)$$

If we consider this along with the distributive laws, we can regard composition as a ''bilinear'' operation on L, that is,

$$(af + bg) \circ h = a(f \circ h) + b(g \circ h)$$

and

$$h \circ (af + bg) = a(h \circ f) + b(h \circ g)$$

for all f, g, and h in L and all scalars a and b. Thus L is a vector space on which an internal multiplication is defined that is associative and bilinear.

LINEAR ALGEBRA

A vector space with an associative and bilinear multiplicative operation is called a *linear algebra*.

It is pretty obvious that $M_n(\mathfrak{F})$, like $L(S)$, is a linear algebra. (See the problems.) Moreover, $L(S)$ and $M_n(\mathfrak{F})$ are isomorphic *as linear algebras,* for the map $\phi: L \to M$ defined by $\phi(f) = [f]$ preserves everything! Sums, scalar products, and compositions in L are sent into the corresponding sums, scalar products, and matrix products in M:

$$\phi(f + g) = \phi(f) + \phi(g) \qquad \phi(kf) = k\phi(f) \qquad \phi(f \circ g) = \phi(f)\phi(g)$$

where f and g are linear operators and k is a scalar. Such an invertible map from one linear algebra to another is of course called an isomorphism (between linear algebras).

You need not pay much attention to these remarks, for we have no plans to launch a study of "algebraic varieties" (groups, rings, fields, linear algebras, . . .). But it is worthwhile to observe how the term "isomorphism" takes on different meanings depending on the context. Generally speaking, an isomorphism is an invertible map from one mathematical system to another, with the crucial property that whatever structure the first system has is carried over intact to the second system. This idea is at the heart of modern algebra.

Problems

23. Define $f: \mathbb{R}^3 \to \mathbb{R}^2$ and $g: \mathbb{R}^2 \to \mathbb{R}^3$ by

$$f(x,y,z) = (x + y - 2z, 3y + z) \qquad \text{and} \qquad g(x,y) = (x + y, 2x, -y)$$

a. Using the standard bases, find the matrix representations A, B, C, and D of f, g, $f \circ g$, and $g \circ f$, respectively (without using matrix multiplication).

b. We should have $AB = C$ and $BA = D$. Use matrix multiplication to check this result.

24. Let

$$A = \begin{bmatrix} 2 & -1 \\ 1 & 3 \end{bmatrix} \qquad B = \begin{bmatrix} 3 & -3 & 0 \\ -1 & 2 & 4 \end{bmatrix} \qquad C = \begin{bmatrix} 0 & 4 \\ -1 & -2 \\ 1 & 3 \end{bmatrix}$$

Carry out the matrix multiplication involved in checking the associative law, $(AB)C = A(BC)$.

25. Name two square matrices that don't commute.

26. Suppose that A is $p \times q$, B is $q \times r$, and that $AB = BA$. Explain why $p = q = r$. (In other words, the idea of two matrices commuting is necessarily restricted to square matrices of the same size.)

27. Suppose that A is $p \times q$ and B and C are $q \times r$. Prove the left-distributive law $A(B + C) = AB + AC$ in the following ways.

a. Use distributivity of composition of maps with respect to addition.
b. Appeal directly to the definitions of matrix addition and multiplication.
Does the right-distributive law, $(B + C)A = BA + CA$, follow by the commutative law as in elementary algebra? Or must it be derived independently?

28. Explain why multiplication on $M_n(\mathfrak{F})$ is bilinear. [Since it is also associative, you have confirmed that $M_n(\mathfrak{F})$ is a linear algebra.]

29. Prove that if A and B are $n \times n$ matrices that commute, then $(A + B)^2 = A^2 + 2AB + B^2$. (Note the context of the question! Since addition, scalar multiplication, and matrix multiplication are all involved, we are regarding $M_n(\mathfrak{F})$ as a linear algebra, not merely as a vector space or ring.)

30. Finish the proof of Theorem 5.11 by showing that $I_mA = A$ for each $m \times n$ matrix A.

31. Prove Theorem 5.11 by introducing appropriate linear maps and using Theorem 5.4.

32. Prove that the unit element in $M_n(\mathfrak{F})$ is unique; that is, if a matrix U exists satisfying $AU = UA = A$ for every $A \in M_n(\mathfrak{F})$, then $U = I$.

★ **33.** Let $C = AB$, where A is $p \times q$ and B is $q \times r$.
a. Prove that the ith row of C is the product of the ith row of A and the matrix B.†
b. Prove that the jth column of C is the product of the matrix A and the jth column of B.†

34. If f is a rotation of the plane through the angle θ, then

$$[f] = \begin{bmatrix} \cos\theta & -\sin\theta \\ \sin\theta & \cos\theta \end{bmatrix} \quad \text{and} \quad [f^{-1}] = \begin{bmatrix} \cos\theta & \sin\theta \\ -\sin\theta & \cos\theta \end{bmatrix}$$

(See Prob. 5, Sec. 4.1.)
a. Use matrix multiplication to confirm that the product of these matrices (in either order) is I.
b. How could this be shown without doing any matrix multiplication?

35. If A and B are $n \times n$ matrices whose product (in either order) is I, we call each the "inverse" of the other.
a. Use matrix multiplication to find a matrix B serving as the inverse of

$$A = \begin{bmatrix} 2 & 3 \\ 3 & 5 \end{bmatrix}$$

b. Let f: $\mathbb{R}^2 \to \mathbb{R}^2$ be the linear map whose matrix is A. Find f^{-1} and use the fact that $[f][f^{-1}] = [f^{-1}][f] = [i] = I$ to construct B.

† In order for these statements to make sense, we must interpret "row" (of A) to mean $1 \times q$ matrix, and "column" (of B) to mean $q \times 1$ matrix. Otherwise the products referred to would be undefined.

★ **36.** Suppose that A and B are compatible for multiplication, say $p \times q$ and $q \times r$, respectively.

a. Prove that $(AB)^T = B^T A^T$. *Hint:* Let $C = B^T$ and $D = A^T$. The i,j entry of $CD = B^T A^T$ is $\Sigma_{k=1}^{q} c_{ik} d_{kj}$, where $c_{ik} = b_{ki}$ and $d_{kj} = a_{jk}$. (See Sec. 4.3.)

b. Prove that $(AB)^* = B^* A^*$.

37. Confirm the formulas in Prob. 36 in the following cases.

a. $A = \begin{bmatrix} 2 & -1 & 1 \\ 0 & 5 & 1 \end{bmatrix} \qquad B = \begin{bmatrix} 4 & 0 \\ 1 & -3 \\ 2 & 1 \end{bmatrix}$

b. $A = \begin{bmatrix} 1-i & 2 \\ -1 & i \end{bmatrix} \qquad B = \begin{bmatrix} 1+i & 0 & 2 \\ i & 1 & 2+i \end{bmatrix}$

38. The map $f\colon M_n(\mathfrak{F}) \to M_n(\mathfrak{F})$ defined by $f(A) = A^T$ is an isomorphism between vector spaces (Prob. 29, Sec. 4.3). Explain why it is not an isomorphism between rings.

39. In general, the ring of $n \times n$ matrices with entries in $\mathfrak{F}$ is not a field. If $n = 1$, however, it *is*. Why? Define an isomorphism which identifies $M_1(\mathfrak{F})$ with $\mathfrak{F}$.

★ **40.** Let $M_2(\mathfrak{R})$ be the ring of 2×2 matrices with real entries, and let $\mathfrak{C}'$ be the subset consisting of matrices of the form

$$\begin{bmatrix} a & b \\ -b & a \end{bmatrix}$$

a. Explain why $\mathfrak{C}'$ is a commutative ring with unit element.

b. Let $\phi\colon \mathfrak{C}' \to \mathfrak{C}$ be the map defined by $\phi(Z) = z$, where

$$Z = \begin{bmatrix} a & b \\ -b & a \end{bmatrix}$$

and $z = a + bi$. Explain why ϕ is invertible and show that

$$\phi(U + V) = \phi(U) + \phi(V) \qquad \text{and} \qquad \phi(UV) = \phi(U)\phi(V)$$

for all U and V in $\mathfrak{C}'$. (Thus ϕ is an isomorphism between rings. If u and v are the images in $\mathfrak{C}$ of U and V, the matrix sum $U + V$ and product UV map into $u + v$ and uv.)

c. Let $\mathfrak{R}'$ be the subset of $\mathfrak{C}'$ consisting of matrices of the form

$$\begin{bmatrix} a & 0 \\ 0 & a \end{bmatrix}$$

Explain why $\mathfrak{R}'$ is a commutative ring with unit element and show that when ϕ is restricted to the domain $\mathfrak{R}'$ it is an isomorphism between $\mathfrak{R}'$ and $\mathfrak{R}$ sending

$$\begin{bmatrix} a & 0 \\ 0 & a \end{bmatrix}$$

into the real number a.

Actually $\mathcal{C}'$ and $\mathcal{R}'$ are *fields* (as we'll see after the discussion of matrix inverses in the next section). The point of the problem is that complex numbers need not be written in a notation involving i. They may be regarded as a field in the ring of real 2×2 matrices; there is nothing "imaginary" about them! Moreover, they contain the real numbers as a subfield. A visitor from outer space might have discovered complex numbers in this way. He would regard our difficulties with the symbol $i = \sqrt{-1}$ as trivial, for in his world the equation $i^2 = -1$ would read

$$\begin{bmatrix} 0 & 1 \\ -1 & 0 \end{bmatrix}\begin{bmatrix} 0 & 1 \\ -1 & 0 \end{bmatrix} = \begin{bmatrix} -1 & 0 \\ 0 & -1 \end{bmatrix}$$

and he would see no problem. The recognition of this fact in the 19th century (by Gauss and others) was a watershed in the history of mathematics. Of course our notation is simpler. The visitor from outer space would probably adopt it forthwith, with the comment that we have had the good sense to paste labels over

$$\begin{bmatrix} a & 0 \\ 0 & a \end{bmatrix} \quad \begin{bmatrix} b & 0 \\ 0 & b \end{bmatrix} \quad \text{and} \quad \begin{bmatrix} 0 & 1 \\ -1 & 0 \end{bmatrix}$$

with a, b, and i written on the labels. Thus the typical "complex number" is

$$\begin{bmatrix} a & b \\ -b & a \end{bmatrix} = \begin{bmatrix} a & 0 \\ 0 & a \end{bmatrix} + \begin{bmatrix} b & 0 \\ 0 & b \end{bmatrix}\begin{bmatrix} 0 & 1 \\ -1 & 0 \end{bmatrix}$$

which reduces to $a + bi$.

5.4 The Inverse of a Matrix

Following the characterization of $L(S)$ as a ring with unit element (Secs. 5.1 and 5.2), we went on to discuss the algebraic properties of inverses. Not every linear operator has an inverse, but when it does we showed that $f \circ f^{-1} = f^{-1} \circ f = i$, where i is the identity map on S. We also proved that if f and g are invertible, so is $f \circ g$, and $(f \circ g)^{-1} = g^{-1} \circ f^{-1}$; this means the set of invertible linear operators is a group relative to composition. (Why?)

The same ideas can be developed in connection with matrices. We know from the last section that $M_n(\mathfrak{F})$, the set of $n \times n$ matrices with entries in $\mathfrak{F}$, is a ring with unit element (the $n \times n$ identity matrix I). The *inverse* of a matrix

A (relative to multiplication) is a matrix B satisfying $AB = BA = I$.† We could discuss its existence, uniqueness, and algebraic properties by a direct attack based on the definition of matrix multiplication, but for the present it is easier to use what we know about maps.

Let $f\colon S \to S$ be the linear operator defined by $[f] = A$ (where S is a vector space over $\mathfrak{F}$ of dimension n and all matrices are computed relative to a fixed basis for S). If f is invertible, we know that $f \circ f^{-1} = f^{-1} \circ f = i$; it seems reasonable to suppose that $B = [f^{-1}]$ is the matrix we seek, since

$$AB = [f][f^{-1}] = [f \circ f^{-1}] = [i] = I$$

and (similarly) $BA = I$. To find the inverse of a matrix, we need only find the inverse of a linear map!

EXAMPLE 5.6

Let

$$A = \begin{bmatrix} 3 & 5 \\ -1 & 1 \end{bmatrix}$$

The map $f\colon \mathcal{R}^2 \to \mathcal{R}^2$ defined by $[f] = A$ is $f(x,y) = (3x + 5y, -x + y)$. Since

$$\begin{aligned} \ker f &= \{(x,y) \mid f(x,y) = (0,0)\} \\ &= \{(x,y) \mid 3x + 5y = -x + y = 0\} = \{(0,0)\} \end{aligned}$$

we know that f is invertible; its inverse is defined by

$$f^{-1}(x,y) = (s,t) \qquad \text{where } f(s,t) = (x,y)$$

To find it explicitly, we solve the system

$$\begin{aligned} 3s + 5t &= x \\ -s + t &= y \end{aligned}$$

to obtain

$$s = \tfrac{1}{8}(x - 5y) \qquad \text{and} \qquad t = \tfrac{1}{8}(x + 3y)$$

that is,

$$f^{-1}(x,y) = \tfrac{1}{8}(x - 5y, x + 3y)$$

The matrix of f^{-1} is

$$B = [f^{-1}] = \frac{1}{8}\begin{bmatrix} 1 & -5 \\ 1 & 3 \end{bmatrix}$$

You can check by matrix multiplication that $AB = BA = I$, so we have found the inverse of A.

† There is only one such matrix, as we shall prove momentarily. Also note that A and B must be square matrices of the same size in order to commute (Prob. 26, Sec. 5.3). That is why we confine the present discussion to $M_n(\mathfrak{F})$.

EXAMPLE 5.7

A more efficient way to do Example 5.6 is to let $[f] = A$ and $g = f^{-1}$. Then

$$f(\mathbf{E}_1) = 3\mathbf{E}_1 - \mathbf{E}_2 \qquad \text{and} \qquad f(\mathbf{E}_2) = 5\mathbf{E}_1 + \mathbf{E}_2 \qquad \text{(why?)}$$

from which

$$g(3\mathbf{E}_1 - \mathbf{E}_2) = \mathbf{E}_1 \qquad \text{and} \qquad g(5\mathbf{E}_1 + \mathbf{E}_2) = \mathbf{E}_2$$

LECTOR *You're assuming that f^{-1} exists.*

AUCTOR *Right. If it didn't, we'd discover that fact, too.*

Now use the linearity of g to write

$$\begin{aligned} 3g(\mathbf{E}_1) - g(\mathbf{E}_2) &= \mathbf{E}_1 \\ 5g(\mathbf{E}_1) + g(\mathbf{E}_2) &= \mathbf{E}_2 \end{aligned}$$

and solve for

$$g(\mathbf{E}_1) = \tfrac{1}{8}(\mathbf{E}_1 + \mathbf{E}_2) \qquad \text{and} \qquad g(\mathbf{E}_2) = \tfrac{1}{8}(-5\mathbf{E}_1 + 3\mathbf{E}_2)$$

This yields

$$B = [f^{-1}] = \frac{1}{8}\begin{bmatrix} 1 & -5 \\ 1 & 3 \end{bmatrix}$$

as before.

EXAMPLE 5.8

To find the inverse of

$$A = \begin{bmatrix} 3 & 5 \\ -1 & 1 \end{bmatrix}$$

without appealing to linear maps at all, let

$$B = \begin{bmatrix} x & u \\ y & v \end{bmatrix}$$

and multiply out $AB = I$ to get

$$\begin{aligned} 3x + 5y &= 1 \qquad & 3u + 5v &= 0 \\ -x + y &= 0 \qquad & -u + v &= 1 \end{aligned}$$

Then solve for x, y, u, v to find

$$B = \frac{1}{8}\begin{bmatrix} 1 & -5 \\ 1 & 3 \end{bmatrix}$$

Having beaten this problem to death, we are ready to present some formal results.

THEOREM 5.12

The inverse of a matrix is unique when it exists.

Proof

Suppose that A is given and there are two matrices B and C with the inverse property $AB = BA = I$ and $AC = CA = I$. Multiplying each side of $AC = I$ by B, we obtain

$$\begin{aligned} B(AC) &= BI \\ (BA)C &= B \\ IC &= B \\ C &= B \ \blacksquare \end{aligned}$$

This result enables us to make the following formal definition.

INVERSE OF A MATRIX

Given the matrix $A \in M_n(\mathfrak{F})$, suppose that there is a matrix $B \in M_n(\mathfrak{F})$ satisfying $AB = BA = I$. Then B is called the *inverse* of A (relative to multiplication), and is denoted by A^{-1}. When it exists, we say that A is *nonsingular;* a square matrix not having an inverse is called *singular.*†

THEOREM 5.13

An $n \times n$ matrix A with entries in $\mathfrak{F}$ is nonsingular if and only if the linear operator f: $S \to S$ defined by $[f] = A$ is invertible (where S is an n-dimensional vector space over $\mathfrak{F}$ and matrices are computed relative to some fixed basis for S).

Proof

If f is invertible, let $B = [f^{-1}]$. Then (as we have already seen) $AB = BA = I$, so A is nonsingular. Conversely, if A^{-1} exists, let g: $S \to S$ be the linear operator defined by $[g] = A^{-1}$. Then

$$[f \circ g] = [f][g] = AA^{-1} = I = [i]$$

from which $f \circ g = i$. (Why?) Similarly, $g \circ f = i$. It follows from Corollary 5.6a that f is invertible. $\blacksquare$

† Note that nonsquare matrices are not classified in these categories at all. We also agree to use the unqualified term "inverse" to mean "inverse relative to multiplication." The inverse of A relative to addition is $-A$; it will be of little interest in the sequel.

COROLLARY 5.13a

If f is an invertible linear operator on a finite-dimensional space, then $[f^{-1}] = [f]^{-1}$. (In words: The matrix of the inverse map is the inverse of the matrix of the map, where it is understood that a fixed basis is used.)

EXAMPLE 5.9

Let $f: \mathcal{R}^2 \to \mathcal{R}^2$ be a rotation of the plane through the angle θ. To find f^{-1} by matrix methods, we compute

$$[f] = \begin{bmatrix} \cos\theta & -\sin\theta \\ \sin\theta & \cos\theta \end{bmatrix}$$

(See Prob. 5, Sec. 4.1.) In the problem section we'll ask you to prove that a 2×2 matrix

$$A = \begin{bmatrix} a & b \\ c & d \end{bmatrix}$$

is nonsingular if and only if $\det A = ad - bc \neq 0$, and that in this case its inverse is

$$A^{-1} = \frac{1}{\det A}\begin{bmatrix} d & -b \\ -c & a \end{bmatrix}$$

Assuming that this result has been shown, the inverse of $[f]$ is

$$[f]^{-1} = \begin{bmatrix} \cos\theta & \sin\theta \\ -\sin\theta & \cos\theta \end{bmatrix}$$

But $[f]^{-1} = [f^{-1}]$, so f^{-1} is the map defined by

$$f^{-1}(x,y) = (x\cos\theta + y\sin\theta, -x\sin\theta + y\cos\theta)$$

Note that this checks with what we did in Sec. 3.3, where we found the formula for f^{-1} by a direct attack.

THEOREM 5.14

If A and B are nonsingular $n \times n$ matrices, so is AB, and $(AB)^{-1} = B^{-1}A^{-1}$.

Proof

Let $f: S \to S$ and $g: S \to S$ be the invertible linear operators defined by $[f] = A$ and $[g] = B$ (where S is an n-dimensional space with fixed basis relative to which matrices are computed). Then $AB = [f][g] = [f \circ g]$; since $f \circ g$ is invertible, AB is nonsingular. Moreover,

$$\begin{aligned}(AB)^{-1} &= [f \circ g]^{-1} = [(f \circ g)^{-1}] = [g^{-1} \circ f^{-1}] \\ &= [g^{-1}][f^{-1}] = [g]^{-1}[f]^{-1} = B^{-1}A^{-1}\end{aligned}$$

(See Theorem 5.7.) ▮

A proof not involving linear maps can be based on the fact that

$$(AB)(B^{-1}A^{-1}) = [(AB)B^{-1}]A^{-1} = [A(BB^{-1})]A^{-1} = (AI)A^{-1} = AA^{-1} = I$$

Similarly, $(B^{-1}A^{-1})(AB) = I$. Hence the matrix $C = B^{-1}A^{-1}$ has the property that $(AB)C = C(AB) = I$, which makes it the inverse of AB. Thus AB is nonsingular and $(AB)^{-1} = B^{-1}A^{-1}$. ∎

COROLLARY 5.14a

If $A_1, \ldots, A_r$ are nonsingular $n \times n$ matrices, so is their product, and $(A_1 \cdots A_r)^{-1} = A_r^{-1} \cdots A_1^{-1}$.

Theorem 5.14 says that the subset of $M_n(\mathfrak{F})$ consisting of nonsingular matrices is closed relative to multiplication. Since I is nonsingular (why?) and the inverse of a nonsingular matrix is nonsingular (why?), this set is a group relative to multiplication.† It corresponds to the group of invertible linear operators mentioned at the beginning of this section. (Like that group it is not abelian, since in general $AB \neq BA$.)

THEOREM 5.15

Given the matrix $A \in M_{m,n}(\mathfrak{F})$, define $f\colon \mathfrak{F}^n \to \mathfrak{F}^m$ by $f(\mathbf{X}) = A\mathbf{X}$, where $\mathbf{X}$ is regarded as a "column vector" (an $n \times 1$ matrix) and $A\mathbf{X}$ is understood to be a matrix product. Then f is linear and $[f] = A$ (relative to the standard bases).

Proof

This is Prob. 26, Sec. 4.2. However, it is sufficiently important to be repeated. To prove that f is linear, let $\mathbf{X}$ and $\mathbf{Y}$ be any elements of $\mathfrak{F}^n$ and suppose that a and b are scalars. Then (regarding $\mathbf{X}$ and $\mathbf{Y}$ as $n \times 1$ matrices) we have

$$\begin{aligned} f(a\mathbf{X} + b\mathbf{Y}) &= A(a\mathbf{X} + b\mathbf{Y}) \\ &= A(a\mathbf{X}) + A(b\mathbf{Y}) && \text{(distributive law)} \\ &= a(A\mathbf{X}) + b(A\mathbf{Y}) && \text{(Prob. 24}c\text{, Sec. 4.2)} \\ &= af(\mathbf{X}) + bf(\mathbf{Y}) \end{aligned}$$

To show that $[f] = A$, observe that if $\mathbf{E}_j = (\delta_{1j}, \ldots, \delta_{nj})$ is the jth element of the standard basis for $\mathfrak{F}^n$, then

$$f(\mathbf{E}_j) = A\mathbf{E}_j = \begin{bmatrix} a_{11} & a_{12} & \cdots & a_{1n} \\ a_{21} & a_{22} & \cdots & a_{2n} \\ \cdots & \cdots & \cdots & \cdots \\ a_{m1} & a_{m2} & \cdots & a_{mn} \end{bmatrix} \begin{bmatrix} \delta_{1j} \\ \delta_{2j} \\ \vdots \\ \delta_{nj} \end{bmatrix} = \begin{bmatrix} a_{1j} \\ a_{2j} \\ \vdots \\ a_{mj} \end{bmatrix}$$

† Associativity is inherited from $M_n(\mathfrak{F})$.

where $j = 1, \ldots, n$. (Confirm this!) Hence $f(\mathbf{E}_j)$ is the jth column of A. By definition of the matrix representation of a linear map, it follows that $[f] = A$. ▮

THEOREM 5.16

Given the matrix $A \in M_{m,n}(\mathfrak{F})$, let $f\colon \mathfrak{F}^n \to \mathfrak{F}^m$ be the linear map whose matrix (relative to the standard bases) is A. Then $f(\mathbf{X}) = A\mathbf{X}$ for each $\mathbf{X} \in \mathfrak{F}^n$.

Proof

According to Theorem 5.15 the map which sends $\mathbf{X}$ into $A\mathbf{X}$ is linear and has A as its matrix. Call this map g. Then $[f] = A = [g]$, from which $f = g$. Hence $f(\mathbf{X}) = g(\mathbf{X}) = A\mathbf{X}$ for each $\mathbf{X} \in \mathfrak{F}^n$. ▮

Thus the matrix A and the map $f(\mathbf{X}) = A\mathbf{X}$ are explicitly identified. Given the matrix, the corresponding map is $f(\mathbf{X}) = A\mathbf{X}$ (Theorem 5.16); given the map, its matrix is A (Theorem 5.15). While these results do not involve the notion of inverse (they have been in the air ever since we first mentioned matrix multiplication in Prob. 24, Sec. 4.2), they lead directly to important characterizations of nonsingularity, as follows.

THEOREM 5.17

An $n \times n$ matrix A with entries in $\mathfrak{F}$ is nonsingular if and only if the linear operator $f\colon \mathfrak{F}^n \to \mathfrak{F}^n$ defined by $f(\mathbf{X}) = A\mathbf{X}$ is invertible.

Proof

Since $[f] = A$, this follows immediately from Theorem 5.13. ▮

THEOREM 5.18

An $n \times n$ matrix A with entries in $\mathfrak{F}$ is nonsingular if and only if the only solution of the equation $A\mathbf{X} = \mathbf{0}$ in $\mathfrak{F}^n$ is $\mathbf{X} = \mathbf{0}$.

Proof

This is left for the problems.

This theorem has profound implications, as we shall see. The only comment we make now is that the equation $A\mathbf{X} = \mathbf{0}$ is a concise version of a system of n linear equations in n unknowns. Indeed the equation reads

$$\begin{bmatrix} a_{11} & a_{12} & \cdots & a_{1n} \\ a_{21} & a_{22} & \cdots & a_{2n} \\ \cdots & \cdots & \cdots & \cdots \\ a_{n1} & a_{n2} & \cdots & a_{nn} \end{bmatrix} \begin{bmatrix} x_1 \\ x_2 \\ \vdots \\ x_n \end{bmatrix} = \begin{bmatrix} 0 \\ 0 \\ \vdots \\ 0 \end{bmatrix}$$

which is equivalent to the system

$$\begin{aligned} a_{11}x_1 + a_{12}x_2 + \cdots + a_{1n}x_n &= 0 \\ a_{21}x_1 + a_{22}x_2 + \cdots + a_{2n}x_n &= 0 \\ \cdots\cdots\cdots\cdots\cdots\cdots\cdots\cdots& \\ a_{n1}x_1 + a_{n2}x_2 + \cdots + a_{nn}x_n &= 0 \end{aligned}$$

in the sense that an n-tuple $\mathbf{X} = (x_1, \ldots, x_n)$ is a solution of $A\mathbf{X} = \mathbf{0}$ if and only if its components satisfy each equation of the system. Later on (Chap. 7) we'll develop methods for solving such systems.

Problems

41. Why is I nonsingular? Name its inverse.

42. Prove that if A is a nonsingular matrix, so is A^{-1}, and also prove that $(A^{-1})^{-1} = A$.

43. Prove that

$$A = \begin{bmatrix} 4 & 2 \\ 6 & 3 \end{bmatrix}$$

is singular as follows.

a. Use matrix multiplication to show that no 2×2 matrix B exists satisfying $AB = I$.

b. Show that an appropriate linear operator having A as its matrix is not invertible.

c. Show that the equation $A\mathbf{X} = \mathbf{0}$ has more than one solution.

44. Let A be an $n \times n$ matrix with entries in $\mathfrak{F}$ and suppose that f: $S \to T$ is a linear map having A as its matrix (where S and T are n-dimensional vector spaces over $\mathfrak{F}$ with specified bases). Prove that A is nonsingular if and only if f is invertible. (This is a more general version of Theorem 5.13.)

45. The matrix of f: $\mathfrak{R}^3 \to \mathfrak{R}^3$ defined by $f(x,y,z) = (x - y + z, 3y, x + 2z)$ is

$$A = \begin{bmatrix} 1 & -1 & 1 \\ 0 & 3 & 0 \\ 1 & 0 & 2 \end{bmatrix}$$

Confirm that

$$A^{-1} = \frac{1}{3}\begin{bmatrix} 6 & 2 & -3 \\ 0 & 1 & 0 \\ -3 & -1 & 3 \end{bmatrix}$$

and use the result to find the formula for $f^{-1}(x,y,z)$.

46. The linear operator $f: \mathcal{R}^2 \to \mathcal{R}^2$ defined by $f(x,y) = (y,x)$ is a reflection of the plane in the line $y = x$. (Why?) The inverse of f is clearly the same reflection, that is, $f^{-1} = f$. Confirm that if $A = [f]$, then $A^{-1} = A$.

47. Prove that if f is an invertible linear operator on a finite-dimensional space, then $[f^{-1}] = [f]^{-1}$ (Corollary 5.13a).

48. Complete the second proof of Theorem 5.14 by showing that $(B^{-1}A^{-1})(AB) = I$ (where A and B are nonsingular $n \times n$ matrices).

49. Prove Theorem 5.18: The $n \times n$ matrix A is nonsingular if and only if the only solution of $A\mathbf{X} = \mathbf{0}$ is $\mathbf{X} = \mathbf{0}$.

★ **50.** Suppose that A and B are $n \times n$ matrices whose product AB is nonsingular. It would be nice to say that $(AB)^{-1} = B^{-1}A^{-1}$ (as in Theorem 5.14), but this requires A and B to be nonsingular too. Show that they *are*, as follows.

a. First prove that B is nonsingular by arguing that otherwise the equation $B\mathbf{X} = \mathbf{0}$, and hence also the equation $(AB)\mathbf{X} = \mathbf{0}$, has more than one solution.

b. Then prove that A is nonsingular. *Hint:* $A = ABB^{-1}$.

Thus if AB is nonsingular, so are A and B. This is the converse of Theorem 5.14. (Note, however, that A and B are *square*. See Prob. 19, Sec. 5.2, for an example of AB nonsingular with A and B not even classifiable as singular or nonsingular.)

★ **51.** Prove that if A and B are $n \times n$ matrices satisfying $AB = I$, then they commute and each is the inverse of the other. *Hint:* AB is nonsingular; use Prob. 50.

This result cuts in half the labor involved in checking that a matrix B is the inverse of a matrix A. We need only multiply A and B in one order and obtain I; the other computation is automatic. Moreover, it follows that if f and g are linear operators on the finite-dimensional space S and if $f \circ g = i$, then f and g commute and each is the inverse of the other. (Why?) This is an improvement of Corollary 5.6a. However, we cannot guarantee that a "one-sided" inverse is an inverse unless S is finite-dimensional. (See Prob. 16, Sec. 5.2.)

★ **52.** Let

$$A = \begin{bmatrix} a & b \\ c & d \end{bmatrix}$$

be a 2×2 matrix.

a. Prove that if the determinant of A is not zero, then A is nonsingular and

$$A^{-1} = \frac{1}{\det A}\begin{bmatrix} d & -b \\ -c & a \end{bmatrix}$$

(This is the 2×2 case of a general formula for the inverse of an $n \times n$ matrix. To prove that it is correct, you need only multiply by A and get I.)

b. Conversely, prove that if A is nonsingular, then $\det A \neq 0$. *Hint:* Suppose that $ad - bc = 0$ and show that the matrix equation

$$\begin{bmatrix} a & b \\ c & d \end{bmatrix}\begin{bmatrix} x & u \\ y & v \end{bmatrix} = \begin{bmatrix} 1 & 0 \\ 0 & 1 \end{bmatrix}$$

is impossible.

53. Let

$$A = \begin{bmatrix} 3 & -1 \\ 2 & 1 \end{bmatrix} \quad \text{and} \quad B = \begin{bmatrix} 0 & -1 \\ 2 & 1 \end{bmatrix}$$

Find A^{-1}, B^{-1}, and $(AB)^{-1}$, and check the formula $(AB)^{-1} = B^{-1}A^{-1}$.

54. Prove that if $A_1, \ldots, A_r$ are nonsingular $n \times n$ matrices, then so is their product, and $(A_1 \cdots A_r)^{-1} = A_r^{-1} \cdots A_1^{-1}$ (Corollary 5.14a).

★ **55.** Prove that if α and β are any bases for the vector space S, then $A = [i]_{\alpha^\beta}$ and $B = [i]_{\beta^\alpha}$ are inverses. *Hint:* Recall the formula

$$[f \circ g]_{\alpha^\gamma} = [f]_{\beta^\gamma}[g]_{\alpha^\beta}$$

in the last section.

Note that A and B are not the identity matrix unless $\beta = \alpha$. (See Probs. 14 and 15, Sec. 4.1.) In the next chapter we'll show that any nonsingular matrix can be regarded as a representation of the identity map relative to appropriate bases. This problem shows that its inverse is the representation obtained by interchanging the bases.

56. Let α and β be the bases $\{(1,0), (0,1)\}$ and $\{(1,1), (-1,2)\}$ for $\mathcal{R}^2$. Find $A = [i]_{\alpha^\beta}$ and $B = [i]_{\beta^\alpha}$ and confirm Prob. 55.

57. Show that a square matrix with a zero row or column is singular. *Hint:* Use Prob. 33, Sec. 5.3.

58. Let A be an $n \times n$ matrix with entries in $\mathfrak{F}$.

a. Prove that if A is nonsingular, the equation $AB = 0$ implies that $B = 0$ (where B is any matrix compatible with A for multiplication).

b. Prove that if A is singular, there exists a nonzero $n \times n$ matrix B satisfying $AB = 0$. *Hint:* The equation $A\mathbf{X} = \mathbf{0}$ has nonzero solutions in $\mathfrak{F}^n$. (Why?) Use any of these as the columns of B and appeal to Prob. 33, Sec. 5.3.

c. Illustrate part *b* in the case

$$A = \begin{bmatrix} 4 & 2 \\ 6 & 3 \end{bmatrix}$$

by constructing a nonzero 2×2 matrix B satisfying $AB = 0$.

Note that this cannot happen in a field; the equation $ab = 0$ implies that $a = 0$ or $b = 0$. But $M_n(\mathfrak{F})$ is not a field (unless $n = 1$).

★ **59.** Suppose that the $n \times n$ matrix A is nonsingular.

a. Prove that A^T is nonsingular, with inverse given by $(A^T)^{-1} = (A^{-1})^T$. *Hint:* Multiply A^T and $(A^{-1})^T$, using Prob. 36 of Sec. 5.3 to show that the result is I.

b. Prove that A^* is nonsingular, with inverse given by $(A^*)^{-1} = (A^{-1})^*$.

60. Confirm Prob. 59 in the following cases.

a. $A = \begin{bmatrix} 2 & 3 \\ 3 & 5 \end{bmatrix}$

b. $A = \begin{bmatrix} 2 - i & 1 \\ 2 + i & 1 + i \end{bmatrix}$

61. Let $\mathcal{C}'$ be the set of 2×2 matrices of the form

$$\begin{bmatrix} a & b \\ -b & a \end{bmatrix}$$

where a and b are real, and let $\mathcal{R}'$ be the subset consisting of matrices of the form

$$\begin{bmatrix} a & 0 \\ 0 & a \end{bmatrix}$$

(Each of these sets is a commutative ring with unit element; see Prob. 40, Sec. 5.3.)

a. Show that if A is a nonzero element of $\mathcal{C}'$, then A^{-1} exists and is an element of $\mathcal{C}'$.

b. Repeat part *a* with $\mathcal{C}'$ replaced by $\mathcal{R}'$.

Thus $\mathcal{C}'$ and $\mathcal{R}'$ are fields; they are isomorphic to $\mathcal{C}$ and $\mathcal{R}$, respectively, by virtue of the map which sends

$$\begin{bmatrix} a & b \\ -b & a \end{bmatrix}$$

into $a + bi$.

Review Quiz

True or false?

1. The set of nonsingular $n \times n$ matrices with entries in $\mathcal{F}$ is a group relative to multiplication.
2. The i,j entry of AB is the dot product of the ith row of A and the jth column of B.
3. If A and B are square matrices satisfying $AB = I$, then A and B commute.
4. If f and g are linear operators on S, then ker f is a subspace of ker $(f \circ g)$.
5. Any matrix representation of an identity map is nonsingular.
6. If A is a square matrix satisfying $A^2 = 0$, then $A = 0$.
7. If A is a matrix representation of the linear operator $f: S \to S$ and ker f is the zero subspace of S, then A is nonsingular.
8. If A is nonsingular, then $(A^*)^{-1} = (A^{-1})^*$.
9. If α and β are bases for the vector space S, then $[i]_{\alpha^\beta}$ and $[i]_{\beta^\alpha}$ are inverses.
10. The i,j entry of AB, where A is $p \times q$ and B is $q \times r$, is $\Sigma_{t=1}^{q} a_{it}b_{tj}$.
11. Nonsingular $n \times n$ matrices commute.
12. The set of nonsingular $n \times n$ matrices is a vector space.
13. If f and g are linear operators satisfying $f \circ g = g \circ f = i$, then each is invertible and the other is its inverse.
14. If the only solution of $A\mathbf{X} = \mathbf{0}$ is $\mathbf{X} = \mathbf{0}$ (where A is an $n \times n$ matrix and $\mathbf{X}$ is a vector in n-space), then A is nonsingular.
15. If A and B are $n \times n$ matrices, then $(A + B)^2 = A^2 + 2AB + B^2$.
16. If P is a nonsingular $n \times n$ matrix and $f: M_n(\mathcal{F}) \to M_n(\mathcal{F})$ is defined by $f(A) = PAP^{-1}$, then f is an isomorphism.
17. $(AB)^T = A^TB^T$

18. The matrices

$$\begin{bmatrix} 5 & 3 \\ 3 & 2 \end{bmatrix} \quad \text{and} \quad \begin{bmatrix} 2 & -3 \\ -3 & 5 \end{bmatrix}$$

are inverses.

19. If A and B are $p \times q$ and C is $q \times r$, then $(aA + bB)C = a(AC) + b(BC)$ for all scalars a and b.

20. If $f\colon \mathfrak{F}^n \to \mathfrak{F}^m$ is defined by $f(\mathbf{X}) = A\mathbf{X}$, where A is $m \times n$, then A is the matrix of f relative to the standard bases.

21. Composition of linear maps is an associative operation.

22. The ith row of AB is equal to the product of the ith row of A and the matrix B.

23. If $f\colon T \to U$, $g\colon S \to T$, and $h\colon S \to T$ are maps (not necessarily linear), then $f \circ (g + h) = (f \circ g) + (f \circ h)$.

24. If S is n-dimensional over $\mathfrak{F}$, then $L(S)$ and $M_n(\mathfrak{F})$ are isomorphic as vector spaces, as rings, and as linear algebras.

25. Every nonzero linear operator is invertible.

26. If either one of the $n \times n$ matrices A and B is singular, then AB is singular.

27. If f and g are invertible linear operators on S, then $(f \circ g)^{-1} = f^{-1} \circ g^{-1}$.

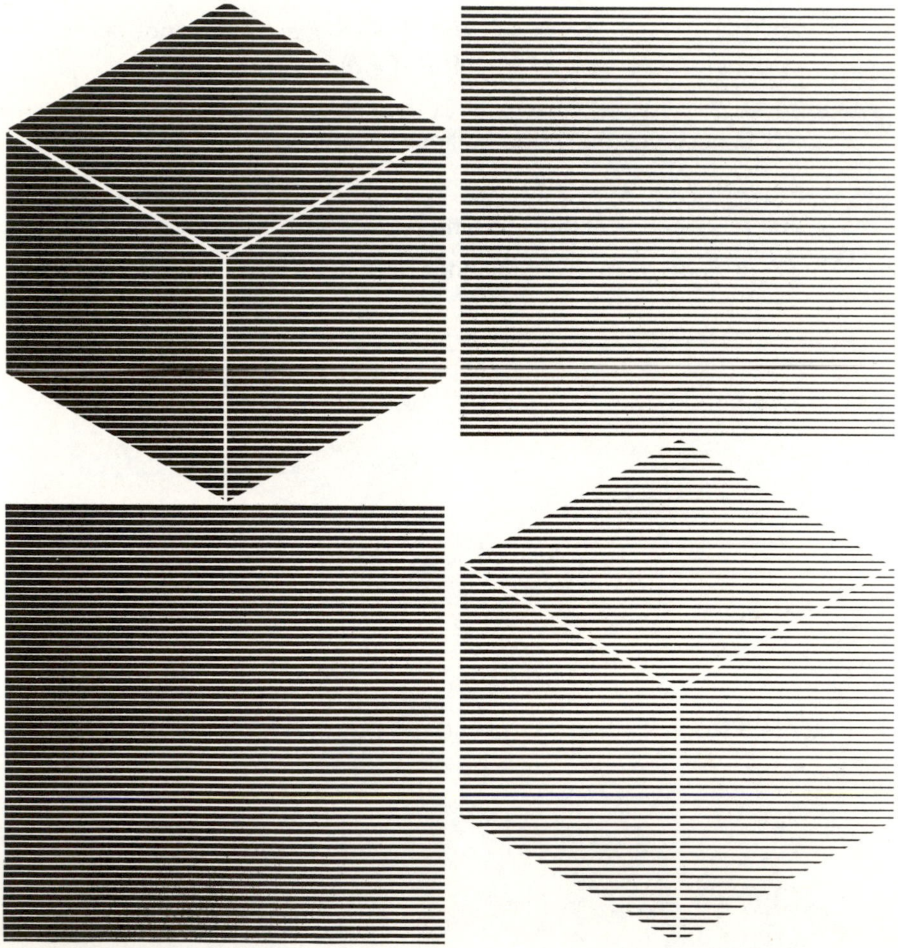

CHAPTER 6

Equivalence and Similarity

Given a linear map $f: S \to T$, where S and T are vector spaces with specified bases, we know that there is only one matrix representing f relative to these bases. However, a different matrix may be obtained if the bases are changed; what we are going to do in this chapter is investigate the relation between the various representations of f. This leads to the concept of ''equivalent'' matrices; when $T = S$ (that is, when f is a linear operator), the idea specializes to a relation called ''similarity.''

6.1 Matrix Representations Revisited

Our first theorem says that any matrix representation of a linear map $f: S \to T$ may be obtained from any other by ''premultiplication'' and ''postmultiplication'' of the other by nonsingular matrices. It turns out that the premultiplication corresponds to a change of basis in T, while the postmultiplication represents a change of basis in S. However, the details of this are rather more complicated than one might expect; we postpone discussion of changing bases to Sec. 6.3.

THEOREM 6.1

If A and B are matrix representations of the same linear map, there are nonsingular matrices P and Q such that $B = PAQ$.

Proof

Let $f: S \to T$ be the map. We are given that

$$A = [f]_{\alpha^\beta} \qquad \text{and} \qquad B = [f]_{\gamma^\delta}$$

where α and γ are bases for S and β and δ are bases for T. It follows that

$$B = [f]_{\gamma^\delta} = [i_T \circ f \circ i_S]_{\gamma^\delta} = [i_T]_{\beta^\delta}[f]_{\alpha^\beta}[i_S]_{\gamma^\alpha} = PAQ$$

where

$$P = [i_T]_{\beta^\delta} \qquad \text{and} \qquad Q = [i_S]_{\gamma^\alpha}$$

Since P and Q are matrix representations of identity maps, they are nonsingular. (See Prob. 55, Sec. 5.4.) ∎

EXAMPLE 6.1

Define $f\colon \mathcal{R}^2 \to \mathcal{R}^3$ by $f(x,y) = (x + y, 2x, x - y)$. The matrix of f relative to the standard bases for $\mathcal{R}^2$ and $\mathcal{R}^3$ is

$$A = \begin{bmatrix} 1 & 1 \\ 2 & 0 \\ 1 & -1 \end{bmatrix}$$

while its matrix relative to the bases

$$\{(1,1), (0,-1)\} \qquad \text{and} \qquad \{(1,-1,0), (0,2,1), (2,1,1)\}$$

is

$$B = \begin{bmatrix} -6 & 5 \\ -4 & 4 \\ 4 & -3 \end{bmatrix}$$

(See Example 4.2.) Let $S = \mathcal{R}^2$ and $T = \mathcal{R}^3$ and label the above bases (in the order named) by α, β, γ, and δ. To find $P = [i_T]_{\beta^\delta}$, we compute

$$\begin{aligned} i_T(1,0,0) &= (1,0,0) = -(1,-1,0) - (0,2,1) + (2,1,1) \\ i_T(0,1,0) &= (0,1,0) = -2(1,-1,0) - (0,2,1) + (2,1,1) \\ i_T(0,0,1) &= (0,0,1) = 4(1,-1,0) + 3(0,2,1) - 2(2,1,1) \end{aligned}$$

Hence

$$P = \begin{bmatrix} -1 & -2 & 4 \\ -1 & -1 & 3 \\ 1 & 1 & -2 \end{bmatrix}$$

The work involved here is considerable, as you can see if you go through it yourself. An alternative procedure is to find $P^{-1} = [i_T]_{\delta^\beta}$ first and then compute its inverse P. Since β is the standard basis for $\mathcal{R}^3$, the columns of P^{-1} are simply the elements of δ (why?); hence

$$P^{-1} = \begin{bmatrix} 1 & 0 & 2 \\ -1 & 2 & 1 \\ 0 & 1 & 1 \end{bmatrix}$$

However, the work involved in finding the inverse of this is equivalent to what we did before.†

†We are assuming that you don't know any method but brute force for finding the inverse of an $n \times n$ matrix (unless $n \leq 2$). One may always insert letters for the entries of its unknown inverse and use matrix multiplication to write down a system of equations they satisfy. (See Example 5.8.) But this is gruesome if n is large.

To find $Q = [i_S]_{\gamma^\alpha}$, we observe that α is the standard basis for $\mathcal{R}^2$. Hence the columns of Q are the elements of γ:

$$Q = \begin{bmatrix} 1 & 0 \\ 1 & -1 \end{bmatrix}$$

You can check by matrix multiplication that $B = PAQ$.

EXAMPLE 6.2

Let A and B be as in Example 6.1. The matrices

$$P = \begin{bmatrix} -1 & -2 & 0 \\ -1 & -1 & -1 \\ 1 & 1 & 0 \end{bmatrix} \quad \text{and} \quad Q = \begin{bmatrix} 1 & -1 \\ 1 & 0 \end{bmatrix}$$

are nonsingular (confirm this!) and $B = PAQ$, as you can check. *Thus the P and Q of Theorem 6.1 are not unique.*

This lack of uniqueness can be confusing if the reason for it is not understood. The matrices $P = [i_T]_{\beta^\delta}$ and $Q = [i_S]_{\gamma^\alpha}$ in the *proof* of Theorem 6.1 are uniquely determined by the bases α, β, γ, and δ. But these bases are not the only ones relative to which A and B are matrix representations of f. For example,

$$A = \begin{bmatrix} 1 & 1 \\ 2 & 0 \\ 1 & -1 \end{bmatrix}$$

is not only the matrix of $f(x,y) = (x + y, 2x, x - y)$ relative to the standard bases for $\mathcal{R}^2$ and $\mathcal{R}^3$, but also relative to the bases

$$\alpha = \{(0,1), (1,0)\} \quad \text{and} \quad \beta = \{(1,0,0), (0,1,0), (0,-2,-1)\}$$

If

$$\gamma = \{(1,1), (0,-1)\} \quad \text{and} \quad \delta = \{(1,-1,0), (0,2,1), (2,1,1)\}$$

are as before, the matrices $P = [i_T]_{\beta^\delta}$ and $Q = [i_S]_{\gamma^\alpha}$ turn out to be those of Example 6.2, as you can check.

EQUIVALENT MATRICES

Suppose that the $m \times n$ matrices A and B are related by an equation of the form $B = PAQ$, where P and Q are nonsingular ($m \times m$ and $n \times n$, respectively). Then we say that A is *equivalent* to B and write $A \sim B$.

AUCTOR *Go back to Theorem 3.7.*

LECTOR *I have it. It is not easy to watch the book in two places at once, but please continue.*

AUCTOR *Note the discussion of "equivalence relation."*

LECTOR *Why?*

AUCTOR *It will make what follows seem less abrupt.*

THEOREM 6.2

Equivalence on the set $M_{m,n}(\mathfrak{F})$ is

a. Reflexive: $A \sim A$;

b. Symmetric: $A \sim B \Rightarrow B \sim A$;

c. Transitive: $A \sim B$ and $B \sim C \Rightarrow A \sim C$.

Proof

We'll leave the proof of *a* and *b* for the problems. To prove *c*, suppose that $A \sim B$ and $B \sim C$. Then there are nonsingular matrices P_1, Q_1, P_2, and Q_2 such that

$$B = P_1AQ_1 \qquad \text{and} \qquad C = P_2BQ_2$$

From this it follows that

$$C = P_2(P_1AQ_1)Q_2 = (P_2P_1)A(Q_1Q_2) = PAQ$$

where $P = P_2P_1$ and $Q = Q_1Q_2$ are nonsingular by Theorem 5.14. Hence $A \sim C$. ▮

LECTOR *Turn to page 239 of the Image edition of* The Path to Rome, *by Hilaire Belloc.*

AUCTOR *I know what you're going to say.*

LECTOR *Have you no shame?*

AUCTOR *Everything is explained in the introduction to this book.*

Thus matrix equivalence is an equivalence relation (Sec. 3.4) which classifies $n \times n$ matrices together if they are related by an equation of the form $B = PAQ$ (where P and Q are nonsingular). Theorem 6.1 may be rendered in another way by saying that the matrix representations of a given linear map are all in the same equivalence class. We can do better than this, however; it turns out that the entire class consists of matrix representations of this same map! To prove this, we need a preliminary theorem.

THEOREM 6.3

Let P be a nonsingular $n \times n$ matrix and let S be an n-dimensional vector space with basis β. Then there is a basis α such that $P = [i]_{\alpha^\beta}$ and a basis γ such that $P = [i]_{\beta^\gamma}$.

Proof

Let $f: S \to S$ be the linear operator whose matrix relative to β is P. Since P is nonsingular, f is invertible and therefore sends any basis into another basis (Prob. 47, Sec. 3.3). Thus if $u_j = f(v_j)$, where $\beta = \{v_1, \ldots, v_n\}$, it follows that $\alpha = \{u_1, \ldots, u_n\}$ is a basis for S. We claim that $P = [i]_{\alpha^\beta}$. To confirm this, we need only look at the effect of i on the elements of α:

$$i(u_j) = u_j = f(v_j) = \sum_{i=1}^{n} p_{ij} v_i$$

where p_{ij} is the i,j entry of P.

To prove the second part of the theorem, apply the first part to the nonsingular matrix P^{-1}; that is, find a basis γ such that $P^{-1} = [i]_{\gamma^\beta}$. By Prob. 55, Sec. 5.4, the inverse of this is $P = [i]_{\beta^\gamma}$. ▮

THEOREM 6.4

Equivalent matrices represent the same linear map.

Proof

Suppose that A and B are equivalent and that A is the matrix representation of a linear map $f: S \to T$ (relative to the bases α and β). The problem is to prove that B also represents f. We know there are nonsingular matrices P and Q such that $B = PAQ$. By Theorem 6.3 there are bases γ and δ for S and T, respectively, such that $P = [i_T]_{\beta^\delta}$ and $Q = [i_S]_{\gamma^\alpha}$. We then have

$$B = PAQ = [i_T]_{\beta^\delta}[f]_{\alpha^\beta}[i_S]_{\gamma^\alpha} = [f]_{\gamma^\delta}$$

which means that B represents f relative to γ and δ. ▮

Theorems 6.1 and 6.4 may be combined to yield an elegant statement about linear maps and matrices:

THEOREM 6.5

Two matrices are equivalent if and only if they represent the same linear map.

Proof

The "if" part is Theorem 6.1; the "only if" part is Theorem 6.4. ▮

Thus a linear map gives rise to a unique equivalence class of matrices. All the matrix representations of the map are in this class (Theorem 6.1); every matrix in the class is a representation of f (Theorem 6.4). The *class itself* may therefore be regarded as a single representative of f which is independent of bases; we rise above such mundane considerations as coordinate vectors and loftily state that a linear map and an equivalence class of matrices are the same thing.

LECTOR *Levitation is for mystics. I prefer one matrix at a time.*
AUCTOR *I don't blame you. But the edelweiss up here is pretty.*
LECTOR *Pass me an ice axe.*

The class of matrices representing a given map must somehow reflect the properties of the map which do not depend on the choice of bases. These are the "invariants" of the map which do not change when the coordinate systems do; they are "essential" as opposed to "accidental" properties of the map. Since they are easier to read off from one matrix than another, one seeks some sort of "canonical form" to represent the equivalence class.

LECTOR *This seems to be degenerating into a lecture on Aristotle.*
AUCTOR *Or Thomas Aquinas.*
LECTOR *Existence precedes essence.*
AUCTOR *Well, maybe.*

EXAMPLE 6.3

Define f: $\mathbb{R}^2 \to \mathbb{R}^2$ by $f(x,y) = (x + y, 2x, x - y)$. The "rank" of this map (defined in Sec. 3.2 to be the dimension of its range) is 2; eventually we'll define the rank of a matrix in such a way as to make all the matrix representatives of a given map have the same rank as the map. Hence the rank of the matrices

$$A = \begin{bmatrix} 1 & 1 \\ 2 & 0 \\ 1 & -1 \end{bmatrix} \quad \text{and} \quad B = \begin{bmatrix} -6 & 5 \\ -4 & 4 \\ 4 & -3 \end{bmatrix}$$

(both representatives of f) must be 2. But neither A nor B is in a simple form for reading off the rank. The problem is to choose bases for $\mathbb{R}^2$ and $\mathbb{R}^3$ such that the matrix of f is "nice." In the problems we'll ask you to show that the class of matrix representations of f contains

$$\begin{bmatrix} 1 & 0 \\ 0 & 1 \\ 0 & 0 \end{bmatrix}$$

which is certainly a friendly ambassador for the class to put forward. (Its rank is the order of the 2×2 identity matrix appearing as a submatrix.)†

† This discussion is deliberately vague, due to our reluctance to get into the question of rank. (Our only purpose is to supply an example of a property of linear maps that is reflected as an "invariant" by the class of matrices representing the map.) However, see Prob. 8 for a definition of the rank of a square matrix, and Sec. 7.5 for the idea in general.

Problems

1. The matrix of $f(x,y) = (x + y,\ 2x,\ x - y)$ relative to the standard bases is

$$A = \begin{bmatrix} 1 & 1 \\ 2 & 0 \\ 1 & -1 \end{bmatrix}$$

a. Show that A is equivalent to

$$B = \begin{bmatrix} 1 & 0 \\ 0 & 1 \\ 0 & 0 \end{bmatrix}$$

by confirming that

$$P = \begin{bmatrix} 1 & 0 & 0 \\ 0 & 1 & 0 \\ 1 & -1 & 1 \end{bmatrix} \quad \text{and} \quad Q = \frac{1}{2}\begin{bmatrix} 0 & 1 \\ 2 & -1 \end{bmatrix}$$

are nonsingular and that $B = PAQ$.

b. Show that if β is the standard basis for $\mathcal{R}^3$, there is no basis δ for $\mathcal{R}^2$ such that $B = [f]_{\delta\beta}$.

c. Letting α be the standard basis for $\mathcal{R}^2$, find a basis γ for $\mathcal{R}^3$ such that $B = [f]_{\alpha\gamma}$. *Hint:* The first two elements of γ must be $(1,2,1)$ and $(1,0,-1)$. (Why?) The third element may be any vector in $\mathcal{R}^3$ which is not a linear combination of these two. (Why?)

2. Define $f\colon \mathcal{R}^3 \to \mathcal{R}^2$ by $f(x,y,z) = (2x + y - z,\ x - 5z)$ and let

$$\alpha = \{(1,0,0),\ (1,1,1),\ (0,0,1)\} \quad \text{and} \quad \beta = \{(1,-1),\ (1,1)\}$$

Then

$$A = [f]_{\alpha\beta} = \frac{1}{2}\begin{bmatrix} 1 & 6 & 4 \\ 3 & -2 & -6 \end{bmatrix}$$

while the matrix of f relative to the standard bases is

$$B = \begin{bmatrix} 2 & 1 & -1 \\ 1 & 0 & -5 \end{bmatrix}$$

Find nonsingular matrices P and Q such that $B = PAQ$ (guaranteed to exist by Theorem 6.1).

3. Finish the proof of Theorem 6.2 by showing that equivalence on $M_{m,n}(\mathcal{F})$ is reflexive and symmetric.

4. The matrix

$$P = \begin{bmatrix} 2 & -1 \\ 1 & 1 \end{bmatrix}$$

is nonsingular, so according to Theorem 6.3 there is a basis α for $\mathcal{R}^2$ such that $P = [i]_{\alpha\beta}$, where β is the standard basis. Find α.

★ **5.** Prove that if P is a nonsingular $n \times n$ matrix with entries in $\mathfrak{F}$ and β is the standard basis for $\mathfrak{F}^n$, the columns of P constitute a basis α for $\mathfrak{F}^n$ such that $P = [i]_{\alpha\beta}$.

★ **6.** Prove that an $n \times n$ matrix with entries in $\mathfrak{F}$ is nonsingular if and only if its columns are linearly independent n-tuples in $\mathfrak{F}^n$. *Hint:* Problem 5 yields the "only if" part immediately and serves as a suggestion for the "if" part.

★ **7.** Do Prob. 6 with "columns" replaced by "rows." *Hint:* Use the transpose.

8. Let P be an $n \times n$ matrix with entries in $\mathfrak{F}$. The rows of P (considered as n-tuples) span a subspace of $\mathfrak{F}^n$ of dimension at most n; so do the columns. The dimensions of these subspaces are called the "row rank" and "column rank," respectively, of P. Prove that P is nonsingular if and only if both its row rank and column rank are equal to n.

Later we'll prove that the row rank and column rank are always equal, whether or not they have their maximum value n, and the word "rank" will be used to mean either one.

9. Find the row rank and column rank of

$$A = \begin{bmatrix} 2 & 1 & 0 \\ -1 & 5 & 3 \\ 3 & 7 & 3 \end{bmatrix}$$

(Don't assume that they're equal! We haven't proved that yet.) Is this matrix singular or nonsingular?

10. Let P and Q be nonsingular matrices ($m \times m$ and $n \times n$, respectively) and define the map $f\colon M_{m,n}(\mathfrak{F}) \to M_{m,n}(\mathfrak{F})$ by $f(A) = PAQ$.

a. Prove that f is linear.

b. Prove that f is one-to-one (without using the next part of this problem).

c. Prove that f is onto (without using the preceding part of this problem).

According to Prob. 45, Sec. 3.3, it is unnecessary to do both parts *b* and *c*. The reason for our prohibition is that each is worthwhile in its own right.

6.2 Similar Matrices

The $m \times n$ matrices A and B are "equivalent" if $B = PAQ$, where P and Q are nonsingular; we saw in the last section that this condition holds if and only if A and B represent the same linear map. Now we are going to specialize the discussion to linear *operators,* the matrix representations of which are square.

SIMILAR MATRICES

Suppose that the square matrices A and B are related by an equation of the form $B = PAP^{-1}$, where P is nonsingular. Then we say that A is *similar* to B, and write $A \precapprox B$.†

THEOREM 6.6

Similarity on the set $M_n(\mathfrak{F})$ is an equivalence relation.

Proof

We'll do the symmetric law, $A \precapprox B \Rightarrow B \precapprox A$. If $A \precapprox B$, there is a nonsingular matrix P such that $B = PAP^{-1}$. Simply solve for A! ''Premultiplication'' of each side by P^{-1} yields $P^{-1}B = AP^{-1}$; ''postmultiplication'' by P then gives $P^{-1}BP = A$. Letting $Q = P^{-1}$ (note that Q is nonsingular and that $Q^{-1} = P$), we have $A = QBQ^{-1}$, and consequently $B \precapprox A$. ∎

Thus similarity classifies $n \times n$ matrices together if they are related by an equation of the form $B = PAP^{-1}$. When A and B are similar they are also equivalent (why?), but equivalent matrices, even when they are square, are not necessarily similar. (See the problems.) The reason we are interested in similarity is indicated by the following theorem.

THEOREM 6.7

Two matrices are similar if and only if they represent the same linear operator (each relative to a single basis).

Proof

Suppose first that A and B are $n \times n$ matrix representations of the linear operator $f\colon S \rightarrow S$, where S is n-dimensional and each representation is relative to a single basis for S, say $A = [f]_\alpha$ and $B = [f]_\beta$.‡ We leave it to you to name a nonsingular matrix P such that $B = PAP^{-1}$. It follows that A and B are similar.

Conversely, suppose that A and B are similar and that A is the matrix of a linear operator $f\colon S \rightarrow S$ relative to a basis α for S. The problem is to show that B also represents f relative to a single basis. We know that $B = PAP^{-1}$ for some nonsingular P; use Theorem 6.3 to find a basis β such that $P = [i]_{\alpha^\beta}$. Then $P^{-1} = [i]_{\beta^\alpha}$ and we have

$$B = PAP^{-1} = [i]_{\alpha^\beta}[f]_\alpha[i]_{\beta^\alpha} = [f]_\beta \quad ∎$$

†This symbol was also used for ''isomorphic'' in Sec. 3.4. There shouldn't be any occasion for confusion. Notation is not standard here; you will find matrix equivalence and similarity represented by a variety of symbols in different books.

‡See the definition of matrix representation in Sec. 4.1. The notation $[f]_\alpha$ is shorthand for $[f]_{\alpha^\alpha}$.

LECTOR *Did you say there are square matrices A and B that are equivalent but not similar?*

AUCTOR *Yes. See Prob. 14.*

LECTOR *All right, If A is the matrix of f: S → S, so is B.*

AUCTOR *Why?*

LECTOR *Theorem 6.5.*

AUCTOR *Yes, but . . .*

LECTOR *Then Theorem 6.7 says that A and B are similar.*

AUCTOR *Well, hold on . . .*

LECTOR *So how can they be equivalent but not similar?*

AUCTOR *If* $A = [f]_\alpha$, *Theorem 6.5 merely says that* $B = [f]_{\beta^\gamma}$ *for some bases* β *and* γ.

LECTOR *So B represents f.*

AUCTOR *But not relative to a single basis, which Theorem 6.7 requires.*

LECTOR *I just wanted to see what you'd say.*

AUCTOR *What I really said was deleted by the publisher.*

LECTOR *Oh, come on, what was it?*

AUCTOR [*deleted*]

As an example of Theorem 6.7, define the linear operator $f\colon \mathfrak{R}^2 \to \mathfrak{R}^2$ by $f(x,y) = (x - y, y - x)$. The matrix of f relative to the standard basis is

$$A = \begin{bmatrix} 1 & -1 \\ -1 & 1 \end{bmatrix}$$

while its matrix relative to $\{(1,2), (-1,1)\}$ is

$$B = \begin{bmatrix} 0 & 0 \\ 1 & 2 \end{bmatrix}$$

(Confirm this!) A nonsingular matrix P satisfying $B = PAP^{-1}$ is

$$P = \frac{1}{3}\begin{bmatrix} 1 & 1 \\ -2 & 1 \end{bmatrix}$$

as you can check. Note, however, that this is not unique; we could just as well name

$$P = \begin{bmatrix} 1 & 1 \\ 2 & -3 \end{bmatrix}$$

Problems

11. Confirm that the matrices

$$A = \begin{bmatrix} 0 & 1 \\ 1 & 0 \end{bmatrix} \quad \text{and} \quad B = \begin{bmatrix} 1 & 0 \\ 0 & -1 \end{bmatrix}$$

are similar by virtue of the relation $B = PAP^{-1}$, where

$$P = \frac{1}{\sqrt{2}} \begin{bmatrix} 1 & 1 \\ -1 & 1 \end{bmatrix}$$

Then name a linear operator $f\colon \mathcal{R}^2 \to \mathcal{R}^2$, and bases α and β for $\mathcal{R}^2$, such that $A = [f]_\alpha$ and $B = [f]_\beta$.

12. Finish the proof of Theorem 6.6 by showing that similarity on $M_n(\mathcal{F})$ is reflexive and transitive.

13. Explain why similar matrices are also equivalent.

14. Show that the matrices

$$A = \begin{bmatrix} 1 & 0 \\ 0 & 0 \end{bmatrix} \quad \text{and} \quad B = \begin{bmatrix} 0 & 1 \\ 0 & 0 \end{bmatrix}$$

are equivalent but not similar.

15. Finish the proof of Theorem 6.7 by naming a nonsingular matrix P such that $B = PAP^{-1}$, where $A = [f]_\alpha$ and $B = [f]_\beta$ are matrix representations of the linear operator $f\colon S \to S$.

16. The matrix representations of $f(x,y,z) = (x + y, y + z, z + x)$ relative to the bases $\alpha = \{\mathbf{E}_1, \mathbf{E}_2, \mathbf{E}_3\}$ and $\beta = \{\mathbf{E}_2, \mathbf{E}_3, -\mathbf{E}_1\}$ for $\mathcal{R}^3$ are

$$A = \begin{bmatrix} 1 & 1 & 0 \\ 0 & 1 & 1 \\ 1 & 0 & 1 \end{bmatrix} \quad \text{and} \quad B = \begin{bmatrix} 1 & 1 & 0 \\ 0 & 1 & -1 \\ -1 & 0 & 1 \end{bmatrix}$$

respectively. Find a nonsingular matrix P satisfying $B = PAP^{-1}$.

17. Given the scalar k, show that the only $n \times n$ matrix similar to kI is kI itself. (Thus the equivalence class of $n \times n$ matrices similar to kI consists of kI alone; each scalar multiple of I is "in a class by itself.")

18. Suppose that the $n \times n$ matrix A is "in a class by itself" in the sense that the only $n \times n$ matrix similar to A is A.

a. Prove that A commutes with every nonsingular $n \times n$ matrix P.

b. Assuming that $n = 2$, show that part *a* implies that $A = kI$ for some scalar k. *Hint:* By choosing simple nonsingular matrices P, show that

$$A = \begin{bmatrix} a & b \\ c & d \end{bmatrix}$$

must have the form

$$\begin{bmatrix} k & 0 \\ 0 & k \end{bmatrix} = kI$$

The argument in part *b* can be generalized to $n \times n$ matrices; Probs. 17 and 18 then yield the statement that a square matrix is in a similarity class by itself if and only if it is a scalar multiple of I.

19. Show that an equivalence class of similar $n \times n$ matrices cannot have both singular and nonsingular members. (In other words, a nonsingular matrix is never similar to a singular matrix.)

20. Prove that the transposes of two similar matrices are similar. Is this true if "similar" is replaced by "equivalent"?

21. Use mathematical induction to prove that if A and B are similar, so are A^r and B^r, where r is any positive integer.

22. Prove that the inverses of two nonsingular similar matrices are similar.

23. Show that if A is a nonsingular $n \times n$ matrix, then AB and BA are similar for every $n \times n$ matrix B.

24. The "trace" of an $n \times n$ matrix $A = [a_{ij}]$ is the sum of its diagonal entries, that is, $\text{tr } A = \sum_{i=1}^{n} a_{ii}$.
a. Prove that if A and B are $n \times n$, then $\text{tr } AB = \text{tr } BA$.
b. Prove that if A and B are similar, then $\text{tr } A = \text{tr } B$.
c. The trace of a linear operator (on a finite-dimensional space) is defined to be the trace of any one of its matrix representations (relative to a single basis, as usual). Why is this definition unambiguous?
d. Find the trace of $f(x,y,z) = (2x - y + z, x + y, x - y - 3z)$.

25. Let P be a nonsingular $n \times n$ matrix and define the map f: $M_n(\mathfrak{F}) \to M_n(\mathfrak{F})$ by $f(A) = PAP^{-1}$. Since this is a special case of the map defined in Prob. 10, Sec. 6.1, we know that f is an isomorphism on $M_n(\mathfrak{F})$ considered as a vector space. Prove that it is also an isomorphism on $M_n(\mathfrak{F})$ considered as a ring (or for that matter a linear algebra). *Hint:* All that is required is to show that $f(AB) = f(A)f(B)$. Why?

6.3 Change of Basis

To begin this section, we adopt two conventions of notation. Neither of these represents anything new, but is simply an attempt to economize on symbols, which are in danger of proliferating.

COORDINATE VECTOR MAP

Let S be a vector space over $\mathfrak{F}$ with (ordered) basis

$$\alpha = \{u_1, \ldots, u_n\}$$

The invertible linear map which sends each element of S into its coordinate vector relative to α will henceforth be denoted by α. In other words, $\alpha\colon S \to \mathfrak{F}^n$ is defined by $\alpha(x) = \mathbf{X}$, where $x = \sum_{j=1}^{n} x_j u_j$ is any element of S and $\mathbf{X} = (x_1, \ldots, x_n)$. (See the proof of Theorem 4.4, where we have already used this notation.)

STANDARD MAP

Let A be an $m \times n$ matrix with entries in $\mathfrak{F}$. By the *standard map A* we mean the linear map from $\mathfrak{F}^n$ to $\mathfrak{F}^m$ whose matrix relative to the standard bases is A. Since this sends $\mathbf{X}$ into $A\mathbf{X}$ (Theorem 5.16), we shall frequently dispense with the usual functional symbol f and use A instead. We'll write $A(\mathbf{X}) = A\mathbf{X}$, where A plays the dual role of the name of the map (on the left side) and the name of the matrix (on the right side).†

Now suppose that we are looking at coordinate vectors of elements of S relative to different bases $\alpha = \{u_1, \ldots, u_n\}$ and $\beta = \{v_1, \ldots, v_n\}$. A given element $x = \sum_{j=1}^{n} x_j u_j$ has coordinate vector

$$\alpha(x) = \mathbf{X} = (x_1, \ldots, x_n)$$

relative to α. But if x is written in terms of β, say $x = \sum_{j=1}^{n} x'_j v_j$, its coordinate vector changes to

$$\beta(x) = \mathbf{X}' = (x'_1, \ldots, x'_n)$$

Thus the same element of S has different coördinate vectors (in general) relative to different bases.

This suggests that we are dealing with the identity map $i\colon S \to S$, for the elements of S stay put—only their coordinate vectors in n-space change as we change bases. In fact, the matrix $P = [i]_{\alpha^\beta}$ contains the necessary information for describing the change, since its entries p_{ij} appear as coefficients in the equations relating the old basis to the new:

$$u_j = i(u_j) = \sum_{i=1}^{n} p_{ij} v_i \qquad j = 1, \ldots, n$$

† This is frequently done in calculus, where (for example) the function defined by $y = x^2$ may be written as $y(x) = x^2$ instead of $f(x) = x^2$, simply to economize on notation.

If these are written out, they are equations for $u_1, \ldots, u_n$ in terms of $v_1, \ldots, v_n$, namely

$$\begin{aligned} u_1 &= p_{11}v_1 + p_{21}v_2 + \cdots + p_{n1}v_n \\ u_2 &= p_{12}v_1 + p_{22}v_2 + \cdots + p_{n2}v_n \\ &\cdots\cdots\cdots\cdots\cdots\cdots\cdots\cdots \\ u_n &= p_{1n}v_1 + p_{2n}v_2 + \cdots + p_{nn}v_n \end{aligned}$$

This motivates the following definition.

MATRIX OF CHANGE OF BASIS

If α and β are (ordered) bases for the vector space S, then

$$P = [i]_{\alpha^\beta}$$

is called the *matrix relating* α *to* β. This, in turn, implies that

$$P^{-1} = [i]_{\beta^\alpha}$$

is the *matrix relating* β *to* α.

The distinction between these two may be confusing. The thing to remember is that when we refer to a matrix relating one basis to another, the elements of the first basis are written in terms of the second. Note, too, that as a consequence of our earlier definition of the matrix of a map (in this case, the identity map on S), the coefficients in the jth equation

$$u_j = p_{1j}v_1 + p_{2j}v_2 + \cdots + p_{nj}v_n$$

become the jth *column* of P. This is nothing new; we remind you of it to help keep things straight.

EXAMPLE 6.4

$\alpha = \{\mathbf{E}_2, \mathbf{E}_1, \mathbf{E}_3\}$ and $\beta = \{\mathbf{E}_3, \mathbf{E}_2, -\mathbf{E}_1\}$ are bases for $\mathcal{R}^3$; the matrix relating α to β is found by looking at the effect of i on the elements of α in terms of the elements of β:

$$\begin{aligned} i(\mathbf{E}_2) &= \mathbf{E}_2 = 0\mathbf{E}_3 + 1\mathbf{E}_2 + 0(-\mathbf{E}_1) \\ i(\mathbf{E}_1) &= \mathbf{E}_1 = 0\mathbf{E}_3 + 0\mathbf{E}_2 + (-1)(-\mathbf{E}_1) \\ i(\mathbf{E}_3) &= \mathbf{E}_3 = 1\mathbf{E}_3 + 0\mathbf{E}_2 + 0(-\mathbf{E}_1) \end{aligned}$$

Hence

$$P = [i]_{\alpha^\beta} = \begin{bmatrix} 0 & 0 & 1 \\ 1 & 0 & 0 \\ 0 & -1 & 0 \end{bmatrix}$$

Although the elements of S do not change as we go from α to β, their coordinate vectors in n-space do. Hence there is a mapping of $\mathfrak{F}^n$ into itself lurking in the wings. Delightfully enough, it is precisely the standard map P.

THEOREM 6.8

Let S be a vector space over $\mathfrak{F}$ with (ordered) bases α and β, and suppose that P is the matrix relating α to β. Then the standard map P sends α-coordinate vectors into β-coordinate vectors, that is,

$$P\alpha(x) = \beta(x) \qquad \text{for each } x \in S$$

Proof

Let $\alpha = \{u_1, \ldots, u_n\}$. Then for each $x = \sum_{j=1}^{n} x_j u_j$ in S, we have

$$P\alpha(x) = P\mathbf{X} \qquad \text{where } \mathbf{X} = \alpha(x) = (x_1, \ldots, x_n)$$

the coordinate vector of x relative to α.[†] The ith entry of the $n \times 1$ product $P\mathbf{X}$ is

$$\sum_{j=1}^{n} p_{ij}x_j \qquad i = 1, \ldots, n$$

(Why?) We need only compare this to the ith entry of

$$\beta(x) = \mathbf{X}' = (x_1', \ldots, x_n')$$

the coordinate vector of $x = \sum_{i=1}^{n} x_i' v_i$ relative to $\beta = \{v_1, \ldots, v_n\}$. Since

$$u_j = \sum_{i=1}^{n} p_{ij}v_i \qquad j = 1, \ldots, n$$

we have

$$x = \sum_{j=1}^{n} x_j u_j = \sum_{j=1}^{n} x_j\left(\sum_{i=1}^{n} p_{ij}v_i\right) = \sum_{i=1}^{n}\left(\sum_{j=1}^{n} p_{ij}x_j\right)v_i$$

But the expression

$$x = \sum_{i=1}^{n} x_i' v_i$$

in terms of the basis vectors $v_1, \ldots, v_n$ is unique, so it follows that the ith entry of $\beta(x)$ is

[†] As we have noted before, the matrix product $P\mathbf{X}$ is properly defined only if $\mathbf{X}$ is interpreted as an $n \times 1$ matrix (or "column vector"). If this is kept in mind, there is no harm in writing it as an ordinary n-tuple.

$$x_i' = \sum_{j=1}^{n} p_{ij}x_j \qquad i = 1, \ldots, n$$

Hence $P\alpha(x) = \beta(x)$. ▮

COROLLARY 6.8a

The standard map P^{-1} sends β-coordinate vectors into α-coordinate vectors.

EXAMPLE 6.5

Let α and β be as in Example 6.4. The standard map P is defined by

$$P(x,y,z) = \begin{bmatrix} 0 & 0 & 1 \\ 1 & 0 & 0 \\ 0 & -1 & 0 \end{bmatrix}(x,y,z) = (z,x,-y)$$

LECTOR *I presume you mean*

$$\begin{bmatrix} 0 & 0 & 1 \\ 1 & 0 & 0 \\ 0 & -1 & 0 \end{bmatrix}\begin{bmatrix} x \\ y \\ z \end{bmatrix} = \begin{bmatrix} z \\ x \\ -y \end{bmatrix}$$

AUCTOR *Precisely.*

LECTOR *Why don't you write it correctly?*

AUCTOR *There's an isomorphism between* $\mathcal{R}^3$ *and* $M_{3,1}(\mathcal{R})$.

LECTOR *I beg your pardon?*

AUCTOR *Each* (x,y,z) *in* $\mathcal{R}^3$ *can be identified with*

$$\begin{bmatrix} x \\ y \\ z \end{bmatrix}$$

in $M_{3,1}(\mathcal{R})$.

LECTOR *Am I to understand that an isomorphism is a license to be sloppy?*

AUCTOR *Well, not exactly. But we do this sort of thing all the time. For example, Prob. 40, Sec. 5.3, says that* $3 + 2i$ *in* $\mathcal{C}$ *really means*

$$\begin{bmatrix} 3 & 2 \\ -2 & 3 \end{bmatrix}$$

in $M_2(\mathcal{R})$.

LECTOR *I didn't much care for that problem.*

AUCTOR *When it comes down to it, how do you justify writing 5 in the rational numbers when you mean 5/1?*

LECTOR *My arithmetic teacher was 6'5" tall.*

Theorem 6.8 says that P sends α-coordinate vectors into β-coordinate vectors. To see this working here, look at the element $(2,-1,5)$ in $\mathcal{R}^3$. Its coordinate vectors relative to α and β are $(-1,2,5)$ and $(5,-1,-2)$, respectively. Since

$$P(-1,2,5) = (5,-1,-2)$$

Theorem 6.8 checks out. More generally, if (x,y,z) is any element of $\mathcal{R}^3$, its coordinate vectors relative to α and β are (y,x,z) and $(z,y,-x)$, respectively. As you can check, P sends the first into the second.

The standard map P^{-1} is defined by

$$P^{-1}(x,y,z) = \begin{bmatrix} 0 & 1 & 0 \\ 0 & 0 & -1 \\ 1 & 0 & 0 \end{bmatrix}(x,y,z) = (y,-z,x)$$

(Confirm this!) Corollary 6.8a says that this sends β-coordinate vectors back into α-coordinate vectors; since $P^{-1}(z,y,-x) = (y,x,z)$, this is correct.

EXAMPLE 6.6

Consider a rotation of axes through the angle θ in the plane. If a given point has coordinates (x,y) relative to the original axes and coordinates (x',y') relative to the rotated axes, we know from analytic geometry that the equations giving the new coordinates in terms of the old are

$$\begin{aligned} x' &= x\cos\theta + y\sin\theta \\ y' &= -x\sin\theta + y\cos\theta \end{aligned}$$

In the language of this section we are changing the standard basis $\alpha = \{(1,0), (0,1)\}$ to a new basis $\beta = \{(\cos\theta, \sin\theta), (-\sin\theta, \cos\theta)\}$, the elements of β being the unit vectors $(1,0)$ and $(0,1)$ rotated through θ as shown in Fig. 6.1. The coordinate vector (x,y) of a given point relative to α is sent into a new coordinate vector

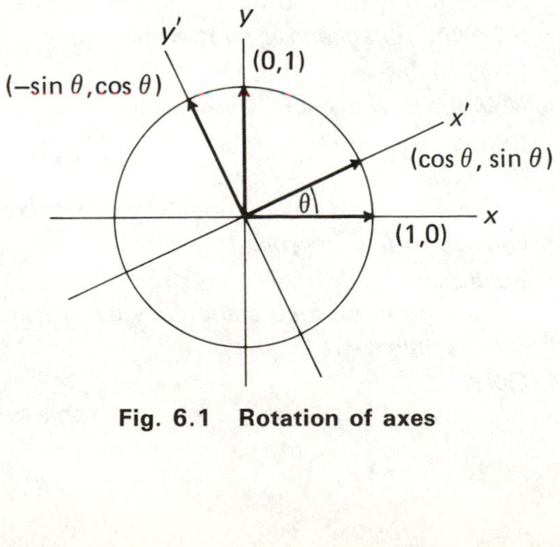

Fig. 6.1 Rotation of axes

$$(x',y') = (x\cos\theta + y\sin\theta,\ -x\sin\theta + y\cos\theta)$$

relative to β. Since this defines the map P, we may write down the matrix

$$P = \begin{bmatrix} \cos\theta & \sin\theta \\ -\sin\theta & \cos\theta \end{bmatrix}$$

immediately, without going to the trouble of computing $[i]_{\alpha^\beta}$. As a check, however, we'll do it anyway:

$$i(1,0) = (1,0) = \cos\theta\,(\cos\theta, \sin\theta) - \sin\theta\,(-\sin\theta, \cos\theta)$$
$$i(0,1) = (0,1) = \sin\theta\,(\cos\theta, \sin\theta) + \cos\theta\,(-\sin\theta, \cos\theta)$$

Thus

$$[i]_{\alpha^\beta} = \begin{bmatrix} \cos\theta & \sin\theta \\ -\sin\theta & \cos\theta \end{bmatrix}$$

as advertised.

Do not confuse this with the matrix

$$\begin{bmatrix} \cos\theta & -\sin\theta \\ \sin\theta & \cos\theta \end{bmatrix}$$

representing a rotation of the plane! (See Example 3.8 and the accompanying footnote.) We are not rotating the plane, but changing its basis. A given point of the plane does not move; its coordinate vector (x,y) relative to α changes to the coordinate vector (x',y') relative to β.

LECTOR *The coordinate vectors are elements of $\mathbb{R}^2$, are they not?*

AUCTOR *Unfortunately.*

LECTOR *They move.*

AUCTOR *Well, yes, but the $\mathbb{R}^2$ you're talking about must be distinguished from the plane whose basis we're changing.*

LECTOR *Perhaps you had better explain that.*

AUCTOR *Start with the cartesian plane S.*

LECTOR *I have it here on the table, equipped with the usual coordinate axes.*

AUCTOR *Stick thumbtacks in it so it won't move.*

LECTOR *Done.*

AUCTOR *Now erase the x and y axes and draw in a new set of axes rotated through the angle θ.*

LECTOR *If you say so.*

AUCTOR *On another table I have supplied you with a cartesian plane $\mathbb{R}^2$ in which you are to find coordinate vectors of the points of S.*

LECTOR *Splendid.*

AUCTOR *Before you erased the standard axes, the coordinate vector of* $(1,-2)$, *for example, was* $(1,-2)$.

LECTOR *Quite so.*

AUCTOR *But relative to the new axes, this same point has coordinate vector*

$(\cos\theta - 2\sin\theta, -\sin\theta - 2\cos\theta)$. It didn't move in S but its coordinate vector in $\mathcal{R}^2$ did.

LECTOR *Fair enough.*

AUCTOR *As a matter of fact, the coordinate vectors on your second table are rotating through the angle $-\theta$, for although we are interpreting*

$$P = \begin{bmatrix} \cos\theta & \sin\theta \\ -\sin\theta & \cos\theta \end{bmatrix}$$

as a matrix of the identity map on S, it may also be written as

$$P = \begin{bmatrix} \cos(-\theta) & -\sin(-\theta) \\ \sin(-\theta) & \cos(-\theta) \end{bmatrix}$$

which is a rotation of $\mathcal{R}^2$ through $-\theta$.

LECTOR *I just dropped the box of thumbtacks.*

AUCTOR *Be careful when you sit down and think it over.*

Now suppose that $f: S \to T$ is a linear map whose matrix relative to the bases α and β for S and T is A. If the bases are changed to γ and δ, we know from Theorem 6.5 that the new matrix of f, namely $B = [f]_{\gamma^\delta}$, is equivalent to A for we can write

$$B = [f]_{\gamma^\delta} = [i_T]_{\beta^\delta}[f]_{\alpha^\beta}[i_S]_{\gamma^\alpha} = PAQ$$

where $P = [i_T]_{\beta^\delta}$ and $Q = [i_S]_{\gamma^\alpha}$. But P is the matrix relating β to δ in T, while Q relates γ to α in S. *Note that Q does not, as you might expect, relate α to γ; instead Q^{-1} does.* This cannot be helped, but in any case we may describe how the matrix of a linear map changes when the bases change, as follows.

THEOREM 6.9

The matrix A of a linear map $f: S \to T$ changes to $B = PAQ$ with a change in bases, where P and Q are the nonsingular matrices relating the old basis to the new in T and the new basis to the old in S, respectively.

Proof

Refer to the above argument.

It is probably hard to remember this theorem (with its references to "old" and "new" bases in two vector spaces). Perhaps you shouldn't try, but should reconstruct the equation

$$B = [f]_{\gamma^\delta} = [i_T]_{\beta^\delta}[f]_{\alpha^\beta}[i_S]_{\gamma^\alpha} = PAQ$$

by which it is obtained. This can be figured out rather than memorized!

COROLLARY 6.9a

If the basis is changed only in T, then $B = PA$, where P relates the old basis to the new.

COROLLARY 6.9b

If the basis is changed only in S, then $B = AQ$, where Q relates the new basis to the old.

These corollaries describe two special cases of equivalence (the general case being $B = PAQ$). In the next chapter we'll show that when $B = PA$ (where P is nonsingular) there is a sequence of "row operations" by which B can be obtained from A. Similarly $B = AQ$ implies the existence of a sequence of "column operations" transforming A into B. For this reason, we call these relations "row equivalence" and "column equivalence," respectively. Thus we may summarize by saying that when the bases are changed, the matrix of a linear map is transformed to an equivalent, row-equivalent, or column-equivalent matrix, depending on whether both bases are changed, or only the basis in the range, or only the basis in the domain.

When we are dealing with linear operators, the situation is simpler, as indicated by the following.

THEOREM 6.10

The matrix A of a linear operator $f: S \to S$ changes to $B = PAP^{-1}$ with a change in basis, where P is the nonsingular matrix relating the old basis to the new.

Proof

This is left for the problems.

Note that Theorems 6.9 and 6.10 don't say anything really new; they have already been given in Secs. 6.1 and 6.2. But now we are interpreting P and Q in Theorem 6.9, and P in Theorem 6.10, as matrices relating one basis to another, an idea we didn't bring up before.

EXAMPLE 6.7

Define $f: \mathcal{R}^2 \to \mathcal{R}^2$ by $f(x,y) = (2x + y, x - y)$. The matrix of f relative to the standard basis α is

$$A = \begin{bmatrix} 2 & 1 \\ 1 & -1 \end{bmatrix}$$

If the basis is changed to $\beta = \{(1,1), (1,-1)\}$, the matrix relating α to β is

$$P = \frac{1}{2}\begin{bmatrix} 1 & 1 \\ 1 & -1 \end{bmatrix}$$

(Confirm this!) Since

$$P^{-1} = \begin{bmatrix} 1 & 1 \\ 1 & -1 \end{bmatrix}$$

we have

$$B = PAP^{-1} = \frac{1}{2}\begin{bmatrix} 1 & 1 \\ 1 & -1 \end{bmatrix}\begin{bmatrix} 2 & 1 \\ 1 & -1 \end{bmatrix}\begin{bmatrix} 1 & 1 \\ 1 & -1 \end{bmatrix} = \frac{1}{2}\begin{bmatrix} 3 & 3 \\ 3 & -1 \end{bmatrix}$$

According to Theorem 6.10, this is the matrix of f relative to β, which you should confirm.

Problems

★ **26.** Suppose that A is the matrix of the linear map $f\colon S \to T$ relative to the bases α and β for S and T.
a. If $\alpha = \{u_1, \ldots, u_n\}$, why is $\beta[f(u_j)]$ the jth column of A?
b. Prove that $\beta[f(x)] = A\alpha(x)$ for each $x \in S$.

27. In each of the following cases we name a vector space S and two bases α and β. Find: (1) the matrix relating α to β, (2) the formula for the linear operator sending α-coordinate vectors into β-coordinate vectors, (3) the matrix relating β to α, and (4) the formula for the linear operator sending β-coordinate vectors into α-coordinate vectors.
a. $S = \mathcal{R}^2$; $\alpha = \{\mathbf{E}_1, \mathbf{E}_2\}$, $\beta = \{\mathbf{E}_1, -\mathbf{E}_2\}$
b. $S = \mathcal{R}^3$; $\alpha = \{(1,1,0), (1,0,1), (0,1,1)\}$, $\beta = \{\mathbf{E}_1, \mathbf{E}_2, \mathbf{E}_3\}$
c. S is the solution space of $y'' + y = 0$ (Example 2.8); $\alpha = \{\phi_1, \phi_2\}$, $\beta = \{-\phi_2, \phi_1\}$, where $\phi_1(x) = \cos x$ and $\phi_2(x) = \sin x$

★ **28.** Let α be any basis for $\mathcal{F}^n$ and suppose that β is the standard basis. Explain why the matrix relating α to β has the elements of α for columns.

29. Let α and β be bases for the n-dimensional space S over $\mathcal{F}$ and suppose that P is the matrix relating α to β. Explain why $P^{-1}\beta(x) = \alpha(x)$ for each $x \in S$.

30. Prove Corollary 6.9a: If A is the matrix of $f\colon S \to T$ and the basis in T is changed, the new matrix of f is $B = PA$, where P relates the old basis to the new.

31. Prove Corollary 6.9b: If A is the matrix of $f\colon S \to T$ and the basis in S is changed, the new matrix of f is $B = AQ$, where Q relates the new basis to the old.

32. Prove Theorem 6.10: If A is the matrix of $f\colon S \to S$ and the basis is changed, the new matrix is $B = PAP^{-1}$, where P relates the old basis to the new.

33. In each of the following cases let A be the matrix of the given linear map $f\colon \mathcal{R}^n \to \mathcal{R}^m$ relative to the standard bases α and β for $\mathcal{R}^n$ and $\mathcal{R}^m$, and suppose that the given bases γ and δ are new. Find the matrices P and Q of Theorem 6.9 and compute the new matrix $B = PAQ$ of f.

a. $f(x,y,z) = (2x - y, x + 3y + 5z)$;
$\gamma = \{(1,0,0), (0,1,0), (-1,-2,7/5)\}$,
$\delta = \{(2,1), (-1,3)\}$

b. $f(x,y) = (2x, x, x - 3y)$; $\gamma = \alpha$, $\delta = \{(2,1,1), (0,0,-3), (1,0,0)\}$

c. $f(x,y) = (2x + y, x + y)$; $\gamma = \{(1,-1), (-1,2)\}$, $\delta = \beta$

34. In each case let A be the matrix of the given linear operator $f\colon \mathcal{R}^n \to \mathcal{R}^n$ relative to the standard basis α, and suppose that the given basis β is new. Find the matrices P and P^{-1} of Theorem 6.10 and compute the new matrix $B = PAP^{-1}$ of f.

a. $f(x,y) = (2x - y, -x + 2y)$; $\beta = \{(1,1), (1,-1)\}$

b. $f(x,y,z) = (3x + 2y + z, -x - z, x + y + 2z)$;
$\beta = \{(1,-1,0), (1,-1,1), (1,0,0)\}$

c. $f(x,y,z) = (2x + z, y, x + 2z)$; $\beta = \{(0,1,0), (-1,0,1), (1,0,1)\}$

Review Quiz

True or false?

1. Every nonsingular $n \times n$ matrix is a representation of the identity map on n-space (relative to appropriate bases).
2. Any matrix which is equivalent to a singular matrix is singular.
3. If $f\colon S \to T$ is a linear map with matrix representations $A = [f]_{\alpha\beta}$ and $B = [f]_{\gamma\delta}$ related by the equation $B = PAQ$, then Q is the matrix relating α to γ.
4. If $\alpha = \{\mathbf{E}_1,\mathbf{E}_2,\mathbf{E}_3\}$ and $\beta = \{\mathbf{E}_3,\mathbf{E}_2,\mathbf{E}_1\}$ in $\mathcal{R}^3$, the matrix relating β to α is
$$\begin{bmatrix} 0 & 0 & 1 \\ 0 & 1 & 0 \\ 1 & 0 & 0 \end{bmatrix}$$
5. If $B = PAQ$, where P and Q are nonsingular, there is a linear map having A and B as matrix representations.
6. An $n \times n$ matrix with linearly independent rows is nonsingular.
7. The matrix relating one basis to another also represents the map sending old coordinate vectors into new.
8. If A and B are similar, so are A^* and B^*.

9. If A and B are matrix representations of the same linear operator (each relative to a single basis), there is a unique matrix P such that $B = PAP^{-1}$.

10. If $\alpha = \{(1,1), (-1,1)\}$ and $\beta = \{(1,0), (0,1)\}$, the linear operator sending α-coordinate vectors into β-coordinate vectors is $P(x,y) = (x - y, x + y)$.

11. The relation $<$ on the set of real numbers is an equivalence relation.

12. Equivalent $n \times n$ matrices represent the same linear operator (each relative to a single basis).

13. The matrices

$$\begin{bmatrix} 2 & 0 \\ 0 & 2 \end{bmatrix} \quad \text{and} \quad \begin{bmatrix} 1 & 0 \\ 0 & -1 \end{bmatrix}$$

are similar.

CHAPTER 7

Linear Systems

At this point we are in a position to say something intelligible about the problem of solving a system of m linear equations in n unknowns. Many books on linear algebra start with this problem, using it to motivate the introduction of matrices and the apparatus of vector spaces and linear maps which we have already developed. This approach has the advantage of exhibiting a concrete (and important) application of linear algebra right away; a student who learns how to solve linear systems may be more easily convinced that he is not wasting his time. The disadvantage is that the theory of linear systems can be cluttered; we have preferred to clear away some of the debris in the hope that your enthusiasm for mathematics will not be seriously injured by what remains.

7.1 Systems of Linear Equations

As an example of the sort of thing we are going to do, consider the 2×2 system

$$\begin{aligned} a_{11}x_1 + a_{12}x_2 &= c_1 \\ a_{21}x_1 + a_{22}x_2 &= c_2 \end{aligned}$$

where a_{11}, a_{12}, a_{21}, a_{22}, c_1, and c_2 are given numbers (real or complex) and x_1, x_2 are "unknowns." This system can be written in the form

$$\begin{bmatrix} a_{11} & a_{12} \\ a_{21} & a_{22} \end{bmatrix} \begin{bmatrix} x_1 \\ x_2 \end{bmatrix} = \begin{bmatrix} c_1 \\ c_2 \end{bmatrix}$$

or, more concisely,

$$A\mathbf{X} = \mathbf{C}$$

where

$$A = \begin{bmatrix} a_{11} & a_{12} \\ a_{21} & a_{22} \end{bmatrix}$$

is the "matrix of coefficients" and $\mathbf{X} = (x_1,x_2)$ and $\mathbf{C} = (c_1,c_2)$ are elements of $\mathfrak{F}^2$ considered as column vectors (2 × 1 matrices). In Prob. 52 of Sec. 5.4 we asked you to prove that if

$$\det A = a_{11}a_{22} - a_{12}a_{21} \neq 0$$

then A is nonsingular, with inverse given by

$$A^{-1} = \frac{1}{\det A}\begin{bmatrix} a_{22} & -a_{12} \\ -a_{21} & a_{11} \end{bmatrix}$$

To solve the system in these circumstances, we need only "premultiply" each side of $A\mathbf{X} = \mathbf{C}$ by A^{-1}:

$$\begin{aligned} A^{-1}(A\mathbf{X}) &= A^{-1}\mathbf{C} \\ (A^{-1}A)\mathbf{X} &= A^{-1}\mathbf{C} \\ I\mathbf{X} &= A^{-1}\mathbf{C} \\ \mathbf{X} &= A^{-1}\mathbf{C} \end{aligned}$$

But

$$\begin{aligned} A^{-1}\mathbf{C} &= \frac{1}{\det A}\begin{bmatrix} a_{22} & -a_{12} \\ -a_{21} & a_{11} \end{bmatrix}(c_1, c_2) \\ &= \frac{1}{\det A}(a_{22}c_1 - a_{12}c_2, -a_{21}c_1 + a_{11}c_2)^{\dagger} \end{aligned}$$

Let

$$A_1 = \begin{bmatrix} c_1 & a_{12} \\ c_2 & a_{22} \end{bmatrix} \quad \text{and} \quad A_2 = \begin{bmatrix} a_{11} & c_1 \\ a_{21} & c_2 \end{bmatrix}$$

be the matrices obtained from A by replacing its first and second columns, respectively, by $\mathbf{C}$. Then the unique solution of the system is

$$\mathbf{X} = \frac{1}{\det A}(\det A_1, \det A_2)$$

and the original unknowns are given by the formulas

$$x_1 = \frac{\det A_1}{\det A} \qquad x_2 = \frac{\det A_2}{\det A}$$

This is the 2 × 2 case of Cramer's rule (Gabriel Cramer, 1704–1752). The $n \times n$ case will be taken up after we have discussed determinants.

You can see how these ideas might be generalized to the $n \times n$ system

$$\begin{aligned} a_{11}x_1 + a_{12}x_2 + \cdots + a_{1n}x_n &= c_1 \\ a_{21}x_1 + a_{22}x_2 + \cdots + a_{2n}x_n &= c_2 \\ \cdots\cdots\cdots\cdots\cdots\cdots\cdots\cdots\cdots & \\ a_{n1}x_1 + a_{n2}x_2 + \cdots + a_{nn}x_n &= c_n \end{aligned}$$

† We remind you again that our notation is mixed; the number pairs should be written as 2 × 1 matrices in order to have compatible matrix products. We are assuming that you translate it this way.

If $A = [a_{ij}]$ is the $n \times n$ coefficient matrix and $\mathbf{X} = (x_1, \ldots, x_n)$ and $\mathbf{C} = (c_1, \ldots, c_n)$ are regarded as column vectors ($n \times 1$ matrices), then the system reduces to $A\mathbf{X} = \mathbf{C}$. Now the problems begin! Is it still correct to say that A is nonsingular if det A is nonzero? Come to think of it, what *is* det A? How do we find A^{-1} when it exists? Does Cramer's rule still hold? If so, how does it read and how is it proved? These questions will be answered as we proceed.

We can say one thing, however, based on our knowledge of matrix multiplication. If A^{-1} exists, then

$$A\mathbf{X} = \mathbf{C} \iff \mathbf{X} = A^{-1}\mathbf{C}$$

(as in the 2×2 case). This establishes the following important result.

> If the coefficient matrix of the $n \times n$ system $A\mathbf{X} = \mathbf{C}$ is nonsingular, then the system has a unique solution, $\mathbf{X} = A^{-1}\mathbf{C}$.

(Existence is the "$\Leftarrow$" part, uniqueness the "$\Rightarrow$" part, of the above double implication.)

EXAMPLE 7.1

The coefficient matrix of the 3×3 system

$$\begin{aligned} 2x - y + z &= 6 \\ x + y + 2z &= 0 \\ y - 3z &= -1 \end{aligned}$$

is

$$A = \begin{bmatrix} 2 & -1 & 1 \\ 1 & 1 & 2 \\ 0 & 1 & -3 \end{bmatrix}$$

Its inverse (see Prob. 9 in Sec. 7.2) is

$$A^{-1} = \frac{1}{12}\begin{bmatrix} 5 & 2 & 3 \\ -3 & 6 & 3 \\ -1 & 2 & -3 \end{bmatrix}$$

Hence the solution is

$$\begin{aligned} \mathbf{X} = A^{-1}\mathbf{C} &= \frac{1}{12}\begin{bmatrix} 5 & 2 & 3 \\ -3 & 6 & 3 \\ -1 & 2 & -3 \end{bmatrix}(6,0,-1) \\ &= \frac{1}{12}(27,-21,-3) = \frac{1}{4}(9,-7,-1) \end{aligned}$$

from which

$$x = 9/4 \qquad y = -7/4 \qquad z = -1/4$$

While the formula $\mathbf{X} = A^{-1}\mathbf{C}$ looks simple, its usefulness in solving $n \times n$ systems is limited by the difficulty in finding A^{-1}; there are better ways to solve such systems. Note, too, that the formula is applicable only to "square" systems in which the number of unknowns and equations is the same.

In general we are interested in solving systems of m equations in n unknowns. These have the form

$$\begin{aligned} a_{11}x_1 + a_{12}x_2 + \cdots + a_{1n}x_n &= c_1 \\ a_{21}x_1 + a_{22}x_2 + \cdots + a_{2n}x_n &= c_2 \\ \cdots\cdots\cdots\cdots\cdots\cdots\cdots\cdots & \\ a_{m1}x_1 + a_{m2}x_2 + \cdots + a_{mn}x_n &= c_m \end{aligned}$$

where the a_{ij} and c_i are given numbers (real or complex) and the x_j are unknowns. The coefficient matrix $A = [a_{ij}]$ is now $m \times n$, while $\mathbf{C} = (c_1, \ldots, c_m)$ is an element of $\mathfrak{F}^m$; the unknowns constitute an n-tuple $\mathbf{X} = (x_1, \ldots, x_n)$ in $\mathfrak{F}^n$. If $\mathbf{X}$ and $\mathbf{C}$ are interpreted to be $n \times 1$ and $m \times 1$ matrices, respectively, the system can be written in the form $A\mathbf{X} = \mathbf{C}$, as before.

EXAMPLE 7.2

Consider the 3×4 system

$$\begin{aligned} 3x_1 - 2x_2 + x_3 &= 0 \\ x_2 - 2x_3 + 2x_4 &= -5 \\ 3x_1 + x_2 + x_3 - x_4 &= 1 \end{aligned}$$

You may never have seen a complete solution of such a system. The operations involved are all elementary; it is just a question of being systematic. In this case it seems natural to eliminate x_4 right away from the second equation by adding twice the third equation to the second, and putting the result in place of the second. This yields the equivalent system†

$$\begin{aligned} 3x_1 - 2x_2 + x_3 &= 0 \\ 6x_1 + 3x_2 &= -3 \\ 3x_1 + x_2 + x_3 - x_4 &= 1 \end{aligned}$$

which can be simplified by dividing the second equation by 3:

$$\begin{aligned} 3x_1 - 2x_2 + x_3 &= 0 \\ 2x_1 + x_2 &= -1 \\ 3x_1 + x_2 + x_3 - x_4 &= 1 \end{aligned}$$

Now suppose that we interchange the first two equations:

† Two systems are "equivalent" if any solution of one is a solution of the other, that is, if their solution sets are the same. Operations of the sort we are now performing always lead to systems equivalent to the original system, as we'll prove. For the present, we are regarding this as obvious.

$$\begin{aligned} 2x_1 + x_2 &= -1 \\ 3x_1 - 2x_2 + x_3 &= 0 \\ 3x_1 + x_2 + x_3 - x_4 &= 1 \end{aligned}$$

Eliminating x_3 from the last equation by adding -1 times the second to the third, we have

$$\begin{aligned} 2x_1 + x_2 &= -1 \\ 3x_1 - 2x_2 + x_3 &= 0 \\ 3x_2 - x_4 &= 1 \end{aligned}$$

Now replace the second equation by the sum of itself and twice the first; similarly, add -3 times the first equation to the third. This eliminates x_2 from the second and third equations:

$$\begin{aligned} 2x_1 + x_2 &= -1 \\ 7x_1 + x_3 &= -2 \\ -6x_1 - x_4 &= 4 \end{aligned}$$

Multiplying the last equation by -1 yields, finally, the system we were after:

$$\begin{aligned} 2x_1 + x_2 &= -1 \\ 7x_1 + x_3 &= -2 \\ 6x_1 + x_4 &= -4 \end{aligned}$$

LECTOR *Great wits are sure to madness near allied.*
AUCTOR *Though this be madness, yet there is method in it.*
LECTOR *I would have done it differently.*
AUCTOR *I rarely do it the same way twice. But attend:*

Each of the unknowns x_2, x_3, and x_4 can now be written in terms of x_1 by inspection. Regarding x_1 as an arbitrary parameter, say $x_1 = t$, we have the solution set of the original system, namely

$$S = \{\mathbf{X} | x_1 = t, x_2 = -2t - 1, x_3 = -7t - 2, x_4 = -6t - 4\}$$

By letting t take on all values in $\mathfrak{F}$, we obtain every solution of the system. For example, $t = 0$ yields $\mathbf{X} = (0, -1, -2, -4)$; $t = 1$ gives $\mathbf{X} = (1, -3, -9, -10)$; and so on.

EXAMPLE 7.3

The sequence of steps by which we arrived at the solution set in Example 7.2 can be carried out in a shorthand version by working with the matrix

$$[A|\mathbf{C}] = \begin{bmatrix} 3 & -2 & 1 & 0 & 0 \\ 0 & 1 & -2 & 2 & -5 \\ 3 & 1 & 1 & -1 & 1 \end{bmatrix}$$

Here we have "augmented" the coefficient matrix A by including $\mathbf{C} = (0, -5, 1)$ as a fifth column. Each operation we carried out with the equations of the system can be imitated by a similar operation on the rows of our matrix. For example, replacing the second equation by the sum of itself and twice the third corresponds to replacing the second row of the matrix by the sum of itself and twice the third:

$$(2) \to (2) + 2(3): \begin{bmatrix} 3 & -2 & 1 & 0 & 0 \\ 0 & 1 & -2 & 2 & -5 \\ 3 & 1 & 1 & -1 & 1 \end{bmatrix} \to \begin{bmatrix} 3 & -2 & 1 & 0 & 0 \\ 6 & 3 & 0 & 0 & -3 \\ 3 & 1 & 1 & -1 & 1 \end{bmatrix}$$

The notation $(2) \to (2) + 2(3)$ is almost self-explanatory; it means "replace row (2) by row (2) + 2 row (3)." Continuing in this fashion, we have the following sequence of "row operations" corresponding to what we did before:

$$(2) \to \tfrac{1}{3}(2): \quad \begin{bmatrix} 3 & -2 & 1 & 0 & 0 \\ 6 & 3 & 0 & 0 & -3 \\ 3 & 1 & 1 & -1 & 1 \end{bmatrix} \to \begin{bmatrix} 3 & -2 & 1 & 0 & 0 \\ 2 & 1 & 0 & 0 & -1 \\ 3 & 1 & 1 & -1 & 1 \end{bmatrix}$$

$$(1) \leftrightarrow (2): \quad \to \begin{bmatrix} 2 & 1 & 0 & 0 & -1 \\ 3 & -2 & 1 & 0 & 0 \\ 3 & 1 & 1 & -1 & 1 \end{bmatrix}$$

$$(3) \to (3) - (2): \quad \to \begin{bmatrix} 2 & 1 & 0 & 0 & -1 \\ 3 & -2 & 1 & 0 & 0 \\ 0 & 3 & 0 & -1 & 1 \end{bmatrix}$$

$$\left.\begin{matrix} (2) \to (2) + 2(1) \\ (3) \to (3) - 3(1) \end{matrix}\right\}: \quad \to \begin{bmatrix} 2 & 1 & 0 & 0 & -1 \\ 7 & 0 & 1 & 0 & -2 \\ -6 & 0 & 0 & -1 & 4 \end{bmatrix}$$

$$(3) \to -(3): \quad \to \begin{bmatrix} 2 & 1 & 0 & 0 & -1 \\ 7 & 0 & 1 & 0 & -2 \\ 6 & 0 & 0 & 1 & -4 \end{bmatrix}$$

Note the submatrix consisting of the second, third, and fourth columns; this is what we were heading for.† For now we can read off the fact that $x_1 = t$ is arbitrary and that

† A purist would object that we don't have things in standard form. More on that in the next section.

$$x_2 = -2x_1 - 1 = -2t - 1 \qquad x_3 = -7x_1 - 2 = -7t - 2$$
$$x_4 = -6x_1 - 4 = -6t - 4$$

The system is "solved."

Row operations on matrices are evidently worthy of study. While you could probably get along in practice without our saying more about them, any general attack on linear systems requires a theoretical treatment. This will come later; meanwhile the next section contains a more systematic description of the "standard form" we're after when the coefficient matrix of a linear system is transformed.

Problems

1. Use the formula $\mathbf{X} = A^{-1}\mathbf{C}$ to solve the following systems.
   a. $3x + 4y = -1$, $x - 2y = 0$ *b.* $2x + y - 2z = 1$, $x + 2y + 2z = -1$, $2x - 2y + z = 2$

   Note: The inverse of the coefficient matrix in part *b* can be found by observing that $AA^T = 9I$.

2. Explain why the formula $\mathbf{X} = A^{-1}\mathbf{C}$ does not apply in the following cases. Then find the solution set of each system.
   a. $2x - y = -3$, $-4x + 2y = 1$ *b.* $2x - y = -3$, $-4x + 2y = 6$

   When A is nonsingular, the $n \times n$ system $A\mathbf{X} = \mathbf{C}$ has a unique solution. This problem suggests that when A is singular, anything can happen.

3. Consider again the linear system solved in Examples 7.2 and 7.3:

$$\begin{aligned} 3x_1 - 2x_2 + x_3 &= 0 \\ x_2 - 2x_3 + 2x_4 &= -5 \\ 3x_1 + x_2 + x_3 - x_4 &= 1 \end{aligned}$$

   a. Use the following sequence of row operations to reduce the augmented matrix of the system to a matrix with the 3×3 identity in the upper left corner:

$$(3) \to (3) - (1)$$
$$(1) \to (1) + 2(2) \quad \text{and} \quad (3) \to (3) - 3(2)$$
$$(1) \to 2(1) \quad \text{and} \quad (2) \to 3(2)$$
$$(1) \to (1) + (3) \quad \text{and} \quad (2) \to (2) + (3)$$
$$(1) \to \tfrac{1}{6}(1) \quad (2) \to \tfrac{1}{3}(2) \quad \text{and} \quad (3) \to \tfrac{1}{6}(3)$$

b. The result in part *a* suggests solving for x_1, x_2, x_3 in terms of the parameter $x_4 = t$. Show that this yields the solution set

$$S = \{\mathbf{X} \mid x_1 = -\tfrac{1}{6}t - \tfrac{2}{3},\ x_2 = \tfrac{1}{3}t + \tfrac{1}{3},\ x_3 = \tfrac{7}{6}t + \tfrac{8}{3},\ x_4 = t\}$$

c. In Example 7.2 we arrived at the solution set

$$S = \{\mathbf{X} \mid x_1 = t,\ x_2 = -2t - 1,\ x_3 = -7t - 2,\ x_4 = -6t - 4\}$$

Show that this set is the same as the set S in part *b*. *Hint:* The parameters are not the same! If the parameter in Example 7.2 is called s, show that the parameter in part *b* is $t = -6s - 4$ and argue that this shows the sets to be identical.

4. Replace $\mathbf{C} = (0, -5, 1)$ in Prob. 3 by $\mathbf{0} = (0,0,0)$, in which case we are dealing with the "homogeneous" system associated with the original system.

a. Show that the solution set of the homogeneous system is

$$S_0 = \{\mathbf{X} \mid x_1 = t,\ x_2 = -2t,\ x_3 = -7t,\ x_4 = -6t\}$$

b. Explain why S_0 is a subspace of 4-space and find a basis. What is dim S_0?

c. Comparing S_0 with S in Prob. 3*c*, conclude that the "general solution" of the nonhomogeneous system is $\mathbf{X} = \mathbf{X}_h + \mathbf{X}_p$, where $\mathbf{X}_h = (t, -2t, -7t, -6t)$ is the general solution of the homogeneous system and $\mathbf{X}_p = (0, -1, -2, -4)$ is a particular solution of the nonhomogeneous system. (This is an application of Prob. 34, Sec. 3.2; see the next problem for a fuller explanation.)

★ **5.** Let $A\mathbf{X} = \mathbf{C}$ be an $m \times n$ linear system and define the map $f\colon \mathfrak{F}^n \to \mathfrak{F}^m$ by $f(\mathbf{X}) = A\mathbf{X}$. (This is the "standard map" associated with A; see Sec. 6.3.)

a. Explain why the solution set of the homogeneous system $A\mathbf{X} = \mathbf{0}$ is ker f, hence a subspace of $\mathfrak{F}^n$.

b. Explain why the existence of a solution of the system $A\mathbf{X} = \mathbf{C}$ is equivalent to the statement $\mathbf{C} \in$ rng f, and why the solution set of this system is the set of vectors in $\mathfrak{F}^n$ which f maps into $\mathbf{C}$. Is this a subspace?

c. Assuming that the system $A\mathbf{X} = \mathbf{C}$ is consistent (that is, a solution $\mathbf{X}_p$ exists), use Prob. 34, Sec. 3.2, to conclude that

$$\mathbf{X} \text{ is a solution} \iff \mathbf{X} = \mathbf{X}_h + \mathbf{X}_p$$

where $\mathbf{X}_h$ is a solution of $A\mathbf{X} = \mathbf{0}$.

Note that since the solution set of $A\mathbf{X} = \mathbf{0}$ is a subspace of $\mathfrak{F}^n$, it has a basis (unless the only solution is $\mathbf{0}$, in which case the system

$AX = \mathbf{C}$ has at most one solution). Hence the general solution of $A\mathbf{X} = \mathbf{0}$ may be written as a linear combination of solutions constituting a basis. To find the general solution of $A\mathbf{X} = \mathbf{C}$, we need only tack on $\mathbf{X}_p$.

6. Find the general solution of the system

$$\begin{aligned} x_1 + 2x_2 + x_3 + x_4 &= 6 \\ 2x_1 + 4x_2 + x_3 - 2x_4 &= 1 \\ 3x_1 + 6x_2 + x_3 &= 4 \end{aligned}$$

as follows.

a. Use row operations to reduce the augmented matrix of the system to the form

$$\begin{bmatrix} 1 & * & 0 & 0 & * \\ 0 & 0 & 1 & 0 & * \\ 0 & 0 & 0 & 1 & * \end{bmatrix}$$

b. Replacing the last column in part *a* by $\mathbf{0}$, read off the solution space of the homogeneous system. Name a basis for this space and write down the general solution of the homogeneous system in terms of this basis.

c. Name a particular solution of the original system and write down its general solution.

7.2 Row Echelon Form

Consider the augmented matrix of the system

$$\begin{aligned} 2x_1 - 4x_2 + 3x_3 + x_4 + 4x_5 &= 5 \\ 3x_1 - 6x_2 + x_3 \phantom{{}+x_4} + 8x_5 &= 2 \\ x_1 - 2x_2 + 3x_3 + x_4 + x_5 &= 5 \\ -x_1 + 2x_2 + 2x_3 + x_4 - 4x_5 &= 3 \end{aligned}$$

namely

$$[A|\mathbf{C}] = \begin{bmatrix} 2 & -4 & 3 & 1 & 4 & 5 \\ 3 & -6 & 1 & 0 & 8 & 2 \\ 1 & -2 & 3 & 1 & 1 & 5 \\ -1 & 2 & 2 & 1 & -4 & 3 \end{bmatrix}$$

Two people working on this independently would be unlikely to reduce it to the same form unless they agreed ahead of time on a specific goal. Suppose that we try to change the first column to $\mathbf{E}_1$, the second to $\mathbf{E}_2$, and so on (having in mind the identity matrix); whenever this is not possible, let us agree to skip the column in question and go on to the next, stopping with the last column of A. (Thus $\mathbf{C}$ is just carried along, changing with each row operation, but not affecting our purpose.)

In this example, we could change the first column to $\mathbf{E}_1$ in several ways; perhaps the easiest is to start with the operation (1) $\leftrightarrow$ (3). This yields

$$\begin{bmatrix} 1 & -2 & 3 & 1 & 1 & 5 \\ 3 & -6 & 1 & 0 & 8 & 2 \\ 2 & -4 & 3 & 1 & 4 & 5 \\ -1 & 2 & 2 & 1 & -4 & 3 \end{bmatrix}$$

Now apply the operations

$$(2) \to (2) - 3(1) \qquad (3) \to (3) - 2(1) \qquad \text{and} \qquad (4) \to (4) + (1)$$

simultaneously:

$$\begin{bmatrix} 1 & -2 & 3 & 1 & 1 & 5 \\ 0 & 0 & -8 & -3 & 5 & -13 \\ 0 & 0 & -3 & -1 & 2 & -5 \\ 0 & 0 & 5 & 2 & -3 & 8 \end{bmatrix}$$

It is impossible to change the second column to $\mathbf{E}_2$ (without ruining the first), so we go on to the third. Again there are several ways to proceed. For example, we could apply the operation (2) $\to -\frac{1}{8}$(2) to obtain 1 in the 2,3 position. To avoid fractions, however, we'll apply (2) $\to$ (2) $-$ 3(3):

$$\begin{bmatrix} 1 & -2 & 3 & 1 & 1 & 5 \\ 0 & 0 & 1 & 0 & -1 & 2 \\ 0 & 0 & -3 & -1 & 2 & -5 \\ 0 & 0 & 5 & 2 & -3 & 8 \end{bmatrix}$$

The simultaneous operations

$$(1) \to (1) - 3(2) \qquad (3) \to (3) + 3(2) \qquad \text{and} \qquad (4) \to (4) - 5(2)$$

reduce this to

$$\begin{bmatrix} 1 & -2 & 0 & 1 & 4 & -1 \\ 0 & 0 & 1 & 0 & -1 & 2 \\ 0 & 0 & 0 & -1 & -1 & 1 \\ 0 & 0 & 0 & 2 & 2 & -2 \end{bmatrix}$$

which fixes up the third column to be $\mathbf{E}_2$. The fourth column may be changed to $\mathbf{E}_3$ by

$$(1) \to (1) + (3) \qquad (4) \to (4) + 2(3) \qquad \text{and} \qquad (3) \to -(3):$$

$$\begin{bmatrix} 1 & -2 & 0 & 0 & 3 & 0 \\ 0 & 0 & 1 & 0 & -1 & 2 \\ 0 & 0 & 0 & 1 & 1 & -1 \\ 0 & 0 & 0 & 0 & 0 & 0 \end{bmatrix}$$

Since the fifth column cannot be changed to $\mathbf{E}_4$ without ruining the earlier work, we are done. The original system has been transformed to the equivalent system

$$\begin{aligned} x_1 - 2x_2 + 3x_5 &= 0 \\ x_3 - x_5 &= 2 \\ x_4 + x_5 &= -1 \end{aligned}$$

Note the last row of the final augmented matrix! The corresponding equation is

$$0x_1 + 0x_2 + 0x_3 + 0x_4 + 0x_5 = 0$$

which is of course true but tells us nothing about the unknowns. On the other hand, if the last entry of this row were nonzero, we would conclude that the system is inconsistent; the solution set is then empty. (A linear system is called "consistent" if it has a solution.)

It is now an easy matter to write down the solution set of the system. The "leading entry" of each nonzero row (the first nonzero entry) is 1; the corresponding unknowns are x_1, x_3, and x_4. These may be expressed in terms of the parameters $x_2 = t_1$ and $x_5 = t_2$ by "solving" the last system:

$$\begin{aligned} x_1 &= 2t_1 - 3t_2 \\ x_3 &= t_2 + 2 \\ x_4 &= -t_2 - 1 \end{aligned}$$

The solution set is therefore

$$S = \{\mathbf{X} \mid x_1 = 2t_1 - 3t_2,\ x_2 = t_1,\ x_3 = t_2 + 2,\ x_4 = -t_2 - 1,\ x_5 = t_2\}$$

Solutions are obtained by assigning values to the pair of parameters (t_1,t_2); for example, the pair (1,0) yields the solution $\mathbf{X} = (2,1,2,-1,0)$.

The original coefficient matrix in this example was

$$A = \begin{bmatrix} 2 & -4 & 3 & 1 & 4 \\ 3 & -6 & 1 & 0 & 8 \\ 1 & -2 & 3 & 1 & 1 \\ -1 & 2 & 2 & 1 & -4 \end{bmatrix}$$

and the last coefficient matrix was

$$B = \begin{bmatrix} 1 & -2 & 0 & 0 & 3 \\ 0 & 0 & 1 & 0 & -1 \\ 0 & 0 & 0 & 1 & 1 \\ 0 & 0 & 0 & 0 & 0 \end{bmatrix}$$

We call B the "row echelon form" of A (also the "Hermite normal form"). Its properties (in the general $m \times n$ case) may be summarized as follows.

ROW ECHELON FORM
(of a nonzero $m \times n$ matrix)

1. The zero rows (if any) come after the nonzero rows, that is, at the bottom of the matrix.
2. The leading entries of the nonzero rows move from left to right down the matrix.
3. The columns in which the leading entries of the nonzero rows occur are $\mathbf{E}_1, \ldots, \mathbf{E}_r$, where r is the number of nonzero rows and $\{\mathbf{E}_1, \ldots, \mathbf{E}_m\}$ is the standard basis for m-space.

It can be proved that every nonzero $m \times n$ matrix can be reduced to row echelon form by elementary row operations and that this form is unique. (Thus two people applying different operations to A will nevertheless get the same B.) But we are not interested in such theoretical questions now.

In the system $A\mathbf{X} = \mathbf{C}$ with which we began this section we had $\mathbf{C} = (5,2,5,3)$; in the process of reducing A to row echelon form $\mathbf{C}$ got changed to $\mathbf{D} = (0,2,-1,0)$. The whole thing may be summarized by writing

$$[A|\mathbf{C}] \to [B|\mathbf{D}]$$

to indicate that the original system $A\mathbf{X} = \mathbf{C}$ has been transformed to the equivalent system $B\mathbf{X} = \mathbf{D}$. The arrow between the augmented matrices is intended to represent the sequence of elementary row operations by which A is changed to B.

HOMOGENEOUS

A linear system $A\mathbf{X} = \mathbf{C}$ is called *homogeneous* if $\mathbf{C} = \mathbf{0}$.

EXAMPLE 7.4

The homogeneous system associated with the system at the beginning of this section is

$$\begin{aligned}
2x_1 - 4x_2 + 3x_3 + x_4 + 4x_5 &= 0\\
3x_1 - 6x_2 + x_3 \qquad\quad + 8x_5 &= 0\\
x_1 - 2x_2 + 3x_3 + x_4 + x_5 &= 0\\
-x_1 + 2x_2 + 2x_3 + x_4 - 4x_5 &= 0
\end{aligned}$$

To solve this system, it is clearly legitimate to work with the coefficient matrix alone, since the sequence of row operations we used before transforms the augmented matrix $[A|\mathbf{0}]$ into $[B|\mathbf{0}]$. The extra column is excess baggage. You can see that the solution set is

$$S_0 = \{\mathbf{X} | x_1 = 2t_1 - 3t_2, x_2 = t_1, x_3 = t_2, x_4 = -t_2, x_5 = t_2\}$$

But this is a subspace of $\mathfrak{F}^5$. (See Probs. 4 and 5 in Sec. 7.1). We can easily name a basis for it by taking the pair of parameters (t_1,t_2) to be (1,0) and (0,1) in turn. This guarantees linearly independent answers (why?), namely

$$\mathbf{U}_1 = (2,1,0,0,0) \qquad \text{and} \qquad \mathbf{U}_2 = (-3,0,1,-1,1)$$

Since every solution of $A\mathbf{X} = \mathbf{0}$ can be written in the form

$$\mathbf{X}_h = t_1\mathbf{U}_1 + t_2\mathbf{U}_2 = (2t_1 - 3t_2,\ t_1,\ t_2,\ -t_2,\ t_2)$$

(the so-called "general solution"), we conclude that $\{\mathbf{U}_1,\mathbf{U}_2\}$ is a basis for S_0 and dim $S_0 = 2$.

Moreover, the general solution of the original system $A\mathbf{X} = \mathbf{C}$ can be written in terms of this general solution and the "particular solution" $\mathbf{X}_p = (0,0,2,-1,0)$ obtained from

$$S = \{\mathbf{X} | x_1 = 2t_1 - 3t_2, x_2 = t_1, x_3 = t_2 + 2, x_4 = -t_2 - 1, x_5 = t_2\}$$

by taking $t_1 = t_2 = 0$.† For according to Prob. 34, Sec. 3.2, every solution of $A\mathbf{X} = \mathbf{C}$ can be written in the form $\mathbf{X} = \mathbf{X}_h + \mathbf{X}_p$ for some $\mathbf{X}_h \in S_0$ and every sum of this form is a solution of $A\mathbf{X} = \mathbf{C}$. Hence the general solution is

$$\begin{aligned} \mathbf{X} = \mathbf{X}_h + \mathbf{X}_p &= t_1\mathbf{U}_1 + t_2\mathbf{U}_2 + \mathbf{X}_p \\ &= (2t_1 - 3t_2,\ t_1,\ t_2,\ -t_2,\ t_2) + (0,0,2,-1,0) \\ &= (2t_1 - 3t_2,\ t_1,\ t_2 + 2,\ -t_2 - 1,\ t_2) \end{aligned}$$

Of course we knew this as soon as we found S. Our purpose is not to gild the lily but to display the form of the general solution, and its relation to the general solution of the homogeneous system.

EXAMPLE 7.5

Suppose that we have reduced the coefficient matrix of a 5×7 homogeneous system $A\mathbf{X} = \mathbf{0}$ to the row echelon form

$$B = \begin{bmatrix} 0 & 1 & -1 & 0 & 0 & 2 & 5 \\ 0 & 0 & 0 & 1 & 0 & 3 & -3 \\ 0 & 0 & 0 & 0 & 1 & -2 & 2 \\ 0 & 0 & 0 & 0 & 0 & 0 & 0 \\ 0 & 0 & 0 & 0 & 0 & 0 & 0 \end{bmatrix}$$

(by appropriate row operations on A). The number of nonzero rows in this case is $r = 3$; their leading entries occur in the columns corresponding to the

† Any other choice of (t_1,t_2) would do as well; this is merely the simplest.

unknowns x_2, x_4, x_5. Hence we expect to solve for these unknowns in terms of the four parameters

$$x_1 = t_1 \qquad x_3 = t_2 \qquad x_6 = t_3 \qquad x_7 = t_4$$

The solution set is the subspace of $\mathfrak{F}^7$ defined by

$$S_0 = \{\mathbf{X} \mid x_1 = t_1,\ x_2 = t_2 - 2t_3 - 5t_4,\ x_3 = t_2,\ x_4 = -3t_3 + 3t_4,\ x_5 = 2t_3 - 2t_4,\ x_6 = t_3,\ x_7 = t_4\}$$

(Confirm this!) Four linearly independent solutions, obtained by letting the 4-tuple (t_1,t_2,t_3,t_4) take on the values (1,0,0,0), (0,1,0,0), (0,0,1,0), (0,0,0,1) in turn, are

$$\begin{aligned}\mathbf{U}_1 &= (1,0,0,0,0,0,0)\\ \mathbf{U}_2 &= (0,1,1,0,0,0,0)\\ \mathbf{U}_3 &= (0,-2,0,-3,2,1,0)\\ \mathbf{U}_4 &= (0,-5,0,3,-2,0,1)\end{aligned}$$

and the general solution is

$$\begin{aligned}\mathbf{X}_h &= t_1\mathbf{U}_1 + t_2\mathbf{U}_2 + t_3\mathbf{U}_3 + t_4\mathbf{U}_4\\ &= (t_1,\ t_2 - 2t_3 - 5t_4,\ t_2,\ -3t_3 + 3t_4,\ 2t_3 - 2t_4,\ t_3,\ t_4)\end{aligned}$$

Note that $\{\mathbf{U}_1,\mathbf{U}_2,\mathbf{U}_3,\mathbf{U}_4\}$ is a basis for S_0 and that

$$\dim S_0 = 7 - 3 = 4$$

the difference between the number of unknowns and the number of nonzero rows of B. This is always the case, as we'll prove in Sec. 7.6.

EXAMPLE 7.6

If the original system in Example 7.5 were nonhomogeneous, say $A\mathbf{X} = \mathbf{C}$, the row operations reducing A to B would change the augmented matrix $[A \mid \mathbf{C}]$ to $[B \mid \mathbf{D}]$. This can then be used to determine whether the system is consistent, and if so, what its solution set is. For example, if $\mathbf{D}$ turns out to be $(3,1,0,-1,0)$, then

$$[A \mid \mathbf{C}] \rightarrow [B \mid \mathbf{D}] = \begin{bmatrix} 0 & 1 & -1 & 0 & 0 & 2 & 5 & 3\\ 0 & 0 & 0 & 1 & 0 & 3 & -3 & 1\\ 0 & 0 & 0 & 0 & 1 & -2 & 2 & 0\\ 0 & 0 & 0 & 0 & 0 & 0 & 0 & -1\\ 0 & 0 & 0 & 0 & 0 & 0 & 0 & 0\end{bmatrix}$$

which clearly represents an inconsistent system. (The fourth equation reads $0x_1 + \cdots + 0x_7 = -1$.) Thus, in general, $\mathbf{D}$ cannot have any nonzero entries beyond the rth position, where r is the number of nonzero rows in B, or the system is inconsistent. On the other hand, if

$$[A|\mathbf{C}] \to [B|\mathbf{D}] = \begin{bmatrix} 0 & 1 & -1 & 0 & 0 & 2 & 5 & 1 \\ 0 & 0 & 0 & 1 & 0 & 3 & -3 & -1 \\ 0 & 0 & 0 & 0 & 1 & -2 & 2 & 5 \\ 0 & 0 & 0 & 0 & 0 & 0 & 0 & 0 \\ 0 & 0 & 0 & 0 & 0 & 0 & 0 & 0 \end{bmatrix}$$

the system $A\mathbf{X} = \mathbf{C}$ is consistent, with solution set

$$S = \{\mathbf{X} | x_1 = t_1, x_2 = t_2 - 2t_3 - 5t_4 + 1, x_3 = t_2,$$
$$x_4 = -3t_3 + 3t_4 - 1, x_5 = 2t_3 - 2t_4 + 5, x_6 = t_3, x_7 = t_4\}$$

A convenient particular solution is $\mathbf{X}_p = (0,1,0,-1,5,0,0)$, obtained by taking $t_1 = t_2 = t_3 = t_4 = 0$; the general solution is

$$\begin{aligned} \mathbf{X} &= \mathbf{X}_h + \mathbf{X}_p = t_1\mathbf{U}_1 + t_2\mathbf{U}_2 + t_3\mathbf{U}_3 + t_4\mathbf{U}_4 + \mathbf{X}_p \\ &= (t_1, t_2 - 2t_3 - 5t_4 + 1, t_2, -3t_3 + 3t_4 - 1, 2t_3 - 2t_4 + 5, t_3, t_4) \end{aligned}$$

These examples should give you the idea. In succeeding sections we'll formally state and prove the theorems needed to justify the methods, but for practical purposes you already know enough to solve any linear systems you will encounter.

Problems

7. In each of the following, let $[A|\mathbf{C}]$ be the augmented matrix of the system. Use row operations to reduce it to $[B|\mathbf{D}]$, where B is the row echelon form of A. Then (1) find a basis (if possible) for the solution space of $A\mathbf{X} = \mathbf{0}$, (2) find a particular solution of $A\mathbf{X} = \mathbf{C}$, and (3) write down the general solution of $A\mathbf{X} = \mathbf{C}$.

a.
$$\begin{aligned} x_1 + 6x_2 + 3x_3 + 4x_4 &= 1 \\ x_1 + 2x_2 + x_3 + x_4 &= 1 \\ -x_1 + 2x_2 + x_3 + 2x_4 &= -1 \end{aligned}$$

b.
$$\begin{aligned} 2x - 3y + z &= 2 \\ y - 2z &= -1 \\ 4x + 3y + 2z &= 5 \end{aligned}$$

c.
$$\begin{aligned} 2x_1 - 2x_2 + x_3 + 3x_4 - x_5 &= 0 \\ -x_1 + x_2 + x_3 - 6x_4 + 2x_5 &= 0 \\ x_1 - x_2 + 3x_4 - x_5 &= 0 \\ 5x_1 - 5x_2 + x_3 + 12x_4 - 4x_5 &= 0 \end{aligned}$$

d.
$$\begin{aligned} 5x + y - z &= 0 \\ x - 3y + z &= -3 \end{aligned}$$

e.
$$\begin{aligned} 2x - 3y &= 1 \\ x + y &= 8 \\ 3x - 5y &= 0 \\ 2x - y &= 7 \end{aligned}$$

8. Reduce the augmented matrix $[A|\mathbf{C}]$ of the following system to $[B|\mathbf{D}]$, where B is the row echelon form of A, and use the result to show that the system is inconsistent.

$$\begin{aligned} 4x - y + z &= -1 \\ x - y - 3z &= 2 \\ 2x + 5y - z &= 0 \\ x + y + z &= 0 \end{aligned}$$

★ **9.** Let

$$A = \begin{bmatrix} 2 & -1 & 1 \\ 1 & 1 & 2 \\ 0 & 1 & -3 \end{bmatrix}$$

The inverse of A (if it exists) is a matrix

$$A^{-1} = \begin{bmatrix} x_1 & y_1 & z_1 \\ x_2 & y_2 & z_2 \\ x_3 & y_3 & z_3 \end{bmatrix}$$

satisfying $AA^{-1} = I$.

a. Confirm that the columns **X**, **Y**, **Z** of A^{-1} must satisfy the linear systems $A\mathbf{X} = \mathbf{E}_1$, $A\mathbf{Y} = \mathbf{E}_2$, $A\mathbf{Z} = \mathbf{E}_3$, respectively.

b. Use row operations to reduce the matrix $[A|I]$ to the form $[I|P]$ (if possible), thus solving all three systems together. (The symbol $[A|I]$ means the 3×6 matrix consisting of A augmented by I.)

c. Explain why it is reasonable to expect that $P = A^{-1}$. Then use matrix multiplication to confirm that this is correct. (This is an efficient method for finding the inverse of a nonsingular matrix.)

10. Use the method described in Prob. 9 to find the inverse of

$$A = \begin{bmatrix} 1 & -1 & 1 \\ 0 & 3 & 0 \\ 1 & 0 & 2 \end{bmatrix}$$

★ **11.** Try the method described in Prob. 9 on the matrix

$$A = \begin{bmatrix} 1 & -2 & 4 \\ 0 & 5 & -5 \\ 3 & 1 & 5 \end{bmatrix}$$

(In view of its failure, it seems reasonable to argue that A is singular. Soon we'll prove that this is always the case; that is, the method is infallible. It produces the inverse when it exists; otherwise it demonstrates singularity.)

12. The formula

$$A^{-1} = \frac{1}{\det A}\begin{bmatrix} d & -b \\ -c & a \end{bmatrix}$$

for the inverse of a nonsingular 2×2 matrix

$$A = \begin{bmatrix} a & b \\ c & d \end{bmatrix}$$

has been presented earlier for checking. Assuming that det A is nonzero, *derive* it as follows.

a. If a is nonzero, apply the sequence of row operations

$$(1) \to a^{-1}(1) \qquad (2) \to (2) - c(1) \qquad (2) \to a(\det A)^{-1}(2)$$

and

$$(1) \to (1) - a^{-1}b(2)$$

to $[A|I]$ to arrive at $[I|A^{-1}]$.

b. If $a = 0$, then b and c are nonzero. (Why?) What sequence of row operations reduces $[A|I]$ to $[I|A^{-1}]$ this time?

7.3 Row Operations

This section has some unavoidably sticky notation. If it were of no more use than to justify the methods we have presented for solving linear systems, we would not bother with it. However, the ideas are of considerably more importance than that; we'll need them in many applications to come.

LECTOR *Do you know what the mountain climber said to his partner who was stuck?*

AUCTOR *No, what did the mountain climber say to his partner who was stuck?*

LECTOR *Let's skip this part and start in again higher up.*

ELEMENTARY ROW OPERATIONS

Let A be an $m \times n$ matrix. The following are called *elementary row operations* on A:

1. The interchange of the ith and jth rows of A, where $j \neq i$, denoted by $(i) \leftrightarrow (j)$.
2. The multiplication of the ith row of A by a nonzero scalar c, denoted by $(i) \to c(i)$.

3. The replacement of the ith row of A by the sum of itself and any scalar multiple of the jth row, where $j \neq i$, denoted by $(i) \to (i) + c(j)$.

It is a remarkable fact that each of these operations can be carried out by applying the operation to the $m \times m$ identity matrix instead, and then "premultiplying" A by the result.

EXAMPLE 7.7

Suppose that

$$A = \begin{bmatrix} 2 & -1 & 5 \\ 1 & 3 & 0 \end{bmatrix}$$

Interchanging the rows of A produces the matrix

$$B = \begin{bmatrix} 1 & 3 & 0 \\ 2 & -1 & 5 \end{bmatrix}$$

If we interchange the rows of

$$I = \begin{bmatrix} 1 & 0 \\ 0 & 1 \end{bmatrix}$$

instead, obtaining the matrix

$$E = \begin{bmatrix} 0 & 1 \\ 1 & 0 \end{bmatrix}$$

then $EA = B$, as you can check. Similarly, the operation $(1) \to (1) - 3(2)$ on A yields the matrix

$$B = \begin{bmatrix} -1 & -10 & 5 \\ 1 & 3 & 0 \end{bmatrix}$$

This same operation applied to I gives

$$E = \begin{bmatrix} 1 & -3 \\ 0 & 1 \end{bmatrix}$$

and again $EA = B$. (Confirm this!)

ELEMENTARY MATRIX

A matrix obtained from I by an elementary row operation is called an *elementary matrix,* denoted by

$$[i,j] \qquad [c(i)] \qquad [(i) + c(j)]$$

depending on which operation is used.

EXAMPLE 7.8

If I is 3×3, the interchange of its first and third rows yields the elementary matrix

$$[1,3] = \begin{bmatrix} 0 & 0 & 1 \\ 0 & 1 & 0 \\ 1 & 0 & 0 \end{bmatrix}$$

The operation $(3) \rightarrow 2(3)$ on I results in

$$[2(3)] = \begin{bmatrix} 1 & 0 & 0 \\ 0 & 1 & 0 \\ 0 & 0 & 2 \end{bmatrix}$$

while $(2) \rightarrow (2) + 5(1)$ gives

$$[(2) + 5(1)] = \begin{bmatrix} 1 & 0 & 0 \\ 5 & 1 & 0 \\ 0 & 0 & 1 \end{bmatrix}$$

Obviously the meaning of this notation depends on the context! For example, $[1,3]$ and $[2(3)]$ would not even occur in the 2×2 case, while

$$[(2) + 5(1)] = \begin{bmatrix} 1 & 0 \\ 5 & 1 \end{bmatrix}$$

In Example 7.7, we suggested that premultiplication of an $m \times n$ matrix A by an elementary $(m \times m)$ matrix accomplishes the corresponding row operation on A. To see why this is true, we need a preliminary result concerning the rows of a matrix product. There is a similar result for columns, so we'll state them together; to avoid confusion, we introduce some notation.

ROW AND COLUMN NOTATION

The ith row of an $m \times n$ matrix $A = [a_{ij}]$ is denoted by a_{i*}, the jth column by a_{*j}.

The idea of the asterisk in a_{i*} (in place of the column subscript) is to indicate that we are not talking about an individual entry a_{ij}, but a whole row of entries, namely the n-tuple

$$(a_{i1}, a_{i2}, \ldots, a_{in})$$

or (in some contexts) the $1 \times n$ matrix

$$[a_{i1} \quad a_{i2} \quad \cdots \quad a_{in}]$$

Similarly, the jth column a_{*j} is regarded as the m-tuple

$$(a_{1j}, a_{2j}, \ldots, a_{mj})$$

or the $m \times 1$ matrix

$$\begin{bmatrix} a_{1j} \\ a_{2j} \\ \vdots \\ a_{mj} \end{bmatrix}$$

Some writers use $A_{(i)}$ and $A^{(j)}$ to stand for the ith row and jth column of A, but notation is not uniform.

THEOREM 7.1

Let $C = AB$, where A and B are $p \times q$ and $q \times r$, respectively. Then

a. For each $i = 1, \ldots, p$,

$$c_{i*} = \sum_{k=1}^{q} a_{ik} b_{k*}$$

that is, the ith row of a matrix product is a linear combination of the rows of its second factor, the coefficients being the entries of the ith row of its first factor.

b. For each $j = 1, \ldots, r$,

$$c_{*j} = \sum_{k=1}^{q} a_{*k} b_{kj}$$

that is, the jth column of a matrix product is a linear combination of the columns of its first factor, the coefficients being the entries of the jth column of its second factor.†

† The notation $a_{*k}b_{kj}$, in which a vector is *postmultiplied* by a scalar, has not been used before. However, we define it to mean the same as the vector premultiplied by the scalar (the usual order in which scalar multiplication is written). For example, $(3, -1, 1)2$ means $2(3, -1, 1)$. The reason for the unusual order is given in the remark following the proof.

Proof

To show that

$$c_{i*} = \sum_{k=1}^{q} a_{ik} b_{k*}$$

we compare the jth entries of the two sides. On the left-hand side this is c_{ij}, while on the right-hand side it is

$$\sum_{k=1}^{q} a_{ik} b_{kj}$$

(Why?) By the definition of matrix multiplication (Sec. 5.3) these are the same for each $i = 1, \ldots, p$ and $j = 1, \ldots, r$. We leave the formula in part b for the problems. ▮

Both these formulas are easy to remember (and distinguish) because of the formula

$$c_{ij} = \sum_{k=1}^{q} a_{ik} b_{kj}$$

One merely replaces j by $*$ or i by $*$, depending on whether the formula for rows or columns is wanted.

EXAMPLE 7.9

Let

$$A = \begin{bmatrix} 1 & -2 & 2 \\ 0 & 5 & 3 \end{bmatrix} \quad \text{and} \quad B = \begin{bmatrix} 2 & -1 & 4 \\ 1 & 3 & 1 \\ 1 & 0 & -2 \end{bmatrix}$$

Their product (in the order given) is

$$C = \begin{bmatrix} 2 & -7 & -2 \\ 8 & 15 & -1 \end{bmatrix}$$

What Theorem 7.1a says in this case is

$$\begin{aligned} c_{1*} &= a_{11}b_{1*} + a_{12}b_{2*} + a_{13}b_{3*} \\ c_{2*} &= a_{21}b_{1*} + a_{22}b_{2*} + a_{23}b_{3*} \end{aligned}$$

that is,

$$\begin{aligned} (2,-7,-2) &= 1(2,-1,4) - 2(1,3,1) + 2(1,0,-2) \\ (8,15,-1) &= 0(2,-1,4) + 5(1,3,1) + 3(1,0,-2) \end{aligned}$$

As you can see, these equations are correct; you might also check out part *b* of the theorem.

THEOREM 7.2

Let A be an $m \times n$ matrix and suppose that B is obtained from A by an elementary row operation. Then $B = EA$, where E is the elementary matrix obtained from I_m by the same row operation.

Proof

Suppose that the ith and jth rows of A are interchanged ($j \neq i$), and let $E = [e_{ij}]$ be the elementary matrix obtained from I_m by interchanging *its* ith and jth rows. To show that $B = EA$, consider the pth row of EA, which by Theorem 7.1 is

$$\sum_{k=1}^{m} e_{pk} a_{k*}$$

If p is neither i nor j, then $e_{pk} = \delta_{pk}$, because the pth row of E in this case is the pth row of I_m. Hence

$$\sum_{k=1}^{m} e_{pk} a_{k*} = \sum_{k=1}^{m} \delta_{pk} a_{k*} = a_{p*}$$

which means that the pth row of EA is the pth row of A. In other words, EA and A are identical except for the ith and jth rows. On the other hand, if $p = i$, we have

$$\sum_{k=1}^{m} e_{pk} a_{k*} = \sum_{k=1}^{m} e_{ik} a_{k*} = \sum_{k=1}^{m} \delta_{jk} a_{k*} = a_{j*}$$

because the ith row of E is the jth row of I_m. Hence the ith row of EA is the jth row of A. Similarly, the jth row of EA is the ith row of A. Therefore $EA = B$.

Next, suppose that the ith row of A is multiplied by the nonzero scalar c. Then if $E = [e_{ij}]$ is the elementary matrix obtained from I_m by multiplying *its* ith row by c, we have (for $p \neq i$)

$$\sum_{k=1}^{m} e_{pk} a_{k*} = \sum_{k=1}^{m} \delta_{pk} a_{k*} = a_{p*}$$

while

$$\sum_{k=1}^{m} e_{ik} a_{k*} = \sum_{k=1}^{m} c\delta_{ik} a_{k*} = ca_{i*}$$

Hence EA and A are identical except that the ith row of EA is c times the ith row of A. Again we conclude that $EA = B$.

We leave the third part of the proof for the problems. ∎

THEOREM 7.3

Elementary matrices are nonsingular, with inverses given by

$$[i,j]^{-1} = [i,j]$$
$$[c(i)]^{-1} = [c^{-1}(i)]$$
$$[(i) + c(j)]^{-1} = [(i) - c(j)]$$

Proof

Premultiplication of a matrix by $[i,j]$ interchanges its ith and jth rows (Theorem 7.2). Hence $[i,j][i,j]$ must be $[i,j]$ with its ith and jth rows interchanged. Therefore

$$[i,j][i,j] = I$$

that is,

$$[i,j]^{-1} = [i,j]$$

Similarly, $[c^{-1}(i)][c(i)] = I$ because premultiplication of $[c(i)]$ by $[c^{-1}(i)]$ multiplies its ith row by c^{-1}. Hence $[c(i)]^{-1} = [c^{-1}(i)]$.

We leave it to you to show that $[(i) + c(j)][(i) - c(j)] = I$. ∎

EXAMPLE 7.10

If I is 2×2, then

$$[1,2] = \begin{bmatrix} 0 & 1 \\ 1 & 0 \end{bmatrix} \qquad [3(1)] = \begin{bmatrix} 3 & 0 \\ 0 & 1 \end{bmatrix} \qquad [(2) + 7(1)] = \begin{bmatrix} 1 & 0 \\ 7 & 1 \end{bmatrix}$$

Using Prob. 12 in Sec. 7.2 (or direct multiplication), you can check that the corresponding inverses are

$$[1,2]^{-1} = \begin{bmatrix} 0 & 1 \\ 1 & 0 \end{bmatrix} = [1,2]$$

$$[3(1)]^{-1} = \begin{bmatrix} \frac{1}{3} & 0 \\ 0 & 1 \end{bmatrix} = [3^{-1}(1)]$$

$$[(2) + 7(1)]^{-1} = \begin{bmatrix} 1 & 0 \\ -7 & 1 \end{bmatrix} = [(2) - 7(1)]$$

THEOREM 7.4

If B is obtained from A by a sequence of elementary row operations, then $B = PA$, where P is nonsingular.

Proof

Let $E_1, \ldots, E_k$ be the elementary matrices corresponding to the row operations performed on A to get B. By repeated application of Theorem 7.2 we have $B = E_k \cdots E_1 A$, where each E_i is nonsingular by Theorem 7.3. Let $P = E_k \cdots E_1$, which is nonsingular by Corollary 5.14a.†

THEOREM 7.5

Let $A\mathbf{X} = \mathbf{C}$ be a system of m linear equations in n unknowns. If B is obtained from A by a sequence of elementary row operations (so that $B = PA$, where P is nonsingular), then the system $B\mathbf{X} = P\mathbf{C}$ is equivalent to the system $A\mathbf{X} = \mathbf{C}$. Moreover, the same sequence of row operations applied to the augmented matrix $[A|\mathbf{C}]$ yields the matrix $[B|P\mathbf{C}]$.

Proof

The systems $A\mathbf{X} = \mathbf{C}$ and $B\mathbf{X} = P\mathbf{C}$ are equivalent because

$$A\mathbf{X} = \mathbf{C} \quad \Leftrightarrow \quad P(A\mathbf{X}) = P\mathbf{C} \quad \Leftrightarrow \quad (PA)\mathbf{X} = P\mathbf{C} \quad \Leftrightarrow \quad B\mathbf{X} = P\mathbf{C}$$

(Note that the reverse implication $\Leftarrow$ is correct because P is nonsingular.)

To prove that $[A|\mathbf{C}]$ is transformed into $[B|P\mathbf{C}]$ by the same sequence of row operations that changes A into B, we need only show that $P[A|\mathbf{C}] = [PA|P\mathbf{C}]$. (Why?) We do this by comparing their i,j entries. If $j \leq n$, the i,j entry of $P[A|\mathbf{C}]$ is $\Sigma_{k=1}^{m} p_{ik}a_{kj}$, while the i, $n + 1$ entry is $\Sigma_{k=1}^{m} p_{ik}c_k$, where $\mathbf{C} = (c_1, \ldots, c_m)$. You can see that these are the same as the corresponding entries of $[PA|P\mathbf{C}]$. ∎

EXAMPLE 7.11

Given the system

$$\begin{aligned} 2x + 3y - 8z &= -1 \\ x + y - 3z &= 1 \end{aligned}$$

the sequence of row operations

$$(1) \leftrightarrow (2) \qquad (2) \rightarrow (2) - 2(1) \qquad (1) \rightarrow (1) - (2)$$

reduces the coefficient matrix

$$A = \begin{bmatrix} 2 & 3 & -8 \\ 1 & 1 & -3 \end{bmatrix}$$

to its row echelon form

$$B = \begin{bmatrix} 1 & 0 & -1 \\ 0 & 1 & -2 \end{bmatrix}$$

† Note that P is not unique, but depends on the sequence of row operations used to get from A to B.

The matrix P in this case is

$$P = E_3E_2E_1 = \begin{bmatrix} -1 & 3 \\ 1 & -2 \end{bmatrix}$$

where E_1, E_2, E_3 are the elementary matrices obtained from I by the same operations. *Note that it is unnecessary to compute these separately and then multiply them together to get P!* Since $P = (E_3E_2E_1)I$, we may regard P as I transformed by the same sequence of row operations that changes A to B. Hence we need only write

$$\begin{bmatrix} 1 & 0 \\ 0 & 1 \end{bmatrix} \to \begin{bmatrix} 0 & 1 \\ 1 & 0 \end{bmatrix} \to \begin{bmatrix} 0 & 1 \\ 1 & -2 \end{bmatrix} \to \begin{bmatrix} -1 & 3 \\ 1 & -2 \end{bmatrix} = P$$

What Theorem 7.5 says is that the system $B\mathbf{X} = P\mathbf{C}$, namely

$$\begin{bmatrix} 1 & 0 & -1 \\ 0 & 1 & -2 \end{bmatrix}(x,y,z) = \begin{bmatrix} -1 & 3 \\ 1 & -2 \end{bmatrix}(-1,1)$$

or

$$\begin{aligned} x - z &= 4 \\ y - 2z &= -3 \end{aligned}$$

is equivalent to the original system, and that the sequence

$$\begin{bmatrix} 2 & 3 & -8 & -1 \\ 1 & 1 & -3 & 1 \end{bmatrix} \to \begin{bmatrix} 1 & 1 & -3 & 1 \\ 2 & 3 & -8 & -1 \end{bmatrix} \to$$

$$\begin{bmatrix} 1 & 1 & -3 & 1 \\ 0 & 1 & -2 & -3 \end{bmatrix} \to \begin{bmatrix} 1 & 0 & -1 & 4 \\ 0 & 1 & -2 & -3 \end{bmatrix}$$

changes $[A|\mathbf{C}]$ to $[B|P\mathbf{C}]$. As you can see, this is correct.

For an illustration of Theorem 7.4, you might also check that

$$PA = \begin{bmatrix} -1 & 3 \\ 1 & -2 \end{bmatrix}\begin{bmatrix} 2 & 3 & -8 \\ 1 & 1 & -3 \end{bmatrix} = \begin{bmatrix} 1 & 0 & -1 \\ 0 & 1 & -2 \end{bmatrix} = B$$

Note that we can get back where we started in this example by reversing the operations; applying

$$(1) \to (1) + (2) \qquad (2) \to (2) + 2(1) \qquad (1) \leftrightarrow (2)$$

to B yields A. (Confirm this!) Since $B = PA$, we have $A = P^{-1}B$; a nonsingular matrix which accomplishes the reversal in one step is

$$P^{-1} = (E_3E_2E_1)^{-1} = E_1^{-1}E_2^{-1}E_3^{-1} = \begin{bmatrix} 2 & 3 \\ 1 & 1 \end{bmatrix}$$

Again, one should not find this by multiplying E_1^{-1}, E_2^{-1}, and E_3^{-1}! Since $P^{-1} = (E_1^{-1}E_2^{-1}E_3^{-1})I$, we may obtain P^{-1} by applying the reversed sequence of row operations to I:

$$\begin{bmatrix} 1 & 0 \\ 0 & 1 \end{bmatrix} \rightarrow \begin{bmatrix} 1 & 1 \\ 0 & 1 \end{bmatrix} \rightarrow \begin{bmatrix} 1 & 1 \\ 2 & 3 \end{bmatrix} \rightarrow \begin{bmatrix} 2 & 3 \\ 1 & 1 \end{bmatrix} = P^{-1}$$

Thus the point of Theorem 7.5 is that elementary row operations are legitimate tools in the process of solving a linear system of equations. Each new system obtained from a preceding system by such an operation is equivalent to the original system. The corresponding relation between matrices is called "row equivalence"; more precisely, we have the following.

ROW-EQUIVALENT MATRICES

If B is obtained from A by a sequence of elementary row operations, we say that A is *row-equivalent* to B, and write $A \rightarrow B$.

THEOREM 7.6

Row equivalence on the set $M_{m,n}(\mathfrak{F})$ is an equivalence relation.

Proof

To prove the reflexive law, $A \rightarrow A$, we need only observe that the operation $(1) \rightarrow 1(1)$ transforms A into A. To prove the symmetric law, suppose that $A \rightarrow B$. Then B can be obtained from A by a sequence of elementary row operations; that is, $B = E_k \cdots E_1A$, where E_i is the elementary matrix that accomplishes the ith operation. It follows that

$$A = E_1^{-1} \cdots E_k^{-1}B \qquad \text{(Why?)}$$

Since each E_i^{-1} is itself an elementary matrix (reversing the operation represented by E_i), we conclude that $B \rightarrow A$. The transitive law is left for the problems. ▮

If $A \rightarrow B$, we know from Theorem 7.4 that $B = PA$, where P is nonsingular. Hence row equivalence is a special case of equivalence, as we suggested in Sec. 6.3. (See the comment following Corollary 6.9b.) However, we said then that row equivalence is *defined* by the relation $B = PA$, where P is nonsingular. Our present definition implies such a relation, but we have not yet shown the converse—that if $B = PA$, where P is nonsingular, then B can be obtained from A by a sequence of elementary row operations. To do this, we need a theorem to the effect that *every nonsingular matrix can be expressed as a product of elementary matrices*. Then we can write $P = E_k \cdots E_1$ (where each E_i is elementary) and a sequence of row operations transforming A into B is clear from the expression $B = E_k \cdots E_1A$. However, the required theorem will be delayed until some other things are done.

THEOREM 7.7

If the $n \times n$ matrix A is row-equivalent to I, then A is nonsingular. Moreover, any sequence of row operations which reduces A to I changes the augmented matrix $[A|I]$ to $[I|A^{-1}]$.

Proof

Since $A \to I$, we may write $I = PA$, where P is a nonsingular product of elementary matrices corresponding to row operations transforming A into I. But $I = PA$ implies that A is nonsingular. (Why?) Moreover, Theorem 7.5 says that these same row operations change $[A|\mathbf{E}_j]$ to $[I|P\mathbf{E}_j]$, where $\mathbf{E}_1, \ldots, \mathbf{E}_n$ are the columns of I. Since $P\mathbf{E}_j$ is the jth column of P (why?), this sequence of operations transforms $[A|I]$ into $[I|P] = [I|A^{-1}]$. ▮

Theorem 7.7 justifies the procedure suggested in Prob. 9 of Sec. 7.2:

> To find the inverse of a matrix A, transform $[A|I]$ into $[I|A^{-1}]$ by elementary row operations (if possible).

Note the phrase "if possible." If all attempts fail, that is, if A is not row-equivalent to I, we cannot yet assert that A is singular (as suggested in Prob. 11 of Sec. 7.2). This would be equivalent to the assertion that if A is nonsingular, it is row-equivalent to I. (Note that Theorem 7.7 does not say this!) The assertion is true, but we won't prove it until we show that every nonsingular matrix is a product of elementary matrices.

Problems

13. In the elementary row operation $(i) \to c(i)$ we specified that c is nonzero. Why is this necessary? What happens if $c = 0$ in the row operation $(i) \to (i) + c(j)$?

14. Let

$$A = \begin{bmatrix} 2 & 3 & -2 \\ 1 & 2 & -1 \end{bmatrix}$$

a. Carry out the sequence of row operations

$$(1) \leftrightarrow (2) \qquad (2) \to (2) - 2(1) \qquad (1) \to (1) + 2(2)$$

and

$$(2) \to -(2)$$

to transform A into

$$B = \begin{bmatrix} 1 & 0 & -1 \\ 0 & 1 & 0 \end{bmatrix}$$

b. Write down the elementary matrices E_1, E_2, E_3, E_4 corresponding to these operations and compute E_1A, E_2E_1A, $E_3E_2E_1A$, $E_4E_3E_2E_1A$.

c. Apply the same sequence of operations to I_2 to produce the matrix $P = E_4E_3E_2E_1$. Then compute PA.

d. Write down the matrices E_1^{-1}, E_2^{-1}, E_3^{-1}, E_4^{-1} and compute $E_4^{-1}B$, $E_3^{-1}E_4^{-1}B$, $E_2^{-1}E_3^{-1}E_4^{-1}B$, $E_1^{-1}E_2^{-1}E_3^{-1}E_4^{-1}B$.

e. Reverse the operations in part *a* to change I_2 to the matrix $E_1^{-1}E_2^{-1}E_3^{-1}E_4^{-1}$. Confirm that this is P^{-1} and compute $P^{-1}B$.

15. The sequence of row operations described in Prob. 3, Sec. 7.1, transforms

$$A = \begin{bmatrix} 3 & -2 & 1 & 0 \\ 0 & 1 & -2 & 2 \\ 3 & 1 & 1 & -1 \end{bmatrix}$$

into

$$B = \begin{bmatrix} 1 & 0 & 0 & \tfrac{1}{6} \\ 0 & 1 & 0 & -\tfrac{1}{3} \\ 0 & 0 & 1 & -\tfrac{7}{6} \end{bmatrix}$$

Find a nonsingular matrix P satisfying $PA = B$. *Note:* This is not the only such P. A different sequence of row operations transforming A into B would yield a different P in general.

16. Complete the proof of Theorem 7.1 by showing that if $C = AB$, where A is $p \times q$ and B is $q \times r$, then for each $j = 1, \ldots, r$ the jth column of C is

$$c_{*j} = \sum_{k=1}^{q} a_{*k} b_{kj}$$

17. Suppose that A is $m \times n$ and B is obtained from A by the operation $(i) \to (i) + c(j)$. Prove that $B = EA$, where $E = [(i) + c(j)]$. (This completes the proof of Theorem 7.2.)

18. Finish the proof of Theorem 7.3 by showing that

$$[(i) + c(j)]^{-1} = [(i) - c(j)]$$

19. In the 3×3 case find the following.

a. $[2(3)]^{-1}$

b. $[(1) - 4(2)]^{-1}$

20. Finish the proof of Theorem 7.6 by showing that row equivalence is transitive.

★ **21.** Let E_{ij} be the $m \times n$ matrix with i,j entry 1 and 0s elsewhere. (See the proof of Theorem 4.3.)

a. Prove that $[i,j] = I_m - E_{ii} - E_{jj} + E_{ij} + E_{ji}$. *Hint:* Compare the rows of the matrices on each side of the equation, noting that the ith row of E_{ij} is $\mathbf{E}_j$ while the other rows are zero.

b. Prove that $[c(i)] = I_m + (c - 1)E_{ij}$.

c. Prove that $[(i) + c(j)] = I_m + cE_{ij}$.

★ **22.** Use Prob. 21 (noting that $E_{ij}^T = E_{ji}$) to prove the following.

a. $[i,j]^T = [i,j]$

b. $[c(i)]^T = [c(i)]$

c. $[(i) + c(j)]^T = [(j) + c(i)]$

7.4 Column Operations

Since linear systems are solved by row operations alone, the corresponding theory of column operations is not of any immediate interest. However, we'll need it later and the logical time to develop it is now.

ELEMENTARY COLUMN OPERATIONS

Let A be an $m \times n$ matrix. The following are called *elementary column operations* on A.

1. The interchange of the ith and jth columns of A, where $j \neq i$, denoted by $(i) \leftrightarrow (j)$.
2. The multiplication of the ith column of A by a nonzero scalar c, denoted by $(i) \rightarrow c(i)$.
3. The replacement of the ith column of A by the sum of itself and any scalar multiple of the jth column, where $j \neq i$, denoted by $(i) \rightarrow (i) + c(j)$.

Any matrix obtained from I by an elementary column operation is called an *elementary matrix*.

Two things should be observed in this definition. First, we use the same notation for column operations as for row operations, trusting the context to make the meaning clear. Second, the last sentence of the definition appears to need sharpening, since we agreed in the last section to call an elementary matrix the result of a *row* operation on I. However, we are going to show that the elementary matrices corresponding to column operations are nothing new, but can be expressed in terms of the same symbols introduced in the last section.

THEOREM 7.8

The elementary matrices corresponding to the column operations

$$(i) \leftrightarrow (j) \qquad (i) \to c(i) \qquad (i) \to (i) + c(j)$$

are

$$[i,j] \qquad [c(i)] \qquad [(j) + c(i)]$$

respectively.

Proof

Any column operation on a matrix can be carried out by performing the corresponding row operation on its transpose and then transposing back. Hence the elementary matrix obtained from I by the column operation $(i) \leftrightarrow (j)$ is

$$\begin{aligned}([i,j]I^T)^T &= (I^T)^T[i,j]^T && \text{(Prob. 36, Sec. 5.3)}\\ &= I[i,j] && \text{(Prob. 22, Sec. 7.3)}\\ &= [i,j]\end{aligned}$$

Similarly, the elementary matrix obtained from I by the column operation $(i) \to c(i)$, where c is nonzero, is

$$([c(i)]I^T)^T = I[c(i)]^T = I[c(i)] = [c(i)]$$

We leave the third part of the proof for the problems. ∎

EXAMPLE 7.12

The column operation $(2) \to (2) + 5(3)$ on I_3 yields

$$\begin{bmatrix} 1 & 0 & 0 \\ 0 & 1 & 0 \\ 0 & 5 & 1 \end{bmatrix}$$

According to Theorem 7.8 this should be the same as $[(3) + 5(2)]$, the elementary matrix obtained from I_3 by the row operation $(3) \to (3) + 5(2)$. As you can check, this is correct.

THEOREM 7.9

Let A be an $m \times n$ matrix and suppose that B is obtained from A by an elementary column operation. Then $B = AE$, where E is the elementary matrix obtained from I_n by the same column operation.

Proof

We leave the first two column operations for the problems. To do the third, suppose that B is obtained from A by the column operation $(i) \to (i) + c(j)$. Then

$$B = ([(i) + c(j)]A^T)^T = A[(i) + c(j)]^T = A[(j) + c(i)]$$

But according to Theorem 7.8 the matrix $E = [(j) + c(i)]$ corresponds to the column operation $(i) \to (i) + c(j)$. ▮

Thus column operations are accomplished by *postmultiplication* (rather than premultiplication) by elementary matrices.

EXAMPLE 7.13

The column operation $(3) \to (3) - (1)$ on

$$A = \begin{bmatrix} 3 & -1 & 5 \\ 1 & 0 & 3 \end{bmatrix}$$

yields

$$B = \begin{bmatrix} 3 & -1 & 2 \\ 1 & 0 & 2 \end{bmatrix}$$

The same column operation on I_3 results in the elementary matrix

$$E = \begin{bmatrix} 1 & 0 & -1 \\ 0 & 1 & 0 \\ 0 & 0 & 1 \end{bmatrix}$$

As you can check, $B = AE$.

THEOREM 7.10

If B is obtained from A by a sequence of elementary column operations, then $B = AQ$, where Q is nonsingular.

Proof

This is left for the problems.

COLUMN-EQUIVALENT MATRICES

If B is obtained from A by a sequence of elementary column operations, we say that A is *column-equivalent* to B, and write $A \to B$. (This is the same symbol used for row equivalence in the last section. There won't be any occasion for confusion.)

THEOREM 7.11

Column equivalence on the set $M_{m,n}(\mathfrak{F})$ is an equivalence relation.

Proof

This is left for the problems.

As in the last section, note that although

$$A \to B \quad \Rightarrow \quad B = AQ \qquad \text{where } Q \text{ is nonsingular}$$

(and hence column equivalence is a special case of equivalence), we cannot as yet assert the converse. When we do, column equivalence (as defined above) will be the same as the relation $B = AQ$, where Q is nonsingular.

Problems

23. Finish the proof of Theorem 7.8 by showing that the elementary matrix corresponding to the column operation $(i) \to (i) + c(j)$ is $[(j) + c(i)]$.

24. In each of the following, carry out the given column operation on I_3 and confirm that the result is the given elementary matrix.
a. $(2) \leftrightarrow (3)$; $[2,3]$
b. $(3) \to 4(3)$; $[4(3)]$
c. $(2) \to (2) - (1)$; $[(1) - (2)]$

25. Suppose that A is an $m \times n$ matrix and that B is obtained from A by the given column operation. Prove that $B = AE$, where E is the elementary matrix obtained from I_n by the same column operation. (This completes the proof of Theorem 7.9.)
a. $(i) \leftrightarrow (j)$
b. $(i) \to c(i)$, $c \neq 0$

26. Carry out the column operation $(1) \to (1) + 3(2)$ on

$$A = \begin{bmatrix} 3 & -1 \\ 0 & 1 \\ 5 & -2 \end{bmatrix}$$

to get B, and on I_2 to get E, and confirm that $B = AE$.

27. Prove Theorem 7.10: If B is obtained from A by a sequence of elementary column operations, then $B = AQ$, where Q is nonsingular. Is Q unique?

28. Prove the following properties of column equivalence on $M_{m,n}(\mathfrak{F})$.
a. reflexive: $A \to A$
b. symmetric: $A \to B \quad \Rightarrow \quad B \to A$
c. transitive: $A \to B$ and $B \to C \quad \Rightarrow \quad A \to C$

★ **29.** Show that if B is obtained from A by a sequence of elementary row and/or column operations, then A is equivalent to B. (The converse is also true, but the proof awaits the theorem that every nonsingular matrix is a product of elementary matrices.)

30. Let

$$A = \begin{bmatrix} 1 & 1 \\ 2 & 0 \\ 1 & -1 \end{bmatrix} \quad \text{and} \quad B = \begin{bmatrix} 1 & 0 \\ 0 & 1 \\ 0 & 0 \end{bmatrix}$$

a. Use elementary row operations to transform A into B, and name a nonsingular matrix P satisfying $B = PA$.

b. Try changing A into B by using elementary column operations.

c. Prove that no nonsingular matrix Q exists satisfying $B = AQ$. Why does this show the attempt in part *b* to be hopeless?

d. Let f: $\mathcal{R}^2 \to \mathcal{R}^3$ be the linear map whose matrix relative to the standard bases is A, namely the standard map $f(\mathbf{X}) = A\mathbf{X}$. In Prob. 1 of Sec. 6.1 we asked you to show that if α is the standard basis for $\mathcal{R}^2$, then there is a basis γ for $\mathcal{R}^3$ such that $B = [f]_{\alpha^\gamma}$. Explain why this follows from part *a*. *Hint:* Use Theorem 6.3.

e. We also asked you to show (in Prob. 1, Sec. 6.1) that if β is the standard basis for $\mathcal{R}^3$, then there is no basis δ for $\mathcal{R}^2$ such that $B = [f]_{\delta^\beta}$. Why does this follow from part *c*?

7.5 Rank

Sylvester's law of nullity (Sec. 3.2) says that if S is finite-dimensional and f: $S \to T$ is linear, then

$$\dim (\ker f) + \dim (\text{rng } f) = \dim S$$

We mentioned in Sec. 3.2 that ker f is sometimes called the "null space" of f and its dimension the "nullity" of f, while dim (rng f) is called the "rank" of f. Now we are going to use similar terminology in connection with matrices.

ROW RANK AND COLUMN RANK

Let $A \in M_{m,n}(\mathfrak{F})$. The subspace of $\mathfrak{F}^n$ spanned by the rows of A (regarded as n-tuples) is called the *row space*, and its dimension the *row rank*, of A, denoted by $R(A)$ and $r(A)$ respectively. Similarly, the subspace of $\mathfrak{F}^m$ spanned by the columns of A (regarded as m-tuples) is the *column space*, and its dimension the *column rank*, of A, denoted by $C(A)$ and $c(A)$ respectively.

EXAMPLE 7.14

If

$$A = \begin{bmatrix} 2 & -3 & 1 & -1 \\ 4 & -6 & 2 & -2 \\ 1 & 0 & -2 & 5 \end{bmatrix}$$

then the row space is

$$R(A) = \langle (2,-3,1,-1), (4,-6,2,-2), (1,0,-2,5) \rangle$$

a subspace of $\mathfrak{F}^4$. (See Theorem 2.15.) Since the second row is twice the first, the rows are linearly dependent. But the first and third are linearly independent, so the row space is 2-dimensional and

$$r(A) = \dim R(A) = 2$$

The column space, on the other hand, is

$$C(A) = \langle (2,4,1), (-3,-6,0), (1,2,-2), (-1,-2,5) \rangle$$

a subspace of $\mathfrak{F}^3$. These vectors must be linearly dependent in $\mathfrak{F}^3$ (why?); as a matter of fact, the first two are linear combinations of the last two, namely

$$(2,4,1) = {}^{11}\!/_{3}(1,2,-2) + {}^{5}\!/_{3}(-1,-2,5)$$

and

$$(-3,-6,0) = -5(1,2,-2) - 2(-1,-2,5)$$

Since the last two are linearly independent, we have

$$c(A) = \dim C(A) = 2$$

The equality of $r(A)$ and $c(A)$ in this example is no accident; it is true in general. However, the proof is not simple.

THEOREM 7.12

The row rank and column rank of a matrix are equal.

Proof

This will be deferred at present. However, see Prob. 33 for a sketch of an elegant proof based on Sylvester's law and the idea of orthogonal projection. For an argument not involving linear maps (less elegant), see Philip Gillett, *Linear Mathematics,* Boston: Prindle, Weber, and Schmidt, 1970.

RANK OF A MATRIX

The common value of the row rank and column rank of a matrix A is called the *rank* of A, and is denoted by $r(A)$.

THEOREM 7.13

Multiplication of a matrix by another matrix cannot increase its rank.

Proof

Let $C = AB$ be any matrix product. Since the rows of C are linear combinations of the rows of B (Theorem 7.1), the row space of C is a subspace of the row space of B. Hence $r(C) \leq r(B)$, which means that the rank of a matrix product cannot exceed the rank of the second factor. We leave it to you to show (similarly, or by using transposes) that the rank of a product cannot exceed the rank of the first factor either. ▮

THEOREM 7.14

Multiplication of a matrix by a nonsingular matrix preserves rank.

Proof

Let A be $m \times n$ and suppose that P is a nonsingular $m \times m$ matrix. By Theorem 7.13, $r(PA) \leq r(A)$ and (again by Theorem 7.13) $r[P^{-1}(PA)] \leq r(PA)$. But $P^{-1}(PA) = A$, so this second inequality reads $r(A) \leq r(PA)$. Hence $r(A) = r(PA)$, which shows that premultiplication of A by a nonsingular matrix does not change its rank.

To show that postmultiplication of A by a nonsingular matrix also preserves its rank, we may simply imitate the above argument. It is more economical, however, to write

$$r(AQ) = r[(AQ)^T] = r(Q^TA^T)$$

where Q is a nonsingular $n \times n$ matrix, and use what we just proved about premultiplication, for Q^T is nonsingular and hence

$$r(Q^TA^T) = r(A^T) = r(A) \quad ▮$$

COROLLARY 7.14a

The rank of a matrix is not changed by elementary row or column operations.

COROLLARY 7.14b

Equivalent matrices have the same rank.

COROLLARY 7.14c

Matrix representations of a given linear map are all of the same rank.

See Example 6.3 for a preview of this last statement. We said then that we would define the rank of a matrix in such a way as to make all the matrix representations of a given map have the same rank as the map. Corollary 7.14c says that they all have the same rank; what remains is to prove that this is the same as the rank of the map.

THEOREM 7.15

The rank of a linear map and the rank of any one of its matrix representations are the same.

Proof

Let $f\colon S \to T$ be the map and let $r = \dim(\text{rng } f)$ be its rank. We need only look at one matrix representation of f; if its rank turns out to be r, we are done.

Assume that $r > 0$ (otherwise f is the zero map and the theorem is trivial). The idea of the proof is to choose bases for S and T in such a way as to make the matrix of f simple and its rank obvious. Sylvester's law of nullity says that

$$\dim(\ker f) = \dim S - \dim(\text{rng } f) = n - r$$

where $n = \dim S$. Let $\{u_{r+1}, \ldots, u_n\}$ be a basis for $\ker f$ and build it up to a basis

$$\alpha = \{u_1, \ldots, u_n\}$$

for S.† Then the elements $v_1 = f(u_1), \ldots, v_r = f(u_r)$ constitute a basis for rng f. (See the proof of Sylvester's law in Sec. 3.2.) Build this up (if necessary) to a basis

$$\beta = \{v_1, \ldots, v_m\}$$

for T, where $m = \dim T$. The matrix of f relative to α and β is found by writing $f(u_j)$ in terms of the v_i:

$$f(u_j) = v_j = \sum_{i=1}^{m} \delta_{ij} v_i \qquad j = 1, \ldots, r$$

$$f(u_j) = 0 = \sum_{i=1}^{m} 0 v_i \qquad j = r + 1, \ldots, n \quad \text{(Why?)}$$

Hence $[f]_{\alpha^\beta} = A = [a_{ij}]$, where

$$a_{ij} = \delta_{ij} \quad j = 1, \ldots, r \qquad \text{and} \qquad a_{ij} = 0 \quad j = r + 1, \ldots, n$$

In other words, the ith row of A is

$$(\delta_{i1}, \ldots, \delta_{ir}, 0, \ldots, 0)$$

where there are $n - r$ zeros following δ_{ir}.

† If $\ker f$ is the zero space, let α be any basis for S. In this case, $r = n$ and the elements $u_{r+1}, \ldots, u_n$ do not appear; their occurrence in the rest of the proof is vacuous.

This means that A has the $r \times r$ identity matrix as a submatrix in its upper left corner and zeros elsewhere.† The row space of A is clearly r-dimensional (the first r rows of A span it and are linearly independent). Hence the rank of A is r. ▮

NULLITY OF A MATRIX

The *null space* of a matrix $A \in M_{m,n}(\mathfrak{F})$ is the solution set of the linear homogeneous system $A\mathbf{X} = \mathbf{0}$; its dimension is the *nullity of A*. (This is the kernel of the standard map $f: \mathfrak{F}^n \to \mathfrak{F}^m$ defined by $f(\mathbf{X}) = A\mathbf{X}$. In Sec. 3.2 we called it the null space of f and referred to its dimension as the nullity of f.)

THEOREM 7.16

The nullity plus the rank of a matrix is equal to the number of columns.

Proof

Let $A \in M_{m,n}(\mathfrak{F})$ and define the standard map $f: \mathfrak{F}^n \to \mathfrak{F}^m$ by $f(\mathbf{X}) = A\mathbf{X}$. Since $A = [f]$, Theorem 7.15 says that dim (rng f) $= r(A)$. As observed in the above definition, dim (ker f) $=$ nullity of A. The theorem follows from Sylvester's law of nullity. ▮

Problems

31. Find a basis for the row space of the matrix

$$A = \begin{bmatrix} 1 & -1 & 3 & 0 & -2 \\ -2 & 2 & -6 & 0 & 4 \\ 0 & 2 & 5 & -1 & 0 \\ 2 & -6 & -4 & 2 & -4 \end{bmatrix}$$

What is the row rank of A?

†This is the "canonical form" of the matrix of f (see the remarks preceding Example 6.3); an example is

$$\begin{bmatrix} 1 & 0 & 0 & 0 \\ 0 & 1 & 0 & 0 \\ 0 & 0 & 0 & 0 \end{bmatrix}$$

in which the rank is 2.

32. Find a basis for the column space of the matrix in Prob. 31. What is the column rank of A?

★ **33.** Let $A \in M_{m,n}(\mathfrak{F})$ and define the standard map $f\colon \mathfrak{F}^n \to \mathfrak{F}^m$ by $f(\mathbf{X}) = A\mathbf{X}$.

a. Explain why ker f consists of those vectors in $\mathfrak{F}^n$ whose dot product with each of the rows of A is 0.

b. If U is a subspace of $\mathfrak{F}^n$ and we define

$$U^{\perp} = \{\mathbf{Y} \in \mathfrak{F}^n \mid \mathbf{X} \cdot \mathbf{Y} = 0 \text{ for every } \mathbf{X} \in U\}$$

why does it follow from part *a* that ker $f = [R(A)]^{\perp}$? *Note:* In $\mathcal{R}^2$ and $\mathcal{R}^3$ this is the "orthogonal complement" as described in Prob. 53, Sec. 2.6, and Example 3.11.

c. Prove that rng $f = C(A)$. *Hint:* Use Theorem 7.1.

d. Why does it follow from parts *b* and *c* that dim $[R(A)]^{\perp}$ + dim $C(A) = n$?

e. It can be proved that if U is a subspace of $\mathfrak{F}^n$, then dim U + dim $U^{\perp} = n$. Assuming this to be true, conclude from part *d* that dim $R(A) =$ dim $C(A)$; that is, the row rank and column rank of A are equal. (This constitutes a proof of Theorem 7.12; we can adopt it as soon as the formula dim U + dim $U^{\perp} = n$ is established.)

34. Why can't the rank of an $m \times n$ matrix exceed the smaller of m and n?

35. Why is the rank of a matrix the same as the rank of its transpose?

36. Confirm that multiplication of the nonsingular matrix

$$A = \begin{bmatrix} 3 & -1 \\ 1 & 2 \end{bmatrix}$$

by the singular matrix

$$B = \begin{bmatrix} 3 & 6 \\ 2 & 4 \end{bmatrix}$$

decreases its rank. (This is true in general; see Prob. 55 in Sec. 7.6.)

37. Complete the proof of Theorem 7.13 by showing that the rank of a matrix product cannot exceed the rank of its first factor in the following ways.

a. Imitate the first part of the proof.

b. Use transposes and appeal to the part of the theorem already proved.

38. Do the second part of the proof of Theorem 7.14 (postmultiplication by nonsingular matrices preserves rank) without appealing to transposes.

39. Prove that the rank of a matrix is not changed by elementary row or column operations (Corollary 7.14a).

40. Prove that equivalent matrices have the same rank (Corollary 7.14b). Do similar matrices always have the same rank?

41. Prove that matrix representations of a given linear map are all of the same rank (Corollary 7.14c).

42. Prove that row-equivalent matrices have the same row space. Do they also have the same column space?

43. Define $f: \mathcal{R}^3 \to \mathcal{R}^2$ by $f(x,y,z) = (x - 2y + z, 2y)$.

a. Without using Theorem 7.15, confirm that the rank of f is $r = 2$.

b. Show that $\mathbf{U}_3 = (1,0,-1)$ constitutes a basis for ker f.

c. Check that $\mathbf{U}_1 = \mathbf{E}_1$, $\mathbf{U}_2 = \mathbf{E}_2$, and $\mathbf{U}_3$ constitute a basis α for $\mathcal{R}^3$.

d. Find $\mathbf{V}_1 = f(\mathbf{U}_1)$ and $\mathbf{V}_2 = f(\mathbf{U}_2)$ and show that they constitute a basis β for rng $f = \mathcal{R}^2$.

e. At this point in the proof of Theorem 7.15 we claimed that $A = [f]_{\alpha^\beta}$ has the $r \times r$ identity matrix in its upper left corner and zeros elsewhere. Confirm that this is the case by finding A.

44. Without referring to linear maps, prove that the null space of a matrix $A \in M_{m,n}(\mathfrak{F})$ is a subspace of $\mathfrak{F}^n$.

45. Regarding

$$A = \begin{bmatrix} 2 & 0 & 4 \\ -1 & 1 & -2 \end{bmatrix}$$

as an element of $M_{2,3}(\mathcal{R})$, find its null space and describe it geometrically.

Note that A could just as well be regarded as an element of $M_{2,3}(\mathcal{C})$. Then its null space is in $\mathcal{C}^3$ and can no longer be visualized. For example, $(-2i,0,i)$ is in the null space. This emphasizes again the importance of specifying what the base field is.

46. Suppose that A is a matrix representation of the linear map $f: S \to T$. If $r(A) = \dim S$, why is f one-to-one?

47. Find the rank and nullity of the nonsingular matrix

$$A = \begin{bmatrix} 2 & -1 & 1 \\ 1 & 1 & 2 \\ 0 & 1 & -3 \end{bmatrix}$$

(See Prob. 9, Sec. 7.2.) Why does the system

$$\begin{aligned} 2x - y + z &= 0 \\ x + y + 2z &= 0 \\ y - 3z &= 0 \end{aligned}$$

have only the trivial solution $(x,y,z) = (0,0,0)$?

48. How can you be sure (without solving it) that the system

$$\begin{aligned} x - 2y + z &= 0 \\ 3x + 5y + z &= 0 \end{aligned}$$

has nontrivial solutions? Prove in general that a linear homogeneous system with more unknowns than equations must have a nontrivial solution.

49. Explain why the system $A\mathbf{X} = \mathbf{0}$, where A is the matrix in Prob. 31, has three linearly independent solutions.

★ **50.** Let A be an $n \times n$ matrix and suppose we consider the following list of statements.

(1) A is row-equivalent to I.

(2) A is a product of elementary matrices.

(3) A is nonsingular.

(4) The system $A\mathbf{X} = \mathbf{0}$ has only the trivial (zero) solution.

(5) The rows of A are linearly independent.

(6) The columns of A are linearly independent.

(7) The rank of A is n.

a. Show that (1) $\Rightarrow$ (2), that (2) $\Rightarrow$ (3), and that (3) $\Rightarrow$ (4).

b. Show that (4) $\Rightarrow$ (7) and that (7) $\Rightarrow$ (5), thus establishing that (4) $\Rightarrow$ (5).

c. Use Theorem 5.18 and Prob. 7, Sec. 6.1, to explain why (4) $\Rightarrow$ (3) and (3) $\Rightarrow$ (5), thus providing a different argument for the implication (4) $\Rightarrow$ (5).

d. Show that (5) $\Rightarrow$ (7) and (7) $\Rightarrow$ (6), thus establishing the fact that (5) $\Rightarrow$ (6).

e. Use Probs. 6 and 7, Sec. 6.1, to explain why (5) $\Rightarrow$ (3) and (3) $\Rightarrow$ (6), thus providing a different argument for the implication (5) $\Rightarrow$ (6).

f. Show that (6) $\Rightarrow$ (7). (At this point we have established that each of the first six statements implies the next.)

g. Explain why (7) $\Rightarrow$ (3), thus establishing the equivalence of (3), (4), (5), (6), (7).

h. Show that (2) $\Rightarrow$ (1), thus establishing the equivalence of statements (1) and (2).

Note that at this point we could establish the equivalence of all seven statements by showing that any one of (3), (4), (5), (6), (7) implies (1) or (2). In the next section we shall prove that (7) $\Rightarrow$ (1), but there are obviously other possibilities.

7.6 Theory of Linear Systems

In Secs. 7.1 and 7.2 we had an intuitive look at the problems involved in solving linear systems. Now we can justify the methods (which you should review before reading on).

THEOREM 7.17

Let $A\mathbf{X} = \mathbf{0}$ be a linear homogenous system of m equations in n unknowns. Its solution set is a subspace of $\mathfrak{F}^n$ of dimension $n - r$, where r is the rank of A.

Proof

This is left for the problems.

COROLLARY 7.17a

If the rank of A is n, the system $A\mathbf{X} = \mathbf{0}$ has only the trivial solution $\mathbf{X} = \mathbf{0}$.

Note that in this case $m \geq n$ (why?); that is, there are at least as many equations as unknowns.

COROLLARY 7.17b

If the rank of A is less than n, the system $A\mathbf{X} = \mathbf{0}$ has $n - r$ linearly independent solutions. The general solution is a linear combination of these.

In particular, a homogeneous system in which there are more unknowns than equations always has a nontrivial solution.

THEOREM 7.18

Let $A\mathbf{X} = \mathbf{C}$ be a linear system of m equations in n unknowns. If $\mathbf{X}_p$ is a solution, the solution set consists of all sums of the form $\mathbf{X} = \mathbf{X}_h + \mathbf{X}_p$, where $\mathbf{X}_h$ is a solution of the homogeneous system $A\mathbf{X} = \mathbf{0}$.

Proof

Let $\mathbf{X}$ be any solution of $A\mathbf{X} = \mathbf{C}$. Then $\mathbf{X}_h = \mathbf{X} - \mathbf{X}_p$ is a solution of the homogeneous system. (Why?) Hence $\mathbf{X} = \mathbf{X}_h + \mathbf{X}_p$, a sum of the form advertised. Conversely, every such sum is a solution of $A\mathbf{X} = \mathbf{C}$, since

$$A(\mathbf{X}_h + \mathbf{X}_p) = A\mathbf{X}_h + A\mathbf{X}_p = \mathbf{0} + \mathbf{C} = \mathbf{C} \blacksquare$$

LECTOR *Why not define $f\colon \mathfrak{F}^n \to \mathfrak{F}^m$ by $f(\mathbf{X}) = A\mathbf{X}$ and apply Prob. 34, Sec. 3.2?*

AUCTOR *I wish I'd said that.*

LECTOR *You probably will in the second edition.*

COROLLARY 7.18a

The general solution of a consistent linear system $A\mathbf{X} = \mathbf{C}$ is $\mathbf{X} = \mathbf{X}_h + \mathbf{X}_p$, where $\mathbf{X}_h$ is the general solution of the homogeneous system $A\mathbf{X} = \mathbf{0}$ and $\mathbf{X}_p$ is a particular solution of $A\mathbf{X} = \mathbf{C}$.

THEOREM 7.19

The linear system $A\mathbf{X} = \mathbf{C}$ has a unique solution if and only if it is consistent and the rank of A is equal to the number of unknowns.

Proof

Suppose that the system is consistent and that $r(A) = n$, where n is the number of unknowns. Then there is a solution $\mathbf{X}_p$ (by definition of "consistent"). Moreover, every solution has the form $\mathbf{X} = \mathbf{X}_h + \mathbf{X}_p$, where $\mathbf{X}_h$ is a solution of $A\mathbf{X} = \mathbf{0}$. But Corollary 7.17a says that the only solution of $A\mathbf{X} = \mathbf{0}$ is $\mathbf{X}_h = \mathbf{0}$. Hence the only solution of $A\mathbf{X} = \mathbf{C}$ is $\mathbf{X} = \mathbf{X}_p$.

The "only if" part is left for the problems. ∎

In practice one can tell from the augmented matrix $[B|P\mathbf{C}]$ whether the system $A\mathbf{X} = \mathbf{C}$ is consistent (where $B = PA$ is the row echelon form of A and P is a nonsingular product of elementary matrices corresponding to row operations that change A to B). If r is the rank of A (the number of nonzero rows of B), then $P\mathbf{C}$ cannot have any nonzero entries beyond the rth position, or the system is inconsistent. (See Example 7.6.) Nonetheless it is worthwhile to have some theoretical criteria.

THEOREM 7.20

Each of the following statements is equivalent to each of the others.

1. The linear system $A\mathbf{X} = \mathbf{C}$ is consistent.
2. $\mathbf{C}$ is a linear combination of the columns of A.
3. The augmented matrix $[A|\mathbf{C}]$ has the same rank as the coefficient matrix A.
4. $\mathbf{C} \in \operatorname{rng} f$, where $f(\mathbf{X}) = A\mathbf{X}$.

Proof

This is left for the problems.

Since our general method for solving $A\mathbf{X} = \mathbf{C}$ depends on the reduction of A to its row echelon form, it is important to know that this is always possible (and has a unique outcome). We state the theorem without proof, not because the argument is deep, but because it is dull.†

THEOREM 7.21

Every nonzero matrix can be reduced to row echelon form by elementary row operations. Moreover, this form is unique, and its rank is equal to the number of nonzero rows.

THEOREM 7.22

Let A be an $n \times n$ matrix. Then each of the following statements is equivalent to each of the others.

(1) A is row-equivalent to I.

(2) A is a product of elementary matrices.

(3) A is nonsingular.

(4) The system $A\mathbf{X} = \mathbf{0}$ has only the trivial (zero) solution.

(5) The rows of A are linearly independent.

(6) The columns of A are linearly independent.

(7) The rank of A is n.

Proof

The implications (1) $\Rightarrow$ (2) $\Rightarrow$ (3) $\Rightarrow$ (4) $\Rightarrow$ (5) $\Rightarrow$ (6) $\Rightarrow$ (7) are already established in Prob. 50 of Sec. 7.5. To prove that (7) $\Rightarrow$ (1), suppose that the rank of A is n. By Theorem 7.21 we can reduce A to a matrix B in row echelon form by elementary row operations; we need only show that $B = I$. Since $r(B) = r(A) = n$ (why?), all the rows of B must be nonzero. But then every row of B has a leading entry, and these move from left to right down the matrix. Hence the main diagonal consists of leading entries. As a result every column contains a leading entry, which means the columns are the standard basis vectors in $\mathfrak{F}^n$. (See the definition of row echelon form in Sec. 7.2.) Therefore $B = I$. ▮

This is a remarkable theorem, for if each implication of the form $(i) \Rightarrow (j)$ is counted as a separate statement (where i and j are distinct integers from

† Except for the fundamental theorem of algebra (which we'll use later without proof), this is the only undefended theorem in the book. Bertrand Russell once said that it is impossible to be both intelligible and complete, an observation which serves as an excuse for omitting one or two things.

the list 1, . . . , 7), there are 42 theorems! In the next chapter we'll add another item to the list, namely

$$\det A \neq 0$$

at which point we'll have 56 distinct implications.

Problems

51. Prove that the solution set of the $m \times n$ system $A\mathbf{X} = \mathbf{0}$ is a subspace of $\mathfrak{F}^n$ of dimension $n - r$, where r is the rank of A (Theorem 7.17).

52. If the rank of the coefficient matrix of an $m \times n$ homogeneous system is n, why must there be at least n equations?

53. Complete the proof of Theorem 7.19 by showing that if the $m \times n$ system $A\mathbf{X} = \mathbf{C}$ has a unique solution, the rank of A is n.

54. Prove Theorem 7.20.

55. Prove that multiplication of a nonsingular matrix by a singular matrix reduces its rank. (See Prob. 36 in Sec. 7.5.)

★ **56.** Suppose that the $n \times n$ matrix A is reduced to its row echelon form B by elementary row operations, and $B \neq I$.

a. Why does it follow that *no* sequence of row operations will reduce A to I?

b. Prove that A is singular. (This justifies the procedure suggested in Prob. 11, Sec. 7.2.)

★ **57.** In Sec. 7.3 we proved that if A is row-equivalent to B, there is a nonsingular matrix P such that $B = PA$. Prove the converse.

★ **58.** In Sec. 7.4 we proved that if A is column-equivalent to B, there is a nonsingular matrix Q such that $B = AQ$. Prove the converse.

Problems 57 and 58 justify the definitions of row equivalence and column equivalence given in Sec. 6.3 (following Corollary 6.9b). In other words, A is row-equivalent to B if and only if $B = PA$ for some nonsingular P, and column-equivalent to B if and only if $B = AQ$ for some nonsingular Q.

★ **59.** Prove that A and B are equivalent if and only if one can be obtained from the other by elementary row and/or column operations.

60. Show that the matrices

$$A = \begin{bmatrix} 2 & 5 & 3 & 0 \\ -1 & 1 & 2 & -2 \\ 5 & 9 & 4 & 2 \end{bmatrix} \quad \text{and} \quad B = \begin{bmatrix} 1 & 0 & 0 & 0 \\ 0 & 1 & 0 & 0 \\ 0 & 0 & 0 & 0 \end{bmatrix}$$

are equivalent but not row-equivalent. What is the rank of A?

Note the 2×2 identity matrix in the upper left corner of B. This is an example of "canonical form." Every $m \times n$ matrix of rank $r > 0$ is equivalent to a canonical form with the $r \times r$ identity in its upper left corner and zeros elsewhere (the same as the matrix A in the proof of Theorem 7.15). Hence the equivalence classes in $M_{m,n}(\mathfrak{F})$ are $C_0, C_1, \ldots, C_p$, where C_r is the class of all $m \times n$ matrices of rank r and p is the smaller of m and n. (See the discussion of equivalence classes in Sec. 6.1.)

★ **61.** Explain why two $m \times n$ matrices are equivalent if and only if they have the same rank. (Note that the "only if" part of this is already done in Sec. 7.5, Corollary 7.14b.)

Review Quiz

True or false?

1. $r(A) = r(A^T)$
2. If A is $n \times n$ and $[A|I]$ cannot be reduced to the form $[I|P]$ by row operations, then A is singular.
3. If B is obtained from A by a sequence of elementary column operations, the systems $A\mathbf{X} = \mathbf{0}$ and $B\mathbf{X} = \mathbf{0}$ are equivalent.
4. If $A\mathbf{X} = A\mathbf{Y}$, then $\mathbf{X} = \mathbf{Y}$.
5. The leading entries of the nonzero rows of a matrix in row echelon form are 1s.
6. If $B = AQ$, where Q is nonsingular, then A can be transformed into B by elementary column operations.
7. If $B = PA$, then $r(B) = r(A)$.
8. If the linear system $A\mathbf{X} = \mathbf{C}$ is consistent, then $\mathbf{C}$ is an element of the column space of A.
9. If A is a matrix representation of the linear map $f\colon S \to T$ and $r(A) = \dim T$, then f is onto.
10. Each row of AB is a linear combination of the rows of B.

11. If $A\mathbf{X} = \mathbf{C}$ is a linear system of m equations in n unknowns and $r(A) = n$, the system has a unique solution.

12. Homogeneous linear systems are always consistent.

13. The canonical form of

$$\begin{bmatrix} 2 & 1 & -3 \\ 1 & -1 & 0 \end{bmatrix}$$

is

$$\begin{bmatrix} 1 & 0 & 0 \\ 0 & 1 & 0 \end{bmatrix}$$

14. Every square matrix can be written as a product of elementary matrices.

15. If A is row-equivalent to B and B is nonsingular, then A is nonsingular.

16. The nullity plus the rank of

$$\begin{bmatrix} 1 & 5 & 0 \\ 2 & -1 & 1 \end{bmatrix}$$

is 2.

17. Elementary matrices resulting from column operations on I can also be obtained by row operations on I.

18. If the rank of A is r, the solution space of $A\mathbf{X} = \mathbf{0}$ is r-dimensional.

19. The rank of

$$\begin{bmatrix} 2 & 6 \\ 1 & 3 \end{bmatrix}$$

is 2.

20. If E is obtained from I_3 by adding the third row to the second, then E^{-1} is obtained from I_3 by subtracting the third row from the second.

21. If B is obtained from A by a sequence of elementary row operations, then $B = PA$, where P is the matrix obtained from I by the same sequence of operations.

22. If the row echelon form of the $n \times n$ matrix A is I, the system $A\mathbf{X} = \mathbf{C}$ has a unique solution.

23. If A is a 4×3 matrix in row echelon form, with two nonzero rows, the system $A\mathbf{X} = \mathbf{0}$ has only the trivial solution.

24. If $\mathbf{X}$ and $\mathbf{Y}$ are solutions of a given linear system, then $\mathbf{X} - \mathbf{Y}$ is a solution of the associated homogeneous system.

25. Any two matrix representations of a given linear map are row-equivalent.

26. If the $n \times n$ matrix A is singular, then $r(A) < n$.

27. The system

$$\begin{aligned} x - 2y + 5z &= 1 \\ 2x + y + z &= 0 \\ x - y - z &= 2 \end{aligned}$$

has a unique solution.

28. The set of solutions of $A\mathbf{X} = \mathbf{C}$ is a vector space.

29. Two $m \times n$ matrices of equal rank are equivalent.

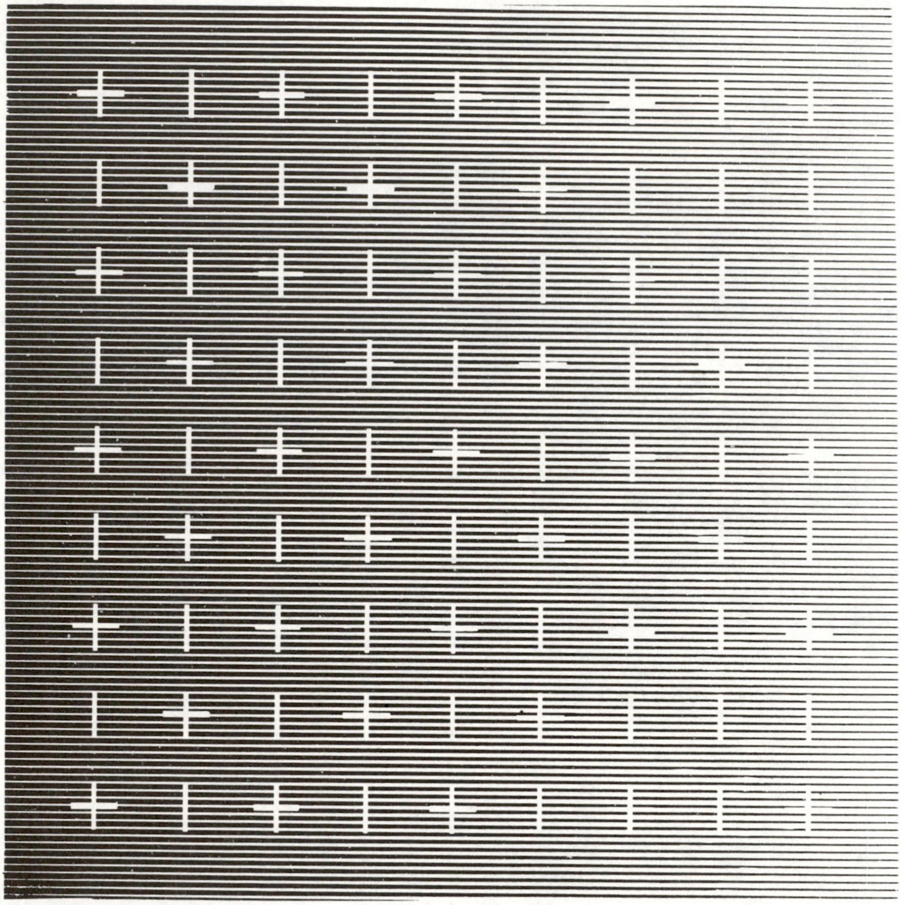

CHAPTER 8

Determinants

The theory of determinants has little to recommend it as a part of linear algebra. Of the several approaches to the subject that have been devised by mathematicians, none is very appealing (although one or two are theoretically elegant), and none has much to do with the subject matter of this book. We have therefore adopted a definition of determinants that is artificial but concise, the idea being to get to the heart of the matter quickly. Don't be surprised if you miss the point for a while! All will become clear after some twenty theorems (eighteen of which come in dual pairs, so it isn't as bad as it sounds).

8.1 Definitions

DETERMINANT

Let $A = [a_{ij}]$ be an $n \times n$ matrix with entries in $\mathfrak{F}$. The *determinant* of A, denoted by $\det A$, is defined recursively as follows.

1. If $n = 1$, in which case $A = [a_{11}]$ has only one entry, then $\det A = a_{11}$.
2. If $n > 1$, then

$$\det A = \sum_{j=1}^{n} (-1)^{i+j} a_{ij} \det A_{ij}$$

where i is any one of the integers $1, \ldots, n$ and A_{ij} is the $(n-1) \times (n-1)$ matrix obtained from A by deleting its ith row and jth column.

This definition is called "recursive" because when $n > 1$ the determinant of A is defined in terms of determinants of $(n-1) \times (n-1)$ matrices, which in turn are given in terms of determinants of $(n-2) \times (n-2)$ matrices, and so on. At the end of this process, the determinant of A is expressed in terms of determinants of 1×1 matrices and these are explicitly defined by condition 1.

EXAMPLE 8.1

If

$$A = \begin{bmatrix} a_{11} & a_{12} \\ a_{21} & a_{22} \end{bmatrix}$$

then

$$\det A = \sum_{j=1}^{2} (-1)^{i+j} a_{ij} \det A_{ij}$$

where i is either 1 or 2. If $i = 1$, we get

$$\det A = a_{11} \det A_{11} - a_{12} \det A_{12} = a_{11}a_{22} - a_{12}a_{21}$$

On the other hand, if $i = 2$, we find that

$$\det A = -a_{21} \det A_{21} + a_{22} \det A_{22} = -a_{21}a_{12} + a_{22}a_{11}$$

The result is the same in either case, so the definition of det A when $n = 2$ is unambiguous.

EXAMPLE 8.2

If

$$A = \begin{bmatrix} a_{11} & a_{12} & a_{13} \\ a_{21} & a_{22} & a_{23} \\ a_{31} & a_{32} & a_{33} \end{bmatrix}$$

then

$$\det A = \sum_{j=1}^{3} (-1)^{i+j} a_{ij} \det A_{ij}$$

where i is 1, 2, or 3. If $i = 1$, we get

$$\begin{aligned} \det A &= a_{11} \det A_{11} - a_{12} \det A_{12} + a_{13} \det A_{13} \\ &= a_{11}(a_{22}a_{33} - a_{23}a_{32}) - a_{12}(a_{21}a_{33} - a_{23}a_{31}) + a_{13}(a_{21}a_{32} - a_{22}a_{31}) \\ &= a_{11}a_{22}a_{33} - a_{11}a_{23}a_{32} - a_{12}a_{21}a_{33} + a_{12}a_{23}a_{31} \\ &\qquad + a_{13}a_{21}a_{32} - a_{13}a_{22}a_{31} \end{aligned}$$

You might check that the same result is obtained when $i = 2$ or $i = 3$; the definition of det A when $n = 3$ is therefore also unambiguous.

These examples raise the question of whether the definition of det A is unambiguous in general. *We are going to assume that it is,* in order to develop the terminology and prove some preliminary theorems. Then we'll show that what we have done is actually independent of this assumption.

EXPANSION BY A ROW

The expression

$$\det A = \sum_{j=1}^{n} (-1)^{i+j} a_{ij} \det A_{ij}$$

is called the *expansion of det A by the ith row of A* (because the coefficients $a_{i1}, \ldots, a_{in}$ are the entries in the ith row of A). The determinant of A_{ij} is called the *minor* of a_{ij}; the expression $(-1)^{i+j} \det A_{ij}$ is called the *cofactor* of a_{ij}.

Note that in Example 8.2 we expanded det A by the first row of A. The expressions in parentheses, namely

$$a_{22}a_{33} - a_{23}a_{32} \qquad a_{21}a_{33} - a_{23}a_{31} \qquad a_{21}a_{32} - a_{22}a_{31}$$

are the minors of a_{11}, a_{12}, and a_{13}, respectively; prefixed by the "signs"

$$(-1)^{1+1} \qquad (-1)^{1+2} \qquad (-1)^{1+3}$$

they become the cofactors of a_{11}, a_{12}, a_{13}.

Thus the expansion of det A by a given row of A may be described as the

sum of the entries of this row multiplied by their cofactors

This should help fix the definition in mind.†

The "sign" $(-1)^{i+j}$ can be quickly determined by mentally superimposing a checkerboard pattern of signs on the matrix, starting with a plus in the upper left corner. The 3×3 case looks like this:

$$\begin{bmatrix} + & - & + \\ - & + & - \\ + & - & + \end{bmatrix}$$

from which one can see that the cofactor of a_{23} (for example) is $-\det A_{23}$.

† Obviously we are talking about determinants of $n \times n$ matrices with $n > 1$. The 1×1 case is trivial; its occurrence in many of the theorems we'll state is vacuous.

EXPANSION BY A COLUMN

Eventually we'll prove that

$$\det A = \sum_{i=1}^{n}(-1)^{i+j}a_{ij} \det A_{ij}$$

where j is any one of the integers $1, \ldots, n$. This is called the *expansion of det A by the* j*th column of A*. Note that it is the same as our original definition, except that the index of summation is now i instead of j. The coefficients $a_{1j}, \ldots, a_{nj}$ are the entries of the jth *column* of A.

EXAMPLE 8.3

Let A be the 3×3 matrix of Example 8.2. The expansion of det A by the second column of A is

$$\begin{aligned}\det A &= \sum_{i=1}^{3}(-1)^{i+2}a_{i2} \det A_{i2}\\ &= -a_{12} \det A_{12} + a_{22} \det A_{22} - a_{32} \det A_{32}\\ &= -a_{12}(a_{21}a_{33} - a_{23}a_{31}) + a_{22}(a_{11}a_{33} - a_{13}a_{31}) - a_{32}(a_{11}a_{23} - a_{13}a_{21})\end{aligned}$$

As you can see, this is the same as the expansion by the first row that we worked out in Example 8.2.

The determinant of a fully written-out matrix is often indicated by simply replacing the brackets with vertical bars. Thus the determinant of

$$A = \begin{bmatrix} 1 & -3 \\ 2 & 1 \end{bmatrix}$$

is

$$\det A = \begin{vmatrix} 1 & -3 \\ 2 & 1 \end{vmatrix} = 7$$

Do not be misled by this notation into thinking that a determinant has rows and columns! *A determinant is a scalar;* the only reason this notation is used is that it refers us back to the matrix from which the scalar is derived. Note that the determinant of

$$\begin{bmatrix} 4 & 5 \\ 1 & 3 \end{bmatrix}$$

is also 7; thus

$$\begin{vmatrix} 1 & -3 \\ 2 & 1 \end{vmatrix} \quad \text{and} \quad \begin{vmatrix} 4 & 5 \\ 1 & 3 \end{vmatrix}$$

are the same determinants derived from different matrices.

EXAMPLE 8.4

The vertical-bar notation sometimes saves writing. Thus we can expand the following determinant by the third column of its parent matrix without explicitly putting down the matrix (or the 2×2 matrices whose determinants are involved in the expansion):

$$\begin{aligned} \begin{vmatrix} 3 & 2 & -1 \\ -1 & 2 & 3 \\ -3 & 1 & 3 \end{vmatrix} &= (-1)\begin{vmatrix} -1 & 2 \\ -3 & 1 \end{vmatrix} - (3)\begin{vmatrix} 3 & 2 \\ -3 & 1 \end{vmatrix} + (3)\begin{vmatrix} 3 & 2 \\ -1 & 2 \end{vmatrix} \\ &= -(-1+6) - 3(3+6) + 3(6+2) \\ &= -8 \end{aligned}$$

Problems

1. We expanded

$$\begin{vmatrix} a_{11} & a_{12} \\ a_{21} & a_{22} \end{vmatrix}$$

by the first and second rows and obtained consistent results. Expand by the first and second *columns* and see what happens.

2. We expanded

$$\begin{vmatrix} a_{11} & a_{12} & a_{13} \\ a_{21} & a_{22} & a_{23} \\ a_{31} & a_{32} & a_{33} \end{vmatrix}$$

by the first row and also by the second column. See if you get the same result expanding by (*a*) the second row; (*b*) the third column.

3. Assuming that det A may be expanded by any row or column, find

$$\begin{vmatrix} 2 & 0 & -1 & 0 \\ 3 & -5 & 1 & 7 \\ 2 & 0 & 3 & -1 \\ 2 & 0 & 2 & 0 \end{vmatrix}$$

in the most economical way.

★ 4. Prove by induction on n that no matter which row or column is used in the expansion, det $I_n = 1$. *Hint:* The $(n-1) \times (n-1)$ matrix obtained from I_n by deleting the ith row and jth column is I_{n-1} when $j = i$. (Why?)

8.2 Expansion by a Row

In the last section we defined

$$\det A = \sum_{j=1}^{n} (-1)^{i+j} a_{ij} \det A_{ij}$$

the expansion of det A by the ith row of A. As yet we haven't determined whether this definition is unambiguous (except in the cases $n = 2$ and $n = 3$), which means that any theorems we state must be cautiously investigated to see how much they depend on our assumption that it is.

> In all the theorems of this section the matrices are understood to be $n \times n$, with entries in $\mathfrak{F}$.

THEOREM 8.1

If two adjacent columns of A are equal, then $\det A = 0$.

Proof

Induction on n. The theorem is vacuous if $n = 1$ and is easily checked if $n = 2$. Hence assume that A is $n \times n$ with $n > 2$ and that the theorem is true for $(n - 1) \times (n - 1)$ matrices. Let a_{*k} and a_{*k+1} be the adjacent columns of A that are equal. Our general assumption in this section (as yet unwarranted) is that if i is any one of the integers $1, \ldots, n$, then

$$\det A = \sum_{j=1}^{n} (-1)^{i+j} a_{ij} \det A_{ij}$$

Since A_{ij} has two equal adjacent columns when j is neither k nor $k + 1$, the induction hypothesis says that $\det A_{ij} = 0$ for each such j. Hence the above sum reduces to

$$\det A = (-1)^{i+k} a_{ik} \det A_{ik} + (-1)^{i+k+1} a_{ik+1} \det A_{ik+1}{}^{\dagger}$$

However, $a_{ik} = a_{ik+1}$ and $\det A_{ik} = \det A_{ik+1}$. (Why?) Therefore $\det A = 0$. ∎

It is important to note that the above argument does not depend on i; that is, we get $\det A = 0$ regardless of what row is chosen for the expansion of det A. Hence *Theorem* 8.1 *is independent of the assumption that det A may be expanded by any row.*

† The notation a_{ik+1} should probably read $a_{i,k+1}$ to distinguish the subscripts properly. But we have been casual about this throughout the book. How do you tell, for example, whether a_{11} means "a sub one-one" or "a sub eleven"? Of course the answer is that the context requires two subscripts, not one. In the present instance the only way to make sense of a_{ik+1} is to read it $a_{i,k+1}$.

THEOREM 8.2

Suppose that A, B, and C are identical except that one column of A is a linear combination of the corresponding columns of B and C. Then $\det A$ is the same linear combination of $\det B$ and $\det C$.†

Proof

Suppose that $a_{*k} = xb_{*k} + yc_{*k}$; we'll prove that

$$\det A = x \det A + y \det C$$

by induction on n. If $n = 1$, the theorem is trivial, so assume that A is $n \times n$ with $n > 1$ and that the theorem is true for $(n-1) \times (n-1)$ matrices. Let i be any one of the integers $1, \ldots, n$ and look at the typical term of the sum

$$\det A = \sum_{j=1}^{n} (-1)^{i+j} a_{ij} \det A_{ij}$$

If $j = k$, this term is

$$(-1)^{i+k} a_{ik} \det A_{ik} = (-1)^{i+k}(xb_{ik} + yc_{ik}) \det A_{ik}$$

because $a_{*k} = xb_{*k} + yc_{*k}$. However, because $\det A_{ik} = \det B_{ik} = \det C_{ik}$ (why?), we find that

$$(-1)^{i+k} a_{ik} \det A_{ik} = (-1)^{i+k}(xb_{ik} \det B_{ik} + yc_{ik} \det C_{ik})$$

On the other hand, if $j \neq k$, we have $a_{ij} = b_{ij} = c_{ij}$. (Why?) Applying the induction hypothesis to the $(n-1) \times (n-1)$ matrix A_{ij} (identical to B_{ij} and C_{ij} except for the column that is the given linear combination of the corresponding columns of B_{ij} and C_{ij}), we have

$$\det A_{ij} = x \det B_{ij} + y \det C_{ij}$$

Thus our typical term in this case is

$$(-1)^{i+j} a_{ij} \det A_{ij} = (-1)^{i+j}(xb_{ij} \det B_{ij} + yc_{ij} \det C_{ij})$$

Looking at both of these cases, we see that the terms of $\det A$ may all be written in the same form, whether $j = k$ or $j \neq k$. Hence

† For example, the first column of

$$A = \begin{bmatrix} -7 & -1 \\ 8 & 3 \end{bmatrix}$$

is $a_{*1} = 3b_{*1} - 5c_{*1}$, where

$$B = \begin{bmatrix} 1 & -1 \\ 6 & 3 \end{bmatrix} \quad \text{and} \quad C = \begin{bmatrix} 2 & -1 \\ 2 & 3 \end{bmatrix}$$

while the second columns of A, B, and C are equal. As you can see, $\det A = 3 \det B - 5 \det C$.

$$\begin{aligned}\det A &= \sum_{j=1}^{n}(-1)^{i+j}a_{ij}\det A_{ij}\\ &= \sum_{j=1}^{n}(-1)^{i+j}(xb_{ij}\det B_{ij} + yc_{ij}\det C_{ij})\\ &= x\sum_{j=1}^{n}(-1)^{i+j}b_{ij}\det B_{ij} + y\sum_{j=1}^{n}(-1)^{i+j}c_{ij}\det C_{ij}\\ &= x\det B + y\det C \ \blacksquare\end{aligned}$$

This argument, like that for Theorem 8.1, goes through regardless of what i is. Expanding det A by one row or another may give different results (this remains to be settled), but *for a given choice of row the linear relation stated in Theorem 8.2 is correct.*

COROLLARY 8.2a

Suppose that A and B are identical except that one column of A is a multiple of the corresponding column of B. Then det A is the same multiple of det B.

EXAMPLE 8.5

Sometimes this corollary is rendered by saying that if the entries of a column of a determinant have a common factor, it may be "factored out" of the determinant. Thus

$$\begin{vmatrix}3 & 2 & -1\\ 5 & 6 & 4\\ 0 & 4 & 4\end{vmatrix} = 2\begin{vmatrix}3 & 1 & -1\\ 5 & 3 & 4\\ 0 & 2 & 4\end{vmatrix}$$

where we removed the factor 2 from the second column. While this sounds simple, it is inexact. As we pointed out in the last section, a determinant is a *number* and can hardly be said to "have columns." Worse yet, matrices (which do have columns) do not behave this way; put brackets in place of the vertical bars and the statement is false. (See the problems.)

Now we are ready for some rapid progress. Keeping in mind that Theorem 8.1 is independent of the row used to expand det A, and that the linear relation in Theorem 8.2 is correct even if different rows give different results, note that the next seven theorems are all based on these two. Hence we don't have much to keep track of; the only unsettled question is whether Theorem 8.2 can be made entirely independent of the row used.

THEOREM 8.3

If B is obtained from A by adding a multiple of one column to an adjacent column, then $\det B = \det A$.[†]

Proof

Suppose that the kth column of A is replaced by the sum of itself and x times the next column. (The argument for the preceding column is the same.) Then B is identical to A except that

$$b_{*k} = a_{*k} + xa_{*k+1}$$

Let C be the matrix which is identical to A except that $c_{*k} = a_{*k+1}$. Then A, B, and C are identical except that

$$a_{*k} = b_{*k} - xc_{*k}$$

By Theorem 8.2 it follows that

$$\det A = \det B - x \det C$$

and since c_{*k} and c_{*k+1} are the same, Theorem 8.1 says that $\det C = 0$. Hence $\det A = \det B$. ▮

COLUMN FORM OF A MATRIX

If $A = [a_{ij}]$ is an $m \times n$ matrix, we sometimes write

$$A = [a_{*1}, \ldots, a_{*n}]$$

This displays A in *column form.*

EXAMPLE 8.6

If

$$A = \begin{bmatrix} 2 & 0 & 3 \\ 1 & -1 & 5 \end{bmatrix}$$

then

$$a_{*1} = \begin{bmatrix} 2 \\ 1 \end{bmatrix} \qquad a_{*2} = \begin{bmatrix} 0 \\ -1 \end{bmatrix} \qquad a_{*3} = \begin{bmatrix} 3 \\ 5 \end{bmatrix}$$

The notation

$$A = [a_{*1}, a_{*2}, a_{*3}]$$

[†] More precisely, a given column is replaced by the sum of itself and a scalar multiple of an adjacent column. This is a column operation of type 3. (See Sec. 7.4.)

merely indicates that we are thinking of A as being made up of these three columns. Although we'll have little occasion to use such notation in the sequel, it is helpful in two or three theorems of this chapter.

THEOREM 8.4

If B is obtained from A by interchanging two adjacent columns, then $\det B = -\det A$.

Proof

Let a_{*k} and a_{*k+1} be the columns that are interchanged. Applying Theorem 8.3 repeatedly, we have

$$\begin{aligned}\det B &= \det [a_{*1}, \ldots, a_{*k+1}, a_{*k}, \ldots, a_{*n}] \\ &= \det [a_{*1}, \ldots, a_{*k+1} + a_{*k}, a_{*k}, \ldots, a_{*n}] \\ &= \det [a_{*1}, \ldots, a_{*k+1} + a_{*k}, -a_{*k+1}, \ldots, a_{*n}] \\ &= \det [a_{*1}, \ldots, a_{*k}, -a_{*k+1}, \ldots, a_{*n}]\end{aligned}$$

The last matrix is identical to A except that its $k + 1$ column is -1 times the $k + 1$ column of A. By Corollary 8.2a its determinant is $-\det A$. Hence $\det B = -\det A$. ∎

THEOREM 8.5

If A has two equal columns, then $\det A = 0$.

Proof

By interchanging adjacent columns (if necessary), we can make the equal columns adjacent. Each such interchange (according to Theorem 8.4) merely introduces a factor of -1. But in the end (according to Theorem 8.1) the determinant is 0. Hence $\det A = 0$. ∎

THEOREM 8.6

If B is obtained from A by adding a multiple of one column to another column, then $\det B = \det A$.

Proof

Suppose that a_{*k} is replaced by $a_{*k} + xa_{*s}$. We assume that $k < s$ (the proof when $k > s$ is the same). Then

$$\begin{aligned}\det B &= \det [a_{*1}, \ldots, a_{*k} + xa_{*s}, \ldots, a_{*s}, \ldots, a_{*n}] \\ &= \det [a_{*1}, \ldots, a_{*k}, \ldots, a_{*s}, \ldots, a_{*n}] \\ &\qquad + x \det [a_{*1}, \ldots, a_{*s}, \ldots, a_{*s}, \ldots, a_{*n}] \\ &= \det A + x \cdot 0 \\ &= \det A \quad ∎\end{aligned}$$

This theorem is the basis for simplifying the evaluation of determinants, for it says that we may apply elementary column operations of type 3 to the matrix A without affecting det A. Appropriate operations of this type will reduce the number of nonzero entries in a given (nonzero) row to one; the expansion of the determinant by this row will then involve only one term.

EXAMPLE 8.7

The determinant

$$\begin{vmatrix} 3 & 2 & -1 \\ -1 & 2 & 3 \\ -3 & 1 & 3 \end{vmatrix}$$

(Example 8.4) may be reduced to

$$\begin{vmatrix} 0 & 0 & -1 \\ 8 & 8 & 3 \\ 6 & 7 & 3 \end{vmatrix}$$

by the column operations $(1) \to (1) + 3(3)$ and $(2) \to (2) + 2(3)$. Expanding this by the first row yields

$$(-1)\begin{vmatrix} 8 & 8 \\ 6 & 7 \end{vmatrix} = -(56 - 48) = -8$$

THEOREM 8.7

If B is obtained from A by interchanging two columns, then det $B = -$det A.

Proof

Use Theorem 8.6 to construct an argument similar to the one presented in the proof of Theorem 8.4.

THEOREM 8.8

If E is an elementary matrix and det E is expanded by any row, then det E is equal to -1, c, or 1 depending on whether E is of type 1 (interchange of columns), type 2 (multiplication of a column by a nonzero scalar c), or type 3 (addition of a multiple of one column to another column).

Proof

Use Prob. 4 in Sec. 8.1, Theorem 8.7, Corollary 8.2a, and Theorem 8.6. ∎

Note that the results in Theorem 8.8 are independent of the row used in the expansion of det E. We get -1, c, or 1 in any case.

THEOREM 8.9

If E is an elementary matrix and the determinants involved are expanded by any row, then for every A, $\det AE = (\det A)(\det E)$.

Proof

By Theorem 8.8 we know that $\det E$ is equal to -1, c, or 1 depending on whether E is of type 1, 2, or 3. But AE may be regarded as the matrix obtained from A by performing the column operation on A corresponding to post-multiplication of A by E. (See Theorem 7.9.) By Theorem 8.7, Corollary 8.2a, or Theorem 8.6, we have

$$\det AE = -\det A,\ c \det A, \text{ or } \det A$$

depending on whether the operation is of type 1, 2, or 3. Hence

$$\det AE = (\det A)(\det E) \ \blacksquare$$

EXAMPLE 8.8

The elementary matrix obtained from I_2 by the column operation $(2) \to (2) + 2(1)$ is

$$E = \begin{bmatrix} 1 & 2 \\ 0 & 1 \end{bmatrix}$$

If

$$A = \begin{bmatrix} 1 & -1 \\ 5 & 2 \end{bmatrix}$$

then

$$AE = \begin{bmatrix} 1 & 1 \\ 5 & 12 \end{bmatrix}$$

As you can see,

$$\det AE = (\det A)(\det E) = 7$$

It turns out that there is no need to confine Theorem 8.9 to products whose second factor is an elementary matrix; it is true in general. (See Theorem 8.22.) *The determinant of a product is the product of the determinants of the factors.*

Problems

5. Show that a 2×2 matrix with equal columns has a zero determinant. (This confirms Theorem 8.1 in the case $n = 2$.)

6. The matrices

$$A = \begin{bmatrix} 1 & 5 & 0 \\ 2 & 16 & 2 \\ -1 & 9 & 0 \end{bmatrix} \qquad B = \begin{bmatrix} 1 & 2 & 0 \\ 2 & 1 & 2 \\ -1 & 3 & 0 \end{bmatrix} \qquad C = \begin{bmatrix} 1 & 1 & 0 \\ 2 & 5 & 2 \\ -1 & 2 & 0 \end{bmatrix}$$

are identical except that $a_{*2} = b_{*2} + 3c_{*2}$. Confirm Theorem 8.2, which says that $\det A = \det B + 3 \det C$.

7. Suppose that A and B are identical except that one column of A is x times the corresponding column of B. Explain why $\det A = x \det B$ (Corollary 8.2a).

★ 8. If A is $n \times n$ and c is a scalar, what is $\det cA$? (Note that it is *not* $c \det A$!)

9. Suppose that B is obtained from A by replacing the kth column of A by the sum of itself and x times the preceding column. Show that $\det B = \det A$. (This completes the proof of Theorem 8.3.)

10. Supply reasons for the steps in the proof of Theorem 8.4.

11. Supply reasons for the steps in the proof of Theorem 8.6.

12. In each of the following cases, use Theorem 8.6 to reduce the given determinant to a single second-order determinant, and evaluate.

$$a. \begin{vmatrix} 6 & -3 & 5 \\ -1 & 4 & -2 \\ 2 & 2 & 4 \end{vmatrix} \qquad b. \begin{vmatrix} 2 & -5 & 0 & 4 \\ 0 & 3 & 2 & -4 \\ 1 & -1 & 2 & 5 \\ 1 & 3 & 1 & 0 \end{vmatrix}$$

13. Prove that if B is obtained from A by interchanging two columns, then $\det B = -\det A$ (Theorem 8.7).

14. Fill in the details of the proof of Theorem 8.8.

15. Let

$$A = \begin{bmatrix} 5 & -1 \\ 1 & 1 \end{bmatrix} \quad \text{and} \quad B = \begin{bmatrix} 3 & -3 \\ 1 & 2 \end{bmatrix}$$

Find AB and confirm that

$$\det AB = (\det A)(\det B)$$

(This is an illustration of a theorem we haven't proved yet, except in the special case where B is an elementary matrix.)

8.3 Expansion by a Column

The theory we have developed so far is contingent on our assumption that det A may be expanded by any row. If we assume instead that det A may be expanded by any *column*, that is,

$$\det A = \sum_{i=1}^{n} (-1)^{i+j} a_{ij} \det A_{ij}$$

where j is any one of the integers $1, \ldots, n$, then everything we did in the last section can be repeated with the words "row" and "column" interchanged. Hence each theorem of Sec. 8.2 has a "dual," which we now state. The proofs of these theorems proceed exactly as before; we merely note that in the proofs of the first two the determinant of A should be expanded by the jth column (instead of the ith row).

THEOREM 8.10

If two adjacent rows of A are equal, then det $A = 0$.

As in the proof of Theorem 8.1, the argument for this theorem goes through regardless of what column we use for the expansion of det A. Hence det $A = 0$ in any case if A has two equal adjacent rows.

THEOREM 8.11

Suppose that A, B, and C are identical except that one row of A is a linear combination of the corresponding rows of B and C. Then det A is the same linear combination of det B and det C.

As in the proof of Theorem 8.2 in the last section, the argument for this does not depend on which column we use to expand the determinants. The expansion by one column or another may give different results (this remains to be settled), but for a given choice of column the linear relation stated in Theorem 8.11 is correct.

COROLLARY 8.11a

Suppose that A and B are identical except that one row of A is a multiple of the corresponding row of B. Then det A is the same multiple of det B.

THEOREM 8.12

If B is obtained from A by adding a multiple of one row to an adjacent row, then det $B =$ det A.

THEOREM 8.13

If B is obtained from A by interchanging two adjacent rows, then det $B = -$det A.

THEOREM 8.14

If A has two equal rows, then $\det A = 0$.

THEOREM 8.15

If B is obtained from A by adding a multiple of one row to another row, then $\det B = \det A$.

THEOREM 8.16

If B is obtained from A by interchanging two rows, then $\det B = -\det A$.

THEOREM 8.17

If E is an elementary matrix and $\det E$ is expanded by any column, then $\det E$ is equal to -1, c, or 1 depending on whether E is of type 1 (interchange of rows), type 2 (multiplication of a row by a nonzero scalar c), or type 3 (addition of a multiple of one row to another row).

Note that in this theorem we interpret E as representing a row operation, whereas in Theorem 8.8 of the last section it was regarded as representing a column operation. Recall from Theorem 7.8 that elementary matrices can be interpreted either way.

THEOREM 8.18

If E is an elementary matrix and the determinants involved are expanded by any column, then for every A, $\det EA = (\det E)(\det A)$.

This theorem should be contrasted with Theorem 8.9, where the product rule reads $\det AE = (\det A)(\det E)$. The reason for the difference is that row operations and column operations are accomplished by premultiplication and postmultiplication, respectively, by elementary matrices. In the next section we'll remove all restrictions and prove that

$$\det AB = (\det A)(\det B)$$

where A and B are arbitrary.

Problems

16. Prove Theorem 8.10.

17. Prove Theorem 8.11.

18. In each of the following cases, use Theorem 8.15 to reduce the given determinant to a single second-order determinant, and evaluate. (Compare with Prob. 12, Sec. 8.2.)

a. $\begin{vmatrix} 6 & -3 & 5 \\ -1 & 4 & -2 \\ 2 & 2 & 4 \end{vmatrix}$ b. $\begin{vmatrix} 2 & -5 & 0 & 4 \\ 0 & 3 & 2 & -4 \\ 1 & -1 & 2 & 5 \\ 1 & 3 & 1 & 0 \end{vmatrix}$

19. Prove Theorem 8.18.

8.4 Uniqueness of Determinants

Now we are ready to justify the assumptions on which everything rests, namely that for each $i = 1, \ldots, n$,

$$\det A = \sum_{j=1}^{n} (-1)^{i+j} a_{ij} \det A_{ij}$$

and for each $j = 1, \ldots, n$,

$$\det A = \sum_{i=1}^{n} (-1)^{i+j} a_{ij} \det A_{ij}$$

that is, det A may be expanded by any row or column. In other words, det A is uniquely defined by any of these $2n$ expressions. (If it were not, we couldn't even talk about it in the singular! There would not be *one* determinant of A, but several different determinants, depending on the row or column used.)

If you are still with us, you will want to pay particular attention to whether we fall into circular reasoning, for we have built up a structure of eighteen theorems based on unproved assumptions. We cannot now use these theorems to justify the assumptions except insofar as they are independent of the assumptions.

THEOREM 8.19

The determinant of a nonsingular matrix is independent of the row or column used to expand it.

Proof

Suppose that A is nonsingular, in which case it can be written as a product of elementary matrices, say $A = E_1 \cdots E_s$. (See Theorem 7.22.) If det A is expanded by any row, we may write $A = (E_1 \cdots E_{s-1})E_s$ and use Theorem 8.9 (expanding each determinant by this same row) to obtain

$$\det A = (\det E_1 \cdots E_{s-1})(\det E_s)$$

Now repeat the process. After $s - 1$ steps[†] we have

$$\det A = (\det E_1) \cdots (\det E_s)$$

But according to Theorem 8.8 each determinant on the right is independent of the row used to expand it. Hence so is det A. The argument for the expansion of det A by any column is similar (using the dual theorems in Sec. 8.3).

Note that this argument does not depend on the assumptions by which we got the eighteen theorems of Secs. 8.2 and 8.3! The only gap in those theorems was that Theorems 8.2 and 8.11 did not settle the question of whether expansion by one row or another (or one column or another) gives different results. The above argument bypasses that question.

One question remains, however. Is the expansion of det A by a row the same as its expansion by a column? In each case we obtain

$$\det A = (\det E_1) \cdots (\det E_s)$$

so what it boils down to is whether the expansion of det E, where E is an elementary matrix, is the same by a row as by a column. Suppose, for example, that E is of type 3, having been obtained from I by adding c times the jth column to the ith column. Then by Theorem 8.8 the expansion of det E by a row yields 1. But E may also be obtained from I by adding c times the ith row to the jth row. (See Theorem 7.8.) Hence by Theorem 8.17 the expansion of det E by a column also yields 1. To complete the proof, type 1 and type 2 elementary matrices should be analyzed in the same way. ▮

THEOREM 8.20

If A is singular, then $\det A = 0$ (regardless of the row or column used in its expansion).

Proof

Suppose that det A is to be expanded by a row. Since A is singular, its columns are linearly dependent, so one is a linear combination of the others.[‡] By interchanging columns (which by Theorem 8.7 merely changes det A by a factor of -1) we may arrange it so that the nth column is a linear combination of the preceding columns, say

$$a_{*n} = \sum_{j=1}^{n-1} x_j a_{*j}$$

By Theorem 8.2 this means that

$$\det A = \sum_{j=1}^{n-1} x_j \det [a_{*1}, \ldots, a_{*n-1}, a_{*j}] \qquad \text{(Why?)}$$

[†] Of course s might be 1, in which case A itself is elementary and no steps are needed.
[‡] We are assuming that A is $n \times n$ with $n > 1$. The theorem is trivial if $n = 1$.

Since each determinant in this sum is zero (Theorem 8.5), we conclude that det $A = 0$. When det A is expanded by a column, the argument is similar (using the dual theorems in Sec. 8.3). ∎

We have now shown that the determinant of any $n \times n$ matrix is independent of the row or column used in its expansion. Hence Theorems 8.1 through 8.9, and their duals in Sec. 8.3, may be adopted without any qualifications concerning how the expansion of a determinant is carried out. Note, too, that we have another item to add to the list in Theorem 7.22:

THEOREM 8.21

An $n \times n$ matrix A is nonsingular if and only if det $A \neq 0$.

Proof

If det A is nonzero, then A is nonsingular by Theorem 8.20. Conversely, if A is nonsingular, then (see the proof of Theorem 8.19)

$$\det A = (\det E_1) \cdots (\det E_s)$$

where $E_1, \ldots, E_s$ are elementary matrices whose product is A. Since the determinant of an elementary matrix is nonzero (why?), it follows that det A is nonzero. ∎

THEOREM 8.22

If A and B are $n \times n$, then det AB = (det A)(det B)

Proof

If AB is singular, then either A or B is singular (why?) and the theorem is trivial. If AB is nonsingular, then both A and B are nonsingular (Prob. 50, Sec. 5.4). Write

$$A = E_1 \cdots E_s \qquad \text{and} \qquad B = E_{s+1} \cdots E_t$$

where each E_i is an elementary matrix. Repeated application of Theorem 8.9 yields

$$\begin{aligned} \det AB &= \det (E_1 \cdots E_t) \\ &= (\det E_1) \cdots (\det E_t) \\ &= (\det E_1 \cdots E_s)(\det E_{s+1} \cdots E_t) \\ &= (\det A)(\det B) \end{aligned}$$ ∎

Problems

20. Finish the proof of Theorem 8.19 by showing that the expansion of the determinant of a type 1 or type 2 elementary matrix is the same by a row as by a column.

21. Why is the determinant of an elementary matrix nonzero? (This question is asked in the proof of Theorem 8.21 before it is established that the determinant of a nonsingular matrix is nonzero. Hence you cannot answer by merely quoting the fact that elementary matrices are nonsingular!)

★ **22.** Prove that if A is nonsingular, then $\det A^{-1} = 1/\det A$.

23. Confirm Prob. 22 in the case

$$A = \begin{bmatrix} 1 & -1 & 0 \\ 0 & 1 & 2 \\ 5 & 3 & -2 \end{bmatrix}$$

★ **24.** Let A be an $n \times n$ matrix with entries in $\mathfrak{F}$. Prove the following.
a. $\det A^T = \det A$ *b.* $\det A^* = \overline{\det A}$

★ **25.** Prove that similar matrices have the same determinant.
This permits an unambiguous definition of the *determinant of a linear operator:* simply compute the determinant of any one of its matrix representations (relative to a single basis). See Theorem 6.7.

26. Find the determinant of the linear operator $f: \mathfrak{R}^2 \to \mathfrak{R}^2$ defined by $f(x,y) = (4x + 2y, 6x + 3y)$. Is this operator invertible?

★ **27.** An $n \times n$ matrix with entries in $\mathfrak{F}$ is called "unitary" if $A^* = A^{-1}$ (also "orthogonal" when $\mathfrak{F} = \mathfrak{R}$, in which case the defining property is $A^T = A^{-1}$).
a. Prove that $\det A$ is a complex number whose absolute value is 1.
b. What are the possible values of the determinant of an orthogonal matrix?

28. Confirm that each of the following matrices is unitary and check Prob. 27.

a. $\dfrac{1}{\sqrt{3}}\begin{bmatrix} 1 & 1-i \\ 1+i & -1 \end{bmatrix}$ *b.* $\begin{bmatrix} \cos\theta & -\sin\theta \\ \sin\theta & \cos\theta \end{bmatrix}$

c. $\dfrac{1}{2}\begin{bmatrix} 1 & i & 1-i \\ i\sqrt{2} & \sqrt{2} & 0 \\ 1 & i & -1+i \end{bmatrix}$ *d.* $\dfrac{1}{3}\begin{bmatrix} 2 & 1 & -2 \\ 1 & 2 & 2 \\ 2 & -2 & 1 \end{bmatrix}$

†8.5 Cramer's Rule

In Sec. 7.1 we proved that if the coefficient matrix of

$$\begin{aligned} a_{11}x_1 + a_{12}x_2 &= c_1 \\ a_{21}x_1 + a_{22}x_2 &= c_2 \end{aligned}$$

† This section can be omitted without loss of continuity.

is nonsingular, the system has a unique solution given by

$$x_1 = \frac{\det A_1}{\det A} \qquad x_2 = \frac{\det A_2}{\det A}$$

where A_1 and A_2 are the matrices obtained from

$$A = \begin{bmatrix} a_{11} & a_{12} \\ a_{21} & a_{22} \end{bmatrix}$$

by replacing its first and second columns, respectively, by $\mathbf{C} = (c_1, c_2)$. That is,

$$x_1 = \frac{\begin{vmatrix} c_1 & a_{12} \\ c_2 & a_{22} \end{vmatrix}}{\begin{vmatrix} a_{11} & a_{12} \\ a_{21} & a_{22} \end{vmatrix}} \qquad x_2 = \frac{\begin{vmatrix} a_{11} & c_1 \\ a_{21} & c_2 \end{vmatrix}}{\begin{vmatrix} a_{11} & a_{12} \\ a_{21} & a_{22} \end{vmatrix}}$$

This is the 2×2 case of Cramer's rule, which we now propose to develop in general. While we're at it, we'll derive a formula for computing the inverse of a nonsingular $n \times n$ matrix (promised in Prob. 52, Sec. 5.4).

THEOREM 8.23

If i and p are among the integers $1, \ldots, n$, then

$$\sum_{j=1}^{n} (-1)^{p+j} a_{ij} \det A_{pj} = \begin{cases} \det A & \text{if } i = p \\ 0 & \text{if } i \neq p \end{cases}$$

Similarly, if j and q are among the integers $1, \ldots, n$, then

$$\sum_{i=1}^{n} (-1)^{i+q} a_{ij} \det A_{iq} = \begin{cases} \det A & \text{if } j = q \\ 0 & \text{if } j \neq q \end{cases}$$

Proof

To establish the first assertion, note that if $i = p$, the given sum is simply the expansion of $\det A$ by the ith row. On the other hand, suppose that $i \neq p$. Let B be the matrix which is identical to A except that its pth row is the ith row of A. Then B has two equal rows and hence $\det B = 0$. But expanding $\det B$ by the pth row yields

$$\det B = \sum_{j=1}^{n} (-1)^{p+j} b_{pj} \det B_{pj} = \sum_{j=1}^{n} (-1)^{p+j} a_{ij} \det A_{pj}$$

The proof of the second assertion is similar. ▮

What this theorem says is that if the entries of a given row (column) of A are multiplied by the cofactors of the corresponding entries of any row (column), and the products are added, the result is zero, unless the two rows (columns) are the same, in which case $\det A$ is obtained.

EXAMPLE 8.9

Suppose that we use the second row of

$$A = \begin{bmatrix} 2 & -1 & 5 \\ 1 & 2 & 2 \\ 3 & -2 & 1 \end{bmatrix}$$

together with the cofactors of the entries in the *first* row:

$$1\begin{vmatrix} 2 & 2 \\ -2 & 1 \end{vmatrix} - 2\begin{vmatrix} 1 & 2 \\ 3 & 1 \end{vmatrix} + 2\begin{vmatrix} 1 & 2 \\ 3 & -2 \end{vmatrix} = 6 + 10 - 16 = 0$$

But if the cofactors of the entries in the second row are used, we are expanding $\det A$ by the second row and the result is $\det A$.

THEOREM 8.24 Cramer's Rule

If A is nonsingular and $\mathbf{C} \in \mathfrak{F}^n$, the unique solution of the linear system $A\mathbf{X} = \mathbf{C}$ is the vector $\mathbf{X}$ with components

$$x_q = \frac{\det A_q}{\det A} \qquad q = 1, \ldots, n$$

where A_q is the matrix obtained from A by replacing its qth column by $\mathbf{C}$.

Proof

If the equations of the system $A\mathbf{X} = \mathbf{C}$ are written out individually, the ith equation is

$$\sum_{j=1}^{n} a_{ij}x_j = c_i \qquad i = 1, \ldots, n$$

Multiplying each side by $(-1)^{i+q} \det A_{iq}$, where q is any one of the integers $1, \ldots, n$, and adding the equations, we have

$$\sum_{i=1}^{n}\left[(-1)^{i+q} \det A_{iq}\sum_{j=1}^{n} a_{ij}x_j\right] = \sum_{i=1}^{n}(-1)^{i+q}c_i \det A_{iq}$$

or (reversing the sums on the left)

$$\sum_{j=1}^{n}\left[\sum_{i=1}^{n}(-1)^{i+q}a_{ij}\det A_{iq}\right]x_j = \sum_{i=1}^{n}(-1)^{i+q}c_i \det A_{iq}$$

By Theorem 8.23 the sum in the brackets reduces to det A if $j = q$ and to 0 if $j \neq q$. Hence the left-hand side collapses to (det A)x_q. The right-hand side is the expansion of det A_q by the qth column. Thus we have

$$(\det A)x_q = \det A_q \qquad q = 1, \ldots, n$$

Since det A is nonzero, the result follows. ∎

This is one of the more elegant theorems of classical mathematics. It is also virtually useless except for theoretical purposes; to solve square linear systems, it is usually easier to work out the row echelon form of the coefficient matrix. H. Schneider and G. P. Barker (*Matrices and Linear Algebra,* New York: Holt, Rinehart and Winston, 1968) have an entertaining comparison of two methods of evaluating det A (cofactors versus row operations, essentially equivalent to Cramer's rule versus row echelon form in solving square systems). When $n = 3$, the number of steps involved is about the same for each method (they estimate 17 versus 15). When $n = 10$, the comparison is 36,288,000 to 667. They go on to point out that modern computers can handle matrices as large as 100 × 100; the comparison of the two methods in that case boggles the imagination.

LECTOR *It is a terrible thing to be boggled.*

AUCTOR *There are more things in heaven and earth, good Lector, than are dreamt of in your philosophy. I boggle every morning before breakfast.*

LECTOR *Does it hurt?*

AUCTOR *Not in the correct Yoga position.*

We end this section with a remarkable formula for the inverse of a nonsingular matrix A. Given such a matrix, we know that finding $A^{-1} = [x_{ij}]$ is equivalent to solving the n linear systems.

$$A\mathbf{X} = \mathbf{E}_j \qquad j = 1, \ldots, n \qquad \text{(Why?)}$$

Since A is nonsingular, Cramer's rule tells us that the solution of the jth system is the vector $\mathbf{X}_j$ with components

$$x_{ij} = \frac{\det A_i(j)}{\det A} \qquad i = 1, \ldots, n$$

where $A_i(j)$ is the matrix obtained from A by replacing its ith column by $\mathbf{E}_j$. Expanding by the ith column, we find that

$$\det A_i(j) = \sum_{k=1}^{n} (-1)^{k+i}\delta_{kj} \det A_{ki} = (-1)^{j+i} \det A_{ji}$$

from which

$$x_{ij} = \frac{1}{\det A}(-1)^{j+i} \det A_{ji} \qquad i = 1, \ldots, n$$

This holds for each $j = 1, \ldots, n$, so the i,j entry of A^{-1} is

$$\frac{1}{\det A}(\text{cofactor of } a_{ji})$$

a formula which motivates the following definition.

ADJOINT OF A MATRIX

Let $A = [a_{ij}]$ be an $n \times n$ matrix. The *adjoint* of A, denoted by adj A, is the $n \times n$ matrix whose i,j entry is the cofactor of the j,i entry of A. That is,

$$\text{adj } A = [(-1)^{i+j} \det A_{ij}]^T$$

the transpose of the matrix whose i,j entry is the cofactor of a_{ij}.

EXAMPLE 8.10

If

$$A = \begin{bmatrix} -2 & -1 & 1 \\ 1 & 3 & -1 \\ 0 & 2 & 1 \end{bmatrix}$$

then

$$\text{adj } A = \begin{bmatrix} 5 & 3 & -2 \\ -1 & -2 & -1 \\ 2 & 4 & -5 \end{bmatrix}$$

We found the 2,3 entry (for example) by computing the cofactor of the 3,2 entry of A, namely

$$-\begin{vmatrix} -2 & 1 \\ 1 & -1 \end{vmatrix} = -1$$

THEOREM 8.25

If A is nonsingular, then

$$A^{-1} = \frac{1}{\det A}(\operatorname{adj} A)$$

Proof

We have already shown that the i,j entry of A^{-1} is

$$\frac{1}{\det A}(\text{cofactor of } a_{ji})$$

The theorem follows from the definition of adj A. ▮

A different proof (assuming that we know the formula ahead of time!) is to simply check by matrix multiplication that

$$B = \frac{1}{\det A}(\operatorname{adj} A)$$

is the inverse of A. Since

$$b_{ij} = \frac{1}{\det A}(\text{cofactor of } a_{ji}) = \frac{1}{\det A}(-1)^{j+i} \det A_{ji}$$

the i,j entry of AB is

$$\sum_{k=1}^{n} a_{ik} b_{kj} = \frac{1}{\det A} \sum_{k=1}^{n} (-1)^{j+k} a_{ik} \det A_{jk}$$

According to the first part of Theorem 8.23, this expression is 1 if $i = j$ and 0 if $i \neq j$; thus the i,j entry of AB is δ_{ij} and hence $AB = I$. ▮

EXAMPLE 8.11

If A is the matrix of Example 8.10, then $\det A = -7$. (Confirm this!) Hence A is nonsingular; by Theorem 8.25

$$A^{-1} = \frac{1}{-7}(\operatorname{adj} A) = \frac{1}{7}\begin{bmatrix} -5 & -3 & 2 \\ 1 & 2 & 1 \\ -2 & -4 & 5 \end{bmatrix}$$

Problems

29. Finish the proof of Theorem 8.23.

30. Solve the system $AX = \mathbf{C}$, where

$$A = \begin{bmatrix} 2 & 1 & -1 \\ 1 & 3 & 0 \\ 5 & -2 & 1 \end{bmatrix} \quad \text{and} \quad \mathbf{C} = (0,2,4)$$

in the following ways.
a. Use Cramer's rule.
b. Reduce the matrix $[A|\mathbf{C}]$ to $[B|P\mathbf{C}]$, where $B = PA$ is the row echelon form of A.
c. Find A^{-1} and compute $\mathbf{X} = A^{-1}\mathbf{C}$.

31. Prove that if A is any $n \times n$ matrix, then

$$A(\text{adj } A) = (\text{adj } A)A = (\det A)I$$

Hint: Use matrix multiplication, as in the second proof of Theorem 8.25. (Note that Theorem 8.25 itself cannot be used, since A might be singular.)

32. Let A be any $n \times n$ matrix.
a. Prove that if A is nonsingular, adj A is also. What is $(\text{adj } A)^{-1}$?
b. Prove that if A is singular, adj A is also. *Hint:* $A(\text{adj } A) = 0$.

33. Show that if A is $n \times n$, then

$$\det(\text{adj } A) = (\det A)^{n-1}$$

Confirm this in the case of

$$A = \begin{bmatrix} -2 & -1 & 1 \\ 1 & 3 & -1 \\ 0 & 2 & 1 \end{bmatrix}$$

whose adjoint and determinant are given in Examples 8.10 and 8.11.

34. Use Theorem 8.25 to find A^{-1} if

$$A = \begin{bmatrix} 3 & 2 & -1 \\ -1 & 2 & 3 \\ -3 & 1 & 3 \end{bmatrix}$$

35. Use Theorem 8.25 to show that if

$$A = \begin{bmatrix} a_{11} & a_{12} \\ a_{21} & a_{22} \end{bmatrix}$$

is nonsingular, then

$$A^{-1} = \frac{1}{\det A}\begin{bmatrix} a_{22} & -a_{12} \\ -a_{21} & a_{11} \end{bmatrix}$$

Review Quiz

True or false?

1. $\begin{vmatrix} 1 & 0 & 1 \\ 2 & -1 & 2 \\ 5 & 1 & 5 \end{vmatrix} = 0$
2. If the entries of the first row of a square matrix are multiplied by their cofactors and the products are added, the result is the determinant of the matrix.
3. If A is a matrix whose transpose is its inverse, then $(\det A)^2 = 1$.
4. $\det cA = c \det A$
5. If B is obtained from A by adding the first row of A to itself, then $\det B = \det A$.
6. If A is a square matrix whose columns are all different, then $\det A$ is nonzero.
7. $\begin{vmatrix} 2 & 0 & 1 \\ 8 & -3 & -1 \\ 6 & 9 & 2 \end{vmatrix} = 6 \begin{vmatrix} 1 & 0 & 1 \\ 4 & -1 & -1 \\ 3 & 3 & 2 \end{vmatrix}$
8. If $A = [a_{ij}]$ is $n \times n$, the expression

$$\sum_{i=1}^{n} (-1)^{i+j} a_{ij} \det A_{ij}$$

is the expansion of $\det A$ by the ith row.

9. If

$$A = \begin{bmatrix} a_{11} & a_{12} \\ a_{21} & a_{22} \end{bmatrix}$$

the cofactor of a_{11} is a_{22}.

10. $\det I = 1$
11. If E is an elementary matrix, then $\det E$ is nonzero.

12. $\begin{vmatrix} 1 & -3 & 7 \\ 0 & 1 & 1 \\ 5 & -1 & 2 \end{vmatrix} = -\begin{vmatrix} 0 & 1 & 1 \\ 1 & -3 & 7 \\ 5 & -1 & 2 \end{vmatrix}$

13. The map $f\colon M_n(\mathfrak{F}) \to \mathfrak{F}$ defined by $f(A) = \det A$ is linear.

14. If A is nonsingular, then $(\det A)(\det A^{-1}) = 1$.

15. If A and B are square matrices related by an equation of the form $B = PAP^{-1}$, then $\det B = \det A$.

16. The adjoint of

$$\begin{bmatrix} 1 & -1 \\ 2 & -2 \end{bmatrix}$$

is

$$\begin{bmatrix} -2 & 1 \\ -2 & 1 \end{bmatrix}$$

CHAPTER 9

Eigenvalues

Suppose that $f\colon S \to S$ is a linear operator. A problem that comes up astonishingly often in mathematics and in its applications is to find vectors in S that map into scalar multiples of themselves, say $f(x) = rx$, where $x \in S$ and r is a scalar. Such vectors are solutions of the equation $f(x) - rx = 0$, or $(f - ri)(x) = 0$, where i is the identity map on S. Since $g = f - ri$ is itself a linear operator, the problem is to solve the homogeneous equation $g(x) = 0$. If S is finite-dimensional, this equation is equivalent to the homogeneous system $B\mathbf{X} = \mathbf{0}$, where B is a matrix representation of g (relative to a given basis) and $\mathbf{X}$ is the coordinate vector of x relative to this basis. Hence it would appear that we have the apparatus at our disposal to solve the problem, at least in principle. The difficulty is that the scalar r is not known either! In other words, we have to find not only the elements of S satisfying $f(x) = rx$, but also the scalar with which these vectors are associated. Our objective in this chapter is to discuss some ramifications of this problem.

9.1 Diagonal Matrices

It will soon be apparent that to solve the problem described above, one looks for a matrix representation of f having the form

$$[f] = \begin{bmatrix} a_{11} & & \bigcirc \\ & a_{22} & \\ \bigcirc & & \ddots a_{nn} \end{bmatrix}$$

where the zeros mean that every entry not on the "main diagonal" is 0. Hence we discuss the properties of such matrices as a preliminary.

DIAGONAL MATRIX

An $n \times n$ matrix $A = [a_{ij}]$ is called *diagonal* if every entry not on the main diagonal is zero, that is, if

$$a_{ij} = 0 \qquad \text{whenever} \quad i \neq j$$

Its *diagonal entries* are $a_{11}, \ldots, a_{nn}$; since these are the only ones of any interest, we often drop the double subscripts and write them simply as $a_1, \ldots, a_n$. Then A itself is denoted by

$$A = \text{diag}\,(a_1, \ldots, a_n)^\dagger$$

EXAMPLE 9.1

The matrix

$$A = \begin{bmatrix} 1 & 0 & 0 \\ 0 & 0 & 0 \\ 0 & 0 & -2 \end{bmatrix}$$

is diagonal; in simplified notation, we may write

$$A = \text{diag}\,(1, 0, -2)$$

The set D of $n \times n$ diagonal matrices (with entries in $\mathfrak{F}$) has a number of attractive algebraic properties; we'll develop these in the following theorems, the proofs of which are all simple.

THEOREM 9.1

D is an n-dimensional subspace of $M_n(\mathfrak{F})$.

Proof

Since D is nonempty and closed relative to addition and scalar multiplication (confirm this!), it is a subspace of $M_n(\mathfrak{F})$. To show that it is an n-dimensional subspace, let A_i be the diagonal matrix with i,i entry 1 and with 0s elsewhere. (We called this E_{ii} in the proof of Theorem 4.3.) That is,

$$A_i = \text{diag}\,(\delta_{i1}, \ldots, \delta_{in}) \qquad i = 1, \ldots, n$$

We leave it to you to confirm that $\{A_1, \ldots, A_n\}$ is a basis for D. ∎

THEOREM 9.2

D is a commutative ring with unit element.

Proof

The additive properties of a ring (see the definition in Sec. 5.1) follow from Theorem 9.1, in which we showed that D is a vector space. Closure relative to multiplication follows from the fact that if

$$A = \text{diag}\,(a_1, \ldots, a_n) \qquad \text{and} \qquad B = \text{diag}\,(b_1, \ldots, b_n)$$

† In terms of this notation, we can also write $A = [a_i\delta_{ij}]$; that is, the i,j entry of A is $a_i\delta_{ij}$. (Why?)

then

$$AB = \text{diag}\,(a_1b_1, \ldots, a_nb_n)$$

(See the problems.) Associativity and distributivity of multiplication are inherited from $M_n(\mathfrak{F})$. Hence D is a ring. Moreover, diagonal matrices commute (why?), and D contains I. Thus it follows (see the definitions in Sec. 5.2) that D is a commutative ring with unit element. ▮

THEOREM 9.3

Let $A = \text{diag}\,(a_1, \ldots, a_n)$. Then A is nonsingular if and only if each $a_i \neq 0$. Moreover, when A is nonsingular,

$$A^{-1} = \text{diag}\,(a_1^{-1}, \ldots, a_n^{-1})$$

Proof

We leave the "if" part for the problems. Conversely, suppose that A is nonsingular, with inverse $A^{-1} = B = [b_{ij}]$. Since $AB = I$, the i,j entry of AB is δ_{ij}. But $A = [a_i\delta_{ij}]$, so the i,j entry of AB is also

$$\sum_{k=1}^{n} (a_i\delta_{ik})b_{kj} = a_ib_{ij}$$

Thus $a_ib_{ij} = \delta_{ij}$. When $j = i$, this reads $a_ib_{ii} = 1$, so a_i is nonzero and $b_{ii} = a_i^{-1}$. When $j \neq i$, we have $a_ib_{ij} = 0$ and hence $b_{ij} = 0$. Therefore

$$A^{-1} = \text{diag}\,(a_1^{-1}, \ldots, a_n^{-1}) \; ▮$$

EXAMPLE 9.2

The diagonal matrix in Example 9.1 is singular because one of its diagonal entries is 0. On the other hand, if

$$A = \text{diag}\,(1,-1,2) = \begin{bmatrix} 1 & 0 & 0 \\ 0 & -1 & 0 \\ 0 & 0 & 2 \end{bmatrix}$$

then A is nonsingular and

$$A^{-1} = \text{diag}\,(1,-1,\tfrac{1}{2}) = \begin{bmatrix} 1 & 0 & 0 \\ 0 & -1 & 0 \\ 0 & 0 & \tfrac{1}{2} \end{bmatrix}$$

THEOREM 9.4

The determinant of a diagonal matrix is the product of its diagonal entries.

Proof

Let $A = \text{diag}(a_1, \ldots, a_n) = [a_i\delta_{ij}]$. We prove that

$$\det A = a_1 \cdots a_n$$

by induction on n. This is obvious when $n = 1$, so assume that $n > 1$ and that the formula is correct for $(n-1) \times (n-1)$ diagonal matrices. Then (expanding by the first row) we have

$$\det A = \sum_{j=1}^{n} (-1)^{1+j} a_1 \delta_{1j} \det A_{1j} = a_1 \det A_{11}$$

Since $A_{11} = \text{diag}(a_2, \ldots, a_n)$, the induction hypothesis yields

$$\det A = a_1 \cdots a_n \quad \blacksquare$$

EXAMPLE 9.3

The determinants of the diagonal matrices in Examples 9.1 and 9.2 are

$$\det A = (1)(0)(-2) = 0 \qquad \text{and} \qquad \det A = (1)(-1)(2) = -2$$

respectively.

THEOREM 9.5

The rank of a diagonal matrix is the number of nonzero diagonal entries.

Proof

If all the diagonal entries are zero, the matrix is zero and its rank is 0. Assuming that some diagonal entries are nonzero, observe that the row space is spanned by the rows in which these entries occur and that these rows are linearly independent. Hence they constitute a basis for the row space, and the dimension of the row space (the rank of the matrix) is the number of such rows. ▮

EXAMPLE 9.4

The rank of diag $(1, 0, -2)$ is 2.

Theorems 9.1 through 9.5 indicate that the ideas we have discussed in connection with matrices are transparent in the case of diagonal matrices. They are obviously convenient to work with!

Problems

1. Let D be the set of $n \times n$ diagonal matrices with entries in $\mathfrak{F}$.
   a. Show that D is closed relative to addition and scalar multiplication.
   b. Confirm that if $A_i = \text{diag}\,(\delta_{i1}, \ldots, \delta_{in})$, where $i = 1, \ldots, n$, then $\{A_1, \ldots, A_n\}$ is a basis for D.
2. Prove that if $A = \text{diag}\,(a_1, \ldots, a_n)$ and $B = \text{diag}\,(b_1, \ldots, b_n)$, then $AB = \text{diag}\,(a_1b_1, \ldots, a_nb_n)$. Why does it follow that diagonal matrices commute?
3. Explain why D is a linear algebra. (Refer to the definition in Sec. 5.3.)
4. Finish the proof of Theorem 9.3 by showing that if each diagonal entry of $A = \text{diag}\,(a_1, \ldots, a_n)$ is nonzero, then A is nonsingular.
5. Use Prob. 57, Sec. 5.4, to prove that if one of the diagonal entries of $A = \text{diag}\,(a_1, \ldots, a_n)$ is zero, then A is singular. (That is, if A is nonsingular, then each a_i is nonzero. This is a different argument for the "only if" part of Theorem 9.3.)

★ 6. Prove that two diagonal matrices with the same diagonal entries (except possibly for order) are similar. *Hint:* If A and B have the same diagonal entries, but not in the same order, change A into B by stages, using a row interchange and the same column interchange at each stage. Each time this is done the new matrix is $[i,j]$(old matrix)$[i,j]$, which is a similarity transformation. (Why?)

9.2 Eigenvalues of a Linear Operator

What we want to investigate in this chapter is the possibility of representing a linear operator by a diagonal matrix. The idea is to take whatever matrix representation is given and "diagonalize" it by changing to an appropriate basis. This is often spectacular, as you will see, but it is not always possible to pull it off. Hence we want to develop a method for trying it and to provide a statement of the conditions guaranteeing its success.

THEOREM 9.6

Let $f\colon S \to S$ be a linear operator on the n-dimensional space S. Then f has a diagonal matrix representation if and only if there are scalars $r_1, \ldots, r_n$ (not necessarily distinct), and a basis $\{v_1, \ldots, v_n\}$ for S, such that

$$f(v_j) = r_jv_j \qquad j = 1, \ldots, n$$

Proof

Suppose that such scalars, and such a basis, exist. Then

$$f(v_j) = r_j v_j = \sum_{i=1}^{n} r_i \delta_{ij} v_i \qquad j = 1, \ldots, n$$

so the matrix of f relative to $\{v_1, \ldots, v_n\}$ is

$$[r_i \delta_{ij}] = \text{diag}\,(r_1, \ldots, r_n)$$

The "only if" part is proved by reversing the argument. ▮

Thus the problem of diagonalizing the matrix of a linear operator is equivalent to finding a basis with the property that *each of its elements is mapped into a scalar multiple of itself*. This idea is so important that it has acquired a fearsome terminology in two languages.

EIGENVALUES AND EIGENVECTORS OF A LINEAR OPERATOR

Let $f: S \to S$ be a linear operator. A nonzero vector $x \in S$ satisfying $f(x) = rx$ for some scalar r is called an *eigenvector* of f associated with r, while r itself is called an *eigenvalue* of f.[†] Given an eigenvalue r, the subspace of S defined by $\{x \mid f(x) = rx\}$ is called the *eigenspace* associated with r; it consists of all the eigenvectors associated with r, together with the zero vector.

Other terms in current use for eigenvectors and eigenvalues are "characteristic vector" and "characteristic value," "proper vector" and "proper value," "latent vector" and "latent value." Purists would use the original German "Eigenvektor" and "Eigenwert," but it has become common to mix the languages; eigenvector and eigenvalue are now standard. Obviously they must arise in a variety of applications to justify such a luxuriant growth of jargon!

EXAMPLE 9.5

Let S be any (nonzero) vector space and let $i: S \to S$ be the identity map. Then $i(x) = 1x$ for every $x \in S$, so 1 is an eigenvalue of i and every nonzero element of S is an eigenvector associated with it. The corresponding eigenspace is S itself.

[†] Note that $x = 0$ satisfies $f(x) = rx$ no matter what r is. However, this is of little interest, since we are looking for linearly independent vectors with this property. Moreover, the zero vector cannot be associated with a unique eigenvalue (see Prob. 9). Hence we do not classify it among the eigenvectors of f.

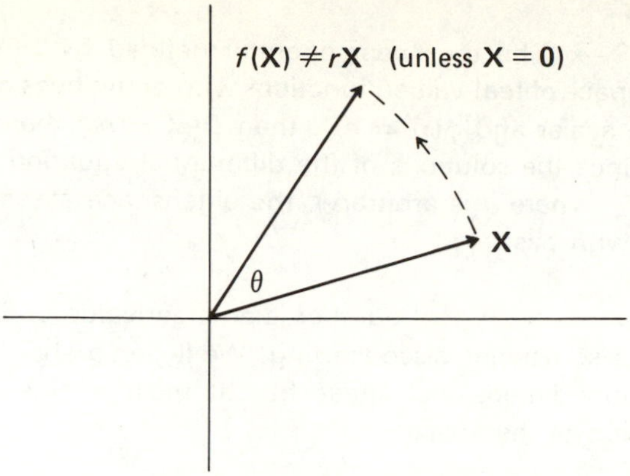

Fig. 9.1 Rotation of the plane

EXAMPLE 9.6

Let f: $\mathcal{R}^2 \to \mathcal{R}^2$ be a rotation of the plane through the angle θ, $0 < \theta < \pi$. Since *no* vector (except **0**) is sent into a scalar multiple of itself, there are no eigenvalues and no eigenvectors. (See Fig. 9.1.)

EXAMPLE 9.7

Define f: $\mathcal{R}^2 \to \mathcal{R}^2$ by $f(x,y) = (x,-y)$. This sends each point of the plane into the image one would expect if the x axis were a mirror. (See Fig. 9.2.) It is called a *reflection in the x axis*. Since every multiple of $\mathbf{E}_1$ is sent into itself, that is, $f(\mathbf{X}) = 1\mathbf{X}$ whenever $\mathbf{X} = a\mathbf{E}_1$, we conclude that $r_1 = 1$ is an eigenvalue of f and that the corresponding eigenspace is the x axis. Similarly, every multiple of $\mathbf{E}_2$ is sent into its negative, that is, $f(\mathbf{X}) = -1\mathbf{X}$ whenever $\mathbf{X} = b\mathbf{E}_2$. Hence $r_2 = -1$ is another eigenvalue of f; the associated eigenspace is the y axis. It is geometrically apparent that r_1 and r_2 are the only eigenvalues of f.

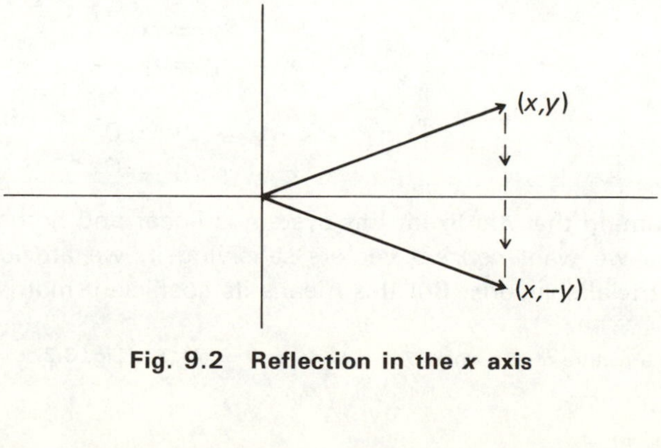

Fig. 9.2 Reflection in the *x* axis

EXAMPLE 9.8

Let $D: S \to S$ be the linear operator defined by $D(\phi) = \phi'$, where S is the vector space of real-valued functions with derivatives of all orders on $(-\infty, \infty)$. If r is a scalar and $\phi(x) = e^{rx}$, then $D(\phi) = r\phi$; therefore r is an eigenvalue of D. Since the solutions of the differential equation $y' = ry$ are of the form $y = ae^{rx}$ (where a is arbitrary), the eigenspace associated with r is 1-dimensional, with basis $\{\phi\}$.

Note that every real number is an eigenvalue of the map in Example 9.8, which is somewhat disconcerting. We'll prove shortly that a linear operator on a finite-dimensional space has at most n eigenvalues, where n is the dimension of the space.

EXAMPLE 9.9

Let S be $\mathcal{R}^2$ or $\mathcal{R}^3$ and define $f: S \to S$ to be the orthogonal projection of S on U, where U is a subspace of S. Recall from Example 3.11 that the image of each $\mathbf{X} \in S$ is found by writing $\mathbf{X} = \mathbf{X}_1 + \mathbf{X}_2$, where $\mathbf{X}_1$ and $\mathbf{X}_2$ are the orthogonal projections of $\mathbf{X}$ on U and $U^\perp$, respectively.† This map sends each element of U into itself and each element of $U^\perp$ into zero. (Why?) Assuming that U is neither the zero subspace of S nor all of S, it follows that $r_1 = 1$ and $r_2 = 0$ are eigenvalues of f. Since no other element of S is sent into a scalar multiple of itself (why?), these are the only eigenvalues; the corresponding eigenspaces are U and $U^\perp$, respectively.

So far these examples have yielded up their eigenvectors and eigenvalues more or less by inspection. The next one is heavier going, and opens up vistas. If you follow it closely, you should have a good idea of where we're heading.

EXAMPLE 9.10

Define $f: \mathcal{R}^2 \to \mathcal{R}^2$ by $f(x,y) = (2x + 2y, x + 3y)$. It is not obvious whether f has any eigenvectors, nor, if so, what they are. We'll attack the problem by brute force, trying to find (x,y) and r satisfying $f(x,y) = r(x,y)$. This leads to the system

$$\begin{aligned} 2x + 2y &= rx \\ x + 3y &= ry \end{aligned}$$

or

$$\begin{aligned} (2 - r)x + 2y &= 0 \\ x + (3 - r)y &= 0 \end{aligned}$$

Assuming that r is fixed, this system is linear and homogeneous in x and y; since we want *nonzero* vectors satisfying it, we are looking for it to have nontrivial solutions. But this means its coefficient matrix should be singular;

† For a picture of this when U is a plane in $S = \mathcal{R}^3$, see Fig. 3.3.

that is, the determinant of the matrix should be 0. Hence we solve the equation

$$\begin{vmatrix} 2-r & 2 \\ 1 & 3-r \end{vmatrix} = 0$$

for r. There are two roots, $r_1 = 1$ and $r_2 = 4$; these are the eigenvalues of f.

However, let's back up a little and use some finesse instead of a sledge hammer. The matrix of f relative to the standard basis for $\mathcal{R}^2$ is

$$A = \begin{bmatrix} 2 & 2 \\ 1 & 3 \end{bmatrix}$$

Writing f in the form $f(\mathbf{X}) = A\mathbf{X}$, we see that the equation we started out to solve, $f(\mathbf{X}) = r\mathbf{X}$, is the same as $A\mathbf{X} = r\mathbf{X}$, that is,

$$(A - rI)\mathbf{X} = \mathbf{0}^{\dagger}$$

This is a concise version of the 2×2 homogeneous system we wrote down before; it has nontrivial solutions if and only if the coefficient matrix

$$A - rI = \begin{bmatrix} 2-r & 2 \\ 1 & 3-r \end{bmatrix}$$

is singular, that is, if and only if

$$\det (A - rI) = \begin{vmatrix} 2-r & 2 \\ 1 & 3-r \end{vmatrix} = 0$$

This is the quadratic equation $r^2 - 5r + 4 = 0$; its roots (the eigenvalues of f) are $r_1 = 1$ and $r_2 = 4$.

With this notation, we are in a position to move on more smoothly. The eigenspace associated with $r_1 = 1$ is the set of solutions of the system $(A - I)\mathbf{X} = \mathbf{0}$. Since

$$A - I = \begin{bmatrix} 1 & 2 \\ 1 & 2 \end{bmatrix}$$

this system is

$$\begin{aligned} x + 2y &= 0 \\ x + 2y &= 0 \end{aligned}$$

Its solution space (the null space of $A - I$) is

$$\{(x,y) \mid x = -2t,\ y = t\}$$

and hence the eigenspace is 1-dimensional (the straight line $x + 2y = 0$ in $\mathcal{R}^2$). A convenient basis is $\{\mathbf{V}_1\}$, where $\mathbf{V}_1 = (-2,1)$.

† Note the insertion of I! It would not make sense to write $(A - r)\mathbf{X} = \mathbf{0}$.

The eigenvectors associated with $r_2 = 4$ are the nonzero solutions of the system $(A - 4I)\mathbf{X} = \mathbf{0}$. Since

$$A - 4I = \begin{bmatrix} -2 & 2 \\ 1 & -1 \end{bmatrix}$$

this is the system

$$\begin{aligned} -2x + 2y &= 0 \\ x - y &= 0 \end{aligned}$$

the solution set of which is

$$\{(x,y) \mid x = t,\ y = t\}$$

Again the dimension is 1; a basis is $\{\mathbf{V}_2\}$, where $\mathbf{V}_2 = (1,1)$.

Lest you lose sight of the original purpose of finding eigenvectors, note that in this example we are now in a position to diagonalize the matrix

$$A = [f]_\alpha = \begin{bmatrix} 2 & 2 \\ 1 & 3 \end{bmatrix}$$

where α is the standard basis for $\mathcal{R}^2$. We have produced two scalars, $r_1 = 1$ and $r_2 = 4$, and two vectors, $\mathbf{V}_1 = (-2,1)$ and $\mathbf{V}_2 = (1,1)$, satisfying

$$f(\mathbf{V}_1) = r_1\mathbf{V}_1 \qquad \text{and} \qquad f(\mathbf{V}_2) = r_2\mathbf{V}_2$$

Since $\mathbf{V}_1$ and $\mathbf{V}_2$ are linearly independent, they constitute a basis β for $\mathcal{R}^2$; recall that we started this section with the observation that A can be diagonalized if and only if such a basis exists. Moreover, the diagonalized representation of f has the r_j as diagonal entries; in this case, it is the matrix

$$B = [f]_\beta = \begin{bmatrix} r_1 & 0 \\ 0 & r_2 \end{bmatrix} = \begin{bmatrix} 1 & 0 \\ 0 & 4 \end{bmatrix}$$

You should also recall that two matrix representations of the same linear operator are similar; more precisely, if $A = [f]_\alpha$ and $B = [f]_\beta$, then $B = PAP^{-1}$, where one choice of P is $P = [i]_{\alpha\beta}$, the nonsingular matrix relating α to β. (See Theorem 6.10.) In the present example,

$$P = \frac{1}{3}\begin{bmatrix} -1 & 1 \\ 1 & 2 \end{bmatrix} \qquad \text{and} \qquad P^{-1} = \begin{bmatrix} -2 & 1 \\ 1 & 1 \end{bmatrix}$$

As you can check,

$$B = \begin{bmatrix} 1 & 0 \\ 0 & 4 \end{bmatrix} = \frac{1}{3}\begin{bmatrix} -1 & 1 \\ 1 & 2 \end{bmatrix}\begin{bmatrix} 2 & 2 \\ 1 & 3 \end{bmatrix}\begin{bmatrix} -2 & 1 \\ 1 & 1 \end{bmatrix} = PAP^{-1}$$

LECTOR *I see two shortcuts.*

AUCTOR *Wonderful.*

LECTOR *If all you want is a diagonal matrix of f, why don't you put down $B = \text{diag}(1,4)$ as soon as you've found the eigenvalues?*

AUCTOR *It doesn't always work.*

LECTOR *Well, we ought to find out when it does.*

AUCTOR *See Sec. 9.4.*

LECTOR *It also appears that P^{-1} can be found by inspection.*

AUCTOR *How?*

LECTOR *Its columns are $\mathbf{V}_1$ and $\mathbf{V}_2$.*

AUCTOR *I'll ask you to explain it in Prob. 21, Sec. 9.3.*

LECTOR *Wonderful*

Problems

7. Suppose that the linear operator $f\colon S \to S$ has a diagonal matrix representation. Name scalars $r_1, \ldots, r_n$, and a basis $\{v_1, \ldots, v_n\}$ for S, such that

$$f(v_j) = r_j v_j \qquad j = 1, \ldots, n$$

where $n = \dim S$. (This is the "only if" part of Theorem 9.6.)

8. Let $f\colon S \to S$ be a linear operator with eigenvalue r and let T be the eigenspace of f associated with r.

a. Prove that T is a subspace of S (as asserted in the definition of eigenspace).

★ *b.* Explain why $f(x) \in T$ for every $x \in T$. (Any subspace of S with this property is said to be "invariant" under f, for f sends each element of T into T.)

9. Prove the following.

a. An eigenvector of a linear operator cannot be associated with two different eigenvalues. Why would this be false if the zero vector were allowed as an eigenvector?

★ *b.* Eigenspaces associated with distinct eigenvalues have only the zero vector in common.

10. Find the eigenvalues and corresponding eigenspaces of each of the following linear operators (without appealing to matrix representations).

a. The rotation of the plane through 180°.

b. The reflection of the plane in the line $y = x$, defined by $f(x,y) = (y,x)$.

c. The orthogonal projection of $\mathcal{R}^3$ on the xy plane, defined by $f(x,y,z) = (x,y,0)$.

d. The differential operator $D\colon S \to S$ defined by $D(\phi) = \phi'$, where S is the vector space (over $\mathcal{R}$) of all linear combinations of $\phi_1(x) = e^x$ and $\phi_2(x) = e^{2x}$.

11. Let f: $S \to S$ be the orthogonal projection of S on U (where S is $\mathcal{R}^2$ or $\mathcal{R}^3$ and U is a subspace of S).
    a. Explain the assertion made in Example 9.9 that each element of U is mapped into itself and each element of $U^\perp$ is mapped into zero.
    b. Why must we restrict U to be neither the zero subspace of S nor all of S before asserting that the eigenvalues of f are 1 and 0?
    c. Explain why no other element of S is sent into a scalar multiple of itself.

★ 12. A linear operator f: $S \to S$ is said to be "idempotent" if $f^2 = f$.
    a. Prove that an idempotent linear operator cannot have any eigenvalues besides 1 and 0.
    b. Explain why the orthogonal projection in Example 9.9 is idempotent.

★ 13. Let f: $S \to S$ be a linear operator.
    a. Prove that r is an eigenvalue of f if and only if the map $f - ri$ is not one-to-one (where i is the identity map on S).
    b. Explain why the eigenspace of f associated with the eigenvalue r is $\ker(f - ri)$.
    c. Why is every eigenspace of f at least 1-dimensional?

14. Define f: $\mathcal{R}^3 \to \mathcal{R}^3$ by $f(\mathbf{X}) = A\mathbf{X}$, where

$$A = \begin{bmatrix} 2/3 & 1/3 & -2/3 \\ 1/3 & 2/3 & 2/3 \\ 2/3 & -2/3 & 1/3 \end{bmatrix}$$

    a. Show that the only eigenvalue of f is $r = 1$ and that the associated eigenspace is a straight line through the origin in $\mathcal{R}^3$.
    b. Why is the line found in part *a* the set of points left fixed by f? (It can be shown that f is a rotation of 3-space; this line is the axis of rotation.)

9.3 Eigenvalues of a Matrix

Let f: $S \to S$ be a linear operator on the n-dimensional space S $(n > 0)$. To find the eigenvalues and eigenvectors of f, we generalize the attack indicated in Example 9.10. The trick is to replace the equation $f(x) = rx$, which involves eigenvectors in S, by an equivalent equation in $\mathfrak{F}^n$. Let α be any (ordered) basis for S and let A be the matrix of f relative to α. We move into n-space by making use of the isomorphism α: $S \to \mathfrak{F}^n$ which sends each element of S into its coordinate vector relative to α. We do this because the equation $f(x) = rx$ is true if and only if $\alpha[f(x)] = \alpha(rx)$. (Why?) But $\alpha[f(x)] = A\alpha(x)$ (Prob.

26, Sec. 6.3) and $\alpha(rx) = r\alpha(x)$, so the equation replacing $f(x) = rx$ is $A\alpha(x) = r\alpha(x)$, or simply $A\mathbf{X} = r\mathbf{X}$ if we let $\mathbf{X} = \alpha(x)$.†

Now we may proceed as in Example 9.10 of the last section. The equation $A\mathbf{X} = r\mathbf{X}$ is equivalent to $(A - rI)\mathbf{X} = \mathbf{0}$, an $n \times n$ linear homogeneous system which has nontrivial solutions if and only if $\det(A - rI) = 0$. The matrix $A - rI$ is

$$\begin{bmatrix} a_{11} - r & a_{12} & \cdots & a_{1n} \\ a_{21} & a_{22} - r & \cdots & a_{2n} \\ \cdots & \cdots & \cdots & \cdots \\ a_{n1} & a_{n2} & \cdots & a_{nn} - r \end{bmatrix}$$

(the same as A except that the diagonal entries are $a_{ii} - r$, where $i = 1, \ldots, n$). Its determinant is a polynomial in r of degree n. (Why?) The problem of finding the eigenvalues of f thus reduces to the problem of solving the equation $\det(A - rI) = 0$ for whatever roots lie in F.‡ This motivates the following definition.

CHARACTERISTIC EQUATION OF A LINEAR OPERATOR

Let $f: S \to S$ be a linear operator on the n-dimensional space S, where $n > 0$. The *characteristic polynomial* of f is $\det(A - rI)$, where A is any one of the matrix representations of f (relative to a single basis for S). The *characteristic equation* of f is

$$\det(A - rI) = 0$$

Of course this raises a question! What if A and B are different matrix representations of f? Then f apparently has two characteristic polynomials,

$$\det(A - rI) \qquad \text{and} \qquad \det(B - rI)$$

We might argue that they are essentially the same because they have the same roots. (The eigenvalues of f are unambiguously defined without reference to matrix representations.) But this is unsatisfactory if $\mathcal{F} = \mathcal{R}$, since different polynomials can have the same *real* roots—for example,

$$(x^2 + 1)(x^2 - 1) \qquad \text{and} \qquad (x^2 + 3)(x^2 - 1)$$

So let us nail it down by a different approach.

† This suggests that nonzero solutions of $A\mathbf{X} = r\mathbf{X}$ should be called eigenvectors of A associated with the eigenvalue r. We'll adopt this terminology shortly.

‡ Some writers convert $A\mathbf{X} = r\mathbf{X}$ to the equation $(rI - A)\mathbf{X} = \mathbf{0}$ and work with $\det(rI - A)$. The roots of this polynomial are the same, but the polynomial itself is $(-1)^n$ times ours. (Why?)

The key to the argument is that A and B are similar, that is, $B = PAP^{-1}$ for some nonsingular matrix P. It follows that

$$B - rI = PAP^{-1} - rPIP^{-1} = P(A - rI)P^{-1}$$

so $A - rI$ and $B - rI$ are also similar. However, similar matrices have the same determinant (Prob. 25, Sec. 8.4), so the polynomials $\det(A - rI)$ and $\det(B - rI)$ are identical.

At this point we are ready to use the same terminology for $n \times n$ matrices as for linear operators.

EIGENVALUES AND EIGENVECTORS OF A MATRIX

Let A be an $n \times n$ matrix with entries in $\mathfrak{F}$. The *characteristic polynomial* of A is $\det(A - rI)$; the *characteristic equation* of A is $\det(A - rI) = 0$. The *eigenvalues* of A are the roots (in $\mathfrak{F}$) of its characteristic equation.† If r is an eigenvalue of A, the nonzero solutions of the system $A\mathbf{X} = r\mathbf{X}$ are called the *eigenvectors* of A associated with r; together with $\mathbf{0}$ they constitute a subspace of $\mathfrak{F}^n$ called the *eigenspace* associated with r.

This proliferation of technical terms is no doubt alarming. However, we haven't really done much yet; the only abstract results we have obtained are summarized in the following three theorems.

THEOREM 9.7

Similar matrices have the same characteristic polynomial (and hence the same eigenvalues).

Proof

This has already been given.

THEOREM 9.8

The eigenvalues of a linear operator (on a nonzero finite-dimensional space) are the eigenvalues of any one of its matrix representations (matrices being computed, as usual, relative to a single basis).

Proof

Refer to the above discussion. Some of the details may be fuzzy, however, so we'll go over it again. Let $f: S \to S$ be the linear operator and suppose that A is any one of its matrix representations, say $A = [f]_\alpha$. Then

† To make the statements of certain theorems simpler, we agree to list multiple eigenvalues repeatedly, as though they were distinct. This is in accordance with the practice in ordinary algebra; for example, the roots of $(x - 1)^2(x - 5) = 0$ are usually given as $r_1 = 1$, $r_2 = 1$, and $r_3 = 5$.

r is an eigenvalue of f

$$\begin{aligned}
&\Leftrightarrow \quad f(x) = rx && \text{for some nonzero } x \in S\\
&\Leftrightarrow \quad \alpha[f(x)] = \alpha(rx) && \text{for some nonzero } x \in S\\
&\Leftrightarrow \quad A\alpha(x) = r\alpha(x) && \text{for some nonzero } x \in S\\
&\Leftrightarrow \quad A\mathbf{X} = r\mathbf{X} && \text{for some nonzero } \mathbf{X} \in \mathfrak{F}^n\\
&\Leftrightarrow \quad (A - rI)\mathbf{X} = \mathbf{0} && \text{for some nonzero } \mathbf{X} \in \mathfrak{F}^n\\
&\Leftrightarrow \quad \det(A - rI) = 0\\
&\Leftrightarrow \quad r \text{ is an eigenvalue of } A \ \blacksquare
\end{aligned}$$

THEOREM 9.9

The eigenvalues of a matrix are the eigenvalues of any linear operator it represents (relative to a single basis).

Proof

Let A be an $n \times n$ matrix with entries in $\mathfrak{F}$. If S is an n-dimensional vector space over $\mathfrak{F}$ and $f\colon S \to S$ is a linear operator whose matrix is A, Theorem 9.8 says that the eigenvalues of f are the eigenvalues of A. ▮

Theorems 9.8 and 9.9 are really the same statement, but with different emphasis. In Theorem 9.8 we start with the map and then identify its eigenvalues with a matrix. In Theorem 9.9 we start with the matrix. Of course if A is given, the simplest space to choose is $S = \mathfrak{F}^n$ and the natural basis is the standard basis. Then f is the standard map defined by $f(\mathbf{X}) = A\mathbf{X}$. In this case we may go further and identify the *eigenspaces* of A and f.

THEOREM 9.10

Let A be an $n \times n$ matrix with entries in $\mathfrak{F}$ and define $f\colon \mathfrak{F}^n \to \mathfrak{F}^n$ by $f(\mathbf{X}) = A\mathbf{X}$. If r is an eigenvalue of A, the eigenspaces of A and f associated with r are the same.

Proof

We need only write $\{\mathbf{X} \mid f(\mathbf{X}) = r\mathbf{X}\} = \{\mathbf{X} \mid A\mathbf{X} = r\mathbf{X}\}$ ▮

There is nothing profound about Theorem 9.10; it is simply the result of planning ahead in the definitions. What is more interesting is the case in which the space S and the map $f\colon S \to S$ represented by A are not so concrete.

THEOREM 9.11

Let A be an $n \times n$ matrix with entries in $\mathfrak{F}$ and suppose that $f\colon S \to S$ is any linear operator represented by A, say $[f]_\alpha = A$. If r is an eigenvalue of A, the eigenspaces of A and f associated with r are isomorphic by virtue of the map which sends each element of S into its coordinate vector relative to α.

Proof

The eigenspaces of f and A associated with r are the subspaces of S and $\mathfrak{F}^n$, respectively, defined by

$$T = \{x \mid f(x) = rx\} \qquad \text{and} \qquad U = \{\mathbf{X} \mid A\mathbf{X} = r\mathbf{X}\}$$

The problem is to name an invertible linear map from one to the other. Why not α? More precisely (since α goes from S to $\mathfrak{F}^n$), let α_T be the restriction of α to the domain T, that is, the map which sends each element of T into its coordinate vector relative to $\alpha = \{u_1, \ldots, u_n\}$. This map is linear and one-to-one because α is; we need only show that its range is U.

Take any $\mathbf{X} = (x_1, \ldots, x_n)$ in U. Then the element $x = \Sigma_{j=1}^n x_j u_j$ in S has the property that $\alpha(x) = \mathbf{X}$; the proof is complete if we show that $x \in T$. Note that

$$\begin{aligned} f(x) &= \sum_{j=1}^n x_j f(u_j) \\ &= \sum_{j=1}^n x_j \left(\sum_{i=1}^n a_{ij} u_i \right) \qquad \text{where } A = [a_{ij}] \\ &= \sum_{i=1}^n \left(\sum_{j=1}^n a_{ij} x_j \right) u_i \end{aligned}$$

But $A\mathbf{X} = r\mathbf{X}$ (because $\mathbf{X} \in U$), and so, for each $i = 1, \ldots, n$, we have

$$\sum_{j=1}^n a_{ij} x_j = r x_i \qquad \text{(Why?)}$$

Hence

$$\begin{aligned} f(x) &= \sum_{i=1}^n (r x_i) u_i \\ &= r \sum_{i=1}^n x_i u_i \\ &= rx \end{aligned}$$

Therefore $x \in T$. ▮

Problems

15. Let A be an $n \times n$ matrix with entries in $\mathfrak{F}$ and suppose that r is an eigenvalue of A. Prove that the eigenspace of A associated with r is a subspace of $\mathfrak{F}^n$ (as asserted in the definition) as follows.
a. Apply Prob. 8, Sec. 9.2, to an appropriate linear operator.
b. Identify the eigenspace with the null space of an appropriate matrix.

★ **16.** Prove that the eigenvalues of a diagonal matrix are its diagonal entries.

17. Why does it follow from Prob. 16 that the determinant of a diagonal matrix is the product of its eigenvalues? (This doesn't sound too impressive until you learn that the same thing is true of *any* $n \times n$ matrix, provided that we take $\mathfrak{F} = \mathbb{C}$ and agree to list multiple eigenvalues as though they were distinct. The proof, however, requires some heavy artillery. See Prob. 30.)

★ **18.** Show that similar diagonal matrices have the same diagonal entries, except possibly for order. (This is the converse of Prob. 6, Sec. 9.1.)

19. Show that if A is an $n \times n$ diagonal matrix, then $\mathbf{E}_1, \ldots, \mathbf{E}_n$ are eigenvectors of A. *Hint:* If A is any $n \times n$ matrix, $A\mathbf{E}_j$ is the jth column of A.

★ **20.** Let A be an $n \times n$ matrix and suppose that a nonsingular matrix P exists such that $B = PAP^{-1}$ is diagonal. Prove that the columns of P^{-1} are eigenvectors of A. *Hint:* Let $B = \text{diag}\,(r_1, \ldots, r_n)$ and look at the jth column of each side of $P^{-1}B = AP^{-1}$.

★ **21.** Suppose that the $n \times n$ matrix A has n linearly independent eigenvectors. Then the matrix having these vectors as columns is nonsingular; call its inverse P. Prove that the matrix $B = PAP^{-1}$ is diagonal. (This justifies Lector's remark at the end of Sec. 9.2.)

★ **22.** Show that the $n \times n$ matrix A is singular if and only if one of its eigenvalues is 0. (This is trivial if you use the unproved remark following Prob. 17. Don't!)
Now we have another item to add to the list in Theorem 7.22:

> An $n \times n$ matrix is nonsingular if and only if none of its eigenvalues is 0.

23. Prove that an $n \times n$ matrix which is similar to a diagonal matrix with nonzero diagonal entries is nonsingular.

24. Show that the nonsingular matrix

$$A = \begin{bmatrix} 1 & 1 \\ 0 & 1 \end{bmatrix}$$

is *not* similar to a diagonal matrix with nonzero diagonal entries. (Thus the converse of Prob. 23 is false.)

25. Show that the matrix in Prob. 24 has the same eigenvalues as I. (We know that similar matrices have the same eigenvalues; this shows the converse to be false.)

26. Prove that an $n \times n$ matrix and its transpose have the same eigenvalues. Then give an example showing that they need not have the same eigenvectors.

★ **27.** Show that if r is an eigenvalue of the nonsingular matrix A, then r^{-1} is an eigenvalue of A^{-1}.

28. Why can't a linear operator on an n-dimensional space have more than n eigenvalues?

29. Let

$$A = \begin{bmatrix} a_{11} & a_{12} \\ a_{21} & a_{22} \end{bmatrix}$$

a. Show that the characteristic polynomial of A is

$$p(r) = r^2 - (\operatorname{tr} A)r + \det A$$

where tr A (the "trace" of A) is the sum of the diagonal entries. (See Prob. 24, Sec. 6.2.)

b. Why does it follow from part *a* that the determinant of A is the product of its eigenvalues? (This is the 2×2 case of the remark following Prob. 17.)

c. If $P(x) = a_0 + a_1x + \cdots + a_mx^m$ is any polynomial, the symbol $P(A)$ means $a_0I + a_1A + \cdots + a_mA^m$. Show that A satisfies its own characteristic equation, that is, $p(A) = 0$. (This is the 2×2 case of the celebrated *Cayley-Hamilton theorem,* which says that every square matrix satisfies its characteristic equation.)

★ **30.** The *fundamental theorem of algebra* (proved by Karl Friedrich Gauss in his doctoral dissertation, 1799) says that every polynomial of degree $n > 0$, with coefficients in $\mathcal{C}$, has at least one root in $\mathcal{C}$. The *factor theorem* (usually proved in intermediate algebra) says that if r is a root of a polynomial in x, then $x - r$ is a factor of the polynomial.

a. Use these theorems to argue that if multiple roots are counted as though they were distinct, the polynomial $P(x) = a_0 + a_1x + \cdots + a_nx^n$ (with coefficients in $\mathcal{C}$ and degree $n > 0$) has exactly n roots in $\mathcal{C}$, and

$$P(x) = a_n(x - r_1) \cdots (x - r_n)$$

where $r_1, \ldots, r_n$ are the roots.

b. Why does it follow that the product of the roots is $(-1)^n(a_0/a_n)$? *Hint:* Set $x = 0$.

c. Now you should be able to prove that the determinant of an $n \times n$ matrix is the product of its eigenvalues in $\mathcal{C}$. (See the remark following Prob. 17.)

9.4 Diagonalization of a Matrix

Let $f: S \to S$ be a linear operator with matrix A relative to a basis α for S. We observed in Sec. 9.2 that A can be diagonalized if and only if there are scalars $r_1, \ldots, r_n$ (not necessarily distinct) and a basis $\beta = \{v_1, \ldots, v_n\}$, such that

$$f(v_j) = r_j v_j \qquad j = 1, \ldots, n$$

where $n = \dim S$. Moreover, the diagonal representation of f in these circumstances is

$$B = [f]_\beta = \text{diag}\,(r_1, \ldots, r_n)$$

Now we are in a position to state conditions for the diagonalization of a matrix, and to describe how it is done when these conditions are met.

THEOREM 9.12

Let A be a matrix representation of the linear operator $f: S \to S$, where S is an n-dimensional vector space over $\mathfrak{F}$. Each of the following statements is equivalent to each of the others.

(1) A can be diagonalized.

(2) A is similar to a diagonal matrix.

(3) $\mathfrak{F}^n$ has a basis consisting of eigenvectors of A.

(4) S has a basis consisting of eigenvectors of f.

Proof

The implication (1) $\Longrightarrow$ (2) is trivial, for if A can be diagonalized, then by definition there is a diagonal matrix B representing f; by Theorem 6.7, A and B are similar.

To show that (2) $\Longrightarrow$ (3), suppose that there is a nonsingular matrix P such that $B = PAP^{-1}$ is diagonal. By Prob. 20 in the last section the columns of P^{-1} are eigenvectors of A; they constitute a basis for $\mathfrak{F}^n$ because they are linearly independent (P^{-1} is nonsingular).

Next, assuming that $\mathfrak{F}^n$ has a basis consisting of eigenvectors of A, let $v_1, \ldots, v_n$ be the elements of S whose coordinate vectors relative to α are these eigenvectors (where α is the basis for S relative to which A is the matrix of f). Then $\beta = \{v_1, \ldots, v_n\}$ is a basis for S. (Why?) Moreover, each v_j is an eigenvector of f (Theorem 9.11). Thus (3) $\Rightarrow$ (4).

Finally, if $\{v_1, \ldots, v_n\}$ is a basis for S consisting of eigenvectors of f, we have

$$f(v_j) = r_j v_j \qquad j = 1, \ldots, n$$

where the r_j are scalars. It follows from Theorem 9.6 that f has a diagonal matrix representation. Thus (4) $\Rightarrow$ (1). ∎

Theorem 9.12 gives necessary and sufficient conditions for diagonalizing the matrix A of a linear operator f. But in practice how does one actually go about doing it (or ascertain that it cannot be done)? The following sequence of steps emerges from what has gone before.

DIAGONALIZATION OF A MATRIX

1. Find the characteristic polynomial of A; that is, compute det $(A - rI)$.
2. Solve the characteristic equation of A; that is, find the roots (in $\mathfrak{F}$) of det $(A - rI) = 0$. These are the eigenvalues of A. (The process may stop right here! For if $\mathfrak{F} = \mathfrak{R}$, the characteristic equation may not have any roots. On the other hand, it always has roots in $\mathfrak{F} = \mathfrak{C}$; recall the fundamental theorem of algebra, quoted in Prob. 30, Sec. 9.3.)
3. For each eigenvalue r, solve the linear homogeneous system $(A - rI)\mathbf{X} = \mathbf{0}$. Its solution set is the eigenspace of A associated with r and is bound to be at least 1-dimensional; find a basis for it. (Note that in complicated cases this may require the apparatus of Chap. 7.)
4. Select n linearly independent eigenvectors of A from the bases for the eigenspaces found in step 3; they constitute a basis for $\mathfrak{F}^n$. (Or ascertain that this cannot be done, in which case the process stops.)
5. Put down the nonsingular matrix whose columns are the eigenvectors found in step 4 and finds its inverse P. Then compute the product $B = PAP^{-1}$; this will be a diagonal matrix representing f.

Note that this process is described entirely in terms of the matrix A; the map f is not explicitly involved. Of course the difficulties may be enormous, even when the process is theoretically guaranteed to work. Among other things, it is rare that a polynomial equation of degree n can be solved; we are likely to get bogged down in step 2 even if the characteristic equation has roots in $\mathfrak{F}$. When $n = 1$ or $n = 2$ everybody knows what to do, and there are formulas for determining the roots when $n = 3$ or $n = 4$, attributed to

Cardano (1501–1576) and Ferrari (1522–1565). But when $n \geq 5$, no general formula exists.[†]

As you can imagine, the diagonalizing we ask you to do in the problems involves carefully chosen matrices. There are certain shortcuts, too, which make the process less gruesome. The most obvious is the case where A has n distinct eigenvalues. Then (as we'll prove shortly) the last three steps outlined above are unnecessary; we need only write down the diagonal matrix which has these eigenvalues as diagonal entries. (See Example 9.10.)

EXAMPLE 9.11

Define $f: \mathcal{R}^3 \to \mathcal{R}^3$ by $f(x,y,z) = (2x + z, y, x + 2z)$. The matrix of f relative to the standard basis is

$$A = \begin{bmatrix} 2 & 0 & 1 \\ 0 & 1 & 0 \\ 1 & 0 & 2 \end{bmatrix}$$

Its eigenvalues are the roots of the characteristic equation

$$\det(A - rI) = \begin{vmatrix} 2-r & 0 & 1 \\ 0 & 1-r & 0 \\ 1 & 0 & 2-r \end{vmatrix} = 0$$

Expanding by the second column, we have

$$\det(A - rI) = (1 - r)\begin{vmatrix} 2-r & 1 \\ 1 & 2-r \end{vmatrix} = (1 - r)(1 - r)(3 - r)$$

so the eigenvalues are $r_1 = 1$, $r_2 = 1$, and $r_3 = 3$. The eigenvectors associated with $r_1 = r_2 = 1$ are the nonzero solutions of $(A - I)\mathbf{X} = \mathbf{0}$, that is, of the homogeneous system whose coefficient matrix is

$$\begin{bmatrix} 1 & 0 & 1 \\ 0 & 0 & 0 \\ 1 & 0 & 1 \end{bmatrix}$$

[†] By "formula" we mean something like the quadratic formula, in which the coefficients of the equation are subjected to the five algebraic operations of addition, subtraction, multiplication, division, and root extraction. The discovery of a "quintic" formula was an outstanding problem of mathematics for more than 250 years after the "cubic" and "quartic" formulas were found; everybody assumed that it was just a matter of being clever enough to find it. There was general disbelief when Abel published a proof that it does not exist. (While still in school, he thought he had found one, but realized his error before publishing it.) Galois (1811–1832) proved that for *every* integer $n \geq 5$, no formula exists for the general equation of degree n. He did this by inventing a body of doctrine known today as the Galois theory of groups; it is still of importance in modern mathematics. The content of this theory, together with an outline of other work he had in progress, was sent in a letter to a friend the night before he was to fight a duel with a provocateur. "There is no time," he repeated, and indeed there was not; not yet 21 years old, he was fatally wounded the next morning.

The eigenspace is therefore

$$\{(x,y,z) \mid x = -t_2,\ y = t_1,\ z = t_2\}$$

a 2-dimensional subspace of $\mathcal{R}^3$ with basis consisting of the eigenvectors $\mathbf{V}_1 = (0,1,0)$ and $\mathbf{V}_2 = (-1,0,1)$. (Confirm this!)

The eigenvectors associated with $r_3 = 3$ are the nonzero solutions of the homogeneous system whose coefficient matrix is

$$A - 3I = \begin{bmatrix} -1 & 0 & 1 \\ 0 & -2 & 0 \\ 1 & 0 & -1 \end{bmatrix}$$

The eigenspace is 1-dimensional, with basis vector $\mathbf{V}_3 = (1,0,1)$. (Confirm this!)

Since $\mathbf{V}_1$, $\mathbf{V}_2$, and $\mathbf{V}_3$ are linearly independent, Theorem 9.12 guarantees that A can be diagonalized. We use these eigenvectors as columns of the matrix

$$P^{-1} = \begin{bmatrix} 0 & -1 & 1 \\ 1 & 0 & 0 \\ 0 & 1 & 1 \end{bmatrix}$$

and compute its inverse,

$$P = \frac{1}{2}\begin{bmatrix} 0 & 2 & 0 \\ -1 & 0 & 1 \\ 1 & 0 & 1 \end{bmatrix}$$

(Confirm this!) The corresponding diagonal matrix representing f is

$$B = [f]_\beta = PAP^{-1} = \begin{bmatrix} 1 & 0 & 0 \\ 0 & 1 & 0 \\ 0 & 0 & 3 \end{bmatrix}$$

where $\beta = \{\mathbf{V}_1, \mathbf{V}_2, \mathbf{V}_3\}$. (Confirm this!) Note that (as predicted by the theory)

$$\begin{aligned} f(\mathbf{V}_1) &= f(0,1,0) = (0,1,0) = r_1\mathbf{V}_1 \\ f(\mathbf{V}_2) &= f(-1,0,1) = (-1,0,1) = r_2\mathbf{V}_2 \\ f(\mathbf{V}_3) &= f(1,0,1) = (3,0,3) = r_3\mathbf{V}_3 \end{aligned}$$

Also note that $B = \text{diag}\,(1,1,3)$ could have been put down as soon as the eigenvalues were found. The next example, however, shows that this is not always the case.

EXAMPLE 9.12

Define $f\colon \mathcal{R}^2 \to \mathcal{R}^2$ by $f(x,y) = (x + 2y,\ -2x - 3y)$. The eigenvalues are $r_1 = r_2 = -1$, as you can check; the corresponding eigenspace is the 1-dimensional subspace of $\mathcal{R}^2$ spanned by $\mathbf{V}_1 = (-1,1)$. This time we cannot span the parent space with eigenvectors; no basis exists relative to which the matrix of f is diagonal. It is no use writing down

$$B = \begin{bmatrix} r_1 & 0 \\ 0 & r_2 \end{bmatrix} = \begin{bmatrix} -1 & 0 \\ 0 & -1 \end{bmatrix}$$

for Theorem 9.12 says that

$$A = [f] = \begin{bmatrix} 1 & 2 \\ -2 & -3 \end{bmatrix}$$

cannot be diagonalized. Hence B cannot be a matrix representation of f.

EXAMPLE 9.13

Define $f\colon \mathcal{R}^2 \to \mathcal{R}^2$ by $f(x,y) = (-x + 5y, -x + 3y)$. The characteristic equation this time is $r^2 - 2r + 2 = 0$; its roots are imaginary and we can't even get off the ground. Hence

$$A = [f] = \begin{bmatrix} -1 & 5 \\ -1 & 3 \end{bmatrix}$$

cannot be diagonalized.

EXAMPLE 9.14

Define f by the same formula as in Example 9.13, but consider it as a linear operator on $\mathcal{C}^2$. Then the characteristic equation *can* be solved; the eigenvalues are $r_1 = 1 + i$ and $r_2 = 1 - i$. The eigenvectors of $A = [f]$ associated with r_1 are the nonzero solutions of

$$[A - (1 + i)I]\mathbf{X} = \mathbf{0}$$

the coefficient matrix of which is

$$\begin{bmatrix} -2 - i & 5 \\ -1 & 2 - i \end{bmatrix}$$

The row operations (1) $\leftrightarrow$ (2), (1) $\to$ $-$(1), (2) $\to$ (2) + $(2 + i)$(1) transform this matrix into its row echelon form

$$\begin{bmatrix} 1 & -2 + i \\ 0 & 0 \end{bmatrix}$$

from which we read off the 1-dimensional eigenspace with basis vector $\mathbf{V}_1 = (2 - i, 1)$.

The eigenspace associated with $r_2 = 1 - i$ is the solution space of the homogeneous system whose coefficient matrix is

$$\begin{bmatrix} -2 + i & 5 \\ -1 & 2 + i \end{bmatrix}$$

The row echelon form is

$$\begin{bmatrix} 1 & -2-i \\ 0 & 0 \end{bmatrix}$$

from which we find the 1-dimensional eigenspace with basis vector $\mathbf{V}_2 = (2+i, 1)$. Since $\mathbf{V}_1$ and $\mathbf{V}_2$ are linearly independent, we can diagonalize A by writing

$$P^{-1} = \begin{bmatrix} 2-i & 2+i \\ 1 & 1 \end{bmatrix} \qquad P = \frac{1}{2}\begin{bmatrix} i & 1-2i \\ -i & 1+2i \end{bmatrix}$$

$$B = PAP^{-1} = \begin{bmatrix} 1+i & 0 \\ 0 & 1-i \end{bmatrix}$$

Much time could be saved by putting B down as soon as $r_1 = 1 + i$ and $r_2 = 1 - i$ are found. As we have already mentioned, this is legitimate whenever the characteristic equation has n distinct roots in $\mathfrak{F}$. We prove this as follows.

THEOREM 9.13

Eigenvectors associated with distinct eigenvalues of a linear operator are linearly independent.

Proof

Suppose that the linear operator $f: S \to S$ has eigenvectors $v_1, \ldots, v_m$ associated with the distinct eigenvalues $r_1, \ldots, r_m$. We proceed by induction on m. If $m = 1$, there is only one eigenvector in the list; being nonzero by definition, it constitutes a linearly independent subset of S by itself.

Now assume that $m > 1$ and that the theorem is true when there are $m - 1$ eigenvectors in the list. Suppose that

$$a_1v_1 + \cdots + a_mv_m = 0 \tag{9.1}$$

The problem is to show that each a_j is equal to 0. Multiplying each side of Eq. (9.1) by r_m yields

$$a_1r_mv_1 + \cdots + a_mr_mv_m = 0 \tag{9.2}$$

while applying f to each side of Eq. (9.1) gives

$$a_1f(v_1) + \cdots + a_mf(v_m) = 0$$

or

$$a_1r_1v_1 + \cdots + a_mr_mv_m = 0 \qquad \text{(Why?)} \tag{9.3}$$

Subtracting Eq. (9.3) from Eq. (9.2), we have

$$a_1(r_m - r_1)v_1 + \cdots + a_{m-1}(r_m - r_{m-1})v_{m-1} = 0$$

But $v_1, \ldots, v_{m-1}$ are linearly independent by the induction hypothesis, so each coefficient in this last equation must be zero, that is,

$$a_j(r_m - r_j) = 0 \qquad j = 1, \ldots, m - 1$$

Since the eigenvalues $r_1, \ldots, r_m$ are distinct, it follows that $a_j = 0$, $j = 1, \ldots, m - 1$. Equation (9.1) therefore reduces to $a_m v_m = 0$, which implies that $a_m = 0$. ∎

COROLLARY 9.13a

If S is n-dimensional and the linear operator $f\colon S \to S$ has n distinct eigenvalues, then S has a basis consisting of eigenvectors of f.

COROLLARY 9.13b

An $n \times n$ matrix with distinct eigenvalues $r_1, \ldots, r_n$ is similar to diag $(r_1, \ldots, r_n)$.

Thus the existence of n distinct eigenvalues is sufficient to guarantee that the $n \times n$ matrix A can be diagonalized. But this is not a necessary condition! In Example 9.11 we diagonalized a 3×3 matrix with eigenvalues 1, 1, 3 by finding two linearly independent eigenvectors associated with the double eigenvalue; that is, the dimension of the eigenspace was the same as the multiplicity of the root.† This condition turns out to be necessary and sufficient for diagonalizing a matrix, as we now show.

THEOREM 9.14

If S is n-dimensional and the linear operator $f\colon S \to S$ has n eigenvalues, then S has a basis consisting of eigenvectors of f if and only if the dimension of each eigenspace is the multiplicity of the corresponding eigenvalue.

Proof

We'll do the "only if" part first. Suppose that S has a basis $\beta = \{v_1, \ldots, v_n\}$ consisting of eigenvectors of f. Then

$$f(v_j) = r_j v_j \qquad j = 1, \ldots, n$$

where $r_1, \ldots, r_n$ are the eigenvalues of f. Let r be any one of these eigenvalues (we drop the subscript to simplify the notation) and suppose that its multiplicity is k. Then the corresponding eigenspace, namely $\{x \mid f(x) = rx\}$, is spanned by the k elements of β which are eigenvectors of f associated with r. (See the problems.) Since these elements are linearly independent, the dimension of the eigenspace associated with r is k.

† A root occurring k times is said to have multiplicity k. Some writers call k the "algebraic" multiplicity, to distinguish it from the "geometric" multiplicity, which is the dimension of the corresponding eigenspace. In Example 9.12 the algebraic multiplicity of the eigenvalue -1 is $k = 2$; the geometric multiplicity is 1.

Conversely, suppose that the dimension of each eigenspace is the multiplicity of the corresponding eigenvalue. Relabel the eigenvalues (if necessary) so that $r_1, \ldots, r_m$ are *distinct,* each r_i having multiplicity k_i, where $k_1 + \cdots + k_m = n$.† Then the eigenspace associated with r_i (where i is any one of the integers $1, \ldots, m$) has a basis consisting of k_i eigenvectors, say $v_{i1}, \ldots, v_{ik_i}$. We claim that the n eigenvectors

$$v_{11}, \ldots, v_{1k_1}, v_{21}, \ldots, v_{2k_2}, \ldots, v_{m1}, \ldots, v_{mk_m}$$

are linearly independent. If so, they constitute a basis for S and the proof is complete.

To prove linear independence, suppose that

$$(c_{11}v_{11} + \cdots + c_{1k_1}v_{1k_1}) + \cdots + (c_{m1}v_{m1} + \cdots + c_{mk_m}v_{mk_m}) = 0$$

and (for each $i = 1, \ldots, m$) let

$$u_i = c_{i1}v_{i1} + \cdots + c_{ik_i}v_{ik_i}$$

Then $u_1 + \cdots + u_m = 0$, where u_i is in the eigenspace associated with r_i. But this implies that each $u_i = 0$ (see the problems) and hence, for each i, the scalars $c_{i1}, \ldots, c_{ik_i}$ are all zero. (Why?) Linear independence follows as advertised. ▮

COROLLARY 9.14a

An $n \times n$ matrix with eigenvalues $r_1, \ldots, r_n$ is similar to diag $(r_1, \ldots, r_n)$ if and only if the dimension of each of its eigenspaces is the multiplicity of the corresponding eigenvalue.

Proof

Let A be the matrix and define the standard map $f\colon \mathfrak{F}^n \to \mathfrak{F}^n$ by $f(\mathbf{X}) = A\mathbf{X}$. Then the eigenvalues (and eigenspaces) of A and f are the same (Theorems 9.9 and 9.10). Suppose that the dimension of each eigenspace is the multiplicity of the corresponding eigenvalue. By Theorem 9.14 $\mathfrak{F}^n$ has a basis consisting of eigenvectors; hence A is similar to a diagonal matrix (Theorem 9.12). By Prob. 6 of Sec. 9.1 we may suppose that this is diag $(r_1, \ldots, r_n)$. The converse is proved by reversing the argument. ▮

Problems

31. Find the eigenvalues of each of the following linear operators by using matrix representations.

a. The identity map on an n-dimensional space. (See Example 9.5.)

† For example, if $n = 3$ and the original list is $r_1 = 1$, $r_2 = 1$, and $r_3 = 3$, we relabel by writing $r_1 = 1$ (multiplicity $k_1 = 2$) and $r_2 = 3$ (multiplicity $k_2 = 1$). Then $m = 2$ and $k_1 + k_2 = 3$.

b. The rotation of the plane through an angle θ, $0 < \theta < \pi$. (See Example 9.6.)

c. The reflection of the plane in the x axis. (See Example 9.7.)

d. The rotation of the plane through 180°. (See Prob. 10*a*, Sec. 9.2.)

e. The reflection of the plane in the line $y = x$. (See Prob. 10*b*, Sec. 9.2.)

f. The orthogonal projection of $\mathcal{R}^3$ on the xy plane. (See Prob. 10*c*, Sec. 9.2.)

g. The differential operator D: $S \to S$ defined by $D(\phi) = \phi'$, where S is the vector space (over $\mathcal{R}$) generated by $\phi_1(x) = e^x$ and $\phi_2(x) = e^{2x}$. (See Prob. 10*d*, Sec. 9.2.)

32. In each case, find the eigenvalues of A and a basis for the eigenspace associated with each eigenvalue. (Assume that the base field is $\mathfrak{F} = \mathcal{C}$.)

a. $A = \begin{bmatrix} 2 & -1 \\ -1 & 2 \end{bmatrix}$ *b*. $A = \begin{bmatrix} 3 & -1 \\ 2 & 1 \end{bmatrix}$

c. $A = \begin{bmatrix} 1 & 1 \\ -1 & 3 \end{bmatrix}$ *d*. $A = \begin{bmatrix} -1 & 3 & -2 \\ 3 & -1 & -2 \\ 0 & 0 & 6 \end{bmatrix}$

e. $A = \begin{bmatrix} 1 & -1 & -1 \\ 1 & 3 & 1 \\ -1 & -1 & 1 \end{bmatrix}$ *f*. $A = \begin{bmatrix} 2 & 0 & 0 \\ 3 & 2 & 0 \\ 5 & -2 & -1 \end{bmatrix}$

g. $A = \begin{bmatrix} 3 & 0 & 1 \\ 2 & 2 & 1 \\ -6 & -1 & -2 \end{bmatrix}$ *h*. $A = \begin{bmatrix} 3 & 2 & 1 \\ -1 & 0 & -1 \\ 1 & 1 & 2 \end{bmatrix}$

33. In each part of Prob. 32 determine whether $\mathfrak{F}^n$ has a basis consisting of eigenvectors of A. Can A be diagonalized?

34. In each part of Prob. 32 (where possible) find a nonsingular matrix P such that $B = PAP^{-1}$ is diagonal, and compute B.

35. In each part of Prob. 32 confirm that the determinant of A is the product of its eigenvalues. (See the remark following Prob. 17, Sec. 9.3.)

36. The matrix

$$A = \begin{bmatrix} 2 & 0 & 0 \\ 3 & 2 & 0 \\ 5 & -2 & -1 \end{bmatrix}$$

in Prob. 32*f* is called a "lower triangular" matrix. (All the entries above the main diagonal are zero.) Its eigenvalues, as in the case of diagonal matrices, are the diagonal entries 2, 2, -1. Prove that this is true in general. *Hint:* If A is a lower triangular matrix, then $a_{ij} = 0$ whenever $i < j$.

37. An $n \times n$ matrix A is called "upper triangular" if all the entries below the main diagonal are zero, that is, if $a_{ij} = 0$ whenever $i > j$. Prove that its eigenvalues are the diagonal entries.

Triangular matrices are not as nice to work with as diagonal matrices. Nevertheless "triangulation" of the matrix of a linear operator is a desirable simplification. Unlike "diagonalization" (which is not always possible), triangulation *is* (provided that the base field is taken to be $\mathfrak{F} = \mathfrak{C}$). See Appendix 2.

38. Let S be the vector space of real-valued functions with derivatives of all orders on $(-\infty, \infty)$ and suppose that $r_1, \ldots, r_n$ are distinct real numbers. Why are the functions $e^{r_1 x}, \ldots, e^{r_n x}$ linearly independent elements of S? *Hint:* See Example 9.8.

39. Prove Corollary 9.13a: If S is n-dimensional and the linear operator $f\colon S \to S$ has n distinct eigenvalues, then S has a basis consisting of eigenvectors of f.

40. Prove Corollary 9.13b: An $n \times n$ matrix with distinct eigenvalues $r_1, \ldots, r_n$ is similar to diag $(r_1, \ldots, r_n)$.

41. Let r be any one of the n eigenvalues of the linear operator $f\colon S \to S$, where S has a basis $\beta = \{v_1, \ldots, v_n\}$ consisting of eigenvectors of f, and suppose that its multiplicity is k. Prove that the corresponding eigenspace is spanned by the k elements of β which are eigenvectors of f associated with r. (This completes the proof of the "only if" part of Theorem 9.14.)

42. Let $r_1, \ldots, r_m$ be the distinct eigenvalues of a linear operator $f\colon S \to S$, where S is n-dimensional and f has n eigenvalues altogether. Prove that if u_i is in the eigenspace associated with r_i, where $i = 1, \ldots, m$, and if $u_1 + \cdots + u_m = 0$, then each $u_i = 0$. *Hint:* Use Theorem 9.13 repeatedly. (This completes the proof of the "if" part of Theorem 9.14.)

43. Suppose that the $n \times n$ matrix A is similar to $B = \text{diag}\,(r_1, \ldots, r_n)$. Prove that the dimension of each of its eigenspaces is the multiplicity of the corresponding eigenvalue, as follows.

a. Reverse the argument for the "if" part of Corollary 9.14a (given in the text).

b. Give an argument not depending on Theorem 9.14. *Hint:* Let r be any one of the eigenvalues (with multiplicity k). Show that the dimension of the corresponding eigenspace is k by deducing the nullity of $A - rI$ from the rank of $B - rI$.

†9.5 An Application to Differential Equations

Suppose that we want to find functions

$$y_1 = \phi_1(x) \qquad y_2 = \phi_2(x) \qquad y_3 = \phi_3(x)$$

satisfying the system of differential equations

$$\begin{aligned} y_1' &= 2y_1 + y_3 \\ y_2' &= y_2 \\ y_3' &= y_1 + 2y_3 \end{aligned}$$

where the primes mean differentiation with respect to x. Introducing the vector $\mathbf{Y} = (y_1, y_2, y_3)$, with derivative $\mathbf{Y}' = (y_1', y_2', y_3')$, we may write this system more concisely in the form

$$\mathbf{Y}' = A\mathbf{Y}$$

where

$$A = \begin{bmatrix} 2 & 0 & 1 \\ 0 & 1 & 0 \\ 1 & 0 & 2 \end{bmatrix}$$

is the matrix of coefficients and $\mathbf{Y}$ and $\mathbf{Y}'$ are regarded as column vectors (3×1 matrices). The unknown functions ϕ_1, ϕ_2, and ϕ_3 may be regarded as components of the vector function $\mathbf{\Phi}$ defined by

$$\mathbf{\Phi}(x) = (\phi_1(x), \phi_2(x), \phi_3(x))$$

The object is to find $\mathbf{\Phi}$.

By analogy with the methods of ordinary (scalar) differential equations (which you may have encountered elsewhere) we propose to try a vector solution of the form

$$\mathbf{\Phi}(x) = e^{rx}\mathbf{V}$$

where r and $\mathbf{V}$ are constants (scalar and vector, respectively) to be determined by substitution in $\mathbf{Y}' = A\mathbf{Y}$. Since $\mathbf{\Phi}'(x) = re^{rx}\mathbf{V}$, the function $\mathbf{\Phi}$ is a solution of $\mathbf{Y}' = A\mathbf{Y}$ if and only if

$$re^{rx}\mathbf{V} = e^{rx}A\mathbf{V}$$

But e^{rx} is never 0, so this is equivalent to

$$r\mathbf{V} = A\mathbf{V}$$

an equation defining $\mathbf{V}$ as an eigenvector of A associated with the eigenvalue r.

We have already found the eigenvalues and eigenspaces of A in Example 9.11: $r_1 = r_2 = 1$, with eigenspace spanned by $\mathbf{V}_1 = (0,1,0)$ and $\mathbf{V}_2 =$

† This section can be omitted without loss of continuity.

$(-1,0,1)$; and $r_3 = 3$, with eigenspace spanned by $\mathbf{V}_3 = (1,0,1)$. Hence each of the vector functions

$$\begin{aligned}\mathbf{\Phi}_1(x) &= e^{r_1x}\mathbf{V}_1 = e^x(0,1,0)\\ \mathbf{\Phi}_2(x) &= e^{r_2x}\mathbf{V}_2 = e^x(-1,0,1)\\ \mathbf{\Phi}_3(x) &= e^{r_3x}\mathbf{V}_3 = e^{3x}(1,0,1)\end{aligned}$$

serves as a candidate for the $\mathbf{\Phi}$ we are seeking. The first one, for example, has components

$$\phi_{11}(x) = 0 \qquad \phi_{21}(x) = e^x \qquad \phi_{31}(x) = 0$$

which (as you can check) satisfy the original system. The second one, with components

$$\phi_{12}(x) = -e^x \qquad \phi_{22}(x) = 0 \qquad \phi_{32}(x) = e^x$$

and the third one, with components

$$\phi_{13}(x) = e^{3x} \qquad \phi_{23}(x) = 0 \qquad \phi_{33}(x) = e^{3x}$$

serve equally well. More important, it can be proved that *every* (vector) solution of the system is a linear combination of these.† Thus the general solution is

$$\begin{aligned}\mathbf{\Phi}(x) &= c_1\mathbf{\Phi}_1(x) + c_2\mathbf{\Phi}_2(x) + c_3\mathbf{\Phi}_3(x)\\ &= c_1e^x(0,1,0) + c_2e^x(-1,0,1) + c_3e^{3x}(1,0,1)\end{aligned}$$

In terms of the components, this says that the general form of the functions we originally set out to find is

$$\phi_1(x) = -c_2e^x + c_3e^{3x} \qquad \phi_2(x) = c_1e^x \qquad \phi_3(x) = c_2e^x + c_3e^{3x}$$

where c_1, c_2, and c_3 are arbitrary constants.

This method applies to any $n \times n$ system $\mathbf{Y}' = A\mathbf{Y}$ whose matrix can be diagonalized. (The reason for the proviso is that we must be able to span $\mathfrak{F}^n$ with eigenvectors of A to guarantee n linearly independent vector solutions.) A modification can be developed to apply to any system whatever, but we won't go into it.‡

Problems

44. Confirm that the functions

$$y_1 = \phi_1(x) = -c_2e^x + c_3e^{3x} \qquad y_2 = \phi_2(x) = c_1e^x$$
$$y_3 = \phi_3(x) = c_2e^x + c_3e^{3x}$$

†This is not obvious! It is a consequence of the fact (proved in differential equations) that the set of vector solutions $\mathbf{\Phi} = (\phi_1,\phi_2,\phi_3)$ is a 3-dimensional space. The functions $\mathbf{\Phi}_1$, $\mathbf{\Phi}_2$, $\mathbf{\Phi}_3$ are linearly independent in this space and hence constitute a basis.

‡See Philip Gillett, *Linear Mathematics*, Boston: Prindle, Weber, and Schmidt, 1970 (Chap. 4).

satisfy the system

$$y_1' = 2y_1 + y_3$$
$$y_2' = y_2$$
$$y_3' = y_1 + 2y_3$$

45. Use the method described in this section to find the general solution of each of the following systems.

a. $y_1' = 2y_1 + 2y_2$
$y_2' = y_1 + 3y_2$ (See Example 9.10.)

b. $y_1' = y_1 - y_2 - y_3$
$y_2' = y_1 + 3y_2 + y_3$
$y_3' = -y_1 - y_2 + y_3$ (See Prob. 32*e*, Sec. 9.4.)

46. Find the solution of the system in Prob. 45*a* whose components satisfy the initial conditions $\phi_1(0) = 1$ and $\phi_2(0) = -2$.

†9.6 The Cayley-Hamilton Theorem

In Prob. 29, Sec. 9.3, we stated the 2×2 case of the Cayley-Hamilton theorem, which says that every square matrix satisfies its characteristic equation. In this section we propose to prove it for $n \times n$ matrices which can be diagonalized; the argument for matrices in general is given in Appendix 2.

Suppose that A is a square matrix with entries in $\mathfrak{F}$ and P is a polynomial with coefficients in $\mathfrak{F}$, say $P(x) = a_0 + a_1x + \cdots + a_mx^m$. We define the symbol $P(A)$ to mean the matrix

$$P(A) = a_0I + a_1A + \cdots + a_mA^m$$

THEOREM 9.15

Let P and Q be polynomials with coefficients in $\mathfrak{F}$ and suppose that A is a square matrix with entries in $\mathfrak{F}$. Then

a. $(P + Q)(A) = P(A) + Q(A)$
b. $(kP)(A) = kP(A)$ for every $k \in \mathfrak{F}$
c. $(PQ)(A) = P(A)Q(A)$

Proof

Suppose that $P(x) = a_0 + a_1x + \cdots + a_mx^m$ and $Q(x) = b_0 + b_1x + \cdots + b_nx^n$. In part *a* assume that $m < n$ and let $a_i = 0$, $i = m + 1, \ldots, n$.

† This section can be omitted without loss of continuity.

(The procedure for $m > n$ is similar, while if $m = n$, no zero terms need be inserted.)
Then

$$\begin{aligned}(P + Q)(A) &= (a_0 + b_0)I + (a_1 + b_1)A + \cdots + (a_n + b_n)A^n \\ &= (a_0I + a_1A + \cdots + a_nA^n) + (b_0I + b_1A + \cdots + b_nA^n) \\ &= P(A) + Q(A)\end{aligned}$$

Part b is left for the problems.

Part c will be omitted. The argument is like that in part a, but the details are messy (and not very enlightening). ∎

Since Theorem 9.15c involves matrix products, it is not an obvious result. Even a simple formula like $(A + B)^2 = A^2 + 2AB + B^2$ is not true in matrix algebra (unless A and B commute).

EXAMPLE 9.15

Let $P(x) = x^2 - 4 = (x - 2)(x + 2)$. If A is any square matrix, Theorem 9.15c says that

$$P(A) = A^2 - 4I = (A - 2I)(A + 2I)$$

Of course in a case this simple we may use the laws of matrix algebra directly:

$$\begin{aligned}(A - 2I)(A + 2I) &= (A - 2I)A + (A - 2I)2I \\ &= A^2 - 2IA + 2AI - 4I^2 = A^2 - 4I\end{aligned}$$

As you can see, the place where one needs the matrices to commute causes no difficulty.

THEOREM 9.16 The Cayley-Hamilton Theorem

Let A be an $n \times n$ matrix with entries in $\mathfrak{F}$ and let p be its characteristic polynomial. Then $p(A) = 0$.

Proof

Suppose that A can be diagonalized. Then $\mathfrak{F}^n$ has a basis $\{\mathbf{V}_1, \ldots, \mathbf{V}_n\}$ consisting of eigenvectors of A. If $r_1, \ldots, r_n$ are the corresponding eigenvalues (roots of p in $\mathfrak{F}$), we can write

$$p(r) = c(r - r_1) \cdots (r - r_n)$$

where c is the coefficient of r^n in $p(r)$. (See the factor theorem stated in Prob. 30, Sec. 9.3.) Apply Theorem 9.15c to write

$$p(A) = c(A - r_1I) \cdots (A - r_nI)$$

and note that, for each $i = 1, \ldots, n$, we have (when $j \neq i$)

$$(A - r_j I)\mathbf{V}_i = A\mathbf{V}_i - r_j\mathbf{V}_i = r_i\mathbf{V}_i - r_j\mathbf{V}_i = (r_i - r_j)\mathbf{V}_i$$

Hence

$$p(A)\mathbf{V}_i = \mathbf{0} \qquad i = 1, \ldots, n \qquad \text{(Write it out!)}$$

The only matrix with this property is the zero matrix. (Why?) Hence $p(A) = 0$.

If A cannot be diagonalized, this argument breaks down. We use instead the fact that A can be triangulated (see Prob. 37 in Sec. 9.4). The proof of this fact is not simple, however, nor is the argument for the Cayley-Hamilton theorem based on it. We have put it in Appendix 2. ▮

Problems

47. Let P be a polynomial and let A be a square matrix. Prove that $(kP)(A) = kP(A)$ for every scalar k.

48. Let A be an $n \times n$ matrix with eigenvalues $r_1, \ldots, r_n$ and corresponding linearly independent eigenvectors $\mathbf{V}_1, \ldots, \mathbf{V}_n$. In the text we argued that

$$(A - r_j I)\mathbf{V}_i = (r_i - r_j)\mathbf{V}_i$$

for each $i = 1, \ldots, n$ and every $j \neq i$.

a. Explain why this implies that $p(A)\mathbf{V}_i = \mathbf{0}$, $i = 1, \ldots, n$, where p is the characteristic polynomial of A.

b. Why does it follow that $p(A) = 0$?

49. Suppose that A satisfies its characteristic equation and that B is similar to A. Show that B must satisfy its characteristic equation. (Thus if the Cayley-Hamilton theorem is true of one matrix, it is true of every similar matrix. Since every square matrix is similar to a triangular matrix, as we show in Appendix 2, the Cayley-Hamilton theorem need only be proved for triangular matrices.)

50. Use the Cayley-Hamilton theorem to find the inverse of each of the following nonsingular matrices. *Hint:* Multiply each side of $p(A) = 0$ by A^{-1}.

a. $A = \begin{bmatrix} 1 & -3 \\ 5 & 2 \end{bmatrix}$ with characteristic polynomial $p(r) = 17 - 3r + r^2$

b. $A = \begin{bmatrix} 3 & 0 & 1 \\ 2 & 2 & 1 \\ -6 & -1 & -2 \end{bmatrix}$ with characteristic polynomial $p(r) = (1 - r)^3$

51. Let A be a nonsingular $n \times n$ matrix with characteristic polynomial $p(r) = c_0 + c_1 r + \cdots + c_n r^n$.

a. Explain why c_0 is nonzero.

b. Show that $A^{-1} = -c_0^{-1}(c_1 I + c_2 A + \cdots + c_n A^{n-1})$.

Thus A^{-1} can be computed in terms of powers of A and the coefficients of the characteristic polynomial of A. While these coefficients are (in general) hard to find directly from the expansion of det $(A - rI)$, there is an alternate method involving the trace of A (and its powers). The resulting calculation of A^{-1} requires fewer than n^4 multiplications and is readily programmed for a computer.†

52. Prove that if A is $n \times n$, the matrices $I, A, A^2, \ldots, A^n$ are linearly dependent in $M_n(\mathfrak{F})$.

53. A square matrix A is called "nilpotent" if $A^k = 0$ for some positive integer k. Prove that A is nilpotent if and only if its eigenvalues are all zero.

Review Quiz

True or false?

1. Abel killed Galois in a duel.

2. A linear operator can have only a finite number of eigenvalues.

3. If the $n \times n$ matrix A has n linearly independent eigenvectors, there is a nonsingular matrix P such that PAP^{-1} is diagonal.

4. Diagonal $n \times n$ matrices commute.

5. The determinant of a square matrix is the product of its eigenvalues.

6. Every eigenspace of a linear operator f is invariant under f.

7. The only eigenvalue of I_n is 1 (with multiplicity n).

8. Eigenvectors associated with distinct eigenvalues of an $n \times n$ matrix are linearly independent in n-space.

9. The eigenspace corresponding to the eigenvalue 3 of

$$\begin{bmatrix} 2 & 4 \\ 1 & -1 \end{bmatrix}$$

is 1-dimensional.

10. Two $n \times n$ matrices are similar if and only if they have the same eigenvalues.

11. If $f: \mathfrak{R}^2 \to \mathfrak{R}^2$ is a reflection of the plane in a line through the origin, the eigenvalues of f are 1 and -1.

12. Every $n \times n$ matrix has eigenvectors spanning $\mathfrak{F}^n$.

13. An $n \times n$ matrix can be diagonalized if and only if it has n distinct eigenvalues.

† See D. T. Finkbeiner, *Introduction to Matrices and Linear Transformations,* 2d ed., San Francisco: Freeman, 1968, pages 173–175.

14. The matrices

$$\begin{bmatrix} 1 & 0 & 0 \\ 0 & 5 & 0 \\ 0 & 0 & -2 \end{bmatrix} \quad \text{and} \quad \begin{bmatrix} 5 & 0 & 0 \\ 0 & 1 & 0 \\ 0 & 0 & -2 \end{bmatrix}$$

are similar.

15. An $n \times n$ matrix with a zero eigenvalue is singular.

16. A linear operator and its matrix representation have the same eigenvectors.

17. If $p(r) = c_0 + c_1 r + \cdots + c_n r^n$ is the characteristic polynomial of a singular $n \times n$ matrix, then $c_0 = 0$.

18. If r is an eigenvalue of the matrix A, the system $(A - rI)\mathbf{X} = \mathbf{0}$ has nontrivial solutions.

19. The matrix

$$\begin{bmatrix} 1 & 0 \\ 0 & 2 \end{bmatrix}$$

is a representation of the linear operator $f(x,y) = (x - y, 2y)$ relative to a single basis for $\mathcal{R}^2$.

20. If n linearly independent eigenvectors of the $n \times n$ matrix A are used as columns to construct the matrix P, then PAP^{-1} is a diagonal matrix similar to A.

21. The eigenvalues of

$$\begin{bmatrix} 1 & 3 & 1 \\ 0 & -1 & 5 \\ 0 & 0 & 0 \end{bmatrix}$$

are $1, -1, 0$.

22. If $A - 3I$ is nonsingular, then 3 is not an eigenvalue of A.

23. If $r_1, \ldots, r_n$ are eigenvalues of the $n \times n$ matrix A (not necessarily distinct), then diag $(r_1, \ldots, r_n)$ is similar to A.

24. Any matrix similar to diag $(1, -1, 3)$ is nonsingular.

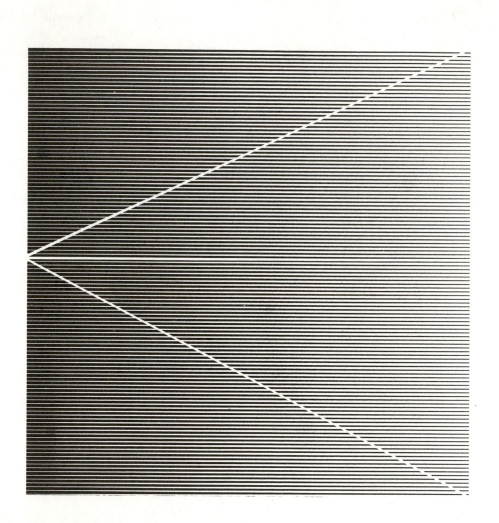

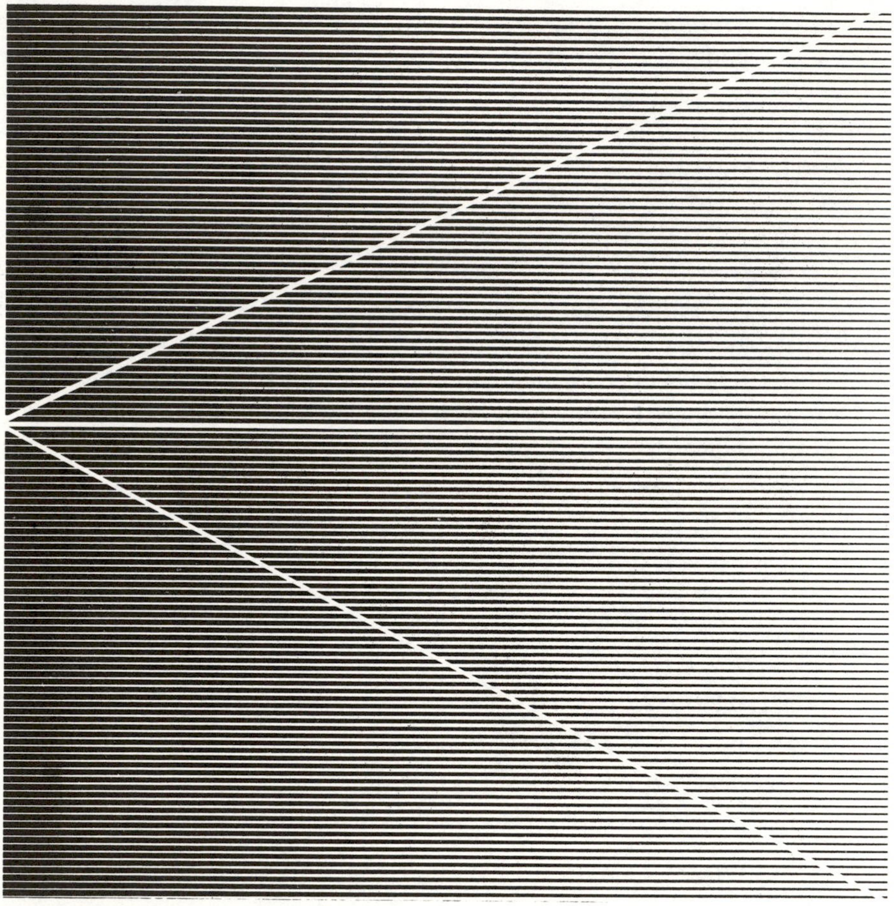

CHAPTER 10

Inner Products

You may recall from calculus that if **X** and **Y** are nonzero vectors in the plane and if θ is the angle between them $(0 \leq \theta \leq \pi)$, their *dot product* is defined to be

$$\mathbf{X} \cdot \mathbf{Y} = \|\mathbf{X}\| \, \|\mathbf{Y}\| \cos \theta$$

where the double bars mean length. This geometric definition (often motivated by physical considerations) is then used to derive the familiar algebraic formula

$$\mathbf{X} \cdot \mathbf{Y} = x_1 y_1 + x_2 y_2 \qquad \text{where} \quad \mathbf{X} = (x_1, x_2) \quad \text{and} \quad \mathbf{Y} = (y_1, y_2)$$

Assuming that this has been done (the argument is based on the law of cosines), it follows that the length of a vector **X** is

$$\|\mathbf{X}\| = \sqrt{\mathbf{X} \cdot \mathbf{X}}$$

and the distance between two points **X** and **Y** is

$$d(\mathbf{X}, \mathbf{Y}) = \|\mathbf{X} - \mathbf{Y}\|$$

Moreover, the angle between two nonzero vectors **X** and **Y** is computed from the formula

$$\cos \theta = \frac{\mathbf{X} \cdot \mathbf{Y}}{\|\mathbf{X}\| \, \|\mathbf{Y}\|}$$

and hence this angle is $\pi/2$ if and only if $\mathbf{X} \cdot \mathbf{Y} = 0$. In other words, two vectors are perpendicular ("orthogonal") if and only if their dot product is 0.

The same discussion in $\mathcal{R}^3$ leads to the same conclusions; our purpose in this chapter is to generalize the ideas of length, distance, and orthogonality to abstract vector spaces. The key to doing this in $\mathcal{R}^n$ is the dot product; more generally, the so-called "standard inner product" on $\mathfrak{F}^n$ is used. (See Sec. 4.3.) This concept leads, in turn, to a definition of "inner product" on abstract vector spaces, at which point we'll be ready to introduce a substantial geometric flavor into linear algebra.

10.1 Geometry in *n*-Space

The *dot product* of two vectors $\mathbf{X} = (x_1, \ldots, x_n)$ and $\mathbf{Y} = (y_1, \ldots, y_n)$ in $\mathfrak{F}^n$ is defined to be the scalar

$$\mathbf{X} \cdot \mathbf{Y} = x_1y_1 + \cdots + x_ny_n = \sum_{j=1}^{n} x_jy_j{}^{\dagger}$$

For example, the dot product of $\mathbf{X} = (1,2,-1,3)$ and $\mathbf{Y} = (0,3,1,1)$ in $\mathcal{R}^4$ is

$$\mathbf{X} \cdot \mathbf{Y} = (1)(0) + (2)(3) + (-1)(1) + (3)(1) = 8$$

while the dot product of $\mathbf{X} = (2, 1 - i)$ and $\mathbf{Y} = (5 + 3i, 2 - 2i)$ in $\mathcal{C}^2$ is

$$\mathbf{X} \cdot \mathbf{Y} = (2)(5 + 3i) + (1 - i)(2 - 2i) = 10 + 2i$$

While this definition is of great importance in some contexts (for example, the definition of matrix multiplication in Sec. 5.3), it is inadequate for a discussion of length and distance in $\mathfrak{F}^n$ (unless $\mathfrak{F} = \mathcal{R}$). Consider, for example, the vector $\mathbf{X} = (1 + i, 1 - i)$ in $\mathcal{C}^2$. If we try to find its length from the formula $\|\mathbf{X}\| = \sqrt{\mathbf{X} \cdot \mathbf{X}}$, we find that

$$\mathbf{X} \cdot \mathbf{X} = (1 + i)(1 + i) + (1 - i)(1 - i) = 0$$

even though $\mathbf{X}$ is nonzero. Worse yet, if $\mathbf{X} = (2i, 1)$, we get

$$\mathbf{X} \cdot \mathbf{X} = (2i)(2i) + (1)(1) = -3$$

from which $\|\mathbf{X}\| = \sqrt{-3}$! Obviously a modification is needed to make sense of length in complex *n*-space.

LECTOR *Can you explain why, in 25 words or less, you are going to talk about length in $\mathcal{C}^n$?*

AUCTOR *Only with difficulty. I prefer to approach the subject with the same attitude George Leigh-Mallory had toward Everest.*

LECTOR *Because it is there.*

AUCTOR *Precisely.*

LECTOR *He fell off.*

AUCTOR *Oh, you pragmatists! The world has grown gray with your breath.*

LECTOR *You needn't misquote Swinburne.*

STANDARD INNER PRODUCT

If $\mathbf{X} = (x_1, \ldots, x_n)$ and $\mathbf{Y} = (y_1, \ldots, y_n)$ are elements of $\mathfrak{F}^n$, their *standard inner product* is the scalar

† See Example 3.9, Sec. 3.1, and the corresponding problems in Sec. 3.1.

$$\mathbf{X} * \mathbf{Y} = x_1\bar{y}_1 + \cdots + x_n\bar{y}_n = \sum_{j=1}^{n} x_j\bar{y}_j$$

where the bar means *complex conjugate*. (See Prob. 43, Sec. 1.3.)

If $\mathcal{F} = \mathcal{R}$, the standard inner product is merely the dot product on $\mathcal{R}^n$ (why?), but if $\mathcal{F} = \mathcal{C}$, it is more complicated. For example, the standard inner product of $\mathbf{X} = (1, 2i, 3 - i)$ and $\mathbf{Y} = (i, -1, 2 + i)$ in $\mathcal{C}^3$ is

$$\mathbf{X} * \mathbf{Y} = (1)(-i) + (2i)(-1) + (3 - i)(2 - i) = 5 - 8i$$

while

$$\mathbf{Y} * \mathbf{X} = (i)(1) + (-1)(-2i) + (2 + i)(3 + i) = 5 + 8i$$

Thus the "symmetric law" (which in the case of the dot product reads $\mathbf{X} \cdot \mathbf{Y} = \mathbf{Y} \cdot \mathbf{X}$) fails. (See Prob. 13, Sec. 3.1.) Note, however, that $\mathbf{X} * \mathbf{Y}$ and $\mathbf{Y} * \mathbf{X}$ are conjugates. This is true in general, since the conjugate of

$$\mathbf{X} * \mathbf{Y} = \sum_{j=1}^{n} x_j\bar{y}_j$$

in $\mathcal{F}$ is

$$\begin{aligned}
\overline{\mathbf{X} * \mathbf{Y}} &= \sum_{j=1}^{n} \overline{x_j\,\bar{y}_j} && \text{(Prob. 43}e\text{, Sec. 1.3)}\\
&= \sum_{j=1}^{n} \bar{x}_j\bar{\bar{y}}_j && \text{(Prob. 43}e\text{, Sec. 1.3)}\\
&= \sum_{j=1}^{n} \bar{x}_j y_j && \text{(Why?)}\\
&= \sum_{j=1}^{n} y_j\bar{x}_j\\
&= \mathbf{Y} * \mathbf{X}
\end{aligned}$$

We call this result the "conjugate-symmetric law."

The "bilinear law," which in the case of the dot product is

$$(a\mathbf{X} + b\mathbf{Y}) \cdot \mathbf{Z} = a(\mathbf{X} \cdot \mathbf{Z}) + b(\mathbf{Y} \cdot \mathbf{Z})$$

and

$$\mathbf{Z} \cdot (a\mathbf{X} + b\mathbf{Y}) = a(\mathbf{Z} \cdot \mathbf{X}) + b(\mathbf{Z} \cdot \mathbf{Y})$$

also fails (unless $\mathfrak{F} = \mathfrak{R}$). (See Example 3.9 and Prob. 14, Sec. 3.1.) Although the formula

$$(a\mathbf{X} + b\mathbf{Y}) * \mathbf{Z} = a(\mathbf{X} * \mathbf{Z}) + b(\mathbf{Y} * \mathbf{Z}) \tag{10.1}$$

is correct (see the problems), the formula

$$\mathbf{Z} * (a\mathbf{X} + b\mathbf{Y}) = a(\mathbf{Z} * \mathbf{X}) + b(\mathbf{Z} * \mathbf{Y})$$

must be replaced by

$$\mathbf{Z} * (a\mathbf{X} + b\mathbf{Y}) = \bar{a}(\mathbf{Z} * \mathbf{X}) + \bar{b}(\mathbf{Z} * \mathbf{Y}) \tag{10.2}$$

To see why, note that

$$a\mathbf{X} + b\mathbf{Y} = (ax_1 + by_1, \ldots, ax_n + by_n)$$

Hence

$$\begin{aligned}\mathbf{Z} * (a\mathbf{X} + b\mathbf{Y}) &= \sum_{j=1}^{n} z_j\overline{(ax_j + by_j)} \\ &= \sum_{j=1}^{n} z_j(\bar{a}\bar{x}_j + \bar{b}\bar{y}_j) \\ &= \bar{a}\sum_{j=1}^{n} z_j\bar{x}_j + \bar{b}\sum_{j=1}^{n} z_j\bar{y}_j \\ &= \bar{a}(\mathbf{Z} * \mathbf{X}) + \bar{b}(\mathbf{Z} * \mathbf{Y})\end{aligned}$$

Equations (10.1) and (10.2) are called the "conjugate-bilinear law."

It is worthwhile to record special cases of the conjugate-bilinear law. If $b = 0$ in Eq. (10.1), we have the associative law

$$(a\mathbf{X}) * \mathbf{Z} = a(\mathbf{X} * \mathbf{Z})$$

But Eq. (10.2) reads

$$\mathbf{Z} * (a\mathbf{X}) = \bar{a}(\mathbf{Z} * \mathbf{X})$$

so unless $\mathfrak{F} = \mathfrak{R}$ one has to be careful moving scalars around. A scalar *in the second factor* comes out of the product as a complex conjugate. For example, if we take $a = i$, $\mathbf{X} = (2, 1 - i)$, and $\mathbf{Y} = (5 + 3i, 2 - 2i)$ in $\mathcal{C}^2$, it is incorrect to write

$$\mathbf{X} * (i\mathbf{Y}) = i(\mathbf{X} * \mathbf{Y})$$

Since $\bar{a} = -i$, the correct formula is

$$\mathbf{X} * (i\mathbf{Y}) = -i(\mathbf{X} * \mathbf{Y})$$

as you can check.

If $a = b = 1$, the two parts of the conjugate-bilinear law reduce to the distributive laws

$$(\mathbf{X} + \mathbf{Y}) * \mathbf{Z} = \mathbf{X} * \mathbf{Z} + \mathbf{Y} * \mathbf{Z} \qquad \text{and} \qquad \mathbf{Z} * (\mathbf{X} + \mathbf{Y}) = \mathbf{Z} * \mathbf{X} + \mathbf{Z} * \mathbf{Y}$$

It is interesting to note that both these formulas are correct, despite the failure of the standard inner product to be symmetric.

The most important property of the standard inner product is the "positive-definite law," which says that

$$\mathbf{X} * \mathbf{X} \geq 0 \qquad \text{with equality holding} \quad \Leftrightarrow \quad \mathbf{X} = \mathbf{0}$$

(See Prob. 12, Sec. 3.1.) Indeed if z is a complex number, then $z\bar{z} = |z|^2$ (Prob. 43c, Sec. 1.3). Hence

$$\mathbf{X} * \mathbf{X} = \sum_{j=1}^{n} x_j \bar{x}_j = \sum_{j=1}^{n} |x_j|^2 \geq 0$$

with equality holding if and only if $x_j = 0$, where $j = 1, \ldots, n$, that is, if and only if $\mathbf{X} = \mathbf{0}$.

The dot product does not have this property when $\mathfrak{F} = \mathfrak{C}$, as we have seen. But it is crucial for a reasonable definition of length and distance; the formulas

$$\|\mathbf{X}\| = \sqrt{\mathbf{X} * \mathbf{X}} \qquad \text{and} \qquad d(\mathbf{X},\mathbf{Y}) = \|\mathbf{X} - \mathbf{Y}\|$$

to be introduced shortly, are useless unless $\mathbf{X} * \mathbf{X}$ is a positive real number in every instance but the definite case $\mathbf{X} = \mathbf{0}$. (This is the reason for the term "positive-definite.")

Now we are ready to introduce length, distance, and orthogonality into n-space.

LENGTH

If $\mathbf{X} \in \mathfrak{F}^n$, the *length* (or *norm*) of $\mathbf{X}$ is defined to be

$$\|\mathbf{X}\| = \sqrt{\mathbf{X} * \mathbf{X}}^{\dagger}$$

Even though we may be working with complex scalars in this definition, note that $\|\mathbf{X}\|$ is a positive real number when $\mathbf{X} \neq \mathbf{0}$ and is zero when $\mathbf{X} = \mathbf{0}$ (because of the positive-definite law). For example, the norm of $\mathbf{X} = (2 - i, 1)$ in $\mathfrak{C}^2$ is found by writing

$$\|\mathbf{X}\|^2 = \mathbf{X} * \mathbf{X} = (2 - i)(2 + i) + (1)(1) = 6$$

† The reason for the double bars is to distinguish length in n-space from absolute value in the scalar field. But if $n = 1$, the ideas are the same. (Why?)

Hence $\|\mathbf{X}\| = \sqrt{6}$.

When $\mathcal{F} = \mathcal{R}$ we may use the standard inner product (dot product) to define the angle between two nonzero vectors. The clue is the equivalence of

$$\mathbf{X} \cdot \mathbf{Y} = \sum_{j=1}^{n} x_j y_j \qquad \text{and} \qquad \mathbf{X} \cdot \mathbf{Y} = \|\mathbf{X}\| \, \|\mathbf{Y}\| \cos \theta$$

when $n = 2$ or $n = 3$.

ANGLE IN REAL *n*-SPACE

The *angle* between nonzero vectors **X** and **Y** in $\mathcal{R}^n$ is the real number θ satisfying

$$\cos \theta = \frac{\mathbf{X} \cdot \mathbf{Y}}{\|\mathbf{X}\| \, \|\mathbf{Y}\|} \qquad 0 \leq \theta \leq \pi$$

For example, if $\mathbf{X} = (1,1,-1,1)$ and $\mathbf{Y} = (-2,1,4,2)$ in $\mathcal{R}^4$, then

$$\cos \theta = \frac{-3}{(2)(5)} = -\frac{3}{10}$$

This came out to be a number between -1 and 1, as it must if the definition is to be defended. An obviously critical question is whether

$$-1 \leq \frac{\mathbf{X} \cdot \mathbf{Y}}{\|\mathbf{X}\| \, \|\mathbf{Y}\|} \leq 1$$

in general, since otherwise we cannot find θ satisfying

$$\cos \theta = \frac{\mathbf{X} \cdot \mathbf{Y}}{\|\mathbf{X}\| \, \|\mathbf{Y}\|} \qquad \text{(Why?)}$$

The usual form in which these inequalities are stated is

$$|\mathbf{X} \cdot \mathbf{Y}| \leq \|\mathbf{X}\| \, \|\mathbf{Y}\|$$

the celebrated *Cauchy-Schwarz inequality.*† In terms of the coordinates of $\mathbf{X} = (x_1, \ldots, x_n)$ and $\mathbf{Y} = (y_1, \ldots, y_n)$, this result is equivalent to

$$(x_1 y_1 + \cdots + x_n y_n)^2 \leq (x_1{}^2 + \cdots + x_n{}^2)(y_1{}^2 + \cdots + y_n{}^2)$$

the form in which Cauchy stated it in 1821. A similar inequality for integrals was proved by Bunyakovsky in 1859 and (for double integrals) by Schwarz in 1885. Remarkably enough, it is also true in complex *n*-space (provided that

† The bars on the left mean absolute value, as usual.

the standard inner product is used) and indeed in any space on which an ''inner product'' has been defined. We'll prove it in this general form in Sec. 10.3.

Assuming that the Cauchy-Schwarz inequality is true, it is not hard to prove the *Minkowski inequality,*

$$\|\mathbf{X} + \mathbf{Y}\| \leq \|\mathbf{X}\| + \|\mathbf{Y}\|$$

(See the problems for an argument in $\mathcal{R}^n$, and Sec. 10.3 for the general case.) This inequality in turn helps clarify the following definition.

DISTANCE

If **X** and **Y** are elements of $\mathcal{F}^n$, the *distance* between them is

$$d(\mathbf{X},\mathbf{Y}) = \|\mathbf{X} - \mathbf{Y}\|$$

For example, the distance between $\mathbf{X} = (1,0,-2,5)$ and $\mathbf{Y} = (0,-1,1,3)$ in $\mathcal{R}^4$ is found by noting that $\mathbf{X} - \mathbf{Y} = (1,1,-3,2)$ and then writing

$$d(\mathbf{X},\mathbf{Y}) = \|\mathbf{X} - \mathbf{Y}\| = \sqrt{(\mathbf{X} - \mathbf{Y}) * (\mathbf{X} - \mathbf{Y})} = \sqrt{1 + 1 + 9 + 4} = \sqrt{15}$$

More generally, it is not hard to show that

$$d(\mathbf{X},\mathbf{Y}) = \sqrt{|x_1 - y_1|^2 + \cdots + |x_n - y_n|^2}$$

the usual formula for distance when $\mathcal{F} = \mathcal{R}$ and $n \leq 3$.†

Distance functions in mathematics are supposed to have certain basic properties. For example, the famous ''triangle inequality'' says that if **X, Y, Z** are any three points, then

$$d(\mathbf{X},\mathbf{Y}) \leq d(\mathbf{X},\mathbf{Z}) + d(\mathbf{Z},\mathbf{Y})$$

(See Fig. 10.1 for a picture of this in $\mathcal{R}^2$.) This property is easy to prove on the basis of our definition of distance, for the Minkowski inequality allows us to write

$$\begin{aligned} d(\mathbf{X},\mathbf{Y}) &= \|\mathbf{X} - \mathbf{Y}\| \\ &= \|(\mathbf{X} - \mathbf{Z}) + (\mathbf{Z} - \mathbf{Y})\| \\ &\leq \|\mathbf{X} - \mathbf{Z}\| + \|\mathbf{Z} - \mathbf{Y}\| \\ &= d(\mathbf{X},\mathbf{Z}) + d(\mathbf{Z},\mathbf{Y}) \end{aligned}$$

Another important property of the distance function, following immediately from the positive-definite law of the standard inner product, is

$$d(\mathbf{X},\mathbf{Y}) \geq 0 \qquad \text{with equality holding} \iff \mathbf{X} = \mathbf{Y}$$

† Note that when $\mathcal{F} = \mathcal{C}$ the absolute values are essential; that is, we cannot write $(x_1 - y_1)^2 + \cdots + (x_n - y_n)^2$. Why?

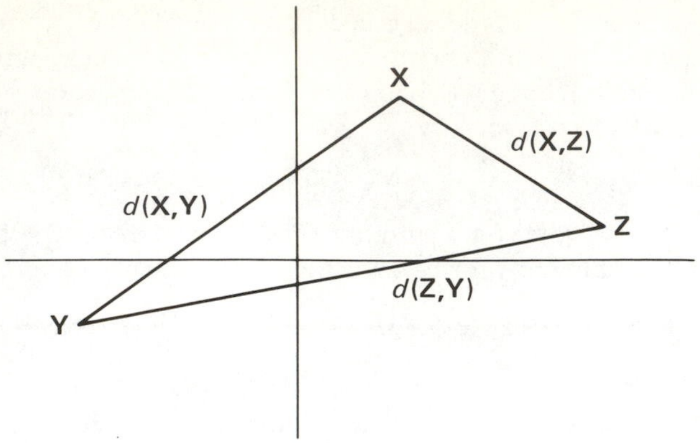

Fig. 10.1 Triangle inequality

(This is sometimes called the positive-definite law, too.) You will find it to be useful from time to time, despite its trivial appearance, for it provides a way of showing that two elements **X** and **Y** are equal when a direct attack may be inconvenient. We simply compute $d(\mathbf{X},\mathbf{Y})$; if it turns out to be zero, then $\mathbf{X} = \mathbf{Y}$.

The third basic property of the distance function is the symmetric law,

$$d(\mathbf{X},\mathbf{Y}) = d(\mathbf{Y},\mathbf{X})$$

which says that the distance between two points does not depend on the order in which they are named. (See the problems.)

Mathematicians usually summarize these properties by saying that the function d is a "metric" on $\mathfrak{F}^n$, and that $\mathfrak{F}^n$ is a "metric space." This concept is of great importance in mathematics, since the measurement of distance (on which so much depends) presupposes a distance function with these properties.†

Our last definition is motivated by the fact that two vectors in $\mathcal{R}^2$ or $\mathcal{R}^3$ are perpendicular ("orthogonal") if and only if their dot product is zero (provided that we agree that the directionless zero vector is orthogonal to every vector).

ORTHOGONAL VECTORS

Two vectors **X** and **Y** in $\mathfrak{F}^n$ are said to be *crthogonal* if $\mathbf{X} * \mathbf{Y} = 0$.

† The whole of elementary calculus, for example, depends on the definition of limit, which involves distances in $\mathcal{R}^1$; $\lim_{x\to a} f(x) = L$ means that $0 < d(x, a) < \delta \Rightarrow d[f(x), L] < \epsilon$, where ϵ is given and δ depends on ϵ.

Note that this definition applies in both real and complex n-space, even though its motivation depends on the idea of angle, which is restricted to $\mathcal{R}^n$. In $\mathcal{C}^n$ we do not distinguish direction except in the special case of orthogonality (which in $\mathcal{R}^n$ corresponds to the angle $\theta = \pi/2$). We'll develop the concept of orthogonality more fully in Secs. 10.4 through 10.6.

Problems

1. Why does the standard inner product on $\mathcal{F}^n$ reduce to the dot product if $\mathcal{F} = \mathcal{R}$?
2. Find the standard inner product of $\mathbf{X} = (2 + i, 3, i)$ and $\mathbf{Y} = (-i, 1 - i, 2)$ in $\mathcal{C}^3$.
3. *a*. If $\mathbf{X} = (2, 1 - i)$ and $\mathbf{Y} = (5 + 3i, 2 - 2i)$, then $\mathbf{X} \cdot \mathbf{Y} = 10 + 2i$. What is $\mathbf{X} * \mathbf{Y}$?
   b. If $\mathbf{X} = (1 + i, 1 - i)$, then $\mathbf{X} \cdot \mathbf{X} = 0$. Show that $\mathbf{X} * \mathbf{X} > 0$.
   c. If $\mathbf{X} = (2i, 1)$, then $\mathbf{X} \cdot \mathbf{X} = -3$. Show that $\mathbf{X} * \mathbf{X} > 0$.
4. Prove the first part of the conjugate-bilinear law,
$$(a\mathbf{X} + b\mathbf{Y}) * \mathbf{Z} = a(\mathbf{X} * \mathbf{Z}) + b(\mathbf{Y} * \mathbf{Z})$$
Then show how this may be used, together with the conjugate-symmetric law, to derive the second part,
$$\mathbf{Z} * (a\mathbf{X} + b\mathbf{Y}) = \bar{a}(\mathbf{Z} * \mathbf{X}) + \bar{b}(\mathbf{Z} * \mathbf{Y})$$
5. Find the norm of $\mathbf{X} = (1 + i, 2, i)$ in $\mathcal{C}^3$.
6. Use the following figure in $\mathcal{R}^2$ to argue that if the dot product of $\mathbf{X} = (x_1, x_2)$ and $\mathbf{Y} = (y_1, y_2)$ is understood to be $\mathbf{X} \cdot \mathbf{Y} = \|\mathbf{X}\|\,\|\mathbf{Y}\| \cos \theta$, the formula $\mathbf{X} \cdot \mathbf{Y} = x_1y_1 + x_2y_2$ follows. *Hint:* Use the law of cosines to obtain $\|\mathbf{X} - \mathbf{Y}\|^2 = \|\mathbf{X}\|^2 + \|\mathbf{Y}\|^2 - 2\|\mathbf{X}\|\,\|\mathbf{Y}\| \cos \theta$.

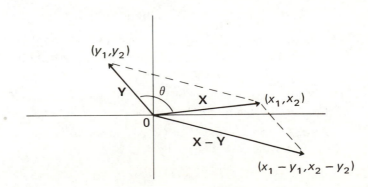

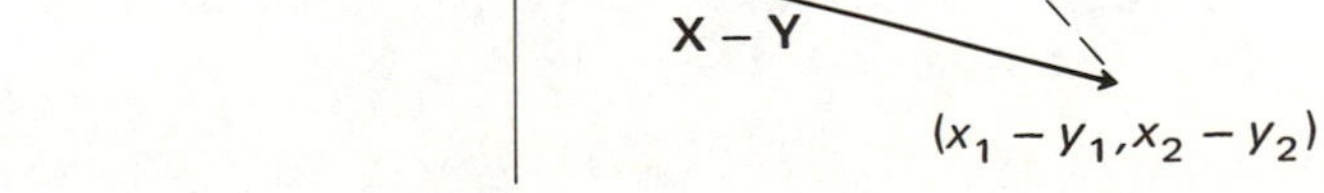

7. In each of the following cases, find the cosine of the angle determined by **X** and **Y**.
a. $\mathbf{X} = (4, -3)$ and $\mathbf{Y} = (-1,7)$
b. $\mathbf{X} = (1,-2,0,2)$ and $\mathbf{Y} = (2,-2,2,-2)$
Why is the definition of $\cos\theta$ restricted to real *n*-space?

8. Prove the Cauchy-Schwarz inequality in $\mathcal{R}^2$. *Hint:* Show that its denial leads to a contradiction of the properties of $\mathcal{R}$. (Do *not* use the formula $\mathbf{X}\cdot\mathbf{Y} = \|\mathbf{X}\|\,\|\mathbf{Y}\|\cos\theta$ for dot product, arguing that the Cauchy-Schwarz inequality follows from the fact that $|\cos\theta| \leq 1$! While this would be legitimate in $\mathcal{R}^2$, it misses the point in general, which is to justify the definition of $\cos\theta$ in $\mathcal{R}^n$.)

9. Try proving the Cauchy-Schwarz inequality in $\mathcal{R}^3$. (If you have trouble, as well you may, you will appreciate the difficulty of this inequality in general!)

★ **10.** Assuming that the Cauchy-Schwarz inequality has been proved, derive the Minkowski inequality in $\mathcal{R}^n$, $\|\mathbf{X} + \mathbf{Y}\| \leq \|\mathbf{X}\| + \|\mathbf{Y}\|$. *Hint:* Show that

$$\|\mathbf{X} + \mathbf{Y}\|^2 = \|\mathbf{X}\|^2 + 2(\mathbf{X} * \mathbf{Y}) + \|\mathbf{Y}\|^2$$

Why doesn't this argument apply in complex *n*-space?

11. Interpreting **X** and **Y** as arrows beginning at the origin, draw a picture in $\mathcal{R}^2$ illustrating why the Minkowski inequality is sometimes called the "triangle inequality."

12. In each of the following cases, find the distance between **X** and **Y**.
a. $\mathbf{X} = (3,1,0,2)$ and $\mathbf{Y} = (5,1,-1,-3)$
b. $\mathbf{X} = (3 - i, 3)$ and $\mathbf{Y} = (i,1)$

★ **13.** Let $\mathbf{X} = (x_1, \ldots, x_n)$ and $\mathbf{Y} = (y_1, \ldots, y_n)$ be elements of $\mathcal{F}^n$.
a. Confirm that $\|\mathbf{X}\| = \sqrt{|x_1|^2 + \cdots + |x_n|^2}$.
b. Show that $d(\mathbf{X},\mathbf{Y}) = \sqrt{|x_1 - y_1|^2 + \cdots + |x_n - y_n|^2}$.
Why are the absolute value signs essential in complex space?

14. Prove the positive-definite law of the distance function:

$$d(\mathbf{X},\mathbf{Y}) \geq 0 \qquad \text{with equality holding} \quad \Leftrightarrow \quad \mathbf{X} = \mathbf{Y}$$

★ **15.** Prove that if *k* is a scalar and $\mathbf{X} \in \mathcal{F}^n$, then $\|k\mathbf{X}\| = |k|\,\|\mathbf{X}\|$. *Hint:* Use properties of the standard inner product already established, avoiding reference to the coordinates of **X**.

16. Why does it follow from Prob. 15 that $\|-\mathbf{X}\| = \|\mathbf{X}\|$? Use this fact to derive the symmetric law of the distance function, $d(\mathbf{X},\mathbf{Y}) = d(\mathbf{Y},\mathbf{X})$.

17. Let $\mathbf{X} = (x_1,x_2)$ and $\mathbf{Y} = (y_1,y_2)$ be points of the plane and define the *urban distance* between them to be

$$t(\mathbf{X},\mathbf{Y}) = |x_1 - y_1| + |x_2 - y_2|$$

a. Draw a picture illustrating why $t(\mathbf{X},\mathbf{Y})$ is called "urban distance."

b. Prove that t is positive-definite, symmetric, and satisfies the triangle inequality. (Thus it is a metric on $\mathcal{R}^2$, like the distance function d. Some people call it the "taxicab metric." In a congested city it is a more meaningful way to measure distance than as the crow flies.)

c. Show that $t(\mathbf{X},\mathbf{Y}) \geq d(\mathbf{X},\mathbf{Y})$ for all $\mathbf{X}$ and $\mathbf{Y}$ in $\mathcal{R}^2$. (Thus "urban distance" is never less than "rural distance.")

★ **18.** Assuming that we are in real n-space, prove the pythagorean theorem and its converse:

$$\|\mathbf{X} + \mathbf{Y}\|^2 = \|\mathbf{X}\|^2 + \|\mathbf{Y}\|^2 \quad \Leftrightarrow \quad \mathbf{X} \text{ and } \mathbf{Y} \text{ are perpendicular}$$

19. Specialize the ideas of this section to $\mathcal{R}^1$, as follows.

a. If x and y are elements of $\mathcal{R}^1$, what is their dot product?

b. What do the positive-definite, symmetric, and bilinear laws reduce to?

c. If $x \in \mathcal{R}^1$, what is the length of x as defined by the dot product?

d. If x and y are nonzero elements of $\mathcal{R}^1$, what are the possibilities for the angle they determine, as defined by the dot product?

e. What does the Cauchy-Schwarz inequality reduce to?

f. What does the Minkowski inequality reduce to?

g. If x and y are elements of $\mathcal{R}^1$, what is the distance between them as defined in terms of the dot product?

10.2 Euclidean and Unitary Spaces

In the last section we saw how to impose metric ideas on n-space by using properties of the standard inner product. The same ideas can be developed in other vector spaces, provided that we can define a product with similar properties.

NOTE TO THE READER

Beginning with Example 10.1 below, and continuing through the rest of this chapter and the next, we present a sequence of examples and problems involving an inner product defined in terms of integration. Do not be intimidated! Virtually no manipulative skills of integration are required; you can fill in the details whenever it pleases you, or skip them entirely. It is important to look at these examples and problems, for they are designed to give substance to the theory (which is hardly worth doing if its applications are confined to n-space).

EXAMPLE 10.1

Let S be the vector space of real-valued functions continuous on the interval $[-\pi,\pi]$. If f and g are elements of S, define their *inner product* to be

$$f * g = \int_{-\pi}^{\pi} f(x)g(x)\,dx$$

It is not hard to show (using well-known properties of the integral) that this product, like the dot product on $\mathcal{R}^n$, is positive-definite, symmetric, and bilinear. Hence we may adopt the same terminology as in $\mathcal{R}^n$. The "norm" of an element $f \in S$, for example, is

$$\|f\| = \sqrt{f * f} = \left[\int_{-\pi}^{\pi} f(x)^2\,dx\right]^{1/2}$$

The "distance" between two elements f and g in S is

$$d(f,g) = \|f - g\|$$

Thus if f and g are the cosine and sine, the distance between them is

$$\left[\int_{-\pi}^{\pi} (\cos x - \sin x)^2\,dx\right]^{1/2} = \sqrt{2\pi}^{\dagger}$$

While it may seem strange to impose euclidean geometry on a space whose elements are functions instead of points, it is important to be able to do so. The reader who is familiar with Fourier[‡] analysis will recognize the connection between that subject and this definition of the inner product of functions.

EXAMPLE 10.2

Let S be the vector space (over $\mathcal{C}$) of "complex-valued" functions continuous on $[-\pi,\pi]$.[¶] For each pair of functions f and g in S, define

$$f * g = \int_{-\pi}^{\pi} f(x)\overline{g(x)}\,dx$$

(Note that this is the same product as in Example 10.1 if we confine our attention to real-valued functions in S and restrict the scalars to be real.) For example, if

$$f(x) = \cos x + i \sin x \qquad \text{and} \qquad g(x) = \cos x - i \sin x$$

[†] Here is an example of what we meant in the Note to the Reader. You can check that the answer is $\sqrt{2\pi}$. But it is unnecessary to do so.

[‡] Jean-Baptiste Joseph Fourier (1768–1830), whose treatise on the conduction of heat (in which he developed the representation of functions by trigonometric series) was a watershed in the history of mathematics and physics.

[¶] A complex-valued function of a real variable on $[a,b]$ is of the form $f(x) = u(x) + iv(x)$, where u and v are real-valued, for example, $f(x) = \cos x + i \sin x$. It is continuous on $[a,b]$ if u and v are; its integral on $[a,b]$ is the complex number $\int_a^b f(x)\,dx = \int_a^b u(x)\,dx + i\int_a^b v(x)\,dx$.

then

$$\begin{aligned} f(x)\overline{g(x)} &= (\cos x + i \sin x)(\cos x + i \sin x) \\ &= (\cos^2 x - \sin^2 x) + i(2 \sin x \cos x) \\ &= \cos 2x + i \sin 2x \end{aligned}$$

and hence

$$\begin{aligned} f * g &= \int_{-\pi}^{\pi} (\cos 2x + i \sin 2x)\, dx \\ &= \int_{-\pi}^{\pi} \cos 2x\, dx + i \int_{-\pi}^{\pi} \sin 2x\, dx \\ &= 0 + 0i = 0 \end{aligned}$$

We claim that this product has the same properties as the standard inner product on $\mathfrak{F}^n$. (If so, it will be reasonable to call f and g orthogonal when $f * g = 0$, as in the above example.)

1. The product $*$ is positive-definite because

$$f * f = \int_{-\pi}^{\pi} f(x)\overline{f(x)}\, dx = \int_{-\pi}^{\pi} |f(x)|^2\, dx \geq 0$$

with equality holding if and only if f is the zero function.†

2. To prove that the product $*$ is conjugate-symmetric, let $f = u_1 + iv_1$ and $g = u_2 + iv_2$ be any elements of S. Then

$$f * g = \int_{-\pi}^{\pi} f\bar{g} = \int_{-\pi}^{\pi} (u_1u_2 + v_1v_2) + i \int_{-\pi}^{\pi} (u_2v_1 - u_1v_2)$$

and

$$\begin{aligned} g * f = \int_{-\pi}^{\pi} g\bar{f} &= \int_{-\pi}^{\pi} (u_1u_2 + v_1v_2) + i \int_{-\pi}^{\pi} (u_1v_2 - u_2v_1) \\ &= \int_{-\pi}^{\pi} (u_1u_2 + v_1v_2) - i \int_{-\pi}^{\pi} (u_2v_1 - u_1v_2) \end{aligned}$$

Hence $f * g$ and $g * f$ are conjugates.

3. The first part of the conjugate-bilinear law,

$$(af + bg) * h = a(f * h) + b(g * h)$$

is confirmed as in the proof of the conjugate-symmetric law above. The second part need not be checked, since it follows from the first part and the conjugate-symmetric law. (See Prob. 4 in Sec. 10.1.)

† The "only if" part of this (the rest is obvious) is correct because if $f = u + iv$, then $|f(x)|^2 = u(x)^2 + v(x)^2$, which is a continuous real-valued function that is never negative. Its integral cannot be zero unless $|f(x)|^2 = 0$ for all x, which implies $f = 0$.

EXAMPLE 10.3

If $\mathbf{X} = (x_1,x_2)$ and $\mathbf{Y} = (y_1,y_2)$ are elements of $\mathcal{R}^2$, define their inner product to be

$$\mathbf{X} \circ \mathbf{Y} = 2x_1y_1 + 3x_2y_2$$

Like the dot product $\mathbf{X} \cdot \mathbf{Y} = x_1y_1 + x_2y_2$, this product is positive-definite, symmetric, and bilinear; if we use it to define length, distance, and angle as in Sec. 10.1, we'll have a consistent geometry in $\mathcal{R}^2$, but it will be distorted compared to the standard geometry. For example, the distance between $\mathbf{X} = (1,-2)$ and $\mathbf{Y} = (3,1)$ will be

$$d(\mathbf{X},\mathbf{Y}) = \|\mathbf{X} - \mathbf{Y}\| = \sqrt{(\mathbf{X} - \mathbf{Y}) \circ (\mathbf{X} - \mathbf{Y})} = \sqrt{35}$$

instead of the normal $\sqrt{13}$. Nevertheless, all the theory of Sec. 10.1 applies: the Cauchy-Schwarz inequality, the Minkowski inequality, and the metric properties of the distance function.

These examples indicate that the standard inner product on $\mathfrak{F}^n$ is a special case of a more general idea, which we now define.

INNER PRODUCT ON A UNITARY SPACE

Let S be a vector space over $\mathfrak{F}$ (where $\mathfrak{F}$ is either $\mathcal{R}$ or $\mathcal{C}$) and suppose that it is possible to define a scalar-valued product $*$ on S which is

1. Positive-definite. If $x \in S$, then $x * x \geq 0$, with equality holding $\Leftrightarrow x = 0$.
2. Conjugate-symmetric. If x and y are elements of S, then $x * y$ and $y * x$ are conjugates.
3. Conjugate-bilinear. If a and b are scalars and x, y, z are elements of S, then

$$(ax + by) * z = a(x * z) + b(y * z)$$

and

$$z * (ax + by) = \bar{a}(z * x) + \bar{b}(z * y)$$

Then $*$ is called an *inner product* on S, and S (with its apparatus of addition, scalar multiplication, and inner product) is called a *unitary space*. When $\mathfrak{F} = \mathcal{R}$ (in which case $*$ is simply positive-definite, symmetric, and bilinear), the term *euclidean space* is also used.

It is worth remarking that many writers use (x,y) or $\langle x,y \rangle$ for the inner product $x * y$. We have adopted the latter notation because it looks more like the usual multiplicative symbol, and to avoid confusion with other meanings of (x,y) and $\langle x,y \rangle$.

Note the simplification of this definition when $\mathfrak{F} = \mathfrak{R}$. In that case $x * y$ is real and the scalars are real; the conjugate-symmetric and conjugate-bilinear laws reduce to the symmetric and bilinear laws, respectively. The term "euclidean" need not be used in this context; we may simply say that S is a unitary space over $\mathfrak{R}$. On the other hand, the unqualified term "unitary" does not specify whether $\mathfrak{F} = \mathfrak{R}$ or $\mathfrak{F} = \mathfrak{C}$. Thus the vector space in Example 10.1 is both euclidean and unitary, while in Example 10.2 it is unitary but not euclidean.†

It is also important to note that when we call $\mathfrak{F}^n$ unitary, we mean its inner product to be

$$\mathbf{X} * \mathbf{Y} = \sum_{j=1}^{n} x_j \bar{y}_j$$

While other inner products on n-space are possible (as in Example 10.3), they are not intended unless specific exception is made. That is why we called this the "standard" inner product in Sec. 10.1. (Of course if $\mathfrak{F} = \mathfrak{R}$, we may also call it the dot product, since that is what it is.)

NORM

If S is a unitary space, the *norm* of an element $x \in S$ is $\|x\| = \sqrt{x * x}$.

EXAMPLE 10.4

For each $x \in \mathfrak{R}$, define

$$e^{ix} = \cos x + i \sin x$$

the celebrated *Euler formula*.‡ Then $f(x) = e^{ix}$ is a complex-valued function with the same exponential properties as e^x in real analysis. (See the problems.) As an element of the unitary space of Example 10.2, its norm is found from

$$\|f\|^2 = \int_{-\pi}^{\pi} f(x)\overline{f(x)}\, dx = \int_{-\pi}^{\pi} e^{ix}e^{-ix}\, dx = \int_{-\pi}^{\pi} 1\, dx = 2\pi$$

Hence we have $\|f\| = \sqrt{2\pi}$.

Note that in terms of Euler's formula we could have computed $f * g$ in Example 10.2 differently. Indeed, if $f(x) = \cos x + i \sin x$ and $g(x) = \cos x - i \sin x$, we have

† There is much variation of terminology in these matters. Some writers call a unitary space (over either $\mathfrak{R}$ or $\mathfrak{C}$) an "inner product space"; others reserve the term inner product for the real case and use "Hermitian product" when complex scalars are involved.

‡ Discovered by the prolific Swiss mathematician Leonhard Euler (1707–1783).

$$f*g = \int_{-\pi}^{\pi} f(x)\overline{g(x)}\,dx = \int_{-\pi}^{\pi} e^{ix}e^{ix}\,dx = \int_{-\pi}^{\pi} e^{2ix}\,dx$$

$$= \frac{1}{2i}e^{2ix}\Big|_{-\pi}^{\pi} = \frac{1}{2i}(e^{2i\pi} - e^{-2i\pi}) = \sin 2\pi = 0$$

DISTANCE

If x and y are elements of the unitary space S, the *distance* between them is $d(x,y) = \|x - y\|$.

To prove that the distance function is a "metric" on S (see Sec. 10.1), we must establish the positive-definite and symmetric properties and the triangle inequality. The first two of these are proved in any unitary space exactly as in $\mathfrak{F}^n$ (see Probs. 14 and 16, Sec. 10.1). But the triangle inequality was the result of the implications

Cauchy-Schwarz inequality $\Longrightarrow$ Minkowski inequality $\Longrightarrow$ Triangle inequality

Since the Cauchy-Schwarz inequality has not been proved, and since the first implication has been done only in $\mathcal{R}^n$ (Prob. 10, Sec. 10.1), we cannot as yet assert that unitary spaces are metric. We'll remedy this in the next section.

Problems

20. Let S be the vector space of real-valued functions continuous on the interval $[a,b]$. If f and g are elements of S, define

$$f*g = \int_a^b f(x)g(x)\,dx$$

Prove that this product is (*a*) positive-definite, (*b*) symmetric, and (*c*) bilinear, and hence that S is a euclidean space. Where in these arguments did you use the fact that the functions in S are continuous?

21. Let $[a,b] = [0,1]$ in Prob. 20. If f and g are defined as follows, what is $d(f,g)$?

a. $f(x) = 2$, $g(x) = 0$
b. $f(x) = x$, $g(x) = x + 3$
c. $f(x) = x$, $g(x) = x^2$

22. For each pair of vectors $\mathbf{X} = (x_1,x_2)$ and $\mathbf{Y} = (y_1,y_2)$ in $\mathcal{R}^2$, define

$$\mathbf{X} \circ \mathbf{Y} = (x_1 - x_2)(y_1 - y_2)$$

Prove that this product is symmetric and bilinear, but not positive-definite.

(Hence $\circ$ is not an inner product on $\mathcal{R}^2$. Some writers call a symmetric bilinear product (on either a real or complex space) a "scalar product." According to this terminology, an inner product on a euclidean space is a positive-definite scalar product.)

23. Let S be a nonzero vector space over $\mathcal{C}$ and suppose that $\circ$ is a scalar-valued bilinear product on S, that is,

$$(ax + by) \circ z = a(x \circ z) + b(y \circ z)$$

and

$$z \circ (ax + by) = a(z \circ x) + b(z \circ y)$$

for all a and b in $\mathcal{C}$ and all x, y, and z in S. Prove that $\circ$ cannot be positive-definite. *Hint:* Look at $(ix) \circ (ix)$, where x is a nonzero element of S. (This shows that the *conjugate*-bilinear law for an inner product on S is not mere caprice; we cannot define a positive-definite bilinear product on a complex space even if we try.)

24. Show that the second part of the conjugate-bilinear law in the definition of inner product is a consequence of the first part and the conjugate-symmetric law. (Thus it could be omitted from the definition and stated as a theorem instead.)

★ **25.** Prove that if S is a unitary space, then $0 * y = 0$ for all $y \in S$. *Hint:* $0 * y = (0 + 0) * y$. Why does it follow that $x * 0 = 0$ for all $x \in S$?

★ **26.** Prove that if $k \in \mathfrak{F}$ and $x \in S$ (where S is a unitary space over $\mathfrak{F}$), then $\|kx\| = |k|\,\|x\|$.

27. Let S be a unitary space with distance function $d(x,y) = \|x - y\|$.

a. Prove that d is positive-definite, that is, $d(x,y) \geq 0$, with equality holding $\Leftrightarrow$ $x = y$.

b. Prove that d is symmetric, that is, $d(x,y) = d(y,x)$ for all x and y in S.

28. Recall from calculus that for all $x \in \mathcal{R}$

$$e^x = 1 + x + \frac{x^2}{2!} + \frac{x^3}{3!} + \frac{x^4}{4!} + \cdots$$

$$\cos x = 1 - \frac{x^2}{2!} + \frac{x^4}{4!} - \frac{x^6}{6!} + \cdots$$

$$\sin x = x - \frac{x^3}{3!} + \frac{x^5}{5!} - \frac{x^7}{7!} + \cdots$$

It is apparent from these series representations that the cosine and sine are somehow "parts" of the exponential function. Substitute ix for x in the first series (not for any good mathematical reason, but just to see what happens), rearrange terms without worrying about convergence, and come up with Euler's formula. (Since e^{ix} is *defined* by the formula, you haven't proved anything; the idea is simply to discover why the formula is intuitively reasonable.)

29. Use Euler's formula to prove the following.

a. $e^{i\pi} + 1 = 0$. (This elegant little equation involves in one place the five most important numbers of mathematics, 0, 1, π, e, and i.)

b. If $x \in \mathcal{R}$, then the conjugate of e^{ix} is e^{-ix}.

c. For all x and y in $\mathcal{R}$,

$$e^{ix}e^{iy} = e^{i(x+y)} \qquad \text{and} \qquad \frac{e^{ix}}{e^{iy}} = e^{i(x-y)}$$

★ *d.* If k is any nonzero integer, $\int_{-\pi}^{\pi} e^{ikx}\,dx = 0$. Why does it follow that if $f(x) = e^{ipx}$ and $g(x) = e^{iqx}$ are regarded as elements of the unitary space in Example 10.2, then $f * g = 0$ when $p \neq q$? What is $f * g$ if $p = q$?

10.3 The Cauchy-Schwarz Inequality

In Sec. 10.1 (Prob. 8) we asked you to prove that if **X** and **Y** are vectors in the plane, then $|\mathbf{X} \cdot \mathbf{Y}| \leq \|\mathbf{X}\|\,\|\mathbf{Y}\|$. Unless you thought up something relatively sophisticated, the argument depended on the coordinates of **X** and **Y**. In Prob. 9 we suggested that the same sort of thing in $\mathcal{R}^3$ is not easy; for if $\mathbf{X} = (x_1,x_2,x_3)$ and $\mathbf{Y} = (y_1,y_2,y_3)$, the inequality reads

$$|x_1y_1 + x_2y_2 + x_3y_3| \leq \sqrt{x_1^2 + x_2^2 + x_3^2}\sqrt{y_1^2 + y_2^2 + y_3^2}$$

which is anything but obvious. One can prove it by assuming that it's false, squaring both sides, and reducing the result to a contradiction of the properties of $\mathcal{R}$, but the details are messy (and not all the steps are apparent).

Even if this succeeds, it is no help in the context of an abstract unitary space. We must come up with an argument that is "coordinate-free," depending only on the properties of an inner product. Because we want to keep it short, we present it in Theorem 10.1 without explanation; its geometrical motivation is given in the problems.

THEOREM 10.1 The Cauchy-Schwarz Inequality

If x and y are elements of the unitary space S, then $|x * y| \leq \|x\|\,\|y\|$.

Proof

If $x = 0$, the theorem is trivial, since $0 * y = 0$ for all y. (See Prob. 25, Sec. 10.2.) Hence we assume that x is nonzero. Then $\|x\|$ is nonzero; let

$$k = \frac{y * x}{\|x\|^2}$$

and define $z = y - kx$. We claim that $x * z = 0$ (see the problems); the idea of the proof is to get the Cauchy-Schwarz inequality from the fact that $z * z \geq 0$. This is admittedly like pulling a rabbit from a hat, but observe how smoothly it works:

$$\begin{aligned} z * z &\geq 0 \\ (y - kx) * z &\geq 0 \\ y * z - k(x * z) &\geq 0 \\ y * z &\geq 0 \\ y * (y - kx) &\geq 0 \\ y * y - \bar{k}(y * x) &\geq 0 \end{aligned}$$

But

$$\bar{k} = \frac{x * y}{\|x\|^2} \qquad \text{(why?)}$$

so we may write the last inequality in the form

$$(x * y)(y * x) \leq \|x\|^2 \|y\|^2$$

Since $x * y$ and $y * x$ are conjugates, this reduces to

$$|x * y|^2 \leq \|x\|^2 \|y\|^2$$

(See Prob. 43*c*, Sec. 1.3.) The Cauchy-Schwarz inequality follows by taking square roots. ▮

Note that when $S = \mathcal{R}^n$ this theorem reads

$$|\mathbf{X} \cdot \mathbf{Y}| \leq \|\mathbf{X}\| \, \|\mathbf{Y}\|$$

which is the Cauchy-Schwarz inequality we stated (but did not prove) in Sec. 10.1. To appreciate the power of the argument, recall the unitary space of complex-valued functions continuous on $[-\pi, \pi]$, with inner product

$$f * g = \int_{-\pi}^{\pi} f(x)\overline{g(x)}\, dx$$

(See Example 10.2.) In this space Theorem 10.1 says that

$$\left| \int_{-\pi}^{\pi} f(x)\overline{g(x)}\, dx \right| \leq \left[\int_{-\pi}^{\pi} |f(x)|^2\, dx \right]^{1/2} \left[\int_{-\pi}^{\pi} |g(x)|^2\, dx \right]^{1/2}$$

While this can be proved as an exercise in calculus, it is more economical to realize that it is nothing but the Cauchy-Schwarz inequality in the context of a certain unitary vector space.

THEOREM 10.2 The Minkowski Inequality

If x and y are elements of the unitary space S, then $\|x + y\| \leq \|x\| + \|y\|$.

Proof

See Prob. 10, Sec. 10.1, for the argument in real n-space. (The same argument is valid in any euclidean space.) If $\mathcal{F} = \mathcal{C}$, however, the inner product is not symmetric and we have to revise the proof.

We begin as in Prob. 10, Sec. 10.1, by writing

$$\begin{aligned}\|x + y\|^2 &= (x + y) * (x + y) \\ &= x * (x + y) + y * (x + y) \\ &= \|x\|^2 + x * y + y * x + \|y\|^2\end{aligned}$$

We cannot replace $x * y + y * x$ by $2(x * y)$ as we did before. However, $x * y$ and $y * x$ are conjugates, so we can write

$$\begin{aligned}\|x + y\|^2 &= \|x\|^2 + x * y + \overline{x * y} + \|y\|^2 \\ &\le \|x\|^2 + 2|x * y| + \|y\|^2 && \text{(Prob. 43}d\text{, Sec. 1.3)} \\ &\le \|x\|^2 + 2\|x\|\,\|y\| + \|y\|^2 && \text{(Cauchy-Schwarz inequality)} \\ &= (\|x\| + \|y\|)^2\end{aligned}$$

The Minkowski inequality follows by taking square roots. ▮

THEOREM 10.3

If S is a unitary space, the distance function $d(x,y) = \|x - y\|$ is a metric on S.

Proof

Since the inner product on S is positive-definite, we know that

$$(x - y) * (x - y) \ge 0 \qquad \text{with equality} \iff x - y = 0$$

Hence

$$d(x,y) = \sqrt{(x - y) * (x - y)} \ge 0 \qquad \text{with equality} \iff x = y$$

Thus d is positive-definite.

To show that d is symmetric, write

$$d(x,y) = \|x - y\| = \|-(y - x)\| = \|y - x\| = d(y,x)$$

(See Prob. 26, Sec. 10.2.) The triangle inequality is proved by writing

$$\begin{aligned}d(x,y) = \|x - y\| = \|(x - z) + (z - y)\| &\le \|x - z\| + \|z - y\| \\ &= d(x,z) + d(z,y)\end{aligned}$$ ▮

Problems

★ **30.** Consider the following figure in $\mathcal{R}^2$, which shows two vectors **X** and **Y**, the projection of **Y** on **X**, and the projection of **Y** perpendicular to **X**:

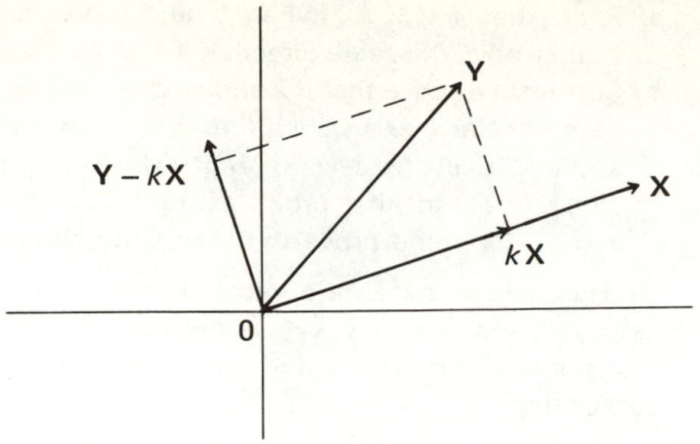

The projection of **Y** on **X** is a scalar multiple of **X**, say $k\mathbf{X}$; the projection of **Y** perpendicular to **X** is then $\mathbf{Y} - k\mathbf{X}$ (since the sum of the projections must be **Y**). Use the fact that $\mathbf{Y} - k\mathbf{X}$ and **X** are perpendicular to show that

$$k = \frac{\mathbf{Y} \cdot \mathbf{X}}{\|\mathbf{X}\|^2}$$

(This is the geometric idea behind the definition of k in the proof of the Cauchy-Schwarz inequality.)

★ **31.** In an arbitrary unitary space we have no prior notion of perpendicularity to fall back on, so we reverse the procedure indicated in Prob. 30. We *define*

$$z = y - kx \qquad \text{where } k = \frac{y * x}{\|x\|^2} \quad \text{and} \quad x \neq 0$$

If the situation in $\mathcal{R}^2$ is any guide, we should have $x * z = 0$. Prove that this is correct. (This justifies the opening move in the proof of the Cauchy-Schwarz inequality.)

32. Give a reason for each step in the proof of the Cauchy-Schwarz inequality (starting with $z * z \geq 0$).

★ **33.** Let x and y be nonzero elements of the euclidean space S. The *angle* between them is defined (as in $\mathcal{R}^n$) by

$$\cos\theta = \frac{x * y}{\|x\|\ \|y\|} \qquad 0 \leq \theta \leq \pi$$

a. Why does this make sense?
b. Why wouldn't it make sense if S were a unitary space over $\mathcal{C}$?

34. Two nonzero elements of the euclidean space S are said to have the "same direction" if the angle between them is 0, and "opposite directions" if it is π.

a. Prove that if $y = kx$ (where x and y are nonzero and $k \in \mathcal{R}$), then x and y have the same direction if $k > 0$, opposite directions if $k < 0$.

b. Conversely, prove that if x and y have the same or opposite directions, then a scalar k exists such that $y = kx$, where $k > 0$ in the first case and $k < 0$ in the second. *Hint:* The problem is to name k. Try the same k as in the proof of the Cauchy-Schwarz inequality, let $z = y - kx$, and prove that $z = 0$ by showing that $z * z = 0$.

★ **35.** To "normalize" a nonzero vector v in a unitary space, we divide v by its norm to obtain $u = v/\|v\|$. Show that u is a unit vector (that is, a vector with norm 1) and that u and v have the same direction if the space is euclidean.

36. Prove that equality holds in $|x * y| \leq \|x\| \, \|y\|$ if and only if x and y are linearly dependent. *Hint:* When $x \neq 0$ (the statement is trivial if $x = 0$), x and y are linearly dependent if and only if $y = kx$ for some $k \in \mathfrak{F}$.

37. Prove that equality holds in $\|x + y\| \leq \|x\| + \|y\|$ if and only if one of the elements x and y is a nonnegative real multiple of the other. *Hint:* Assuming that x is nonzero (the statement is trivial if $x = 0$), prove that

$$\|x + y\| = \|x\| + \|y\| \quad \Leftrightarrow \quad y = kx \qquad \text{for some nonnegative real } k$$

The $\Leftarrow$ part is easy; to do the $\Rightarrow$ part, note that the inequalities in our proof of Theorem 10.2 become equalities.

38. Let S be a unitary space over $\mathfrak{F}$. Derive the following identities (called *polar forms*) for $x * y$.

a. If $\mathfrak{F} = \mathcal{R}$, then $x * y = \frac{1}{4}(\|x + y\|^2 - \|x - y\|^2)$.

b. If $\mathfrak{F} = \mathcal{C}$, then $x * y = \frac{1}{4}[(\|x + y\|^2 - \|x - y\|^2) + i(\|x + iy\|^2 - \|x - iy\|^2)]$.

39. Let x and y be nonzero elements of the euclidean space S and let θ be the angle between them. Show that the polar form in Prob. 38*a* implies that

$$\|x - y\|^2 = \|x\|^2 + \|y\|^2 - 2\|x\| \, \|y\| \cos \theta$$

(This is the law of cosines, with which we introduced this chapter. Also see Prob. 6, Sec. 10.1.)

10.4 Orthogonality

In a euclidean space, the notion of perpendicularity is a specialization of the idea of angle; simply take $\theta = \pi/2$ in the formula

$$\cos \theta = \frac{x * y}{\|x\| \, \|y\|} \qquad \text{(Prob. 33, Sec. 10.3)}$$

This is equivalent to taking the inner product to be 0 (at least in the case of nonzero vectors). Since we can impose that idea on complex spaces as well (where angle is not defined), we adopt it as our point of departure.

ORTHOGONAL

Let S be a unitary space over $\mathfrak{F}$ (with inner product $*$). The elements x and y of S are said to be *orthogonal* (also *perpendicular*) if $x * y = 0$. More generally, $\{v_1, \ldots, v_m\}$ is called *orthogonal* if $v_i * v_j = 0$ whenever $i \neq j$. Thus an orthogonal set consists of mutually perpendicular elements.

EXAMPLE 10.5

Let S be the euclidean space of real-valued functions continuous on $[-\pi,\pi]$, with inner product defined by

$$f * g = \int_{-\pi}^{\pi} f(x)g(x)\, dx$$

(See Example 10.1.) Since

$$\int_{-\pi}^{\pi} \sin mx \cos nx\, dx = 0$$

for all positive integers m and n (see any calculus book), the functions $\sin mx$ and $\cos nx$ are orthogonal elements of S. So are $\sin mx$ and $\sin nx$ ($m \neq n$), as well as $\cos mx$ and $\cos nx$ ($m \neq n$). These "orthogonality relations" of the trigonometric functions are important in Fourier analysis.

THEOREM 10.4

An orthogonal set of nonzero vectors is linearly independent.

Proof

Let $\{v_1, \ldots, v_m\}$ be the set and suppose that

$$c_1v_1 + \cdots + c_mv_m = 0$$

Then if j is any one of the integers $1, \ldots, m$, we have

$$(c_1v_1 + \cdots + c_mv_m) * v_j = 0$$

or (by the conjugate-bilinear law)

$$c_1(v_1 * v_j) + \cdots + c_m(v_m * v_j) = 0$$

Each term of this sum is 0 except the jth, so it collapses to

$$c_j(v_j * v_j) = 0$$

Since v_j is nonzero, the positive-definite law says that $v_j * v_j$ must be nonzero; hence $c_j = 0$, where $j = 1, \ldots, m$. ∎

ORTHONORMAL

Let S be a unitary space. If $\|u\| = 1$, the element $u \in S$ is called *normal* (also a *unit vector*). More generally, $\{u_1, \ldots, u_n\}$ is *normal* if each of its elements is a unit vector. A set which is both orthogonal and normal is called *orthonormal*.

THEOREM 10.5

The set $\{u_1, \ldots, u_m\}$ is orthonormal if and only if $u_i * u_j = \delta_{ij}$ for all i and j.

Proof

Suppose that $u_i * u_j = \delta_{ij}$ for all i and j. Then

$$u_i * u_j = 0 \qquad \text{whenever } i \neq j$$

which means that $\{u_1, \ldots, u_m\}$ is orthogonal. On the other hand,

$$u_i * u_i = 1 \qquad \text{for each } i$$

which implies that $\|u_i\| = 1$, where $i = 1, \ldots, m$. Hence $\{u_1, \ldots, u_m\}$ is normal. The converse is left for the problems. ∎

EXAMPLE 10.6

The standard basis $\{\mathbf{E}_1, \ldots, \mathbf{E}_n\}$ for $\mathfrak{F}^n$ is orthonormal. This is easy to see from the definition, but it is worthwhile to prove it formally by recalling from Example 1.4 that $\mathbf{E}_i$ and $\mathbf{E}_j$ can be written as

$$\mathbf{E}_i = (\delta_{i1}, \ldots, \delta_{in}) \qquad \text{and} \qquad \mathbf{E}_j = (\delta_{1j}, \ldots, \delta_{nj})$$

Hence their standard inner product is

$$\begin{aligned} \mathbf{E}_i * \mathbf{E}_j &= \delta_{i1}\overline{\delta}_{1j} + \cdots + \delta_{in}\overline{\delta}_{nj} \\ &= \sum_{k=1}^{n} \delta_{ik}\delta_{kj} \qquad \text{(because Kronecker's delta is real)} \end{aligned}$$

The factor δ_{ik} in the typical term of this sum is 0 unless $k = i$, so the sum collapses to

$$\mathbf{E}_i * \mathbf{E}_j = \delta_{ii}\delta_{ij} = \delta_{ij}$$

By Theorem 10.5 the conclusion follows.

Note that when $n = 3$ (and $\mathfrak{F} = \mathfrak{R}$), we are making the ordinary geometric observation that the elements $\mathbf{E}_1 = (1,0,0)$, $\mathbf{E}_2 = (0,1,0)$, $\mathbf{E}_3 = (0,0,1)$ are mutually perpendicular unit vectors in space. Similar conditions hold in the case of $n = 2$.

LECTOR *What about when $n = 1$?*

AUCTOR *Then there is only one basis element, $E_1 = 1$.*

LECTOR *Do you call $\{E_1\}$ an orthonormal set?*

AUCTOR *Sure. It satisfies the definition of orthogonality by default, and $\|E_1\| = 1$.*

LECTOR *I don't like it.*

AUCTOR *You agree, don't you, that $E_i * E_j = 0$ whenever $i \neq j$?*

LECTOR *But $i = j = 1$.*

AUCTOR *Precisely. It's like saying "If Christmas is on July 4, then July 5 is the day after Christmas."*

LECTOR *I don't like Christmas in July.*

AUCTOR *You're being difficult.*

EXAMPLE 10.7

Let v_1 and v_2 be the functions defined by $v_1(x) = \cos mx$ and $v_2(x) = \sin mx$, where m is a positive integer. Then in the space S of Example 10.5 we have

$$\|v_1\|^2 = \int_{-\pi}^{\pi} \cos^2 mx\, dx = \pi \qquad \text{and} \qquad \|v_2\|^2 = \int_{-\pi}^{\pi} \sin^2 mx\, dx = \pi$$

Hence

$$u_1 = \frac{v_1}{\sqrt{\pi}} \qquad \text{and} \qquad u_2 = \frac{v_2}{\sqrt{\pi}}$$

are normal vectors in S (see Prob. 35, Sec. 10.3). Since they are also orthogonal (Example 10.5), $\{u_1,u_2\}$ is an orthonormal set. By analogy with the orthonormal vectors $\mathbf{E}_1, \ldots, \mathbf{E}_n$ in $\mathfrak{F}^n$ (which provide a particularly simple type of basis), we should expect to find $\{u_1,u_2\}$ a convenient basis for the subspace of S generated by v_1 and v_2.†

To see how convenient an orthonormal basis is, let us look more closely at the standard basis for $\mathfrak{F}^n$. If $\mathbf{X} = (x_1, \ldots, x_n)$ is any vector in $\mathfrak{F}^n$, its expression as a linear combination of $\mathbf{E}_1, \ldots, \mathbf{E}_n$ is simply

$$\mathbf{X} = x_1\mathbf{E}_1 + \cdots + x_n\mathbf{E}_n = \sum_{i=1}^{n} x_i\mathbf{E}_i$$

† This is $T = \langle v_1,v_2 \rangle$, the set of all linear combinations of $\cos mx$ and $\sin mx$. See Sec. 2.6.

The coefficients are just the coordinates of **X**. There is another way of saying this, however, which generalizes to arbitrary (finite-dimensional) unitary spaces. Observe that for each $j = 1, \ldots, n$, we have

$$\begin{aligned}\mathbf{X} * \mathbf{E}_j &= \left(\sum_{i=1}^{n} x_i \mathbf{E}_i\right) * \mathbf{E}_j \\ &= \sum_{i=1}^{n} x_i (\mathbf{E}_i * \mathbf{E}_j) \qquad \text{(Why?)} \\ &= \sum_{i=1}^{n} x_i \delta_{ij} \qquad \text{(Theorem 10.5)}\end{aligned}$$

But this reduces to

$$\mathbf{X} * \mathbf{E}_j = x_j \qquad j = 1, \ldots, n$$

which means that the expression of **X** as a linear combination of $\mathbf{E}_1, \ldots, \mathbf{E}_n$ can be written

$$\mathbf{X} = (\mathbf{X} * \mathbf{E}_1)\mathbf{E}_1 + \cdots + (\mathbf{X} * \mathbf{E}_n)\mathbf{E}_n = \sum_{i=1}^{n} (\mathbf{X} * \mathbf{E}_i)\mathbf{E}_i$$

Each coefficient is the inner product of **X** and the corresponding element of the standard basis. This observation can be generalized to apply to any orthonormal basis, as we'll see shortly (Theorem 10.7). But first we state a preliminary result for orthogonal (not necessarily orthonormal) bases.

THEOREM 10.6

Let $\{v_1, \ldots, v_n\}$ be an orthogonal basis for the unitary space S. If

$$x = \sum_{i=1}^{n} x_i v_i$$

is any element of S, then

$$x_i = \frac{x * v_i}{\|v_i\|^2} \qquad i = 1, \ldots, n$$

Proof

For each $j = 1, \ldots, n$, we have

$$x * v_j = \left(\sum_{i=1}^{n} x_i v_i\right) * v_j = \sum_{i=1}^{n} x_i (v_i * v_j) = x_j (v_j * v_j) = x_j \|v_j\|^2$$

Since v_j is nonzero (why?), it follows that $\|v_j\|$ is nonzero and hence that

$$x_j = \frac{x * v_j}{\|v_j\|^2} \qquad j = 1, \ldots, n \blacksquare$$

The formula in Theorem 10.6 gives the typical coefficient in the expression of an arbitrary element x in terms of $v_1, \ldots, v_n$; it should remind you of something you've seen before. In the proof of the Cauchy-Schwarz inequality (Sec. 10.3) we defined

$$k = \frac{y * x}{\|x\|^2}$$

where x is nonzero and y is arbitrary; then $z = y - kx$ turned out to be perpendicular to x. In $\mathcal{R}^2$ or $\mathcal{R}^3$ (given nonzero **X** and any **Y**) the geometric interpretation is that $k\mathbf{X}$ is the projection of **Y** parallel to **X** and $\mathbf{Y} - k\mathbf{X}$ is its projection perpendicular to **X**, as in Fig. 10.2. (See Probs. 30 and 31, Sec. 10.3.) In view of this background it appears that we ought to give a name to

$$x_i = \frac{x * v_i}{\|v_i\|^2}$$

in Theorem 10.6. This is the point of the next definition.

FOURIER COEFFICIENT

Let y be any nonzero element of the unitary space S. If $x \in S$, the *Fourier coefficient of x with respect to y* is

$$\frac{x * y}{\|y\|^2}$$

Theorem 10.6 may now be paraphrased by saying that if we want to express an element $x \in S$ as a linear combination of the orthogonal basis

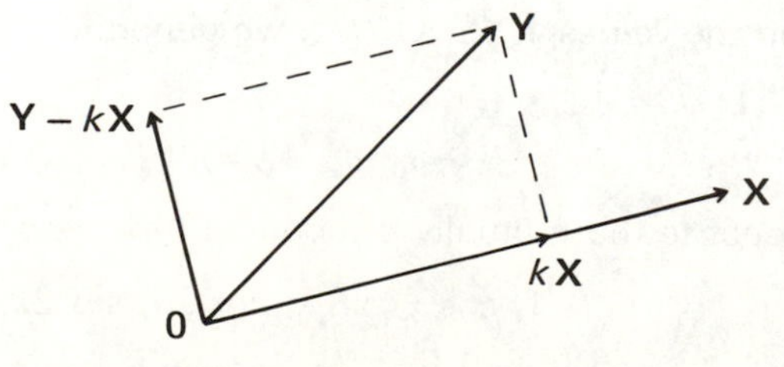

Fig. 10.2

elements $v_1, \ldots, v_n$, we need only compute the Fourier coefficients of x with respect to each v_i. These are the coefficients in the desired expression.

EXAMPLE 10.8

Suppose that $\mathbf{E}_1$ and $\mathbf{E}_2$ in $\mathcal{R}^2$ are replaced by $\mathbf{V}_1 = (1,1)$ and $\mathbf{V}_2 = (-1,1)$. Then $\{\mathbf{V}_1,\mathbf{V}_2\}$ is an orthogonal basis for $\mathcal{R}^2$. Given any vector $\mathbf{X} = (x_1,x_2)$, suppose that we want to write $\mathbf{X}$ as a linear combination of $\mathbf{V}_1$ and $\mathbf{V}_2$. Unaware of Theorem 10.6, we might proceed by putting $\mathbf{X} = c_1\mathbf{V}_1 + c_2\mathbf{V}_2$, where c_1 and c_2 are unknown coefficients to be determined from

$$(x_1,x_2) = c_1(1,1) + c_2(-1,1)$$

This is equivalent to the system

$$\begin{aligned} c_1 - c_2 &= x_1 \\ c_1 + c_2 &= x_2 \end{aligned}$$

the solution of which (as you can check) is

$$c_1 = \tfrac{1}{2}(x_1 + x_2) \qquad c_2 = \tfrac{1}{2}(-x_1 + x_2)$$

But this is the hard way. According to Theorem 10.6 the coefficients are given by

$$c_1 = \frac{\mathbf{X}\cdot\mathbf{V}_1}{\|\mathbf{V}_1\|^2} = \frac{x_1 + x_2}{2} \qquad c_2 = \frac{\mathbf{X}\cdot\mathbf{V}_2}{\|\mathbf{V}_2\|^2} = \frac{-x_1 + x_2}{2}$$

EXAMPLE 10.9

Let f be an arbitrary element of the euclidean space of real-valued functions continuous on $[-\pi, \pi]$. The Fourier coefficients of f with respect to the functions $\cos mx$ and $\sin mx$ are

$$a_m = \frac{1}{\pi}\int_{-\pi}^{\pi} f(x)\cos mx\,dx \qquad \text{and} \qquad b_m = \frac{1}{\pi}\int_{-\pi}^{\pi} f(x)\sin mx\,dx$$

respectively. (See Example 10.7.) Defining

$$a_0 = \frac{1}{2\pi}\int_{-\pi}^{\pi} f(x)\,dx$$

(the Fourier coefficient of f with respect to the constant function 1), and letting m take on the values 1, 2, 3, . . . , we generate a sequence of Fourier coefficients

$$a_0, a_1, b_1, a_2, b_2, \ldots$$

with respect to the mutually orthogonal functions

$$1, \cos x, \sin x, \cos 2x, \sin 2x, \ldots$$

This sequence is then used to construct the *Fourier series*

$$a_0 + (a_1 \cos x + b_1 \sin x) + (a_2 \cos 2x + b_2 \sin 2x) + \cdots$$

As we shall see later, it is natural to investigate whether this series converges to $f(x)$; it was in connection with this question that mathematicians first used the terminology "Fourier coefficients."

THEOREM 10.7

Let $\{u_1, \ldots, u_n\}$ be an orthonormal basis for the unitary space S. If

$$x = \sum_{i=1}^{n} x_i u_i$$

is any element of s, then

$$x_i = x * u_i \qquad i = 1, \ldots, n$$

Proof

Use Theorem 10.6.

Theorem 10.7 expresses the idea we discussed in connection with the standard basis for $\mathfrak{F}^n$. To write an arbitrary element x in terms of the elements of an *orthonormal* basis, simply compute the inner product of x with each basis element. These scalars are the coefficients in the desired expression.

EXAMPLE 10.9

The vectors

$$\mathbf{U}_1 = \frac{1}{\sqrt{3}}(1,-1,1) \qquad \mathbf{U}_2 = \frac{1}{\sqrt{6}}(2,1,-1) \qquad \mathbf{U}_3 = \frac{1}{\sqrt{2}}(0,1,1)$$

are mutually perpendicular unit vectors in $\mathcal{R}^3$. (Confirm!) This means they are linearly independent (Theorem 10.4) and hence constitute a basis for $\mathcal{R}^3$. If $\mathbf{X} = (x_1,x_2,x_3)$ is any element of $\mathcal{R}^3$, it would take a while to write it as a linear combination of $\mathbf{U}_1,\mathbf{U}_2,\mathbf{U}_3$ (using brute force). But according to Theorem 10.7 we need only compute $\mathbf{X}\cdot\mathbf{U}_1$, $\mathbf{X}\cdot\mathbf{U}_2$, $\mathbf{X}\cdot\mathbf{U}_3$ and we have found the coefficients:

$$\mathbf{X} = \frac{1}{\sqrt{3}}(x_1 - x_2 + x_3)\mathbf{U}_1 + \frac{1}{\sqrt{6}}(2x_1 + x_2 - x_3)\mathbf{U}_2 + \frac{1}{\sqrt{2}}(x_2 + x_3)\mathbf{U}_3$$

If $\mathbf{X} = (5,-1,2)$, for example, we can write down by inspection the expression

$$\mathbf{X} = \frac{8}{\sqrt{3}}\mathbf{U}_1 + \frac{7}{\sqrt{6}}\mathbf{U}_2 + \frac{1}{\sqrt{2}}\mathbf{U}_3$$

Problems

40. Explain why $x * y = 0 \iff y * x = 0$. (Thus the definition of orthogonality does not depend on the order in which the inner product is written, despite the fact that in general $x * y$ and $y * x$ are not the same.)

41. Let S be a unitary space. Then $0 * y = 0$ for all $y \in S$ (Prob. 25, Sec. 10.2). Explain why 0 is the only element with this property; that is, prove that

$$x * y = 0 \quad \text{for all } y \quad \Longrightarrow \quad x = 0$$

42. Prove that if $\{u_1, \ldots, u_m\}$ is orthonormal, then $u_i * u_j = \delta_{ij}$ for all i and j (the "only if" part of Theorem 10.5).

43. Show that in a euclidean space x and y are orthogonal if and only if $\|x + y\| = \|x - y\|$. Draw a picture illustrating this theorem in $\mathcal{R}^2$.

44. Find two vectors in $\mathcal{C}^2$ for which $\|\mathbf{X} + \mathbf{Y}\| = \|\mathbf{X} - \mathbf{Y}\|$ while $\mathbf{X} * \mathbf{Y} \neq 0$. (Thus the "if" part of Prob. 43 is false if the space is not euclidean. How about the "only if" part?)

★ **45.** In a euclidean space it is correct to say that

$$\|x + y\|^2 = \|x\|^2 + \|y\|^2 \iff x \text{ and } y \text{ are orthogonal}$$

(This is the pythagorean theorem and its converse; see Prob. 18, Sec. 10.1.) Show that in a unitary space over $\mathcal{C}$ the situation is as follows.

a. $x * y = 0 \implies \|x + y\|^2 = \|x\|^2 + \|y\|^2$. (Thus the pythagorean theorem is still true.)

b. If $y = ix$ and $x \neq 0$, then $\|x + y\|^2 = \|x\|^2 + \|y\|^2$, but $x * y \neq 0$. (The converse of the pythagorean theorem is false.)

c. $\|x + y\|^2 = \|x\|^2 + \|y\|^2 = \|x + iy\|^2 \implies x * y = 0$

46. Let $\mathbf{V}_1 = (0, -1, 1)$, $\mathbf{V}_2 = (-2, 3, 3)$, and $\mathbf{V}_3 = (3, 1, 1)$.

a. Confirm that $\{\mathbf{V}_1, \mathbf{V}_2, \mathbf{V}_3\}$ is an orthogonal set of nonzero vectors in $\mathcal{R}^3$. Why is it therefore a basis?

b. Use Theorem 10.6 to express $\mathbf{X} = (-1, 5, 0)$ as a linear combination of $\mathbf{V}_1, \mathbf{V}_2, \mathbf{V}_3$.

47. In the euclidean space of real-valued functions continuous on $[0, 1]$, with inner product

$$f * g = \int_0^1 f(x)g(x)\,dx$$

let T be the set of all polynomials of the form $ax + b$.

a. Why is T a subspace?

b. Confirm that $v_1(x) = 1$ and $v_2(x) = x - \frac{1}{2}$ are orthogonal elements of T which span T. Why does it follow that $\{v_1, v_2\}$ is a basis for T?

c. "Normalize" v_1 and v_2 to obtain an orthonormal basis $\{u_1, u_2\}$.
d. Express $f(x) = 2x - 3$ as a linear combination of u_1 and u_2 by setting $f = c_1u_1 + c_2u_2$ and solving for c_1 and c_2. By way of contrast, use Theorem 10.7 to write $g(x) = x$ in terms of u_1 and u_2.

48. In the following cases find the Fourier coefficients of f with respect to

$$1, \cos x, \sin x, \cos 2x, \sin 2x, \ldots$$

and write down the Fourier series associated with f. (See Example 10.9.)
a. $f(x) = 1$
b. $f(x) = \cos x$
c. $f(x) = x$

49. Let $\{u_1, \ldots, u_n\}$ be a basis for the unitary space S and define $a_{ij} = u_i * u_j$ for each i and j. Let $x = \Sigma_{i=1}^n x_iu_i$ and $y = \Sigma_{i=1}^n y_iu_i$ be any elements of S.
a. Show that $x * y = \Sigma_{j=1}^n (\Sigma_{i=1}^n a_{ij}x_i)\bar{y}_j$.
b. Show that the result in part *a* reduces to $x * y = \Sigma_{j=1}^n \|u_j\|^2 x_j \bar{y}_j$ if $\{u_1, \ldots, u_n\}$ is orthogonal.
c. What does $x * y$ reduce to if $\{u_1, \ldots, u_n\}$ is orthonormal?

50. Employing the same notation as in Prob. 49, let $A = [a_{ij}]$.
a. Explain why A is symmetric if S is euclidean. (See Sec. 4.3 for a definition of symmetric matrix.)
b. Why is A a diagonal matrix (Sec. 9.1) if $\{u_1, \ldots, u_n\}$ is orthogonal?
c. What is A if $\{u_1, \ldots, u_n\}$ is orthonormal?
d. Taking $\mathbf{U}_1 = (1,2,-1)$, $\mathbf{U}_2 = (3,5,0)$, $\mathbf{U}_3 = (-1,2,2)$ in $S = \mathcal{R}^3$, compute the matrix A. Note that it is symmetric, as advertised in part *a*. (See Example 4.8, where this problem has already been done.)
e. The vectors $\mathbf{U}_1 = (1,2,-1)$, $\mathbf{U}_2 = (5,4,13)$, $\mathbf{U}_3 = (5,-3,-1)$ are orthogonal in $\mathcal{R}^3$. Confirm that A is a diagonal matrix in this case.
f. Find A if the standard basis for $\mathcal{R}^3$ is used.

51. Let $\{\mathbf{U}_1, \mathbf{U}_2\}$ be an arbitrary basis for $\mathcal{R}^2$. Given that $\mathbf{X} = a_1\mathbf{U}_1 + a_2\mathbf{U}_2$ and $\mathbf{Y} = b_1\mathbf{U}_1 + b_2\mathbf{U}_2$, define $\mathbf{X} \circ \mathbf{Y} = a_1b_1 + a_2b_2$.
a. Show that $\circ$ is an inner product on $\mathcal{R}^2$.
b. Prove that with respect to this inner product (instead of the dot product) $\{\mathbf{U}_1, \mathbf{U}_2\}$ is an orthonormal basis for $\mathcal{R}^2$. (Thus the ordinary geometry of the plane has been replaced by one in which $\mathbf{U}_1$ and $\mathbf{U}_2$ are perpendicular unit vectors.)
c. If $\mathbf{U}_1 = (1,0)$ and $\mathbf{U}_2 = (0,1)$, what does $\circ$ reduce to?
d. If $\mathbf{U}_1 = (2,1)$ and $\mathbf{U}_2 = (3,2)$, show that the inner product of $\mathbf{X} = (x_1,x_2)$ and $\mathbf{Y} = (y_1,y_2)$ is $\mathbf{X} \circ \mathbf{Y} = 5x_1y_1 - 8x_1y_2 - 8x_2y_1 + 13x_2y_2$.

e. Use the formula in part *d* to compute $\mathbf{E}_1 \circ \mathbf{E}_2$. Are $\mathbf{E}_1$ and $\mathbf{E}_2$ perpendicular in this new geometry?

f. Use the formula in part *d* to compute the length of $\mathbf{E}_1$. Is $\mathbf{E}_1$ a unit vector in this new geometry?

When the standard inner product (dot product) on $\mathcal{R}^2$ is replaced by another inner product, we are dealing with a different euclidean space. There is a mapping from one to the other, but it does not (in general) preserve angles or distances (as shown by parts *e* and *f* in Prob. 51). However, there are circumstances in which it does, as the following problems indicate.

52. Take $\mathbf{U}_1 = (1,-2)$ and $\mathbf{U}_2 = (2,1)$ in Prob. 51, noting that they are orthogonal.

a. Show that the inner product of $\mathbf{X}$ and $\mathbf{Y}$ is $\mathbf{X} \circ \mathbf{Y} = \frac{1}{5}(\mathbf{X} \circ \mathbf{Y})$.

b. Prove that angles are preserved as we change from the old to the new geometry. *Hint:* This amounts to showing that if $\mathbf{X}$ and $\mathbf{Y}$ are nonzero vectors, then

$$\frac{\mathbf{X} \circ \mathbf{Y}}{\sqrt{\mathbf{X} \circ \mathbf{X}}\sqrt{\mathbf{Y} \circ \mathbf{Y}}} = \frac{\mathbf{X} \cdot \mathbf{Y}}{\|\mathbf{X}\|\,\|\mathbf{Y}\|} \qquad \text{(Why?)}$$

c. Show that lengths are not preserved.

The map that sends $\mathbf{E}_1$ and $\mathbf{E}_2$ into $\mathbf{U}_1$ and $\mathbf{U}_2$ is in this case $f\colon \mathcal{R}^2 \to \mathcal{R}^2$ defined by $f(x,y) = (x + 2y, -2x + y)$. Because it preserves angles it is called "conformal."

53. Take $\mathbf{U}_1 = (1/\sqrt{5})(1,-2)$ and $\mathbf{U}_2 = (1/\sqrt{5})(2,1)$ in Prob. 51, noting that they are orthonormal.

a. Show that the inner product of $\mathbf{X}$ and $\mathbf{Y}$ is $\mathbf{X} \circ \mathbf{Y} = \mathbf{X} \cdot \mathbf{Y}$.

b. Why does it follow that both angles and lengths are preserved as we change from the old to the new geometry?

c. What is the matrix (relative to the standard basis) of the linear map $f\colon \mathcal{R}^2 \to \mathcal{R}^2$ which sends $\mathbf{E}_1$ and $\mathbf{E}_2$ into $\mathbf{U}_1$ and $\mathbf{U}_2$? Confirm that f is a rotation of the plane. (Had we noticed the rotation first, the preservation of angles and lengths would have been obvious. Such a map is called "orthogonal," as is its matrix; we'll discuss orthogonal maps and matrices in detail in Chap. 12.)

10.5 Parseval's Identity

An important implication of our discussion of orthonormal bases in the last section is that the inner product on an abstract unitary space of dimension n can be interpreted as the standard inner product on $\mathcal{F}^n$. To see what we mean, let S be the space (with inner product $*$) and suppose that $\{u_1, \ldots, u_n\}$ is an orthonormal basis for S.[†] Let

$$x = \sum_{j=1}^{n} x_j u_j \qquad \text{and} \qquad y = \sum_{j=1}^{n} y_j u_j$$

be any elements of S. Then

$$\begin{aligned} x * y &= x * \left(\sum_{j=1}^{n} y_j u_j\right) \\ &= \sum_{j=1}^{n} \bar{y}_j (x * u_j) \qquad \text{(Why?)} \\ &= \sum_{j=1}^{n} \bar{y}_j x_j \qquad \text{(Theorem 10.7)} \\ &= \sum_{j=1}^{n} x_j \bar{y}_j{}^{\ddagger} \end{aligned}$$

Now let $\mathbf{X} = (x_1, \ldots, x_n)$ and $\mathbf{Y} = (y_1, \ldots, y_n)$ be the n-tuples of coefficients in the expressions

$$x = \sum_{j=1}^{n} x_j u_j \qquad \text{and} \qquad y = \sum_{j=1}^{n} y_j u_j$$

that is, the coordinate vectors of x and y relative to the given basis. (See Sec. 3.4.) These are vectors in $\mathfrak{F}^n$; their standard inner product (Sec. 10.1) is

$$\mathbf{X} * \mathbf{Y} = \sum_{j=1}^{n} x_j \bar{y}_j$$

Hence

$$x * y = \mathbf{X} * \mathbf{Y}$$

That is, the inner product of x and y in S is the standard inner product of $\mathbf{X}$ and $\mathbf{Y}$ in $\mathfrak{F}^n$.

[†] A question that arises here is whether S *has* an orthonormal basis. In the next section we'll prove that it does.

[‡] See Prob. 49*c*, Sec. 10.4, where this identity has already occurred.

THEOREM 10.8 Parseval's Identity

Let $\{u_1, \ldots, u_n\}$ be an orthonormal basis for the unitary space S. If

$$x = \sum_{j=1}^{n} x_j u_j \quad \text{and} \quad y = \sum_{j=1}^{n} y_j u_j$$

are any elements of S, their inner product is

$$x * y = \sum_{j=1}^{n} x_j \bar{y}_j = \mathbf{X} * \mathbf{Y}$$

where $\mathbf{X} = (x_1, \ldots, x_n)$ and $\mathbf{Y} = (y_1, \ldots, y_n)$ are the coordinate vectors of x and y relative to the given basis.

Proof

Refer to the preceding discussion.

COROLLARY 10.8a

Let $\{u_1, \ldots, u_n\}$ be an orthonormal set in a unitary space (not necessarily finite-dimensional). If

$$x = \sum_{j=1}^{n} x_j u_j \quad \text{and} \quad y = \sum_{j=1}^{n} y_j u_j$$

are elements of the subspace spanned by $u_1, \ldots, u_n$, then

$$x * y = \sum_{j=1}^{n} x_j \bar{y}_j$$

Theorem 10.8 should make abstract (finite-dimensional) unitary spaces seem more concrete. Although nothing explicit is known about S, we can always move into n-space, replacing the elements of S by their coordinate vectors in $\mathfrak{F}^n$. Review Sec. 3.4, where the coordinate vector map that sends

$$x = \sum_{j=1}^{n} x_j u_j \quad \text{into} \quad \mathbf{X} = (x_1, \ldots, x_n)$$

was shown to be an isomorphism between S and $\mathfrak{F}^n$. What we are saying now is that if the abstract space is unitary and *if an orthonormal basis is used,* this isomorphism preserves not only the addition and scalar multiplication of S as it is mapped onto $\mathfrak{F}^n$, but also the inner product.

To be precise, let f be the function defined by $f(x) = \mathbf{X}$, where $x = \sum_{j=1}^{n} x_j u_j$ is any element of S and $\mathbf{X} = (x_1, \ldots, x_n)$ is its coordinate vector relative to the basis $\{u_1, \ldots, u_n\}$. In Sec. 3.4 we proved that f maps S *onto*

$\mathfrak{F}^n$ in a *one-to-one* fashion; thus it is *invertible*. We also showed that it is *linear;* for all x and y in S and every scalar k,

$$f(x+y) = f(x) + f(y) = \mathbf{X} + \mathbf{Y} \qquad \text{and} \qquad f(kx) = kf(x) = k\mathbf{X}$$

Thus f matches $\mathbf{X} + \mathbf{Y}$ with $x + y$ and $k\mathbf{X}$ with kx; the algebraic structure of S is exactly copied in $\mathfrak{F}^n$.

Parseval's identity says that if $\{u_1, \ldots, u_n\}$ is orthonormal, the inner product on S is also duplicated in $\mathfrak{F}^n$:

$$x * y = \mathbf{X} * \mathbf{Y}$$

In other words, it makes no difference whether one finds the product of x and y in S or the product of their images in $\mathfrak{F}^n$. These products are *computed* differently, since S and $\mathfrak{F}^n$ are (in general) different spaces. But the numerical results are the same.

EXAMPLE 10.11

In the euclidean space of real-valued functions continuous on $[-\pi,\pi]$ the functions

$$u_1(x) = \frac{1}{\sqrt{\pi}} \cos x \qquad \text{and} \qquad u_2(x) = \frac{1}{\sqrt{\pi}} \sin x$$

constitute an orthonormal set. (See Example 10.7.) Corollary 10.8a says that the inner product of $f = 2u_1 + 5u_2$ and $g = 2u_1 - u_2$ is

$$f * g = (2,5) \cdot (2,-1) = -1$$

Compare this with the direct computation of $f * g$:

$$\begin{aligned}\int_{-\pi}^{\pi} f(x)g(x)\, dx &= \frac{1}{\pi}\int_{-\pi}^{\pi} (2\cos x + 5\sin x)(2\cos x - \sin x)\, dx \\ &= \frac{4}{\pi}\int_{-\pi}^{\pi} \cos^2 x\, dx + \frac{8}{\pi}\int_{-\pi}^{\pi} \sin x \cos x\, dx - \frac{5}{\pi}\int_{-\pi}^{\pi} \sin^2 x\, dx \\ &= -1\end{aligned}$$

THEOREM 10.9

Let $\{u_1, \ldots, u_n\}$ be an orthonormal basis for the unitary space S. If

$$x = \sum_{j=1}^{n} x_j u_j$$

is any element of S, then

$$\|x\| = \sqrt{|x_1|^2 + \cdots + |x_n|^2}$$

Proof

This is left for the problems.

COROLLARY 10.9a

If

$$x = \sum_{j=1}^{n} x_j u_j \qquad \text{and} \qquad y = \sum_{j=1}^{n} y_j u_j$$

are elements of S, then $d(x,y) = \sqrt{|x_1 - y_1|^2 + \cdots + |x_n - y_n|^2}$.

These results, too, should make matters more concrete. You know that if $\mathbf{X} = (x_1, \ldots, x_n)$ and $\mathbf{Y} = (y_1, \ldots, y_n)$ are vectors in n-space, then

$$\|\mathbf{X}\| = \sqrt{|x_1|^2 + \cdots + |x_n|^2}$$

and

$$d(\mathbf{X},\mathbf{Y}) = \sqrt{|x_1 - y_1|^2 + \cdots + |x_n - y_n|^2}$$

(See Prob. 13, Sec. 10.1.) Theorem 10.9 and Corollary 10.9a say that one may use the same formulas in any unitary space, provided that x and y are expressed in terms of an orthonormal basis.

EXAMPLE 10.12

In terms of the orthonormal vectors

$$\mathbf{U}_1 = \frac{1}{\sqrt{3}}(1,-1,1) \qquad \mathbf{U}_2 = \frac{1}{\sqrt{6}}(2,1,-1) \qquad \mathbf{U}_3 = \frac{1}{\sqrt{2}}(0,1,1)$$

the vector $\mathbf{X} = (5,-1,2)$ is

$$\mathbf{X} = \frac{8}{\sqrt{3}}\mathbf{U}_1 + \frac{7}{\sqrt{6}}\mathbf{U}_2 + \frac{1}{\sqrt{2}}\mathbf{U}_3$$

(See Example 10.10.) Theorem 10.9 says that its norm can be found from

$$\|\mathbf{X}\|^2 = \frac{64}{3} + \frac{49}{6} + \frac{1}{2} = 30$$

LECTOR *Anybody in his right mind would find it from* $\|\mathbf{X}\|^2 = 25 + 1 + 4 = 30$.

AUCTOR *Quite so. I was just trying to illustrate Theorem 10.9.*

LECTOR *Please try again.*

AUCTOR *Right.*

EXAMPLE 10.13

In the unitary space of complex-valued functions continuous on $[-\pi,\pi]$ the functions

$$u_1(x) = \frac{1}{\sqrt{2\pi}}e^{ix} \qquad \text{and} \qquad u_2(x) = \frac{1}{\sqrt{2\pi}}e^{2ix}$$

constitute an orthonormal set. (See Examples 10.2 and 10.4, and Prob. 29*d*, in Sec. 10.2.) If $f = 3u_1 + u_2$ and $g = u_1 - iu_2$, Theorem 10.9 and Corollary 10.9a, applied to $T = \langle u_1, u_2 \rangle$, yield

$$\|f\| = \sqrt{9 + 1} = \sqrt{10} \qquad \|g\| = \sqrt{1 + |-i|^2} = \sqrt{2}$$

$$d(f,g) = \|f - g\| = \sqrt{4 + |1 + i|^2} = \sqrt{4 + 2} = \sqrt{6}$$

LECTOR *I like Example 10.12 better.*

AUCTOR *You would rather bear those ills we have than fly to those that we know not of?*

LECTOR *I am sicklied o'er with the pale cast of thought.*

AUCTOR *In that case you may omit the rest of this section without loss of continuity.*

Parseval's identity suggests something else. Suppose that S is a vector space on which no inner product has been defined, so that geometry has not been imposed on S. Is there any reason to suppose it can be? The definition of inner product in Sec. 10.2 refers to a space as unitary *if it is possible* to define a product with the same properties as the standard inner product on $\mathfrak{F}^n$ (which we know "geometrizes" n-space). But there is no guarantee that such a product exists. In the space of real-valued functions continuous on $[a,b]$, for example, it is not obvious that we should define

$$f * g = \int_a^b f(x)g(x)\,dx$$

(See Prob. 20, Sec. 10.2.) What if this had not occurred to us? Then the existence of an inner product might be in some doubt.

Of course this space is not finite-dimensional. Suppose that S *is,* with basis $\{u_1, \ldots, u_n\}$. Naturally we cannot discuss whether this basis is orthonormal, since that concept is meaningless unless S is unitary. But Parseval's identity suggests that we may "induce" an inner product on S by copying the standard inner product on $\mathfrak{F}^n$. To be precise, we make the following definition.

STANDARD INNER PRODUCT

Let $\{u_1, \ldots, u_n\}$ be a basis for the vector space S. If

$$x = \sum_{j=1}^{n} x_j u_j \qquad \text{and} \qquad y = \sum_{j=1}^{n} y_j u_j$$

are any elements of S, their *standard inner product* relative to the given basis is

$$x \circ y = \sum_{j=1}^{n} x_j \bar{y}_j = \mathbf{X} * \mathbf{Y}$$

where $\mathbf{X} = (x_1, \ldots, x_n)$ and $\mathbf{Y} = (y_1, \ldots, y_n)$ are their coordinate vectors relative to this basis.

Of course this ingenious definition makes sense only if $\circ$ is, in fact, an inner product on S. But this is obvious, since $\circ$ copies the standard inner product on $\mathfrak{F}^n$. The delightful thing about it is that now the basis we started with is orthonormal! We state these results in two theorems, as follows.

THEOREM 10.10

Every nonzero finite-dimensional vector space over $\mathfrak{F}$, together with the standard inner product relative to a given basis, is unitary.

Proof

Refer to the above discussion. If it is not obvious to you that $x \circ y = \mathbf{X} * \mathbf{Y}$ defines an inner product, check it out directly. (See the problems.)

THEOREM 10.11

Any basis for a vector space is orthonormal if the standard inner product relative to this basis is used to make the space unitary.

Proof

Let S be the space, with basis $\{u_1, \ldots, u_n\}$, and let $\circ$ be the standard inner product relative to this basis. The problem is to show that

$$u_i \circ u_j = \delta_{ij} \qquad \text{for all } i \text{ and } j$$

(See Theorem 10.5.) Since

$$u_i = \sum_{k=1}^{n} \delta_{ik} u_k \qquad \text{and} \qquad u_j = \sum_{k=1}^{n} \delta_{kj} u_k$$

we have

$$u_i \circ u_j = \sum_{k=1}^{n} \delta_{ik} \bar{\delta}_{kj} = \delta_{ii} \delta_{ij} = \delta_{ij} \quad \blacksquare$$

EXAMPLE 10.14

Let $\mathbf{U}_1 = (2,1)$ and $\mathbf{U}_2 = (3,2)$ in $\mathfrak{R}^2$. Then $\{\mathbf{U}_1, \mathbf{U}_2\}$ is *not* an orthonormal basis when $\mathfrak{R}^2$ is considered in the usual way (as a euclidean space by virtue of the dot product). But Theorem 10.11 doesn't say it is. Since the typical element $\mathbf{X} = (x_1,x_2)$ is given in terms of $\mathbf{U}_1$ and $\mathbf{U}_2$ by

$$\mathbf{X} = (2x_1 - 3x_2)\mathbf{U}_1 + (-x_1 + 2x_2)\mathbf{U}_2$$

(confirm this!), the standard inner product relative to $\{\mathbf{U}_1,\mathbf{U}_2\}$ is defined by

$$\begin{aligned}\mathbf{X} \circ \mathbf{Y} &= (2x_1 - 3x_2)(2y_1 - 3y_2) + (-x_1 + 2x_2)(-y_1 + 2y_2)\\ &= 5x_1y_1 - 8x_1y_2 - 8x_2y_1 + 13x_2y_2\end{aligned}$$

(See Prob. 51*d*, Sec. 10.4.) Regarding $\mathcal{R}^2$ as unitary by virtue of the inner product $\circ$, we find that $\{\mathbf{U}_1,\mathbf{U}_2\}$ is orthonormal:

$$\begin{aligned}\mathbf{U}_1 \circ \mathbf{U}_2 &= 5(6) - 8(4) - 8(3) + 13(2) = 0\\ \|\mathbf{U}_1\|^2 = \mathbf{U}_1 \circ \mathbf{U}_1 &= 5(4) - 8(2) - 8(2) + 13(1) = 1\\ \|\mathbf{U}_2\|^2 = \mathbf{U}_2 \circ \mathbf{U}_2 &= 5(9) - 8(6) - 8(6) + 13(4) = 1\end{aligned}$$

Of course this is the sledge hammer approach. It is easier to write $\mathbf{U}_1 = 1\mathbf{U}_1 + 0\mathbf{U}_2$ and $\mathbf{U}_2 = 0\mathbf{U}_1 = 1\mathbf{U}_2$ and then use the coordinate vectors $\mathbf{E}_1 = (1,0)$ and $\mathbf{E}_2 = (0,1)$:

$$\begin{aligned}\mathbf{U}_1 \circ \mathbf{U}_2 &= \mathbf{E}_1 \cdot \mathbf{E}_2 = 0\\ \|\mathbf{U}_1\|^2 = \mathbf{U}_1 \circ \mathbf{U}_1 &= \mathbf{E}_1 \cdot \mathbf{E}_1 = 1\\ \|\mathbf{U}_2\|^2 = \mathbf{U}_2 \circ \mathbf{U}_2 &= \mathbf{E}_2 \cdot \mathbf{E}_2 = 1\end{aligned}$$

For a preview of all this, look back to Probs. 51 through 53 in Sec. 10.4. As we observed there, the use of a different inner product on $\mathcal{R}^2$ results in a different euclidean space, with angles and lengths preserved only in special circumstances. Because we don't want to lose the connection between ordinary geometry and *n*-space, the inner product on $\mathfrak{F}^n$ is always understood to be the standard inner product (as agreed in Sec. 10.2) unless otherwise specified. $\mathfrak{F}^n$ is already unitary; there is no point in introducing some other inner product. However, note that the standard inner product can be described in the language of this section as the *standard inner product relative to the standard basis*. Our theory says that in these circumstances the standard basis is orthonormal, which is of course correct. Thus $\mathfrak{F}^n$ fits into the theory. But the discussion is superfluous when the space in question is already unitary.

Problems

54. What is the coordinate vector of $\mathbf{X} = (-1,5,0)$ relative to the basis $\{(0,-1,1), (-2,3,3), (3,1,1)\}$ for $\mathcal{R}^3$? (See Prob. 46, Sec. 10.4.)

55. In the euclidean space of real-valued functions continuous on [0,1] an orthonormal basis for the subspace of polynomials of the form $ax + b$ is $\{u_1,u_2\}$, where $u_1(x) = 1$ and $u_2(x) = 2\sqrt{3}(x - \frac{1}{2})$. (See Prob. 47, Sec. 10.4.)

a. What are the coordinate vectors of $f(x) = 2x - 3$ and $g(x) = x$ relative to this basis?
b. Use Parseval's identity to find $f * g$. Confirm by evaluating $\int_0^1 f(x)g(x)\,dx$.
c. Use Theorem 10.9 to find $\|f\|$ and $\|g\|$.
d. Use Corollary 10.9a to find $d(f,g)$.

56. Let $x = \sum_{j=1}^{n} x_j u_j$ and $y = \sum_{j=1}^{n} y_j u_j$, where $\{u_1, \ldots, u_n\}$ is an orthonormal set in the unitary space S. Prove that $x * y = \sum_{j=1}^{n} x_j \bar{y}_j$ (Corollary 10.8a).

57. Use Corollary 10.8a to find the inner product of $f = 3u_1 + u_2$ and $g = u_1 - iu_2$, where $u_1(x) = (1/\sqrt{2\pi})e^{ix}$ and $u_2(x) = (1/\sqrt{2\pi})e^{2ix}$ are the orthonormal functions in Example 10.13. Confirm by evaluating

$$\int_{-\pi}^{\pi} f(x)\overline{g(x)}\,dx$$

58. Let $\{u_1, \ldots, u_n\}$ be a basis for the vector space S and let $\circ$ be the standard inner product relative to this basis. Prove that $\circ$ is an inner product (Theorem 10.10) in the following ways.
a. Use the definition of $\circ$ directly.
b. Make use of the coordinate vector map $f(x) = \mathbf{X}$, where $x = \sum_{j=1}^{n} x_j u_j$ and $\mathbf{X} = (x_1, \ldots, x_n)$.

59. Let S be the vector space of 2×2 matrices with entries in $\mathfrak{F}$. A basis for S is $\{E_{11}, E_{12}, E_{21}, E_{22}\}$, where

$$E_{11} = \begin{bmatrix} 1 & 0 \\ 0 & 0 \end{bmatrix} \qquad E_{12} = \begin{bmatrix} 0 & 1 \\ 0 & 0 \end{bmatrix} \qquad E_{21} = \begin{bmatrix} 0 & 0 \\ 1 & 0 \end{bmatrix} \qquad E_{22} = \begin{bmatrix} 0 & 0 \\ 0 & 1 \end{bmatrix}$$

a. Given the matrices

$$A = \begin{bmatrix} a & b \\ c & d \end{bmatrix} \quad \text{and} \quad B = \begin{bmatrix} e & f \\ g & h \end{bmatrix}$$

what are their coordinate vectors relative to this basis?
b. Explain why the standard inner product relative to this basis is defined by

$$A \circ B = a\bar{e} + b\bar{f} + c\bar{g} + d\bar{h}$$

c. Use the formula in part *b* to check that E_{11}, E_{12}, E_{21}, E_{22} are mutually orthogonal unit vectors when S is regarded as unitary with inner product $\circ$.

The point of this problem is to illustrate Theorems 10.10 and 10.11. Here we have a vector space which is not obviously unitary, but which can be made unitary by selecting a basis and imposing the standard inner product relative to that basis (which then becomes orthonormal). Of course it is easier to map S onto $\mathfrak{F}^4$ by sending

$$A = \begin{bmatrix} a & b \\ c & d \end{bmatrix}$$

into $\mathbf{X} = (a,b,c,d)$; then $A \circ B = \mathbf{X} * \mathbf{Y}$. (In fact, that is precisely what we are doing when we find coordinate vectors.)

60. Let $\{u_1, \ldots, u_n\}$ be an orthonormal basis for the subspace T of the unitary space S. Given any $x \in S$, let $x_i = x * u_i$ be the Fourier coefficient of x with respect to u_i, where $i = 1, \ldots, n$. Prove *Bessel's inequality,*

$$\sum_{i=1}^{n} |x_i|^2 \leq \|x\|^2$$

Hint: Let $y = x - \sum_{i=1}^{n} x_i u_i$ and look at the inequality $y * y \geq 0$.

61. In Prob. 60 explain why

$$\sum_{i=1}^{n} |x_i|^2 = \|x\|^2 \quad \Leftrightarrow \quad x \in T$$

62. Show that the orthonormal set $\{u_1, \ldots, u_n\}$ is a basis for the unitary space S if and only if $\|x\| = \sqrt{|x_1|^2 + \cdots + |x_n|^2}$ for every $x \in S$, where x_i is the Fourier coefficient of x with respect to u_i, $i = 1, \ldots, n$. *Hint:* The "only if" part has already been shown (where?). Use Prob. 61 for the "if" part.

63. Apply Bessel's inequality in the case where T is the xy plane in $S = \mathcal{R}^3$, with basis $\{\mathbf{E}_1, \mathbf{E}_2\}$. (The result is obvious without the heavy artillery! See the next problem for a nontrivial application.)

64. Let S be the euclidean space of real-valued functions continuous on $[-\pi,\pi]$ and let T be the subspace generated by $\cos x$ and $\sin x$, with orthonormal basis consisting of

$$u_1(x) = \frac{1}{\sqrt{\pi}} \cos x \qquad \text{and} \qquad u_2(x) = \frac{1}{\sqrt{\pi}} \sin x$$

Use Bessel's inequality to show that if f is any function in S, then

$$\frac{1}{\pi}\left[\int_{-\pi}^{\pi} f(x) \cos x \, dx\right]^2 + \frac{1}{\pi}\left[\int_{-\pi}^{\pi} f(x) \sin x \, dx\right]^2 \leq \int_{-\pi}^{\pi} f(x)^2 \, dx$$

(Note that this is not obvious as an inequality out of the blue! Try proving it directly, as a theorem in calculus, to see what we mean.)

65. In the euclidean space of real-valued functions continuous on $[0,1]$ an orthonormal basis for the subspace of polynomials of the form $ax + b$ is $\{u_1,u_2\}$, where $u_1(x) = 1$ and $u_2(x) = 2\sqrt{3}(x - \tfrac{1}{2})$. (See Prob. 47, Sec. 10.4.)

a. Use Bessel's inequality to show that if f is any function in this space, then

$$\left[\int_0^1 f(x)\,dx\right]^2 + 3\left[\int_0^1 (2x-1)f(x)\,dx\right]^2 \leq \int_0^1 f(x)^2\,dx$$

b. Why does equality hold in part a if $f(x) = x$? Confirm this by evaluating the integrals directly. Also see Prob. 55c.

c. Why is the inequality strict if $f(x) = e^x$? (It is not trivial to confirm this by brute force! Try it and see what you get into.)

66. Suppose that S contains infinitely many elements $u_1, u_2, u_3, \ldots$ with the orthonormal property $u_i * u_j = \delta_{ij}$ for all i and j. If $x \in S$ and $x_i = x * u_i$, where $i = 1, 2, 3, \ldots$, prove Bessel's inequality for infinite-dimensional spaces,

$$\sum_{i=1}^{\infty} |x_i|^2 \leq \|x\|^2$$

Hint: Recall from calculus that $\Sigma_{i=1}^{\infty} a_i$ is defined to mean $\lim_{n\to\infty} s_n$, where $s_n = \Sigma_{i=1}^{n} a_i$ is the "nth partial sum" associated with the series. If each $a_i \geq 0$ and the partial sums are bounded, then $\lim_{n\to\infty} s_n$ exists and is no larger than the bound.

67. Let f be a real-valued function continuous on $[-\pi,\pi]$ and let

$$a_0,\ a_1,\ b_1,\ a_2,\ b_2,\ \ldots$$

be its Fourier coefficients with respect to the orthogonal functions

$$1,\ \cos x,\ \sin x,\ \cos 2x,\ \sin 2x,\ \ldots$$

(See Example 10.9.)

a. Normalize the functions $1, \cos x, \sin x, \cos 2x, \sin 2x, \ldots$ to obtain the infinitely many orthonormal elements referred to in Prob. 66.

b. Find (in terms of $a_0, a_1, b_1, a_2, b_2, \ldots$) the Fourier coefficients of f with respect to the orthonormal functions in part a.

c. Apply Prob. 66 to obtain

$$2a_0^2 + (a_1^2 + b_1^2) + (a_2^2 + b_2^2) + \cdots \leq \frac{1}{\pi}\int_{-\pi}^{\pi} f(x)^2\,dx$$

(It can be shown that in this instance equality holds; the result is known as *Parseval's equality*.)

68. Confirm Parseval's equality (Prob. 67) in each of the following cases. (See Prob. 48, Sec. 10.4.)

a. $f(x) = 1$

b. $f(x) = \cos x$

69. Assuming that Parseval's equality is true, show that its application to $f(x) = x$ leads to the equation

$$\sum_{k=1}^{\infty} \frac{1}{k^2} = 1 + \frac{1}{4} + \frac{1}{9} + \frac{1}{16} + \cdots = \frac{\pi^2}{6}$$

(This is one of the most remarkable of the various ways in which π is involved in elementary infinite series.)

†10.6 Gram-Schmidt Orthogonalization

The point of the last two sections has been that the most convenient basis to use in a unitary space is an orthonormal one. In n-space there is no question that such a basis exists; the standard basis is orthonormal. But given an abstract unitary space S, how do we know that there are mutually orthogonal unit vectors spanning S? Even if we convince ourselves that such vectors exist, how do we find them?

To get started on this, observe first that the real question is whether S has an *orthogonal* basis. For if it does, say $\{v_1, \ldots, v_n\}$, we can always ''normalize'' it; simply divide each vector by its norm. (See Prob. 35, Sec. 10.3.) Thus if

$$u_i = \frac{v_i}{\|v_i\|} \qquad i = 1, \ldots, n$$

then $u_1, \ldots, u_n$ are mutually orthogonal unit vectors, which constitute an orthonormal basis for S.

Second, note that S must be nonzero and finite-dimensional for the question to make sense; otherwise no basis exists. What it comes down to, then, is this: Given a basis which may not be orthogonal, is it possible to ''orthogonalize'' it? Can we use the given basis to construct an orthogonal basis?

To get an idea of what is involved, let's look at an example in $\mathcal{R}^3$. Of course we know that $\mathcal{R}^3$ has an orthonormal basis, namely $\{\mathbf{E}_1, \mathbf{E}_2, \mathbf{E}_3\}$. But it will focus the issue if we start with a different basis and see what can be done with it.

Let $\mathbf{W}_1 = (2,0,-1)$, $\mathbf{W}_2 = (1,1,-2)$, $\mathbf{W}_3 = (-1,3,2)$; then $\{\mathbf{W}_1, \mathbf{W}_2, \mathbf{W}_3\}$ is a basis for $\mathcal{R}^3$. To convert it to an orthogonal basis $\{\mathbf{V}_1, \mathbf{V}_2, \mathbf{V}_3\}$, take $\mathbf{V}_1 = \mathbf{W}_1$ (that is, use the first element of the given basis intact). Then replace $\mathbf{W}_2$ by its component perpendicular to $\mathbf{V}_1$, say $\mathbf{V}_2$. (See Fig. 10.3, which is drawn in the plane of $\mathbf{W}_1$ and $\mathbf{W}_2$.)

Since the component of $\mathbf{W}_2$ parallel to $\mathbf{V}_1$ is a multiple of $\mathbf{V}_1$, say $c_1\mathbf{V}_1$, the perpendicular component satisfies

† This section may be omitted. However, the theorems should be noted for the record.

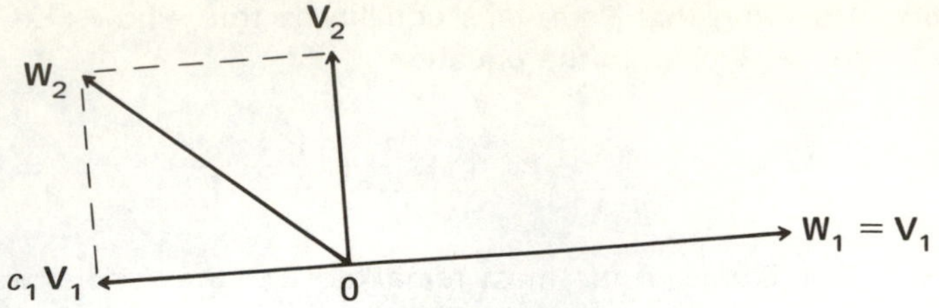

Fig. 10.3

$$c_1\mathbf{V}_1 + \mathbf{V}_2 = \mathbf{W}_2 \qquad \mathbf{V}_1 \cdot \mathbf{V}_2 = 0$$

This means that

$$\mathbf{V}_1 \cdot (\mathbf{W}_2 - c_1\mathbf{V}_1) = 0$$

from which

$$c_1 = \frac{\mathbf{W}_2 \cdot \mathbf{V}_1}{\|\mathbf{V}_1\|^2} = \frac{4}{5}$$

(the Fourier coefficient of $\mathbf{W}_2$ with respect to $\mathbf{V}_1$). Hence

$$\mathbf{V}_2 = \mathbf{W}_2 - c_1\mathbf{V}_1 = \frac{1}{5}(-3,5,-6)$$

The vectors $\mathbf{V}_1 = (2,0,-1)$ and $\mathbf{V}_2 = \frac{1}{5}(-3,5,-6)$ are orthogonal, as you can check directly.

The problem now is to replace $\mathbf{W}_3$ by a vector $\mathbf{V}_3$ which is orthogonal to *both* $\mathbf{V}_1$ and $\mathbf{V}_2$.† The clue to the procedure is the formula $\mathbf{V}_2 = \mathbf{W}_2 - c_1\mathbf{V}_1$, which says that once $\mathbf{V}_1$ is named, a second vector orthogonal to $\mathbf{V}_1$ can be found by subtracting from $\mathbf{W}_2$ its projection parallel to $\mathbf{V}_1$. Why not repeat this in a more general form? That is, find $\mathbf{V}_3$ by subtracting from $\mathbf{W}_3$ its projections parallel to $\mathbf{V}_1$ *and* $\mathbf{V}_2$:

$$\mathbf{V}_3 = \mathbf{W}_3 - c_1\mathbf{V}_1 - c_2\mathbf{V}_2$$

where

$$c_1 = \frac{\mathbf{W}_3 \cdot \mathbf{V}_1}{\|\mathbf{V}_1\|^2} = -\frac{4}{5} \qquad \text{and} \qquad c_2 = \frac{\mathbf{W}_3 \cdot \mathbf{V}_2}{\|\mathbf{V}_2\|^2} = \frac{3}{7}$$

are the Fourier coefficients of $\mathbf{W}_3$ with respect to $\mathbf{V}_1$ and $\mathbf{V}_2$.‡ This yields

$$\mathbf{V}_3 = \frac{6}{7}(1,3,2)$$

† This is no problem in $\mathcal{R}^3$ if you remember from calculus how to compute $\mathbf{V}_1 \times \mathbf{V}_2 = (1,3,2)$. But cross products don't generalize to higher dimensions.

‡ Not the same c_1 as in the preceding step! We use the same label to avoid excessive notation.

which is orthogonal to both $\mathbf{V}_1$ and $\mathbf{V}_2$.

Thus we have produced an orthogonal basis $\{\mathbf{V}_1,\mathbf{V}_2,\mathbf{V}_3\}$. An orthonormal basis is $\{\mathbf{U}_1,\mathbf{U}_2,\mathbf{U}_3\}$, where

$$\mathbf{U}_1 = \frac{\mathbf{V}_1}{\|\mathbf{V}_1\|} = \frac{1}{\sqrt{5}}(2,0,-1) \qquad \mathbf{U}_2 = \frac{\mathbf{V}_2}{\|\mathbf{V}_2\|} = \frac{1}{\sqrt{70}}(-3,5,-6)$$

$$\mathbf{U}_3 = \frac{\mathbf{V}_3}{\|\mathbf{V}_3\|} = \frac{1}{\sqrt{14}}(1,3,2)$$

The process suggested by this example is called *Gram-Schmidt orthogonalization*. It works in general, as we now prove.

THEOREM 10.12

Every nonzero finite-dimensional unitary space has an orthogonal basis.

Proof

Let S be the space, with inner product $*$, and suppose that $\dim S = n$. Let $\{w_1, \ldots, w_n\}$ be a basis for S; we propose to use it to construct an orthogonal basis $\{v_1, \ldots, v_n\}$.

The first step is to name $v_1 = w_1$, using the first element of the given basis intact. If $n = 1$, we are finished, since $\{v_1\}$ is already an orthogonal basis. (See the conversation in Sec. 10.4 concerning this point.) If $n > 1$, we observe that v_1 and w_2 are linearly independent and define

$$v_2 = w_2 - c_1 v_1$$

where

$$c_1 = \frac{w_2 * v_1}{\|v_1\|^2}$$

is the Fourier coefficient of w_2 with respect to v_1. (Note that v_1 is nonzero, so this coefficient makes sense.) Since

$$v_2 * v_1 = (w_2 - c_1 v_1) * v_1 = w_2 * v_1 - c_1(v_1 * v_1) = w_2 * v_1 - w_2 * v_1 = 0$$

we have produced an orthogonal set $\{v_1,v_2\}$. Of course if $v_2 = 0$, nothing has been accomplished. But in that case we have $w_2 = c_1 v_1$, contradicting the linear independence of v_1 and w_2. Thus $\{v_1,v_2\}$ is a set of *nonzero* orthogonal vectors, hence linearly independent. (See Theorem 10.8.)

If $n = 2$, we are finished; if $n > 2$, we observe that v_1,v_2,w_3 are linearly independent. Otherwise w_3 would be a linear combination of v_1 and v_2 (prove!), and hence a linear combination of w_1 and w_2 (why?), in contradiction of the linear independence of w_1, w_2, and w_3. Define

$$v_3 = w_3 - c_1 v_1 - c_2 v_2$$

where

$$c_1 = \frac{w_3 * v_1}{\|v_1\|^2} \quad \text{and} \quad c_2 = \frac{w_3 * v_2}{\|v_2\|^2}$$

are the Fourier coefficients of w_3 with respect to v_1 and v_2.[†] Then for $j = 1,2$ we have

$$\begin{aligned} v_3 * v_j &= (w_3 - c_1 v_1 - c_2 v_2) * v_j \\ &= w_3 * v_j - c_1(v_1 * v_j) - c_2(v_2 * v_j) \\ &= w_3 * v_j - w_3 * v_j \qquad \text{(Confirm this!)} \\ &= 0 \end{aligned}$$

Hence $\{v_1,v_2,v_3\}$ is an orthogonal set. Moreover, v_3 must be nonzero, since otherwise we would have $w_3 = c_1v_1 + c_2v_2$, a result contradicting the linear independence of v_1, v_2, and w_3. Thus $\{v_1,v_2,v_3\}$ is linearly independent and we are done if $n = 3$.

It is clear that this process may be continued until an orthogonal basis $\{v_1, \ldots, v_n\}$ has been constructed.

LECTOR *You're waving your hands.*
AUCTOR *All right, here's the general step.*

Suppose that the orthogonal set $\{v_1, \ldots, v_m\}$ of nonzero vectors has been constructed from the basis $\{w_1, \ldots, w_n\}$ by defining $v_1 = w_1$ and (for each $p = 2, \ldots, m$)

$$v_p = w_p - \sum_{i=1}^{p-1} \frac{w_p * v_i}{\|v_i\|^2} v_i$$

where $2 \le m < n$. Then $v_1, \ldots, v_m, w_{m+1}$ are linearly independent. (See the problems.) Define

$$v_{m+1} = w_{m+1} - \sum_{i=1}^{m} c_i v_i$$

where

$$c_i = \frac{w_{m+1} * v_i}{\|v_i\|^2}$$

is the Fourier coefficient of w_{m+1} with respect to v_i, $i = 1, \ldots, m$. Then for each $j = 1, \ldots, m$, we have

[†] To keep the notation simple, the Fourier coefficients at each stage of this process are labeled as in the preceding stage.

$$\begin{aligned} v_{m+1} * v_j &= \left(w_{m+1} - \sum_{i=1}^{m} c_i v_i\right) * v_j \\ &= w_{m+1} * v_j - \sum_{i=1}^{m} c_i (v_i * v_j) \\ &= w_{m+1} * v_j - c_j (v_j * v_j) \\ &= w_{m+1} * v_j - w_{m+1} * v_j \\ &= 0 \end{aligned}$$

Hence $\{v_1, \ldots, v_{m+1}\}$ is an orthogonal set. Moreover v_{m+1} must be nonzero, because otherwise

$$w_{m+1} = \sum_{i=1}^{m} c_i v_i$$

which contradicts the linear independence of $v_1, \ldots, v_m, w_{m+1}$. Thus $\{v_1, \ldots, v_{m+1}\}$ is linearly independent and we have reached the next stage of the process. Obviously if we can do this in general, we can keep going until $m + 1 = n$ and then we are finished. ▮

COROLLARY 10.12a

Every nonzero finite-dimensional unitary space has an orthonormal basis.

Proof

Normalize the orthogonal basis produced in Theorem 10.12. ▮

THEOREM 10.13

A set of nonzero mutually orthogonal elements of a finite-dimensional unitary space S is either an orthogonal basis for S or can be built up to an orthogonal basis.

Proof

Let $\{w_1, \ldots, w_m\}$ be the set and suppose that $\dim S = n$. If $m = n$, there is nothing to prove, so assume that $m < n$. Then by Theorem 2.7 we can build up $\{w_1, \ldots, w_m\}$ to a basis $\{w_1, \ldots, w_n\}$ for S. Define $v_i = w_i$, where $i = 1, \ldots, m$, and then proceed as in the general step described in the proof of Theorem 10.12. The result is an orthogonal basis $\{v_1, \ldots, v_n\}$ whose first m elements are the original vectors. ▮

COROLLARY 10.13a

A set of orthonormal elements of a finite-dimensional unitary space S either is an orthonormal basis for S or can be built up to an orthonormal basis.

Proof

Use Theorem 10.13 (if necessary) to build up the given set to an orthogonal basis for S. Then normalize each of the additional vectors (if any). The given vectors are normal already. ∎

In practice, it is unnecessary to follow these theoretical procedures exactly. For example, let $\mathbf{U}_1 = (1/\sqrt{2})(1,0,-1)$. To build up $\{\mathbf{U}_1\}$ to an orthonormal basis for $\mathcal{R}^3$, we may take $\mathbf{U}_2$ to be any unit vector in the plane $x - z = 0$ (the plane through the origin perpendicular to $\mathbf{U}_1$), say $\mathbf{U}_2 = (1/\sqrt{2})(1,0,1)$. Then (if we like) we may use $\{\mathbf{U}_1,\mathbf{U}_2,\mathbf{E}_3\}$ as the source of additional vectors in the Gram-Schmidt process to construct the vector

$$\mathbf{E}_3 - (\mathbf{E}_3 \cdot \mathbf{U}_1)\mathbf{U}_1 - (\mathbf{E}_3 \cdot \mathbf{U}_2)\mathbf{U}_2 = (0,0,0)$$

perpendicular to both . . .

AUCTOR *Oops.*
LECTOR *There seems to be some mistake.*
AUCTOR *I thought* $\{\mathbf{U}_1,\mathbf{U}_2,\mathbf{E}_3\}$ *was a basis.*
LECTOR *Try* $\{\mathbf{U}_1,\mathbf{U}_2,\mathbf{E}_2\}$.
AUCTOR *Right.*

The vector

$$\mathbf{E}_2 - (\mathbf{E}_2 \cdot \mathbf{U}_1)\mathbf{U}_1 - (\mathbf{E}_2 \cdot \mathbf{U}_2)\mathbf{U}_2 = (0,1,0)$$

is perpendicular to both $\mathbf{U}_1$ and $\mathbf{U}_2$; normalizing it to obtain $\mathbf{U}_3 = (0,1,0)$, we have produced an orthonormal basis $\{\mathbf{U}_1,\mathbf{U}_2,\mathbf{U}_3\}$.

LECTOR *Good grief. The vector* $\mathbf{E}_2$ *is already perpendicular to* $\mathbf{U}_1$ *and* $\mathbf{U}_2$ *and doesn't need normalizing.*
AUCTOR *One gets caught up in these routines.*
LECTOR *Not if one is paying attention.*
AUCTOR *Well, at least we've illustrated the fact that the Gram-Schmidt process doesn't hurt anything when it's superfluous.*

Problems

70. Use the Gram-Schmidt process to convert

$$\{(1,1,1,1), (2,0,0,0), (1,-1,0,1), (1,2,2,1)\}$$

to an orthogonal basis for $\mathcal{R}^4$. Then normalize it.

71. Complete $\{(1,-2,2)\}$ to an orthogonal basis for $\mathcal{R}^3$. Then normalize it. Explain geometrically why there are infinitely many ways of doing this.

72. Name a vector which will build up $\{(1,0,0,0), (0,1,0,0), (1/\sqrt{2})(0,0,1,1)\}$ to an orthonormal basis for $\mathcal{R}^4$. Then name another.

73. In the vector space of real-valued functions continuous on [0,1], with inner product

$$f * g = \int_0^1 f(x)g(x)\,dx$$

the functions $w_1(x) = 1$ and $w_2(x) = x$ constitute a basis for the subspace of polynomials of the form $ax + b$. Use the Gram-Schmidt process to convert this to an orthogonal basis. (See Prob. 47, Sec. 10.4.)

74. In the vector space of real-valued functions continuous on [−1,1], with inner product

$$f * g = \int_{-1}^1 f(x)g(x)\,dx$$

the functions $w_1(x) = 1$, $w_2(x) = x$, $w_3(x) = x^2$, and $w_4(x) = x^3$ constitute a basis for the subspace of polynomials of the form $ax^3 + bx^2 + cx + d$. Use the Gram-Schmidt process to convert this to an orthogonal basis. (The results, suitably adjusted to take on the value 1 at $x = 1$, are the first four *Legendre polynomials,* which are important in the solution of certain differential equations arising in mathematical physics.)

75. Let $\mathbf{U}$ be the subspace of $\mathcal{C}^3$ generated by $\mathbf{W}_1 = (i,1,0)$ and $\mathbf{W}_2 = (-1, 1 + i, 1)$. Name an orthonormal basis for $\mathbf{U}$.

76. Let T be an m-dimensional subspace of the n-dimensional unitary space S ($m > 0$). Prove that S has an orthonormal basis $\{u_1, \ldots, u_n\}$ with the property that $\{u_1, \ldots, u_m\}$ is an orthonormal basis for T.

77. In the general step of the Gram-Schmidt process we assume that the orthogonal set $\{v_1, \ldots, v_m\}$ of nonzero vectors has been constructed from the basis $\{w_1, \ldots, w_n\}$ by defining $v_1 = w_1$ and

$$v_p = w_p - \sum_{i=1}^{p-1} \frac{w_p * v_i}{\|v_i\|^2} v_i$$

for each $p = 2, \ldots, m$, where $2 \le m < n$. Prove that $v_1, \ldots, v_m, w_{m+1}$ are linearly independent.

78. Suppose that $\{w_1, \ldots, w_n\}$ is already an orthogonal basis for S. Prove that the Gram-Schmidt process just gives this basis back.

79. Suppose that the Gram-Schmidt process is used in $\mathfrak{F}^n$ to build up $\{\mathbf{E}_1, \ldots, \mathbf{E}_m\}$ to an orthonormal basis (using the standard basis as the source of additional vectors). Prove that the result is the standard basis.

80. Show that the Gram-Schmidt process is an effective test for linear dependence by proving that if it is applied to a linearly dependent set $\{w_1, \ldots, w_n\}$, one of the v_i turns out to be zero. *Hint:* Let T be the

m-dimensional space spanned by $w_1, \ldots, w_n$. Relabel (if necessary) to produce a basis $\{w_1, \ldots, w_m\}$ (Theorem 2.13), use the Gram-Schmidt process to convert this to an orthogonal basis $\{v_1, \ldots, v_m\}$. Then $v_{m+1} = 0$.

81. What about the converse of Prob. 80? If one of the v_i turns out to be zero, is $\{w_1, \ldots, w_n\}$ a linearly dependent set?

82. Apply the Gram-Schmidt process to show that $\{(1,1,0,-1), (2,1,0,-3), (0,1,0,1)\}$ is a linearly dependent subset of $\mathcal{R}^4$.

Review Quiz

True or false?

1. In $\mathcal{C}^n$ the formula $\mathbf{X} * (a\mathbf{Y}) = a(\mathbf{X} * \mathbf{Y})$ is correct.

2. A unitary space with a basis has an orthonormal basis.

3. An orthogonal subset of a unitary space is linearly independent.

4. The Cauchy-Schwarz inequality reduces to $|xy| = |x|\,|y|$ in $\mathcal{R}^1$.

5. In a euclidean space $\|kx\| = k\|x\|$.

6. The angle between $\mathbf{X} = (-3,4,0)$ and $\mathbf{Y} = (1,2,2)$ in $\mathcal{R}^3$ is $\pi/3$.

7. In a unitary space $x \neq 0 \Rightarrow x * x > 0$.

8. If $\{\mathbf{U}_1, \mathbf{U}_2\}$ is an orthonormal basis for $\mathcal{R}^2$ and $\mathbf{X} = 2\mathbf{U}_1 + 5\mathbf{U}_2$ and $\mathbf{Y} = \mathbf{U}_1 - \mathbf{U}_2$, then $\mathbf{X} \cdot \mathbf{Y} = -3$.

9. The triangle inequality in a unitary space is a consequence of the Cauchy-Schwarz inequality.

10. Schwarz and Cauchy wrote *My Fair Lady*.

11. The dot product on $\mathcal{C}^2$ satisfies the positive-definite law.

12. Let $\mathbf{U}_1 = (1/\sqrt{2})(1,i)$ and $\mathbf{U}_2 = (1/\sqrt{2})(i,1)$ in $\mathcal{C}^2$. If A is the 2×2 matrix whose i,j entry is $\mathbf{U}_i * \mathbf{U}_j$, then $A = I$.

13. In a unitary space $x * y = 0$ for all $y \Rightarrow x = 0$.

14. If $\mathbf{X} = (2 + i, 3)$, then $\|\mathbf{X}\| = 3 + \sqrt{5}$.

15. The standard inner product on $\mathcal{R}^n$ is symmetric.

16. In a unitary space $d(x,y) = 0 \Leftrightarrow x = y$.

17. If $\{u_1, \ldots, u_n\}$ is an orthonormal set and $c_1, \ldots, c_n$ are scalars, then for each $j = 1, \ldots, n$, we have $\sum_{i=1}^{n} c_i(u_i * u_j) = c_j$.

18. If u is a normal vector in the unitary space S and $x \in S$, then $x * u$ is the Fourier coefficient of x with respect to u.

19. If $f(x) = 3\cos x + \sin x$ and $g(x) = \cos x - \sin x$, then

$$\int_{-\pi}^{\pi} f(x)g(x)\,dx = 2\pi$$

20. If $\mathbf{X} = (3i,5)$ and $\mathbf{Y} = (2 + i,\ 1 - i)$, then $d(\mathbf{X},\mathbf{Y}) = 3\sqrt{5}$.

21. If $\{u_1, \ldots, u_n\}$ is a basis for the vector space S and $*$ is the standard inner product relative to this basis, then $u_i * u_j = \delta_{ij}$ for all i and j.

22. Every euclidean space is unitary.

23. Let $v_1(x) = 1$ and $v_2(x) = x$ in the euclidean space of real-valued functions continuous on $[0,1]$. Then $\{v_1,v_2\}$ is an orthogonal set.

24. If $\mathbf{X} = (x_1, \ldots, x_n)$ and $\mathbf{Y} = (y_1, \ldots, y_n)$ are vectors in n-space, then $d(\mathbf{X},\mathbf{Y}) = \sqrt{|x_1 - y_1|^2 + \cdots + |x_n - y_n|^2}$.

25. $e^{\pi i} = -1$

26. In the euclidean space of real-valued functions continuous on $[-\pi,\pi]$ the function $u(x) = (1/\sqrt{\pi})\cos x$ is a unit vector.

27. In the unitary space of complex-valued functions continuous on $[-\pi,\pi]$ the inner product of $\cos x$ and $i \sin x$ is 0.

28. If $\{\mathbf{U}_1,\mathbf{U}_2,\mathbf{U}_3\}$ is an orthonormal set in $\mathcal{R}^3$ and $\mathbf{X} = 2\mathbf{U}_1 - \mathbf{U}_2 + \mathbf{U}_3$, then $\|\mathbf{X}\| = \sqrt{6}$.

29. If $\mathbf{U}_1$ and $\mathbf{U}_2$ are perpendicular unit vectors in $\mathcal{R}^3$, there is only one vector $\mathbf{U}_3$ with the property that $\{\mathbf{U}_1,\mathbf{U}_2,\mathbf{U}_3\}$ is an orthonormal basis.

30. If S is unitary and x is a nonzero element of S, then $\{0,x\}$ is an orthogonal set.

31. If $\{u_1, \ldots, u_n\}$ is an orthonormal set in the unitary space S and $x \in S$, then $\|x\|^2 = \sum_{i=1}^{n} |x * u_i|^2$.

32. The standard inner product on $\mathcal{C}^2$ is bilinear.

33. The Minkowski inequality reduces to $|x + y| \leq |x| + |y|$ in $\mathcal{R}^1$.

34. In a unitary space $(x * y)(y * x) = |x * y|^2$.

35. It is always possible to define an inner product on a finite-dimensional vector space.

36. The set $\{1,-1,i,-i\}$ is normal in $\mathcal{C}^1$.

37. Every unitary space is a metric space.

38. $\mathbf{X} = (2,-1,5)$ and $\mathbf{Y} = (-2,1,-5)$ have opposite directions in $\mathcal{R}^3$.

39. If v_1 is any nonzero element of the finite-dimensional unitary space S, it is possible to produce an orthogonal basis for S having v_1 as its first element.

40. If $\{v_1, \ldots, v_n\}$ is an orthogonal basis for the unitary space S and $x \in S$, then $x = \sum_{j=1}^{n} x_j v_j$, where x_j is the Fourier coefficient of x with respect to v_j, $j = 1, \ldots, n$.

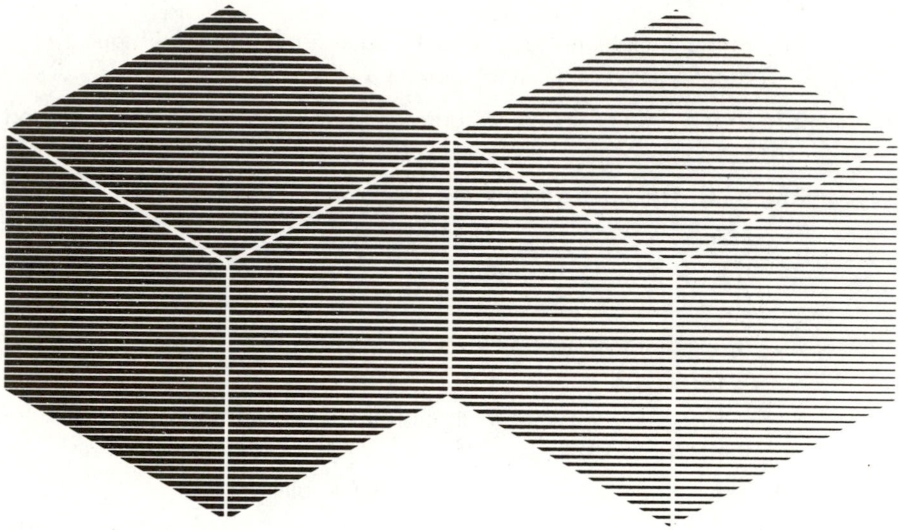

CHAPTER 11

The Projection Theorem

Our purpose in this chapter is to investigate the structure of vector spaces by "decomposing" them into sums of subspaces. This undertaking requires a preliminary discussion of what a sum of subspaces is; then we shall turn our attention to orthogonal subspaces of a unitary space. The heart of the matter is the "projection theorem," which says that every unitary space can be expressed as the sum of an arbitrary (finite-dimensional) subspace and the maximal subspace orthogonal to it. Applications of this idea to geometry and analysis are presented in an optional section at the end of the chapter.

11.1 Sums of Subspaces

SUM AND DIRECT SUM

Let S be a vector space over $\mathfrak{F}$ and suppose that T_1 and T_2 are subspaces of S. Their *sum* is

$$T_1 + T_2 = \{x \in S | x = x_1 + x_2, x_1 \in T_1, x_2 \in T_2\}$$

that is, the set of all sums of elements of T_1 and T_2. If $T_1 \cap T_2 = \{0\}$, that is, if T_1 and T_2 have only the zero element in common, their sum is called a *direct sum* and is written as $T_1 \oplus T_2$.

EXAMPLE 11.1

Let $\mathbf{U}_1 = (1,-2,1)$, $\mathbf{U}_2 = (3,1,0)$, and $\mathbf{U}_3 = (5,-3,2)$ in $\mathcal{R}^3$. Let $T_1 = \langle \mathbf{U}_1,\mathbf{U}_2 \rangle$ and $T_2 = \langle \mathbf{U}_3 \rangle$, the plane generated by $\mathbf{U}_1$ and $\mathbf{U}_2$ and the line generated by $\mathbf{U}_3$, respectively. Since the typical elements of T_1 and T_2 are

$$\mathbf{X}_1 = c_1\mathbf{U}_1 + c_2\mathbf{U}_2 \qquad \text{and} \qquad \mathbf{X}_2 = c_3\mathbf{U}_3$$

respectively, we have

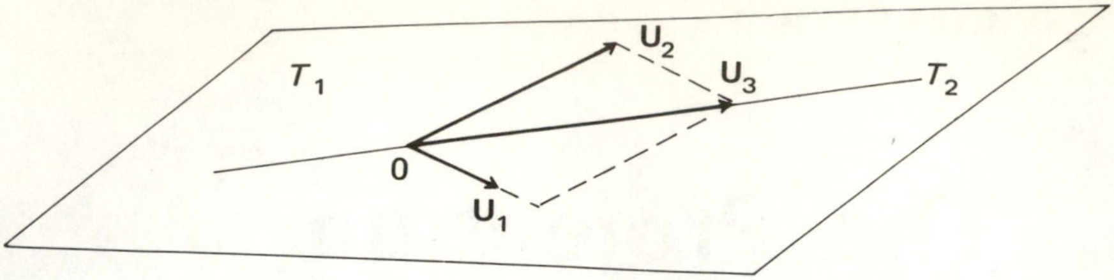

Fig. 11.1 Linear dependence of $\mathbf{U}_1$, $\mathbf{U}_2$, $\mathbf{U}_3$

$$\begin{aligned} T_1 + T_2 &= \{\mathbf{X} \mid \mathbf{X} = \mathbf{X}_1 + \mathbf{X}_2, \mathbf{X}_1 \in T_1, \mathbf{X}_2 \in T_2\} \\ &= \{\mathbf{X} \mid \mathbf{X} = c_1\mathbf{U}_1 + c_2\mathbf{U}_2 + c_3\mathbf{U}_3\} \\ &= \langle \mathbf{U}_1, \mathbf{U}_2, \mathbf{U}_3 \rangle \end{aligned}$$

This is a subspace of $\mathcal{R}^3$ (Theorem 2.15), the whole space if $\mathbf{U}_1$, $\mathbf{U}_2$, $\mathbf{U}_3$ are linearly independent. However, they are not. You can verify that $\mathbf{U}_3 = 2\mathbf{U}_1 + \mathbf{U}_2$; thus the line T_2 lies in the plane T_1. (See Fig. 11.1.) Hence

$$T_1 + T_2 = \langle \mathbf{U}_1, \mathbf{U}_2 \rangle = T_1$$

that is, the sum of the plane and the line is the plane.

Note that in this example T_1 and T_2 have a nontrivial intersection, $T_1 \cap T_2 = T_2$, so $T_1 + T_2$ is not a direct sum. Also note that

$$\dim (T_1 + T_2) < \dim T_1 + \dim T_2 \qquad \text{(Why?)}$$

EXAMPLE 11.2

Let T_1 be as in Example 11.1 and take $T_2 = \langle \mathbf{U}_3 \rangle$, where $\mathbf{U}_3 = (0,0,1)$. (See Fig. 11.2.) This time $\mathbf{U}_1$, $\mathbf{U}_2$, and $\mathbf{U}_3$ are linearly independent and

$$T_1 + T_2 = \langle \mathbf{U}_1, \mathbf{U}_2, \mathbf{U}_3 \rangle = \mathcal{R}^3$$

that is, the sum of the plane and the line is all of 3-space. Since the line T_2 does not lie in the plane T_1, the two subspaces have only the origin in common and their sum is a direct sum. Note that in this example

$$\dim (T_1 \oplus T_2) = \dim T_1 + \dim T_2$$

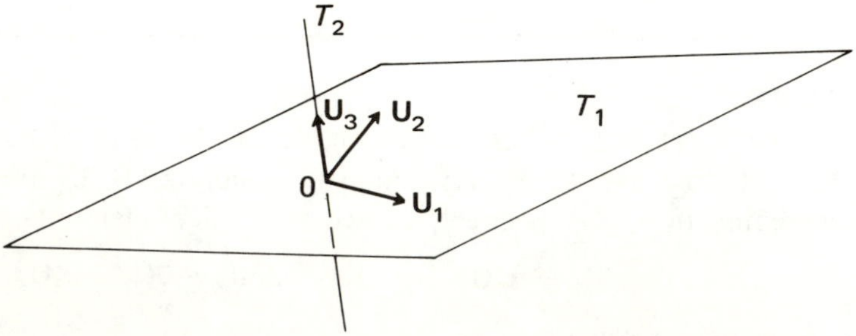

Fig. 11.2 Linear independence of $\mathbf{U}_1$, $\mathbf{U}_2$, $\mathbf{U}_3$

THEOREM 11.1

The sum of two subspaces of a vector space S is a subspace of S.

Proof

Let T_1 and T_2 be the subspaces and let $T = T_1 + T_2$. Since $0 \in T_1$ and $0 \in T_2$ (why?), the element $0 = 0 + 0$ is by definition an element of T; hence T is nonempty. By Theorem 2.14 we need only show that $ax + by \in T$ for all scalars a and b and all elements x and y of T.

Given x and y in T, write $x = x_1 + x_2$ and $y = y_1 + y_2$, where x_1 and y_1 are in T_1 and x_2 and y_2 are in T_2. Then if a and b are scalars, we have

$$ax + by = a(x_1 + x_2) + b(y_1 + y_2) = (ax_1 + by_1) + (ax_2 + by_2)$$

Since T_1 and T_2 are subspaces (hence closed relative to addition and scalar multiplication), we have $ax_1 + by_1 \in T_1$ and $ax_2 + by_2 \in T_2$. Thus $ax + by \in T$. ∎

THEOREM 11.2 The Boolean Formula[†]

If T_1 and T_2 are finite-dimensional subspaces of a vector space, then

$$\dim (T_1 + T_2) = \dim T_1 + \dim T_2 - \dim (T_1 \cap T_2)$$

Proof

Assume that $T_1 \cap T_2$ is neither the zero space nor equal to T_1 or T_2. (See the problems for these special cases.) Then it has a basis, say $\{u_1, \ldots, u_p\}$.[‡] Build this up to a basis $\{u_1, \ldots, u_p, v_1, \ldots, v_q\}$ for T_1 and (separately) to a basis $\{u_1, \ldots, u_p, w_1, \ldots, w_r\}$ for T_2. (Note that each build-up is needed because $T_1 \cap T_2$ is not equal to T_1 or T_2.)

We now have $\dim T_1 = p + q$, $\dim T_2 = p + r$, and $\dim (T_1 \cap T_2) = p$; thus

$$\dim T_1 + \dim T_2 - \dim (T_1 \cap T_2) = p + q + r$$

If the theorem is going to work, we have to find a basis for $T_1 + T_2$ containing $p + q + r$ elements. This requirement would suggest that

$$\{u_1, \ldots, u_p, v_1, \ldots, v_q, w_1, \ldots, w_r\}$$

is the basis we seek; the problem is to show that its elements span $T_1 + T_2$ and are linearly independent.

Let x be an element of $T_1 + T_2$, say $x = x_1 + x_2$, where $x_1 \in T_1$ and $x_2 \in T_2$. Since x_1 is a linear combination of $u_1, \ldots, u_p, v_1, \ldots, v_q$, while x_2 is a linear combination of $u_1, \ldots, u_p, w_1, \ldots, w_r$, it follows that x

†This is named after George Boole (1815–1864), one of the creators of symbolic logic. "Pure mathematics," said Bertrand Russell, "was discovered by Boole, in a work which he called *The Laws of Thought*."

‡Note that $T_1 \cap T_2$ is a vector space; see Prob. 49, Sec. 2.6.

is a linear combination of $u_1, \ldots, u_p, v_1, \ldots, v_q, w_1, \ldots, w_r$. Hence these elements span $T_1 + T_2$.

To prove that they are linearly independent, suppose that

$$\sum_{i=1}^{p} a_i u_i + \sum_{i=1}^{q} b_i v_i + \sum_{i=1}^{r} c_i w_i = 0$$

The problem is to show that the a_i, b_i, and c_i are all zero. Designate these three sums by u, v, and w, respectively; then

$$u \in T_1 \cap T_2 \qquad v \in T_1 \qquad w \in T_2 \qquad \text{and} \qquad u + v + w = 0$$

Now $v = -u - w \in T_2$ (why?), so $v \in T_1 \cap T_2$. This means that v can be written as a linear combination of $u_1, \ldots, u_p$, say $v = \sum_{i=1}^{p} d_i u_i$. Since we already have $v = \sum_{i=1}^{q} b_i v_i$, it follows that

$$d_1 u_1 + \cdots + d_p u_p - b_1 v_1 - \cdots - b_q v_q = 0$$

But $u_1, \ldots, u_p, v_1, \ldots, v_q$ are linearly independent, so all the coefficients in this equation must be zero; in particular, each b_i is zero. Thus $v = 0$ and our original equation reduces to

$$a_1 u_1 + \cdots + a_p u_p + c_1 w_1 + \cdots + c_r w_r = 0$$

Since $u_1, \ldots, u_p, w_1, \ldots, w_r$ are linearly independent, each a_i and c_i is zero. ∎

COROLLARY 11.2a

Let T_1 and T_2 be finite-dimensional subspaces of a vector space. Then $T_1 + T_2$ is a direct sum if and only if $\dim (T_1 + T_2) = \dim T_1 + \dim T_2$.

Note that Theorem 11.2 and Corollary 11.2a do not require the parent space to be finite-dimensional, but only the subspaces T_1 and T_2.

EXAMPLE 11.3

Let S be the vector space of real-valued functions continuous on $(-\infty,\infty)$. Then the solution sets of the differential equations $y'' + y = 0$ and $y'' - y = 0$ are subspaces of S, namely

$$T_1 = \langle f_1, f_2 \rangle \qquad \text{where } f_1(x) = \cos x \quad \text{and} \quad f_2(x) = \sin x$$
$$T_2 = \langle g_1, g_2 \rangle \qquad \text{where } g_1(x) = \cosh x \quad \text{and} \quad g_2(x) = \sinh x$$

(See Example 2.8 and Prob. 23, Sec. 2.3.) Since $\{f_1, f_2\}$ and $\{g_1, g_2\}$ are linearly independent, both T_1 and T_2 are 2-dimensional. Moreover,

$$T_1 + T_2 = \langle f_1, f_2, g_1, g_2 \rangle$$

is 4-dimensional because $\{f_1, f_2, g_1, g_2\}$ is linearly independent. (Confirm this!) Hence

$$\dim (T_1 + T_2) = \dim T_1 + \dim T_2$$

and $T_1 + T_2$ must be a direct sum. It turns out to be the solution set of the differential equation $y'''' - y = 0$; that is, every solution of this equation is of the form

$$\phi(x) = c_1 \cos x + c_2 \sin x + c_3 \cosh x + c_4 \sinh x$$

In differential equations one learns that the "characteristic polynomial" associated with this equation is $x^4 - 1$, which is the product of the characteristic polynomials $x^2 + 1$ and $x^2 - 1$ associated with $y'' + y = 0$ and $y'' - y = 0$. This suggests a connection between *products* of characteristic polynomials and *direct sums* of solution spaces, an idea that is exploited in the theory of differential equations.

THEOREM 11.3

Let T_1 and T_2 be subspaces of a vector space. Then $T_1 + T_2$ is a direct sum if and only if the expression of each of its elements as a sum of elements of T_1 and T_2 is unique.

Proof

Suppose that $T_1 + T_2$ is a direct sum (meaning that $T_1 \cap T_2$ is the zero space). Let x be any element of $T_1 + T_2$ and suppose that $x = x_1 + x_2 = y_1 + y_2$, where x_1 and y_1 are in T_1 and x_2 and y_2 are in T_2. To prove uniqueness of expression, we must show that $x_1 = y_1$ and $x_2 = y_2$. But $x_1 - y_1 = y_2 - x_2$. Since $x_1 - y_1 \in T_1$ and $y_2 - x_2 \in T_2$ (why?), it follows that $x_1 - y_1$ and $y_2 - x_2$ are in $T_1 \cap T_2$. The only element in $T_1 \cap T_2$ is 0, so $x_1 - y_1 = 0$ and $y_2 - x_2 = 0$ and the conclusion follows.

Conversely, suppose that each element of $T_1 + T_2$ has a unique expression as a sum of elements of T_1 and T_2. To prove that $T_1 + T_2$ is a direct sum, we show that the intersection of T_1 and T_2 is the zero space. Suppose that $x \in T_1 \cap T_2$. Since $x \in T_1$ and $0 \in T_2$, we can write $x = x + 0$ as a sum of elements of T_1 and T_2. Another way is $x = 0 + x$, since $0 \in T_1$ and $x \in T_2$. But the expression of x as such a sum is unique, so $x = 0$. Hence $T_1 \cap T_2$ is the zero space. ∎

EXAMPLE 11.4

In Example 11.3 we concluded that $T_1 + T_2$ is a direct sum. It follows from Theorem 11.3 that each of its elements (solutions of $y'''' - y = 0$) can be written in only one way as a sum of elements of T_1 and T_2 (solutions of $y'' + y = 0$ and $y'' - y = 0$, respectively). For example, the solution ϕ of $y'''' - y = 0$ satisfying the initial conditions $\phi(0) = 1$, $\phi'(0) = 0$, $\phi''(0) = -1$, and $\phi'''(0) = 0$ is

$$\phi(x) = \phi_1(x) + \phi_2(x)$$

where

$$\phi_1(x) = \cos x \qquad \text{and} \qquad \phi_2(x) = 0$$

are solutions of $y'' + y = 0$ and $y'' - y = 0$, respectively. (Confirm this!) Having discovered this, we can be certain that there is no other way of writing $\phi = \phi_1 + \phi_2$, where $\phi_1 \in T_1$ and $\phi_2 \in T_2$.

THEOREM 11.4

Let U be a subspace of the finite-dimensional space S. Then S can be written as the direct sum of U and another subspace.

Proof

If $U = \{0\}$ or $U = S$, the theorem is trivial. Assuming that U is a "proper" subspace (neither the zero space nor all of S), let $\{u_1, \ldots, u_m\}$ be a basis for U and build it up to a basis $\{u_1, \ldots, u_n\}$ for S. Then $S = U \oplus V$, where

$$V = \langle u_{m+1}, \ldots, u_n \rangle$$

We leave it to you to explain why. ▮

EXAMPLE 11.5

Let U be the plane $2x - y + z = 0$ in $S = \mathcal{R}^3$. The expression of $\mathcal{R}^3$ as a direct sum of U and another subspace V is not unique; one may choose V to be any line through the origin not lying in the plane. (Why?) Speaking algebraically, a basis for U is $\{(1,2,0), (-1,0,2)\}$; we may build this up to a basis for $\mathcal{R}^3$ in any way we like. One obvious choice is the vector $(2,-1,1)$ perpendicular to the plane; then $V = \langle(2,-1,1)\rangle$ is the "normal line" through the origin. (See Fig. 11.3.) But we could just as well (for example) take $V = \langle(0,0,1)\rangle$.

It is worthwhile in this example to confirm that $\mathcal{R}^3 = U \oplus V$, where

$$U = \langle(1,2,0), (-1,0,2)\rangle \qquad \text{and} \qquad V = \langle(2,-1,1)\rangle$$

Let $\mathbf{X}$ be any element of $\mathcal{R}^3$. Since $\{(1,2,0), (-1,0,2), (2,-1,1)\}$ is a basis for $\mathcal{R}^3$, we may certainly express $\mathbf{X}$ in terms of its elements, say

$$\mathbf{X} = c_1(1,2,0) + c_2(-1,0,2) + c_3(2,-1,1)$$

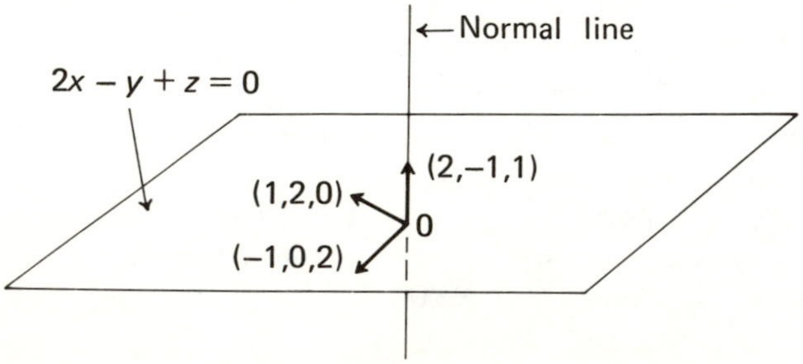

Fig. 11.3

To write this as a sum of elements of U and V, we need only combine the first two terms:

$$\mathbf{X} = \mathbf{X}_1 + \mathbf{X}_2$$

where

$$\mathbf{X}_1 = c_1(1,2,0) + c_2(-1,0,2) \in U \qquad \text{and} \qquad \mathbf{X}_2 = c_3(2,-1,1) \in V$$

This shows that $\mathcal{R}^3 \subset U + V$; the inclusion $U + V \subset \mathcal{R}^3$ is trivial. Hence $\mathcal{R}^3 = U + V$. The sum is direct because U and V have only the origin in common.

Theorem 11.4 is our first "decomposition" theorem. It says that any (finite-dimensional) space may be broken into the sum of an arbitrary subspace and another subspace (with only the zero element in common), in the sense that every element of the space is uniquely expressible as the sum of elements of the subspaces. The second subspace, however, is not uniquely determined by the first, so the theorem is not as precise as we might like. We'll remedy this in the case of unitary spaces by imposing an additional requirement, namely that the second subspace should be "perpendicular" to the first (like the normal line in Example 11.5). The main result in this direction is the projection theorem, which is the objective of the next two sections.

Problems

1. Let $T = T_1 + T_2$, where T_1 and T_2 are subspaces of a vector space S. Explain why T_1 and T_2 are also subspaces of T.
2. Explain why $\mathcal{R}^2 = \langle \mathbf{E}_1 \rangle \oplus \langle \mathbf{E}_2 \rangle$. What is the geometric interpretation of this?
3. Let $\mathbf{U}_1 = (2,0,-1,0)$, $\mathbf{U}_2 = (1,1,0,0)$, $\mathbf{U}_3 = (-1,0,0,0)$ and $\mathbf{U}_4 = (2,-1,-2,0)$ in $\mathcal{R}^4$. If $T_1 = \langle \mathbf{U}_1,\mathbf{U}_2 \rangle$ and $T_2 = \langle \mathbf{U}_3,\mathbf{U}_4 \rangle$, show that $T_1 + T_2$ is not a direct sum. What is its dimension?
4. In Example 11.1 we found that $\dim (T_1 + T_2) < \dim T_1 + \dim T_2$. Confirm the Boolean formula in this case.
5. In the vector space of real-valued functions differentiable on $(-\infty,\infty)$ let $T_1 = \langle \phi_1 \rangle$ and $T_2 = \langle \phi_2 \rangle$, where $\phi_1(x) = e^{2x}$ and $\phi_2(x) = \phi_1'(x)$. Find $T_1 + T_2$ and confirm the Boolean formula.
6. Prove the Boolean formula (Theorem 11.2) in the following special cases.
   a. $T_1 \cap T_2$ is equal to T_1 or T_2. *Hint:* If $T_1 \cap T_2 = T_1$ (for example), then $T_1 \subset T_2$ and $T_1 + T_2 = T_2$. (Why?)
   b. $T_1 \cap T_2$ is the zero space. *Hint:* Suppress the elements $u_1, \ldots, u_p$ in the proof of Theorem 11.2. Note that we may suppose that T_1 and T_2 are nonzero, since otherwise we are back to part *a*.

7. Prove Corollary 11.2a: If T_1 and T_2 are finite-dimensional subspaces of a vector space, then $T_1 + T_2$ is a direct sum if and only if $\dim (T_1 + T_2) = \dim T_1 + \dim T_2$.

8. Let T_1 and T_2 be the lines in $\mathcal{R}^3$ with parametric equations

$$x = t \quad y = 3t \quad z = t \qquad \text{and} \qquad x = -2t \quad y = t \quad z = 5t$$

respectively. Find $T_1 + T_2$ and confirm that it is a direct sum.

9. Let U be the set of all even, and V the set of all odd, functions in the space S of real-valued functions continuous on $[-a,a]$.†
   a. Prove that U and V are subspaces of S. Are they finite-dimensional?
   b. Prove that $U + V$ is a direct sum.
   c. Prove that $S = U \oplus V$. *Hint:* If $f \in S$, then

$$f(x) = \frac{f(x) + f(-x)}{2} + \frac{f(x) - f(-x)}{2}$$

   for all $x \in [-a,a]$. (Thus every function in S can be expressed in one and only one way as the sum of an even function and an odd function.)
   d. Write $f(x) = e^x$ as the sum of an even function and an odd function. Do you recognize the results?

10. Finish the proof of Theorem 11.4.

11. Let U be the line $y = x$ in $\mathcal{R}^2$. Write $\mathcal{R}^2$ in two ways as a direct sum of U and another subspace.

11.2 Orthogonal Subspaces of a Unitary Space

In Example 11.5 of the last section we expressed $\mathcal{R}^3$ as the direct sum of the plane $U = \{(x,y,z) \mid 2x - y + z = 0\}$ and the normal line $V = \langle(2,-1,1)\rangle$. The choice of the normal line suggests that in a unitary space the notion of direct sum is simpler than it is in general. Given a subspace U, perhaps we can come up with a *unique* subspace V, perpendicular to U, with the property that $U + V$ is the whole space. But first we have to define what we mean by "perpendicular" subspaces.

ORTHOGONAL SUBSPACES

Let U and V be subspaces of the unitary space S (with inner product $*$). If $x * y = 0$ for all $x \in U$ and $y \in V$, then U and V are called *orthogonal subspaces* of S. Each element of one space is also said to be *orthogonal* to the other space.

---

† A function f is "even" if $f(-x) = f(x)$, "odd" if $f(-x) = -f(x)$, for all x in its domain. The cosine, for example, is even, while the sine is odd.

EXAMPLE 11.6

Let U and V be the lines in $\mathcal{R}^3$ with parametric equations

$$x = 3t \quad y = t \quad z = -t \qquad \text{and} \qquad x = 2t \quad y = -3t \quad z = 3t$$

respectively. Then

$$U = \{(3t,t,-t) \mid t \in \mathcal{R}\} = \{t(3,1,-1) \mid t \in \mathcal{R}\} = \langle(3,1,-1)\rangle$$

and (similarly) $V = \langle(2,-3,3)\rangle$. Thus U and V are subspaces spanned by $(3,1,-1)$ and $(2,-3,3)$, respectively; their typical elements are

$$\mathbf{X} = a(3,1,-1) \qquad \text{and} \qquad \mathbf{Y} = b(2,-3,3)$$

Since

$$\mathbf{X} \cdot \mathbf{Y} = ab(6 - 3 - 3) = 0$$

for all such $\mathbf{X}$ and $\mathbf{Y}$, we conclude that U and V are orthogonal subspaces of $\mathcal{R}^3$. Each element of U (a vector starting at the origin and lying in the first line) is perpendicular to every element of V, and vice versa. In plain English, U and V are perpendicular lines (through the origin). The definition not only allows us to say this, but also that individual vectors in one line are perpendicular to the other line. This corresponds to ordinary geometrical usage.

Note that in this example $U \oplus V$ is not all of $\mathcal{R}^3$. (Why?) If we regard U as given, V is only part of the set of vectors perpendicular to U, as you can see from Fig. 11.4.

EXAMPLE 11.7

Let $U = \langle(3,1,-1)\rangle$, as in Example 11.6. The set

$$V = \{(x,y,z) \mid 3x + y - z = 0\}$$

is a subspace of $\mathcal{R}^3$; if $\mathbf{X} = a(3,1,-1)$ is any element of U, then for every $\mathbf{Y} = (x,y,z)$ which is an element of V we have $\mathbf{X} \cdot \mathbf{Y} = a(3x + y - z) = 0$ (because $\mathbf{Y}$ satisfies the equation defining V). Hence U and V are orthogonal subspaces of $\mathcal{R}^3$. This time V is the entire set of elements perpendicular to U (namely the plane shown in Fig. 11.4) and, as you can check, $U \oplus V = \mathcal{R}^3$.

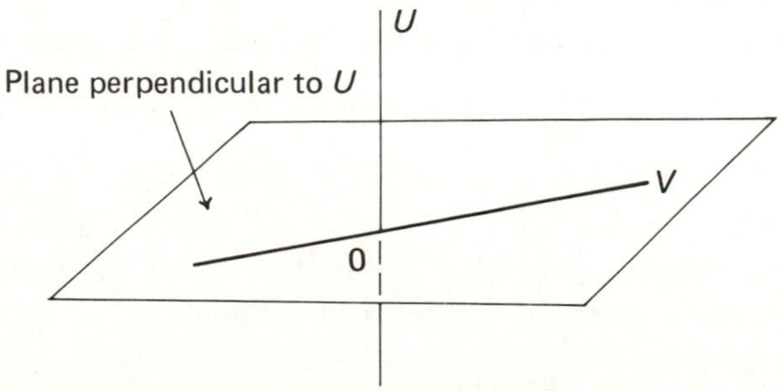

Fig. 11.4 Orthogonal subspaces

These examples suggest that if we are going to decompose a unitary space into the sum of a subspace U and another subspace orthogonal to U, we must include in the second subspace *all* the vectors orthogonal to U. Such a "maximal" orthogonal subspace is given a special name, as in our next definition.

ORTHOGONAL COMPLEMENT

Let U be a (nonempty) subset of the unitary space S. The *orthogonal complement* of U, denoted by $U^\perp$, is the set of elements of S which are orthogonal to U, that is,

$$U^\perp = \{y \in S \mid x * y = 0 \text{ for every } x \in U\}^\dagger$$

EXAMPLE 11.8

Suppose that U consists of $(1,-2,5)$ alone (in $S = \mathcal{R}^3$). Then the orthogonal complement of U is

$$U^\perp = \{(x,y,z) \mid (1,-2,5) \cdot (x,y,z) = 0\} = \{(x,y,z) \mid x - 2y + 5z = 0\}$$

the plane through the origin perpendicular to the vector $(1,-2,5)$.

EXAMPLE 11.9

If U is the subspace of $\mathcal{R}^3$ generated by $(1,-2,5)$, the orthogonal complement is still the plane $x - 2y + 5z = 0$. (Why?) This time, since the elements of U are of the form $k(1,-2,5)$, the definition reads

$$\begin{aligned} U^\perp &= \{(x,y,z) \mid k(1,-2,5) \cdot (x,y,z) = 0 \text{ for every } k \in \mathcal{R}\} \\ &= \{(x,y,z) \mid k(x - 2y + 5z) = 0 \text{ for every } k \in \mathcal{R}\} \\ &= \{(x,y,z) \mid x - 2y + 5z = 0\} \qquad \text{(Why?)} \end{aligned}$$

The complication due to k in this example can be avoided by making use of the following result.

THEOREM 11.5

Let $\{u_1, \ldots, u_m\}$ be a basis for the subspace U of the unitary space S. Then

$$U^\perp = \{y \in S \mid u_i * y = 0,\ i = 1, \ldots, m\}$$

Proof

Let $V = \{y \in S \mid u_i * y = 0, i = 1, \ldots, m\}$. To prove that $V = U^\perp$, we must establish the inclusions $V \subset U^\perp$ and $U^\perp \subset V$. The first of these is left for the problems; the second is proved by writing

† See Prob. 53, Sec. 2.6, and Example 3.11 for a preview of this idea in $\mathcal{R}^2$ and $\mathcal{R}^3$.

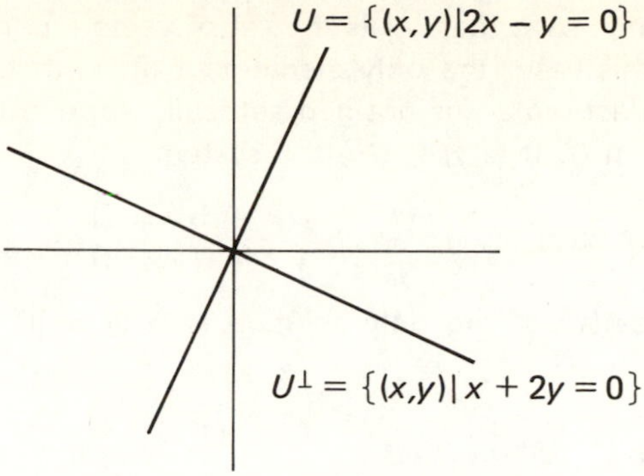

Fig. 11.5 Orthogonal complements

$$\begin{aligned} y \in U^{\perp} &\Rightarrow x * y = 0 \quad \text{for every } x \in U \\ &\Rightarrow u_i * y = 0 \quad i = 1, \ldots, m \\ &\Rightarrow y \in V \ \blacksquare \end{aligned}$$

With this theorem in hand we can do Example 11.9 more easily, for U has basis $\{(1,-2,5)\}$, and we need only write

$$\begin{aligned} U^{\perp} &= \{(x,y,z) | (1,-2,5) \cdot (x,y,z) = 0\} \\ &= \{(x,y,z) | x - 2y + 5z = 0\} \end{aligned}$$

as in Example 11.8.

EXAMPLE 11.10

If U is the line $2x - y = 0$ in $\mathcal{R}^2$, the orthogonal complement is the perpendicular line through the origin, $x + 2y = 0$. (See Fig. 11.5.)

But if U is the line $2x - y + 1 = 0$ (note that this is not a subspace), the orthogonal complement is *not* $x + 2y = 0$ (Fig. 11.6), despite the fact that in elementary geometry these lines are referred to as perpendicular. For

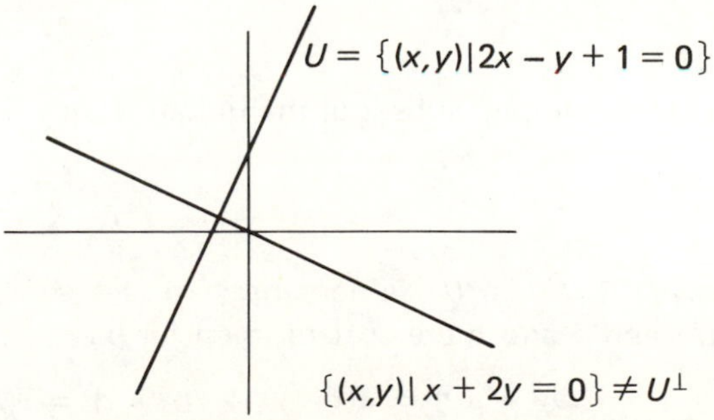

Fig. 11.6 Perpendicular lines but not orthogonal complements

the definition says that $U^\perp$ is the set of vectors perpendicular to *every* vector in U. In this case, the only element of $\mathcal{R}^2$ with that property is $\mathbf{0}$. (Why?)

If this last remark is not geometrically apparent, look at the vectors (1,3) and (0,1) in U. If $(x,y) \in U^\perp$, the system

$$\begin{aligned} 1x + 3y &= 0 \\ 0x + 1y &= 0 \end{aligned}$$

must be satisfied; the only solution is $(x,y) = (0,0)$. Hence $U^\perp$ is the zero subspace of $\mathcal{R}^2$.

EXAMPLE 11.11

Let S be the euclidean space of polynomials of the form $ax^2 + bx + c$, with inner product

$$f * g = \int_{-1}^{1} f(x)g(x)\,dx$$

and let U be the subspace generated by $u_1(x) = 1$ and $u_2(x) = x$. Then we have

$$U^\perp = \left\{ f \in S \,\middle|\, \int_{-1}^{1} f(x)u_1(x)\,dx = \int_{-1}^{1} f(x)u_2(x)\,dx = 0 \right\}$$

The typical element of $U^\perp$ is $f(x) = ax^2 + bx + c$ satisfying

$$\int_{-1}^{1} (ax^2 + bx + c)\,dx = \int_{-1}^{1} (ax^3 + bx^2 + cx)\,dx = 0$$

This implies that $a + 3c = 0$ and $b = 0$ (confirm this!); the typical element of $U^\perp$ is therefore $k(3x^2 - 1)$, where k is arbitrary. Hence $U^\perp = \langle v \rangle$, where $v(x) = 3x^2 - 1$.

These examples indicate that the orthogonal complement of a subset U is always a subspace of S, even when U is not. (You have already proved this in $\mathcal{R}^2$ and $\mathcal{R}^3$; see Prob. 53*f*, Sec. 2.6.)

THEOREM 11.6

Let U be a (nonempty) subset of the unitary space S. Then $U^\perp$ is a subspace of S.

Proof

Obviously $0 \in U^\perp$, so $U^\perp$ is nonempty. It is a subspace because if y and z are in $U^\perp$ and a and b are scalars, then for every $x \in U$

$$x * (ay + bz) = \bar{a}(x * y) + \bar{b}(x * z) = \bar{a}0 + \bar{b}0 = 0$$

Hence $ay + bz \in U^\perp$. ∎

THEOREM 11.7

Let U and V be orthogonal subspaces of the unitary space S. If their sum is S, then each is the orthogonal complement of the other.

Proof

This is left for the problems.

Theorem 11.7 is the uniqueness statement we advertised at the beginning of this section. It does not guarantee that S can be written as a sum of U and another subspace orthogonal to U; that is yet to come. But it does say that such a decomposition (if one exists) is unique. The second subspace is necessarily the orthogonal complement of U.

Thus we have narrowed down the question. Given a subspace U of the unitary space S, can we write $S = U \oplus U^{\perp}$? The "projection theorem" (proved in the next section) says we can if U is finite-dimensional. There are also cases where $S = U \oplus U^{\perp}$ even if U is not finite-dimensional, as the following example shows.

EXAMPLE 11.12

Let U be the set of all even, and V the set of all odd, functions in the euclidean space S of real-valued functions continuous on $[-a,a]$. Then U and V are subspaces of S and $S = U \oplus V$. (See Prob. 9, Sec. 11.1.) Moreover, U and V are orthogonal, as we ask you to show in the problems. It follows from Theorem 11.7 that $V = U^{\perp}$, so we have $S = U \oplus U^{\perp}$. But neither U nor $U^{\perp}$ is finite-dimensional. (Why?)

Problems

12. Explain why the sum of two orthogonal subspaces of a unitary space is always a direct sum. Why does it follow that if U is a subspace, then $U + U^{\perp}$ is a direct sum?

13. In each of the following cases, find $U^{\perp}$ and name a basis for it (if possible). Also describe it geometrically. (See Prob. 53, Sec. 2.6, for additional examples.)

a. $U = \{(x,y) \in \mathcal{R}^2 | x + y = 1\}$

b. U is the z axis in $\mathcal{R}^3$

c. $U = \langle(1,-1)\rangle$ in $\mathcal{R}^2$

d. $U = \{(x,y,z) \in \mathcal{R}^3 | x + y + z = 0\}$

14. Finish the proof of Theorem 11.5 by showing that if $\{u_1, \ldots, u_m\}$ is a basis for U (in the unitary space S) and $V = \{y \in S | u_i * y = 0, i = 1, \ldots, m\}$, then $V \subset U^{\perp}$.

15. Explain why two orthogonal subspaces of a unitary space whose sum is the whole space are orthogonal complements (Theorem 11.7).

16. Let S be the euclidean space of real-valued functions continuous on $[-a,a]$, with inner product

$$f * g = \int_{-a}^{a} f(x)g(x)\,dx$$

In Prob. 9, Sec. 11.1, you proved that S is the direct sum of the subspace of even functions and the subspace of odd functions. Show that these subspaces are orthogonal. *Hint:* Use the theorem from calculus that if h is an odd function continuous on $[-a,a]$, then

$$\int_{-a}^{a} h(x)\,dx = 0$$

17. Let S be the euclidean space $\langle \phi_1, \phi_2 \rangle$, where $\phi_1(x) = \cos x$ and $\phi_2(x) = \sin x$, the inner product being defined by

$$f * g = \int_{-\pi}^{\pi} f(x)g(x)\,dx$$

If $U = \langle \phi_1 \rangle$, what is $U^{\perp}$?

18. Let S be the euclidean space of polynomials of the form $ax + b$, with inner product

$$f * g = \int_{-1}^{1} f(x)g(x)\,dx$$

and let U be the subspace generated by $u(x) = 1$.

a. What is the orthogonal complement of U? Find a basis for it.

b. Given the function $f(x) = ax + b$ in S, show that f can be expressed in one and only one way as the sum of a function in U and a function in $U^{\perp}$. Why does it follow that $S = U \oplus U^{\perp}$?

19. Let U and V be the lines in $S = \mathbb{R}^3$ spanned by $(1,3,-1)$ and $(2,-1,-1)$, respectively.

a. Show that U and V are orthogonal subspaces of S, but that $V \neq U^{\perp}$. Why doesn't this contradict Theorem 11.7?

b. Confirm that the sum of U and V is the plane

$$T = \{(x,y,z) \mid 4x + y + 7z = 0\}$$

c. Since U and V are also orthogonal subspaces of T, Theorem 11.7 says that V is the orthogonal complement of U in T. Explain the relation between this orthogonal complement and $U^{\perp}$ (where $U^{\perp}$, as in part *a*, means the orthogonal complement of U in S).

20. Prove that if U and V are subspaces of a unitary space, then

$$(U + V)^\perp = U^\perp \cap V^\perp$$

[This is similar to *De Morgan's law* in the algebra of sets: If A and B are subsets of a given universal set, then $(A \cup B)' = A' \cap B'$, where A' means the complement of A.]

21. Let U be a subspace of the finite-dimensional unitary space S.
a. If $U = \{0\}$ or $U = S$, why does $S = U \oplus U^\perp$?
b. Assuming that U is a "proper" subspace (neither the zero space nor all of S), let $\{u_1, \ldots, u_m\}$ be an orthonormal basis for U and build it up to an orthonormal basis $\{u_1, \ldots, u_n\}$ for S. (How do you know this can be done?) Prove that $U^\perp = \langle u_{m+1}, \ldots, u_n \rangle$.
c. Why does it follow that $S = U \oplus U^\perp$?
d. Explain why $\dim S = \dim U + \dim U^\perp$.

22. Confirm the formula $\dim S = \dim U + \dim U^\perp$ in Examples 11.9, 11.10, and 11.11. What about Example 11.12?

23. Refer back to Prob. 33, Sec. 7.5, in which we outlined a proof of the fact that the row rank and column rank of a matrix are equal.
a. Why is the proof now complete if $\mathfrak{F} = \mathfrak{R}$?
b. Why is the proof still unfinished if $\mathfrak{F} = \mathfrak{C}$?

24. Let $S = \mathfrak{C}^2$ and $U = \langle (1,i) \rangle$. What is the difference between

$$U^\perp = \{\mathbf{Y} \in \mathfrak{C}^2 \mid \mathbf{X} * \mathbf{Y} = 0 \text{ for all } \mathbf{X} \in U\}$$

(defined in this section) and

$$U^\perp = \{\mathbf{Y} \in \mathfrak{C}^2 \mid \mathbf{X} \cdot \mathbf{Y} = 0 \text{ for all } \mathbf{X} \in U\}$$

(defined in Prob. 33, Sec. 7.5)? Find $\dim U^\perp$ in each case and note that the formula $\dim U + \dim U^\perp = \dim S$ holds both times. (If we knew this to be true in general, we could finish the proof that the row rank and column rank of a matrix are equal.)

11.3 The Projection Theorem

Theorem 11.6 says that if U is any (nonempty) subset of a unitary space S, then $U^\perp$ is a subspace of S. This is true even if U is not itself a subspace, as we have seen (Example 11.8). But now we are going to assume that U *is* a subspace; our objective is to prove that $S = U \oplus U^\perp$, at least when U is finite-dimensional.†

† This has already been done for finite-dimensional unitary spaces. (See Prob. 21, Sec. 11.2.) But in general we do not assume that the parent space is finite-dimensional.

THEOREM 11.8 The Projection Theorem

If U is a finite-dimensional subspace of the unitary space S, then S is the direct sum of U and its orthogonal complement, that is, $S = U \oplus U^{\perp}$.

Proof

If U is the zero subspace of S, then $U^{\perp} = S$ and the theorem is trivial. Hence assume that U is a nonzero subspace and let $\{u_1, \ldots, u_m\}$ be an orthonormal basis for U (guaranteed to exist by the Gram-Schmidt theory of Sec. 10.6). Let x be any element of S. The problem is to name elements $x_1 \in U$ and $x_2 \in U^{\perp}$ such that $x = x_1 + x_2$. If this can be done for every x, then $S = U + U^{\perp}$, which is a direct sum because the only element U and $U^{\perp}$ have in common is 0. (See Prob. 12, Sec. 11.2.)

To get a clue, suppose that x_1 and x_2 exist (we do not yet assert they do). Then we must have

$$x_1 = \sum_{i=1}^{m} c_i u_i \qquad \text{and} \qquad x_2 = x - x_1$$

where the scalars $c_1, \ldots, c_m$ remain to be determined. Since x_2 is to be an element of $U^{\perp}$, we want

$$x_2 * u_j = 0 \qquad \text{for each } j = 1, \ldots, m$$

(See Theorem 11.5.) But this condition is equivalent to

$$\begin{aligned} x_2 * u_j &= (x - x_1) * u_j \\ &= x * u_j - x_1 * u_j \\ &= x * u_j - \left(\sum_{i=1}^{m} c_i u_i\right) * u_j \\ &= x * u_j - \sum_{i=1}^{m} c_i (u_i * u_j) \\ &= x * u_j - c_j \\ &= 0 \qquad j = 1, \ldots, m \end{aligned}$$

Hence each c_j should be taken to be

$$c_j = x * u_j \qquad j = 1, \ldots, m$$

This is sufficient evidence to convict. Given that $x \in S$, *define*

$$x_1 = \sum_{i=1}^{m} c_i u_i \qquad \text{where } c_i = x * u_i \quad i = 1, \ldots, m$$

and then let $x_2 = x - x_1$.[†] All we have to confirm is that

$$x = x_1 + x_2 \qquad \text{where } x_1 \in U \text{ and } x_2 \in U^\perp$$

Obviously $x = x_1 + x_2$ and x_1 is an element of U; the proof that x_2 is an element of $U^\perp$ is contained in the above scratchwork, where we showed that

$$x_2 * u_i = x * u_i - c_i \qquad i = 1, \ldots, m$$

for we have named $c_i = x * u_i$, $i = 1, \ldots, m$. It follows that $x_2 * u_i = 0$ for all i; by Theorem 11.5, this implies that $x_2 \in U^\perp$. ∎

It is worth noting that the expression $x = x_1 + x_2$ in the above argument is unique, for we showed in the scratchwork that if x_1 and x_2 exist, they must be defined as we defined them. If we had not already observed that $U + U^\perp$ is a direct sum, this would prove it. (See Theorem 11.3.)

COROLLARY 11.8a

If U is a finite-dimensional subspace of the unitary space S, then every element of S can be written in exactly one way as the sum of an element of U and an element of $U^\perp$.

Proof

By Theorem 11.3 this is equivalent to Theorem 11.8.[‡]

COROLLARY 11.8b

If U is a subspace of the finite-dimensional unitary space S, then $\dim U + \dim U^\perp = \dim S$.

Note the stronger hypothesis in Corollary 11.8b compared to the projection theorem and its first corollary. In the latter two statements, S need not be finite-dimensional; the only assumption is that U is. But Corollary 11.8b is meaningless unless S has a dimension.

The reason Theorem 11.8 (or its equivalent Corollary 11.8a) is called the "projection theorem" is that each $x \in S$ can be "projected" onto a given (finite-dimensional) subspace U by writing $x = x_1 + x_2$, where $x_1 \in U$ and $x_2 \in U^\perp$ (x_1 being the "projection"). More precisely, we adopt the following definition.

ORTHOGONAL PROJECTION

Let U be a finite-dimensional subspace of the unitary space S and suppose that $x \in S$. In the expression $x = x_1 + x_2$, where $x_1 \in U$ and

[†] We could have shortened the proof by doing this right away. But then you might have wondered where the body was buried.

[‡] For that reason, Corollary 11.8a is often called the projection theorem too.

$x_2 \in U^\perp$, we call x_1 and x_2 the *orthogonal projections* of x on U and $U^\perp$, respectively. Moreover, the map $f: S \to S$ defined by $f(x) = x_1$ is called the *orthogonal projection* of S on U.†

A question that might arise in your mind concerning this definition is what happens if we apply it to $V = U^\perp$, writing $x = y_1 + y_2$, where $y_1 \in V$ and $y_2 \in V^\perp$? Assuming that V is finite-dimensional (otherwise the definition does not apply), we would then be calling y_1 the orthogonal projection of x on V, whereas before it was x_2. Consistency requires that $y_1 = x_2$, a fact that we ask you to establish in the problems.

THEOREM 11.9

Let $\{u_1, \ldots, u_m\}$ be an orthonormal basis for the subspace U of the unitary space S and let x be any element of S. Then the orthogonal projections of x on U and $U^\perp$ are

$$x_1 = \sum_{i=1}^{m} (x * u_i)u_i \qquad \text{and} \qquad x_2 = x - x_1$$

respectively.

Proof

See the definition of x_1 and x_2 in the proof of Theorem 11.8.

EXAMPLE 11.13

Let U be the plane $x - 2y + z = 0$ in $\mathcal{R}^3$ and suppose that $\mathbf{X} = (1,-1,4)$. An orthonormal basis for U is $\{\mathbf{U}_1,\mathbf{U}_2\}$, where

$$\mathbf{U}_1 = \frac{1}{\sqrt{5}}(2,1,0) \qquad \text{and} \qquad \mathbf{U}_2 = \frac{1}{\sqrt{30}}(-1,2,5)$$

By Theorem 11.9 the orthogonal projection of $\mathbf{X}$ on U is

$$\mathbf{X}_1 = (\mathbf{X}\cdot\mathbf{U}_1)\mathbf{U}_1 + (\mathbf{X}\cdot\mathbf{U}_2)\mathbf{U}_2 = \frac{1}{\sqrt{5}}\mathbf{U}_1 + \frac{17}{\sqrt{30}}\mathbf{U}_2 = \frac{1}{6}(-1,8,17)$$

The orthogonal projection of $\mathbf{X}$ on $U^\perp$ (the line in $\mathcal{R}^3$ with parametric equations $x = t$, $y = -2t$, and $z = t$) is

$$\mathbf{X}_2 = \mathbf{X} - \mathbf{X}_1 = (1,-1,4) - \tfrac{1}{6}(-1,8,17) = \tfrac{7}{6}(1,-2,1)$$

One bonus that we may collect from this is that the distance from $(1,-1,4)$ to the plane (see Fig. 3.3.) is the length of $\mathbf{X}_2$, namely

† See Example 3.11 for a special case. Also see Fig. 3.3.

$$\|\mathbf{X}_2\| = \frac{7}{\sqrt{6}}$$

(Actually this is premature, since we haven't defined distance between a point and a subspace. See the next section.)

LECTOR *I know a formula for doing that in 30 seconds.*
AUCTOR *Cherish it. I am after bigger game.*

Problems

25. Suppose that S is a unitary space which is not finite-dimensional. Explain why the orthogonal complement of a finite-dimensional subspace of S cannot be finite-dimensional.

26. Prove that if U is any finite-dimensional subspace of the unitary space S, then $(U^\perp)^\perp = U$. *Hint:* It is easy to show that $U \subset (U^\perp)^\perp$. To prove the reverse inclusion, let $x \in (U^\perp)^\perp$ and use the projection theorem to write $x = x_1 + x_2$, where $x_1 \in U$ and $x_2 \in U^\perp$. Then show that $x_2 * x_2 = 0$.

27. Suppose that U and $V = U^\perp$ are both finite-dimensional (in the unitary space S) and let $x \in S$. Then $x = x_1 + x_2$, where $x_1 \in U$ and $x_2 \in U^\perp$, and also $x = y_1 + y_2$, where $y_1 \in V$ and $y_2 \in V^\perp$. Our definition calls for both x_2 and y_1 to be the orthogonal projection of x on V. Explain why there is no inconsistency by showing that $y_1 = x_2$. (Similarly $y_2 = x_1$, so each is the orthogonal projection of x on U.)

28. Use Theorem 11.9 to find the orthogonal projection of $\mathbf{X} = (4,-1,1)$ on the plane $x - y - z = 0$ in $\mathcal{R}^3$. What is the component of $\mathbf{X}$ which is perpendicular to this plane? What is the distance from $\mathbf{X}$ to the plane?

29. If U is the subspace of $\mathcal{C}^3$ generated by $\mathbf{W}_1 = (i,1,0)$ and $\mathbf{W}_2 = (-1, 1 + i, 1)$, what is the orthogonal projection of $\mathbf{X} = (2 - i, 1, i)$ on U? (See Prob. 75, Sec. 10.6.)

★ **30.** Let f: $S \to S$ be the orthogonal projection of S on U (where S is unitary and U is a finite-dimensional subspace).

a. Prove that f is linear.
b. Explain why rng $f = U$. (Thus f maps S onto U.)
c. Show that ker $f = U^\perp$.
d. Assuming that S is finite-dimensional, what is the connection between the projection theorem and Sylvester's law of nullity (Sec. 3.2)?

31. In each of the following cases, find the formula defining the orthogonal projection of S on U.

a. U is the xy plane in $S = \mathcal{R}^3$.
b. U is the line $2x + 3y = 0$ in $S = \mathcal{R}^2$. (See Prob. 19, Sec. 3.1.)

c. U is the plane $x - y - z = 0$ in $S = \mathbb{R}^3$. (See Prob. 20, Sec. 3.1.)

32. Use the formula in Prob. 31*c* to confirm your answer to the first question in Prob. 28.

33. Let U be a subspace of the finite-dimensional unitary space S and let $f\colon S \to S$ and $g\colon S \to S$ be the orthogonal projections of S on U and $U^\perp$, respectively.

a. Explain why $f + g = i$, the identity map on S.

b. In each part of Prob. 31, use part *a* of this problem to find the orthogonal projection of S on $U^\perp$.

★ **34.** A linear operator $f\colon S \to S$ is said to be "idempotent" if $f^2 = f$. (See Prob. 12, Sec. 9.2.)

a. Why is every orthogonal projection idempotent?

b. Prove that an idempotent linear operator (on a unitary space) is an orthogonal projection if its range is finite-dimensional and its kernel and range are orthogonal complements.

★ **35.** Let U be a subspace of the finite-dimensional euclidean space S. The map $f\colon S \to S$ defined by $f(x) = x_1 - x_2$ (where x_1 and x_2 are the orthogonal projections of x on U and $U^\perp$, respectively) is called a "reflection in U."

a. Draw a picture showing why this terminology is appropriate when U is a line through the origin in $S = \mathbb{R}^2$.

b. What is f if $U = S$?

c. What is f if $U = \{0\}$? (This is called a "reflection in the origin" in analytic geometry.)

d. Prove that f is linear.

e. Show that f preserves the inner product on S, that is, $f(x) * f(y) = x * y$ for all x and y in S. (Such a map is called "orthogonal"; see Prob. 53, Sec. 10.4.)

f. Explain why $f^2 = i$, the identity map on S. What is the geometric interpretation of this? What is f^{-1}?

36. In each of the following cases, find the reflection in U.

a. U is the line $y = x$ in $\mathbb{R}^2$.

b. U is the line $2x + 3y = 0$ in $\mathbb{R}^2$.

c. U is the plane $x - y - z = 0$ in $\mathbb{R}^3$.

37. Regarding $\mathbb{C}$ as a vector space over $\mathbb{R}$ (Prob. 1, Sec. 2.1), define $f\colon \mathbb{C} \to \mathbb{R}^2$ by $f(x) = (a,b)$, where $x = a + bi$ is any element of $\mathbb{C}$.

a. Show that f is linear, one-to-one, and maps $\mathbb{C}$ onto $\mathbb{R}^2$. (Hence it is an isomorphism; see Sec. 3.3.)

b. Let $\circ$ be the standard inner product on $\mathbb{C}$ relative to the basis $\{1,i\}$. Given the complex numbers $x = a + bi$ and $y = c + di$, confirm that $x \circ y = \mathbf{X} \cdot \mathbf{Y}$, where $\mathbf{X} = f(x)$ and $\mathbf{Y} = f(y)$ are the images of x and y in $\mathbb{R}^2$. (See Sec. 10.5, where it is shown that the standard inner product relative to a given basis makes the space unitary and the basis orthonormal, and that this coordinate vector map preserves inner products.)

c. If $U = \langle 1 \rangle$, what is $U^{\perp}$? Given the complex number $x = a + bi$, what are its orthogonal projections on U and $U^{\perp}$? What are the orthogonal projections of $\mathbf{X} = f(x)$ on the usual coordinate axes in $\mathcal{R}^2$?

The point of this problem is to throw some light on the common practice of representing complex numbers graphically by points in the cartesian plane. The horizontal and vertical axes in $\mathcal{R}^2$ are often called the "real axis" and "imaginary axis," respectively (as though one had better standing than the other!). This problem shows that $\mathcal{C}$ and $\mathcal{R}^2$, as euclidean vector spaces, are essentially the same thing. (However, do not confuse $\mathcal{C}$ over $\mathcal{R}$ with $\mathcal{C}^1$! See Prob. 2*d*, Sec. 2.1.)

†11.4 Applications to Geometry and Analysis

In this section we are going to exploit the projection theorem to discuss two problems. The first is comparatively simple: Given a (finite-dimensional) subspace of a unitary space and an arbitrary point of the space, what is the distance from the point to the subspace? The second is much deeper: Given a continuous function f, what "trigonometric series" of the form

$$a_0 + (a_1 \cos x + b_1 \sin x) + (a_2 \cos 2x + b_2 \sin 2x) + \cdots$$

converges to f? While we are in no position to prove anything about this, you will see that our answer to the first question throws considerable light on the second, and indeed provides us with its answer as well.

EXAMPLE 11.14

Suppose that U is the plane $x - 2y + z = 0$ in $\mathcal{R}^3$ and the distance from $\mathbf{X} = (1,-1,4)$ to this plane is wanted. (See Fig. 11.7.) We saw in the last section that the orthogonal projection of $\mathbf{X}$ on U is

$$\mathbf{X}_1 = \frac{1}{\sqrt{5}}\mathbf{U}_1 + \frac{17}{\sqrt{30}}\mathbf{U}_2$$

where $\mathbf{U}_1 = (1/\sqrt{5})(2,1,0)$ and $\mathbf{U}_2 = (1/\sqrt{30})(-1,2,5)$ constitute an orthonormal basis for U; the orthogonal projection of $\mathbf{X}$ on $U^{\perp}$ is

$$\mathbf{X}_2 = \mathbf{X} - \mathbf{X}_1 = \tfrac{7}{6}(1,-2,1)$$

We then said that the required distance is

$$\|\mathbf{X}_2\| = \frac{7}{\sqrt{6}}$$

† This section may be omitted. However, it contains an impressive (and not at all hard) application of algebra to analysis.

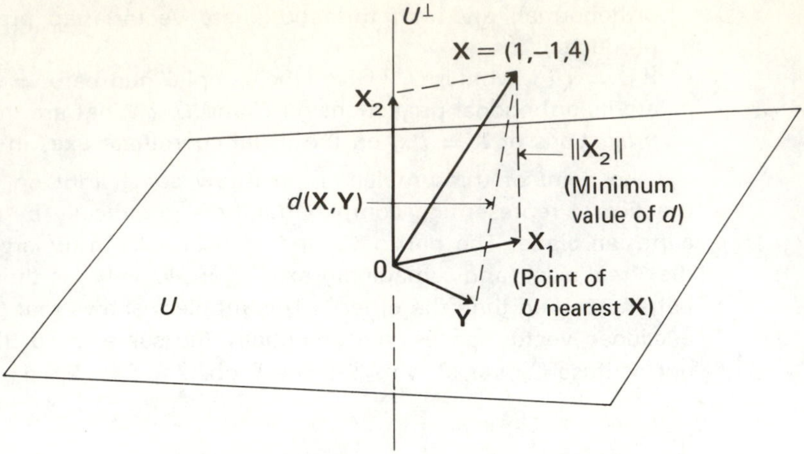

Fig. 11.7 Distance from a point to a plane

An algebraic confirmation of this result may be obtained by noting that the unit vector

$$\mathbf{U}_3 = \frac{\mathbf{X}_2}{\|\mathbf{X}_2\|} = \frac{1}{\sqrt{6}}(1,-2,1)$$

completes $\{\mathbf{U}_1,\mathbf{U}_2\}$ to an orthonormal basis for $\mathcal{R}^3$. Hence

$$\mathbf{X} = \mathbf{X}_1 + \mathbf{X}_2 = \frac{1}{\sqrt{5}}\mathbf{U}_1 + \frac{17}{\sqrt{30}}\mathbf{U}_2 + \frac{7}{\sqrt{6}}\mathbf{U}_3{}^{\dagger}$$

The typical element of U is $\mathbf{Y} = y_1\mathbf{U}_1 + y_2\mathbf{U}_2$; the distance from $\mathbf{X}$ to U (see the definition below) is the minimum value of $d(\mathbf{X},\mathbf{Y})$ as $\mathbf{Y}$ ranges over U. But

$$[d(\mathbf{X},\mathbf{Y})]^2 = \left(\frac{1}{\sqrt{5}} - y_1\right)^2 + \left(\frac{17}{\sqrt{30}} - y_2\right)^2 + \frac{49}{6} \qquad \text{(Corollary 10.9a)}$$

Obviously the minimum value occurs when

$$y_1 = \frac{1}{\sqrt{5}} \qquad \text{and} \qquad y_2 = \frac{17}{\sqrt{30}}$$

† Alternatively, we may use the Gram-Schmidt process to find $\mathbf{U}_3$ directly, bypassing the computation of $\mathbf{X}_2$. The above expression for $\mathbf{X}$ may then be obtained by using Theorem 10.7.

For this choice of **Y**, we find

$$d(\mathbf{X},\mathbf{Y}) = \frac{7}{\sqrt{6}}$$

as before. Also note that we have found the unique point of U which is closest to **X**, namely

$$\mathbf{Y} = \mathbf{X}_1 = \frac{1}{\sqrt{5}}\mathbf{U}_1 + \frac{17}{\sqrt{30}}\mathbf{U}_2 = \frac{1}{6}(-1,8,17)$$

Of course this problem is easily solved by calculus. The distance from $(1,-1,4)$ to a typical point (x,y,z) of the plane $x - 2y + z = 0$ is

$$\sqrt{(x-1)^2 + (y+1)^2 + (z-4)^2} = \sqrt{2x^2 - 4xy + 5y^2 + 6x - 14y + 18}$$

The minimum value occurs when the partial derivatives of this function of x and y are equal to zero; this leads to the system

$$\begin{aligned} 2x - 2y &= -3 \\ 2x - 5y &= -7 \end{aligned}$$

with solution $x = -1/6$, $y = 4/3$. The point of U which is closest to $(1,-1,4)$ is therefore

$$\left(-\frac{1}{6},\frac{4}{3},\frac{17}{6}\right) = \frac{1}{6}(-1,8,17)$$

and the distance from $(1,-1,4)$ to the plane is $7/\sqrt{6}$.

This example should help clarify the following definition.

DISTANCE FROM A POINT TO A SUBSPACE

Let U be a finite-dimensional subspace of the unitary space S and suppose that $x \in S$. The *distance* from x to S is the minimum value of $d(x,y)$ as y ranges over U. The element of U *closest* to x is the $y \in U$ for which $d(x,y)$ has its minimum value.

Of course it is necessary to show that $d(x,y)$ actually has a minimum value as y ranges over U, and that there is exactly one $y \in U$ for which this minimum occurs. Otherwise the definition is meaningless.

THEOREM 11.10

Let U be a finite-dimensional subspace of the unitary space S and let $x = x_1 + x_2$ be any element of S, where x_1 and x_2 are the orthogonal projections of x on U and $U^{\perp}$, respectively. Then the element of U closest to x is x_1 and the distance from x to U is $\|x_2\|$.

Proof

Note first that the projection theorem plays an important role in the very statement of this theorem, for it guarantees that x can be written (uniquely) in the form $x = x_1 + x_2$, where $x_1 \in U$ and $x_2 \in U^{\perp}$. If y is an arbitrary element of U, we have

$$x - y = (x_1 - y) + x_2$$

Since $x_1 - y$ and x_2 are orthogonal (why?), the pythagorean theorem (Prob. 45, Sec. 10.4) yields

$$[d(x,y)]^2 = \|x - y\|^2 = \|x_1 - y\|^2 + \|x_2\|^2$$

Hence $d(x,y)$ has a minimum value; it occurs when $y = x_1$. This choice of y is the element of U closest to x; the corresponding minimum value of d is

$$d(x,y) = \|x_2\| \blacksquare$$

THEOREM 11.11

If $\{u_1, \ldots, u_n\}$ is an orthonormal basis for S whose first m elements span U $(0 < m < n)$, the distance from x to U is

$$\left(\sum_{i=m+1}^{n} |x * u_i|^2\right)^{1/2}$$

Proof

This is left for the problems. Note that this formula for distance applies only in finite-dimensional spaces.

EXAMPLE 11.15

An orthonormal basis for $\mathcal{R}^3$ whose first two elements span

$$U = \{(x,y,z) \,|\, x - 2y + z = 0\}$$

is $\{\mathbf{U}_1,\mathbf{U}_2,\mathbf{U}_3\}$, where

$$\mathbf{U}_1 = \frac{1}{\sqrt{5}}(2,1,0) \qquad \mathbf{U}_2 = \frac{1}{\sqrt{30}}(-1,2,5) \qquad \mathbf{U}_3 = \frac{1}{\sqrt{6}}(1,-2,1)$$

(See Example 11.14.) According to Theorem 11.11 the distance from $\mathbf{X} = (1,-1,4)$ to U is

$$\left(\sum_{i=3}^{3}|\mathbf{X}\cdot\mathbf{U}_i|^2\right)^{1/2} = |\mathbf{X}\cdot\mathbf{U}_3| = \frac{7}{\sqrt{6}}$$

EXAMPLE 11.16

Let S be the euclidean space of real-valued functions continuous on $[-\pi,\pi]$ and let U be the subspace generated by the functions

$$1,\ \cos x,\ \sin x,\ \ldots,\ \cos nx,\ \sin nx$$

where n is a given positive integer. These $2n + 1$ elements of S are mutually orthogonal (Example 10.5), so $\dim U = 2n + 1$; an orthonormal basis for U is

$$\{u_0, u_1, \ldots, u_n, v_1, \ldots, v_n\}$$

where

$$u_0(x) = \frac{1}{\sqrt{2\pi}},\ u_1(x) = \frac{1}{\sqrt{\pi}}\cos x,\ \ldots,\ u_n(x) = \frac{1}{\sqrt{\pi}}\cos nx$$

and

$$v_1(x) = \frac{1}{\sqrt{\pi}}\sin x,\ \ldots,\ v_n(x) = \frac{1}{\sqrt{\pi}}\sin nx$$

(See Example 10.7.)

If f is any function in S, its orthogonal projection on U is (by Theorem 11.9)

$$f_1 = (f * u_0)u_0 + \sum_{k=1}^{n}(f * u_k)u_k + \sum_{k=1}^{n}(f * v_k)v_k$$

But

$$f * u_0 = \frac{1}{\sqrt{2\pi}} \int_{-\pi}^{\pi} f(x)\, dx$$

$$f * u_k = \frac{1}{\sqrt{\pi}} \int_{-\pi}^{\pi} f(x) \cos kx\, dx \qquad k = 1, \ldots, n$$

$$f * v_k = \frac{1}{\sqrt{\pi}} \int_{-\pi}^{\pi} f(x) \sin kx\, dx \qquad k = 1, \ldots, n$$

so we can write

$$\begin{aligned} f_1(x) &= a_0 + a_1 \cos x + \cdots + a_n \cos nx + b_1 \sin x + \cdots + b_n \sin nx \\ &= a_0 + (a_1 \cos x + b_1 \sin x) + \cdots + (a_n \cos nx + b_n \sin nx) \end{aligned}$$

where

$$a_0, a_1, b_1, \ldots, a_n, b_n$$

are the Fourier coefficients of f with respect to the original generators of U,

$$1, \cos x, \sin x, \ldots, \cos nx, \sin nx$$

(See Example 10.9.)

Now according to Theorem 11.10 the function f_1 is the element of U which is *closest to f,* in the sense that $d(f,f_1)$ is the minimum value of $d(f,g)$ as g ranges over U. In other words, if we want to approximate a given function f by a "trigonometric polynomial"

$$a_0 + (a_1 \cos x + b_1 \sin x) + \cdots + (a_n \cos nx + b_n \sin nx)$$

the best approximation we can manage (for a given n) is the one in which the scalars are chosen to be the Fourier coefficients of f with respect to

$$1, \cos x, \sin x, \ldots, \cos nx, \sin nx^{\dagger}$$

Moreover, the error in this approximation is the distance from f to U. Since the orthogonal projection of f on $U^{\perp}$ is $f_2 = f - f_1$, Theorem 11.10 says the error is

$$\|f_2\| = \|f - f_1\| = \left[\int_{-\pi}^{\pi} [f(x) - f_1(x)]^2\, dx\right]^{1/2} \ddagger$$

† If you haven't grasped before why these Fourier coefficients are important, this should clarify matters! They are precisely the coefficients needed to construct the best approximation.

‡ There is no need to mention f_2. Since f has been approximated by f_1, the error is simply $d(f,f_1) = \|f - f_1\|$.

It is at this point that "Fourier analysis" begins, for the question which naturally arises is whether the error approaches zero as n increases. If so, any function continuous on $[-\pi,\pi]$ can be approximated as closely as we please by taking n sufficiently large, and the "Fourier series"

$$a_0 + (a_1 \cos x + b_1 \sin x) + (a_2 \cos 2x + b_2 \sin 2x) + \cdots$$

$$= a_0 + \sum_{k=1}^{\infty} (a_k \cos kx + b_k \sin kx)$$

converges to f. That this is indeed the case is one of the classical results of mathematical analysis.

It is worth mentioning that the convergence referred to here is defined in terms of the integral norm in S; it is not the same as "pointwise" convergence of an infinite series as defined in elementary calculus. There exist continuous functions whose Fourier series do not converge (in the ordinary sense) at every point; this difficulty is overcome by working with a slightly different space of functions in place of S.

Whether a continuous function exists whose Fourier series does not converge (in the ordinary sense) at *any* point of $[-\pi,\pi]$ was still an unsolved problem in 1965. (You will find it labeled as such on page 341 of *An Introduction to Linear Analysis* by D. L. Kreider, R. G. Kuller, D. R. Ostberg, and F. W. Perkins, Reading, Mass.: Addison-Wesley, 1966.) But in 1966 the Swedish mathematician Carleson showed that the Fourier series of any reasonably well-behaved function (including continuous functions) must converge "almost everywhere" in $[-\pi,\pi]$, that is, except for a set of "measure zero."†

EXAMPLE 11.17

Let S be the unitary space of complex-valued functions continuous on $[-\pi,\pi]$, with inner product

$$f * g = \int_{-\pi}^{\pi} f(x)\overline{g(x)}\, dx$$

(See Example 10.2.) According to Prob. 29, Sec. 10.2, the functions

$$v_k(x) = e^{ikx} \qquad k = 0, 1, 2, \ldots$$

† See the article on "Mathematics" by Irving Kaplansky in the 1967 *Encyclopaedia Britannica Yearbook*. There are some interesting statistics in this article on the growth of mathematical research since World War II. The number of pages in the American review journal alone have been as follows: 1941, 419; 1946, 621; 1951, 1004; 1956, 1437; 1961, 2547; 1966, 2900 (approximately). At this rate a mathematician in 1991 who merely wants to read *reviews* of research articles published will have to scan some 20,000 pages annually.

are mutually orthogonal in S. If p and q are nonnegative integers, we can write

$$\begin{aligned} v_p * v_q &= \int_{-\pi}^{\pi} v_p(x)\overline{v_q(x)}\, dx \\ &= \int_{-\pi}^{\pi} e^{ipx}e^{-iqx}\, dx \\ &= \int_{-\pi}^{\pi} e^{i(p-q)x}\, dx \\ &= \begin{cases} 2\pi & \text{if } p = q \\ 0 & \text{if } p \neq q \end{cases} \end{aligned}$$

[Parenthetically, it is interesting to note that these functions, when written out by Euler's formula (Example 10.4) are

$$v_0(x) = 1 \qquad v_1(x) = \cos x + i \sin x \qquad v_2(x) = \cos 2x + i \sin 2x, \ldots$$

Their real and imaginary parts are the functions

$$1, \cos x, \sin x, \cos 2x, \sin 2x, \ldots$$

of the preceding example.]

Since $\|v_k\| = \sqrt{2\pi}$ for each k, the functions

$$u_k(x) = \frac{1}{\sqrt{2\pi}} v_k(x) \qquad k = 0, 1, 2, \ldots, n$$

are orthonormal (where n is an arbitrary nonnegative integer); let U be the subspace they generate. If f is any function in S, its orthogonal projection on U is

$$f_1 = \sum_{k=0}^{n} (f * u_k)u_k$$

where

$$f * u_k = \frac{1}{\sqrt{2\pi}} \int_{-\pi}^{\pi} f(x)e^{-ikx}\, dx \qquad k = 0, 1, \ldots, n$$

Hence

$$f_1(x) = a_0 + a_1 e^{ix} + \cdots + a_n e^{inx}$$

where

$$a_k = \frac{1}{2\pi} \int_{-\pi}^{\pi} f(x)e^{-ikx}\, dx$$

is the Fourier coefficient of f with respect to e^{ikx}, $k = 0, 1, \ldots, n$.

Thus f_1 is the polynomial in e^{ix} which best approximates f; the error is

$$\|f - f_1\| = \left[\int_{-\pi}^{\pi} |f(x) - f_1(x)|^2 \, dx\right]^{1/2}$$

Complex Fourier analysis has this as its point of departure. What is shown is that the error approaches zero as n increases; then it follows that the Fourier series

$$a_0 + a_1 e^{ix} + a_2 e^{2ix} + \cdots$$

converges to f (in the sense defined by the integral norm on S). This is much the same as the real case discussed in Example 11.16, but note that the complex case is less cluttered because of the absorption of sines and cosines into the exponential.

Problems

38. Prove Theorem 11.11.

39. Use Theorem 11.11 to find the distance from $(2,-1,1)$ to the line in $\mathcal{R}^3$ whose parametric equations are $x = 2t$, $y = t$, and $z = -t$.

40. Do Prob. 39 by using calculus to minimize an appropriate function.

41. Let U be the subspace of $\mathcal{C}^3$ generated by $\mathbf{W}_1 = (i,1,0)$ and $\mathbf{W}_2 = (-1, 1 + i, 1)$ and let $\mathbf{X} = (2 - i, 1, i)$. What is the distance from $\mathbf{X}$ to U? (See Prob. 75, Sec. 10.6, and Prob. 29, Sec. 11.3.)

42. Prove that the distance from (x_0,y_0) to the line $ax + by = 0$ in $\mathcal{R}^2$ is

$$d = \frac{|ax_0 + by_0|}{\sqrt{a^2 + b^2}}$$

Hint: A unit vector perpendicular to the line is $(1/\sqrt{a^2 + b^2})(a,b)$. Use this as the second element of an orthonormal basis for $\mathcal{R}^2$.

43. Prove that the distance from (x_0,y_0,z_0) to the plane $ax + by + cz = 0$ in $\mathcal{R}^3$ is

$$d = \frac{|ax_0 + by_0 + cz_0|}{\sqrt{a^2 + b^2 + c^2}}$$

44. In each of the following cases, find the trigonometric polynomial

$$a_0 + (a_1 \cos x + b_1 \sin x) + \cdots + (a_n \cos nx + b_n \sin nx)$$

which best approximates the given function in $[-\pi,\pi]$. (See Prob. 48, Sec. 10.4, for the corresponding Fourier series.)

a. $f(x) = 1$
b. $f(x) = \cos x$
c. $f(x) = x$

45. Apply Bessel's inequality for infinite-dimensional spaces (Prob. 66, Sec. 10.5) in Example 11.17 to show that if f is any complex-valued function continuous on $[-\pi,\pi]$, then

$$\left|\int_{-\pi}^{\pi} f(x)\,dx\right|^2 + \left|\int_{-\pi}^{\pi} f(x)e^{-ix}\,dx\right|^2 + \left|\int_{-\pi}^{\pi} f(x)e^{-2ix}\,dx\right|^2 + \cdots \leq 2\pi \int_{-\pi}^{\pi} |f(x)|^2\,dx$$

(As in the real case given in Prob. 67, Sec. 10.5, this is actually an equality.)

46. Confirm equality in Prob. 45 when $f(x) = e^{ix}$.

Review Quiz

True or false?

1. The orthogonal complement of a finite-dimensional subspace of a unitary space is finite-dimensional.

2. If T_1 and T_2 are distinct lines through the origin in $\mathcal{R}^3$, then $T_1 + T_2$ is a direct sum.

3. If T_1 and T_2 are finite-dimensional subspaces of S, then

$$\dim (T_1 + T_2) = \dim T_1 + \dim T_2$$

4. $(1,1,1)$ is orthogonal to $U = \{(x,y,z) \mid x = t,\ y = -3t,\ z = 2t\}$ in $\mathcal{R}^3$.

5. The orthogonal projection of $(2,0,-3)$ on $U = \{(x,y,z) \mid x = 3t,\ y = -t,\ z = t\}$ in $\mathcal{R}^3$ is $\frac{3}{11}(3,-1,1)$.

6. $\mathcal{R}^3$ is the direct sum of $\langle \mathbf{E}_1 \rangle$, $\langle \mathbf{E}_2 \rangle$, and $\langle \mathbf{E}_3 \rangle$.

7. The distance from $(1, 1 + i)$ to $\langle \mathbf{E}_1 \rangle$ in $\mathcal{C}^2$ is $\sqrt{2}$.

8. If T_1 is the x axis in $\mathcal{R}^3$ and T_2 is the xy plane, then $\dim (T_1 + T_2) = 2$.

9. If $\mathbf{X} \cdot \mathbf{E}_1 = 0$ in $\mathcal{R}^2$, then $\mathbf{X}$ is orthogonal to $U = \{(x,y) \mid y = 0\}$.

10. If each element of S can be written in the form $x = x_1 + x_2$, where $x_1 \in T_1$ and $x_2 \in T_2$, then $S = T_1 \oplus T_2$.

11. If $U = \{(x,y) \mid x + y = 1\}$ in $\mathcal{R}^2$, then $U^\perp = \{(x,y) \mid y = x\}$.

12. If the orthogonal projections of x and y on U (where U is a finite-dimensional subspace of a unitary space) are x_1 and y_1, then $x_1 + y_1$ is the orthogonal projection of $x + y$ on U.

13. If U is a subspace of the unitary space S, then $U + U^\perp$ is a direct sum.
14. If U is a proper subspace of $\mathcal{R}^3$, there are infinitely many subspaces V such that $\mathcal{R}^3 = U \oplus V$.
15. If U is the zero subspace of the unitary space S, then $U^\perp = S$.
16. If U and V are orthogonal subspaces of S, then $V = U^\perp$.

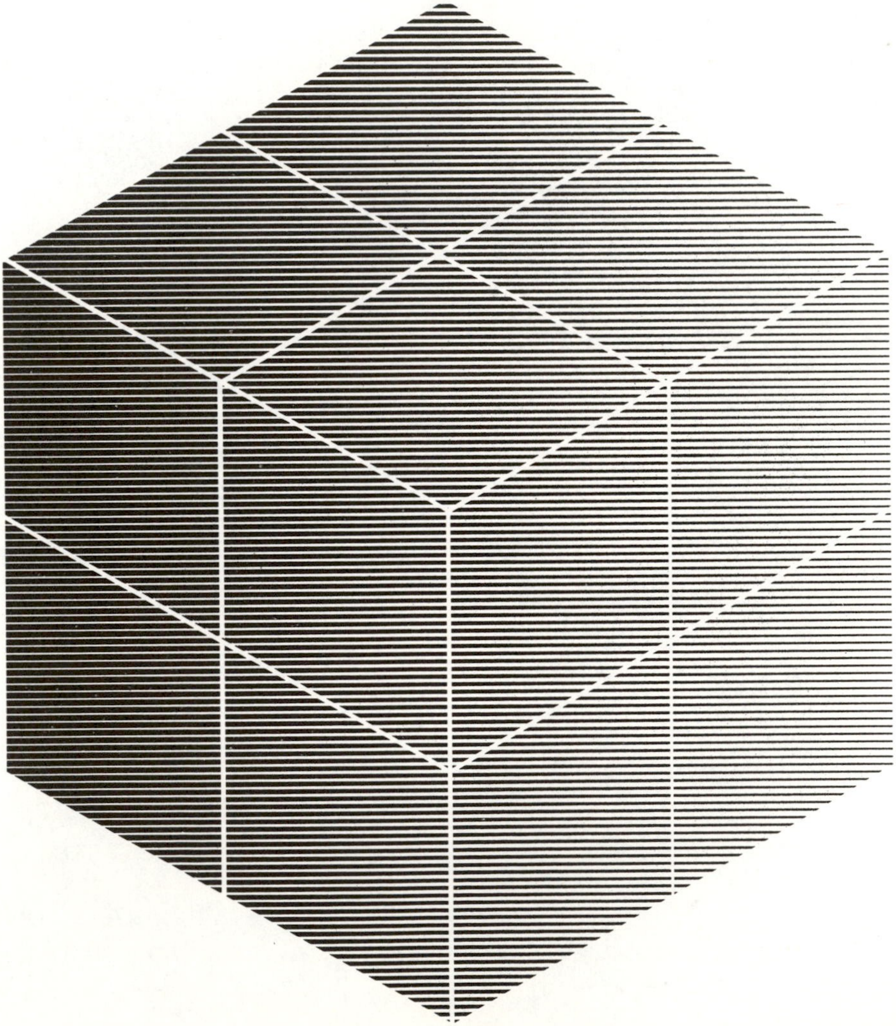

CHAPTER 12

Linear Operators on a Unitary Space

In this chapter and the next we shall try to pull this book together. Linear operators on a unitary space, and the "spectral theory" associated with them, are central to linear algebra and its applications.

12.1 The Adjoint of a Linear Operator

Let $f\colon \mathcal{R}^2 \to \mathcal{R}^2$ be the linear operator whose matrix relative to the standard basis is

$$A = \begin{bmatrix} 1 & -1 \\ 3 & 1 \end{bmatrix}$$

Then for all $\mathbf{X} = (x_1,x_2)$ and $\mathbf{Y} = (y_1,y_2)$ in $\mathcal{R}^2$, we can write

$$\begin{aligned} f(\mathbf{X}) \cdot \mathbf{Y} &= (x_1 - x_2, 3x_1 + x_2) \cdot (y_1,y_2) \\ &= (x_1 - x_2)y_1 + (3x_1 + x_2)y_2 \\ &= x_1(y_1 + 3y_2) + x_2(-y_1 + y_2) \\ &= (x_1,x_2) \cdot (y_1 + 3y_2, -y_1 + y_2) \\ &= \mathbf{X} \cdot g(\mathbf{Y}) \end{aligned}$$

where $g\colon \mathcal{R}^2 \to \mathcal{R}^2$ is the linear operator whose matrix is

$$B = \begin{bmatrix} 1 & 3 \\ -1 & 1 \end{bmatrix}$$

The fascinating thing about this is that $B = A^T$. The manipulations that produced the identity

$$f(\mathbf{X}) \cdot \mathbf{Y} = \mathbf{X} \cdot g(\mathbf{Y})$$

had the effect of changing f to a map whose matrix is $[f]^T$. Thus it seems natural to call g the "transpose" of f. However, most writers prefer to use the term "adjoint."†

Before jumping to any conclusions, let's look at an example in complex space. Let $f: \mathcal{C}^2 \to \mathcal{C}^2$ be the map whose matrix relative to the standard basis is

$$A = \begin{bmatrix} 1 & 2-i \\ i & 3 \end{bmatrix}$$

Then

$$\begin{aligned} f(\mathbf{X}) * \mathbf{Y} &= [x_1 + (2-i)x_2, ix_1 + 3x_2] * (y_1, y_2) \\ &= [x_1 + (2-i)x_2]\bar{y}_1 + (ix_1 + 3x_2)\bar{y}_2 \\ &= x_1(\bar{y}_1 + i\bar{y}_2) + x_2[(2-i)\bar{y}_1 + 3\bar{y}_2] \\ &= x_1\overline{(y_1 - iy_2)} + x_2\overline{[(2+i)y_1 + 3y_2]} \\ &= (x_1, x_2) * [y_1 - iy_2, (2+i)y_1 + 3y_2] \\ &= \mathbf{X} * g(\mathbf{Y}) \end{aligned}$$

where $g: \mathcal{C}^2 \to \mathcal{C}^2$ is the map whose matrix is

$$B = \begin{bmatrix} 1 & -i \\ 2+i & 3 \end{bmatrix}$$

This time we have $B = A^*$ (Sec. 4.3); that is, the identity

$$f(\mathbf{X}) * \mathbf{Y} = \mathbf{X} * g(\mathbf{Y})$$

is satisfied by a map g whose matrix is the conjugate transpose of the matrix of f.

The point of view we are going to adopt is that the key to the map g in both these examples is the identity

$$f(\mathbf{X}) * \mathbf{Y} = \mathbf{X} * g(\mathbf{Y})$$

For the standard inner product on $\mathcal{F}^n$ covers both cases (reducing to the dot product when $\mathcal{F} = \mathcal{R}$); so does the conjugate transpose, since $A^* = A^T$ when A has real entries.

More generally, we make the following definition, which applies not only when matrices can be invoked, but also to linear operators on infinite-dimensional spaces (when no matrix representation is available).

ADJOINT OF A LINEAR OPERATOR

Let S be a unitary space with inner product $*$ and suppose that f is a linear operator on S. If a linear operator $g: S \to S$ exists such that

† See Prob. 39, Sec. 4.3, for a preview of this idea. Do not confuse the term "adjoint" as used here with its use in Sec. 8.5!

$$f(x) * y = x * g(y) \qquad \text{for all } x \text{ and } y \text{ in } S$$

we call g the *adjoint* of f, denoted by f^*.

The first question to settle is whether this definition is unambiguous. What if there is more than one linear operator on S with the property of an adjoint? To show that this is not the case, suppose that g and h satisfy the identities

$$f(x) * y = x * g(y) \qquad \text{and} \qquad f(x) * y = x * h(y)$$

Then for all x and y in S, we have

$$\begin{aligned} x * g(y) &= x * h(y) \\ x * g(y) - x * h(y) &= 0 \\ x * [g(y) - h(y)] &= 0 \\ x * (g - h)(y) &= 0 \end{aligned}$$

In particular, taking $x = (g - h)(y)$ for each $y \in S$, we have

$$(g - h)(y) * (g - h)(y) = 0 \qquad \text{for all } y$$

Since $*$ is positive-definite, this implies that

$$(g - h)(y) = 0 \qquad \text{for all } y$$

which means that $g - h$ is the zero map, that is, $g = h$.

THEOREM 12.1

The adjoint of a linear operator is unique if it exists.

Proof

Refer to the above argument.

The second question is more profound: Given a linear operator $f\colon S \to S$, what guarantee do we have that f^* exists? We are going to show that it does when S is finite-dimensional (by using matrices), but in general the answer is negative. However, it is not easy to give a supportive example; since it would require a digression beyond our present purposes, we leave the question open.

THEOREM 12.2

Let $f\colon S \to S$ be a linear operator on the finite-dimensional unitary space S. Then there exists a unique linear operator $f^*\colon S \to S$ such that

$$f(x) * y = x * f^*(y) \qquad \text{for all } x \text{ and } y \text{ in } S$$

where $*$ is the inner product on S.

Proof

Uniqueness (even for spaces which are not finite-dimensional) is already established by Theorem 12.1. To demonstrate the existence of f^*, let $\alpha = \{u_1, \ldots, u_n\}$ be an orthonormal basis for S and let

$$A = [f]_\alpha = [a_{ij}]$$

We claim that the map $g: S \to S$ whose matrix relative to α is A^* is the adjoint of f, for suppose that

$$B = A^* = [b_{ij}]$$

Then for each i and j, we have

$$\begin{aligned}
f(u_i) * u_j &= \left(\sum_{k=1}^{n} a_{ki} u_k\right) * u_j && \text{(definition of } A = [f]) \\
&= \sum_{k=1}^{n} a_{ki}(u_k * u_j) && (* \text{ is conjugate-bilinear}) \\
&= a_{ji} && (\alpha \text{ is orthonormal})
\end{aligned}$$

On the other hand,

$$\begin{aligned}
u_i * g(u_j) &= u_i * \left(\sum_{k=1}^{n} b_{kj} u_k\right) && \text{(definition of } B = [g]) \\
&= \sum_{k=1}^{n} \overline{b}_{kj}(u_i * u_k) && (* \text{ is conjugate-bilinear}) \\
&= \overline{b}_{ij} && (\alpha \text{ is orthonormal}) \\
&= a_{ji} && (B = A^*)
\end{aligned}$$

Hence

$$f(u_i) * u_j = u_i * g(u_j) \qquad \text{for all } i \text{ and } j$$

It is not hard to extend this formula to any elements x and y of S, for if

$$x = \sum_{i=1}^{n} x_i u_i \qquad \text{and} \qquad y = \sum_{j=1}^{n} y_j u_j$$

we have

$$f(x) * y = \left[\sum_{i=1}^{n} x_i f(u_i)\right] * y = \sum_{i=1}^{n} x_i [f(u_i) * y]$$

But (for each i)

$$f(u_i) * y = f(u_i) * \left(\sum_{j=1}^{n} y_j u_j\right) = \sum_{j=1}^{n} \bar{y}_j[f(u_i) * u_j]$$

so

$$f(x) * y = \sum_{i=1}^{n} \sum_{j=1}^{n} x_i \bar{y}_i [f(u_i) * u_j]$$

Similarly

$$x * g(y) = \sum_{j=1}^{n} \sum_{i=1}^{n} x_i \bar{y}_j [u_i * g(u_j)]$$

(Confirm this!) Since the order of the sums is immaterial, it follows that

$$f(x) * y = x * g(y)$$

and hence $g = f^*$. ▮

Note that the choice of α in this argument does not affect the final result. If a different orthonormal basis is used, the matrix of f is different, say $[f] = C$, and the adjoint of f is the map h: $S \to S$ whose matrix is $[h] = C^*$. Nonetheless we can obtain the identity $f(x) * y = x * h(y)$ as before; Theorem 12.1 guarantees that $h = g$.

EXAMPLE 12.1

Let f: $\mathcal{R}^2 \to \mathcal{R}^2$ be defined by $f(x,y) = (x - y, 3x + y)$. As we saw at the beginning of this section, the adjoint of f is the map g: $\mathcal{R}^2 \to \mathcal{R}^2$ defined by $g(x,y) = (x + 3y, -x + y)$. We used matrices relative to the standard basis to describe f and g; suppose that we had used the orthonormal basis $\{\mathbf{U}_1, \mathbf{U}_2\}$ instead, where

$$\mathbf{U}_1 = (0,1) = \mathbf{E}_2 \qquad \text{and} \qquad \mathbf{U}_2 = (-1,0) = -\mathbf{E}_1$$

Then since the matrix of f relative to the standard basis is

$$A = \begin{bmatrix} 1 & -1 \\ 3 & 1 \end{bmatrix}$$

we have

$$f(\mathbf{U}_1) = f(\mathbf{E}_2) = -\mathbf{E}_1 + \mathbf{E}_2 = \mathbf{U}_1 + \mathbf{U}_2$$
$$f(\mathbf{U}_2) = -f(\mathbf{E}_1) = -\mathbf{E}_1 - 3\mathbf{E}_2 = -3\mathbf{U}_1 + \mathbf{U}_2$$

Hence the matrix of f relative to $\{\mathbf{U}_1,\mathbf{U}_2\}$ is

$$C = \begin{bmatrix} 1 & -3 \\ 1 & 1 \end{bmatrix}$$

Since

$$C^* = C^T = \begin{bmatrix} 1 & 1 \\ -3 & 1 \end{bmatrix}$$

the adjoint of f is the map h: $\mathcal{R}^2 \to \mathcal{R}^2$ defined by

$$h(\mathbf{U}_1) = \mathbf{U}_1 - 3\mathbf{U}_2$$
$$h(\mathbf{U}_2) = \mathbf{U}_1 + \mathbf{U}_2$$

But then

$$h(\mathbf{E}_1) = -h(\mathbf{U}_2) = -\mathbf{U}_1 - \mathbf{U}_2 = \mathbf{E}_1 - \mathbf{E}_2 = (1,-1)$$
$$h(\mathbf{E}_2) = h(\mathbf{U}_1) = \mathbf{U}_1 - 3\mathbf{U}_2 = 3\mathbf{E}_1 + \mathbf{E}_2 = (3,1)$$

from which

$$\begin{aligned} h(x,y) = h(x\mathbf{E}_1 + y\mathbf{E}_2) &= xh(\mathbf{E}_1) + yh(\mathbf{E}_2) \\ &= x(1,-1) + y(3,1) = (x + 3y, -x + y) \end{aligned}$$

This is the same as g!

THEOREM 12.3

Let f: $S \to S$ be a linear operator on the finite-dimensional unitary space S. The linear operator g: $S \to S$ is the adjoint of f if and only if $[g] = [f]^*$, where the matrices are computed relative to an orthonormal basis.

Proof

A little thought should convince you that this is trivial.†

COROLLARY 12.3a

If matrices are computed relative to an orthonormal basis, then $[f^*] = [f]^*$.

Note the role played by orthonormal bases in these statements. The relation between the matrix of a linear operator and the matrix of its adjoint is not this simple when an arbitrary basis is used. (However, see Probs. 4 and 5.)

THEOREM 12.4

If f: $S \to S$ is a linear operator on the unitary space S, and f^* exists, then $(f^*)^* = f$.

†This "proof" is inspired by an anecdote of the British mathematician G. H. Hardy, who once announced a theorem to his class and said, "Gentlemen, the proof is trivial." There was a silence, followed by Hardy's retirement to his office for a few minutes. On his return he said, "Yes, gentlemen, the proof is trivial," and proceeded to the next theorem.

Proof

For all x and y in S, we have

$$\begin{aligned} f^*(x) * y &= \overline{y * f^*(x)} && (* \text{ is conjugate-symmetric}) \\ &= \overline{f(y) * x} && (\text{definition of } f^*) \\ &= x * f(y) \end{aligned}$$

But this means that f is the adjoint of f^*, that is, $(f^*)^* = f$. ∎

When S is finite-dimensional, an argument can be given using matrices. Let $A = [f]$ relative to an orthonormal basis for S. By Corollary 12.3a we have

$$[f^*] = [f]^* = A^* \qquad \text{and hence} \qquad [(f^*)^*] = [f^*]^* = (A^*)^*$$

But $(A^*)^* = A$ (Sec. 4.3), so $(f^*)^*$ and f both have matrix representation A. It follows that $(f^*)^* = f$. (See Theorem 4.2.)

COROLLARY 12.4a

For all x and y in S, $x * f(y) = f^*(x) * y$.

We now have two identities involving the adjoint:

$$f(x) * y = x * f^*(y) \qquad \text{and} \qquad x * f(y) = f^*(x) * y$$

Taken together, they state that when a linear operator is moved from one factor to the other in an inner product, it is changed to its adjoint. (See Prob. 39, Sec. 4.3, for a preview.) The analogous statement for matrices is

$$(A\mathbf{X}) * \mathbf{Y} = \mathbf{X} * (A^*\mathbf{Y}) \qquad \text{and} \qquad \mathbf{X} * (A\mathbf{Y}) = (A^*\mathbf{X}) * \mathbf{Y}$$

where A is an $n \times n$ matrix and $\mathbf{X}$ and $\mathbf{Y}$ are vectors in n-space. (See the problems.) In other words, when a matrix A is moved "across the star," it is changed to A^*.

THEOREM 12.5

If f and g are linear operators on the unitary space S and k is a scalar, then (assuming that f^* and g^* exist)

a. $(f + g)^* = f^* + g^*$

b. $(kf)^* = \bar{k}f^*$

c. $(f \circ g)^* = g^* \circ f^*$

Proof

See the problems. (As in Theorem 12.4, an argument using matrices can be given if S is finite-dimensional, using Theorem 4.6 and Prob. 36, Sec. 5.3.)

Problems

1. Prove Theorem 12.3: $g = f^* \Leftrightarrow [g] = [f]^*$ (relative to an orthonormal basis).
2. In each of the following cases, find the formula for the adjoint of f.
   a. $f\colon \mathcal{R}^3 \to \mathcal{R}^3$ defined by $f(x,y,z) = (x + y - z, 2x, x - z)$
   b. $f\colon \mathcal{C}^2 \to \mathcal{C}^2$ defined by $f(x,y) = (x - iy, ix + y)$
   c. $f\colon \mathcal{R}^2 \to \mathcal{R}^2$ defined by $f(\mathbf{X}) = A\mathbf{X}$, where
   $$A = \begin{bmatrix} 1 & 4 \\ 0 & -1 \end{bmatrix}$$
3. Let $f\colon \mathcal{R}^2 \to \mathcal{R}^2$ be a rotation of the plane through the angle θ. Confirm that $f^* = f^{-1}$.
4. Define $f\colon \mathcal{R}^2 \to \mathcal{R}^2$ and $g\colon \mathcal{R}^2 \to \mathcal{R}^2$ by
   $$f(x,y) = (2x, -x + 5y) \qquad \text{and} \qquad g(x,y) = (2x - y, 5y)$$
   a. Confirm that $g = f^*$.
   b. The matrix representations of f and g relative to the basis $\{(1,0), (0,2)\}$ are
   $$\begin{bmatrix} 2 & 0 \\ -\tfrac{1}{2} & 5 \end{bmatrix} \qquad \text{and} \qquad \begin{bmatrix} 2 & -2 \\ 0 & 5 \end{bmatrix}$$
   respectively; note that the basis is orthogonal but the matrices are not transposes. Why doesn't this contradict Theorem 12.3?
   c. The matrix representations of f and g relative to the basis $\{(1,1), (-1,1)\}$ are
   $$\begin{bmatrix} 3 & 2 \\ 1 & 4 \end{bmatrix} \qquad \text{and} \qquad \begin{bmatrix} 3 & 1 \\ 2 & 4 \end{bmatrix}$$
   respectively; this time the basis is not orthonormal but the matrices *are* transposes. Why doesn't this contradict Theorem 12.3?
   d. In view of parts *b* and *c* it is unnecessary for the basis to be orthonormal to get $[f^*] = [f]^*$, while orthogonality is not sufficient to guarantee $[f^*] = [f]^*$. Can you guess a condition somewhere between orthogonal and orthonormal that is sufficient?
5. Let S be a unitary space with an orthogonal basis $\{v_1, \ldots, v_n\}$ whose elements are all of the same length. Let $f\colon S \to S$ be a linear operator

with matrix A relative to this basis and suppose that $g\colon S \to S$ is the linear operator whose matrix relative to this basis is A^*.

a. Prove that $f(v_i) * v_j = v_i * g(v_j)$ for all i and j.

b. Use part *a* to derive the identity $f(x) * y = x * g(y)$.

c. Why does it follow that, relative to the given basis, $[f^*] = [f]^*$?

6. Let $f\colon S \to S$ be a linear operator on the unitary space S. Explain why $(f^*)^*$ exists if f^* does.

7. Let $f\colon S \to S$ and $g\colon S \to S$ be linear operators on the unitary space S and suppose that k is a scalar. Assuming that f^* and g^* exist, prove the following.

a. $(f + g)^* = f^* + g^*$

b. $(kf)^* = \bar{k}f^*$

c. $(f \circ g)^* = g^* \circ f^*$

8. Assuming that S is finite-dimensional, derive the formulas in Prob. 7 by appealing to matrices.

9. Let A be an $n \times n$ matrix with entries in $\mathfrak{F}$ and suppose that $\mathbf{X}$ and $\mathbf{Y}$ are elements of $\mathfrak{F}^n$ (regarded as "column vectors," that is, $n \times 1$ matrices). Prove the following.

$$(A\mathbf{X}) * \mathbf{Y} = \mathbf{X} * (A^*\mathbf{Y}) \qquad \text{and} \qquad \mathbf{X} * (A\mathbf{Y}) = (A^*\mathbf{X}) * \mathbf{Y}$$

Hint: See Theorem 5.15.

10. Confirm Prob. 9 in each of the following cases.

a. $A = \begin{bmatrix} 1 & -1 & 2 \\ 2 & 0 & -3 \\ 0 & 1 & -4 \end{bmatrix}$

b. $A = \begin{bmatrix} 1 & 1+i \\ 0 & i \end{bmatrix}$

★ **11.** Suppose that $f\colon S \to S$ is a linear operator on the unitary space S and that U is a subspace of S "invariant" under f, that is, $x \in U \Rightarrow f(x) \in U$. Prove that $U^\perp$ is invariant under f^* (assuming that f^* exists).

★ **12.** Let $f\colon S \to S$ be a linear operator on the unitary space S.

a. Assuming that f^* exists, prove that $\ker f^* = (\operatorname{rng} f)^\perp$. *Hint:* Establish the inclusion $(\operatorname{rng} f)^\perp \subset \ker f^*$ by showing that if $z \in (\operatorname{rng} f)^\perp$, then $f^*(z)$ is orthogonal to S. The opposite inclusion is straightforward.

b. Assuming that S is finite-dimensional, conclude that f and f^* have the same rank (and hence the same nullity). *Hint:* Use Sylvester's law of nullity and the projection theorem.

★ **13.** Let $f\colon S \to S$ be an invertible linear operator on the unitary space S. Assuming that f and f^{-1} both have adjoints, prove that f^* is invertible and that its inverse is $(f^*)^{-1} = (f^{-1})^*$. *Hint:* Use Theorem 5.6, Sec. 5.2.

(Note that if S is finite-dimensional, this also follows from the matrix formula $(A^*)^{-1} = (A^{-1})^*$ in Prob. 59, Sec. 5.4.)

14. Let S and T be unitary spaces with inner products $*$ and $\circ$, and orthonormal bases $\alpha = \{u_1, \ldots, u_n\}$ and $\beta = \{v_1, \ldots, v_m\}$, respectively. Suppose that $f: S \to T$ is a linear map with matrix $A = [f]_{\alpha^\beta}$ and define the linear map $g: T \to S$ by $[g]_{\beta^\alpha} = A^*$.

a. Prove that $f(u_i) \circ v_j = u_i * g(v_j)$ for all i and j and then use this to establish that $f(x) \circ y = x * g(y)$ for all $x \in S$ and $y \in T$.

b. Prove that if $h: T \to S$ is any linear map satisfying the identity $f(x) \circ y = x * h(y)$, then $h = g$.

This establishes the existence and uniqueness of a linear map $f^*: T \to S$ such that $f(x) \circ y = x * f^*(y)$ for all $x \in S$ and $y \in T$. Naturally enough, f^* is called the "adjoint" of f. Thus the theory in this section (which applies to linear *operators* on a single unitary space) can be generalized to linear maps from one unitary space to another.

15. Define $f: \mathcal{R}^2 \to \mathcal{R}^3$ by $f(x,y) = (x + y, 2x, x - y)$.

a. Find the formula defining f^*.

b. Confirm that if $\mathbf{X} \in \mathcal{R}^2$ and $\mathbf{Y} \in \mathcal{R}^3$, then $f(\mathbf{X}) \cdot \mathbf{Y} = \mathbf{X} \cdot f^*(\mathbf{Y})$.

c. Confirm that if $A = [f]$, then $(A\mathbf{X}) \cdot \mathbf{Y} = \mathbf{X} \cdot (A^*\mathbf{Y})$ for all $\mathbf{X} \in \mathcal{R}^2$ and $\mathbf{Y} \in \mathcal{R}^3$.

12.2 Orthogonal and Unitary Operators

We have mentioned once or twice before that a linear operator on a euclidean space which preserves inner products is called "orthogonal." (See Prob. 53, Sec. 10.4, and Prob. 35, Sec. 11.3.) Now we are ready to discuss this concept in general.

UNITARY OPERATOR

Let S be a unitary space over $\mathcal{F}$ and suppose that $f: S \to S$ is a linear operator which preserves inner products, that is,

$$f(x) * f(y) = x * y \qquad \text{for all } x \text{ and } y \text{ in } S$$

Then f is called *unitary* (also *orthogonal* if the base field is $\mathcal{F} = \mathcal{R}$).

THEOREM 12.6

Let $f: S \to S$ be a unitary operator. Then

(1) f preserves unit vectors: $\|x\| = 1 \Rightarrow \|f(x)\| = 1$;

(2) f preserves orthogonality: $x * y = 0 \Rightarrow f(x) * f(y) = 0$;

(3) f preserves lengths: $\|f(x)\| = \|x\|$ for all $x \in S$;

(4) f preserves distances: $d[f(x), f(y)] = d(x,y)$ for all x and y in S;

(5) f is one-to-one: $f(x) = f(y) \Rightarrow x = y$.

Proof

We'll do Property (5), leaving the others for the problems:

$$\begin{aligned} f(x) = f(y) &\Rightarrow d[f(x), f(y)] = 0 \\ &\Rightarrow d(x,y) = 0 \qquad \text{[by (2)]} \\ &\Rightarrow x = y \qquad \text{(Why?)} \end{aligned}$$

Thus f is one-to-one. ∎

Note particularly that although a unitary operator is necessarily one-to-one, we do not assert that it is invertible, although of course it is if S is finite-dimensional. (Why?) In general, unitary operators need not be onto, as we shall see.

What Theorem 12.6 says is that a linear operator not only preserves the algebraic structure of its domain (like all linear maps), but also the geometric structure.[†] Because the "metric" properties of the space are left undisturbed by such a map, many writers call a unitary operator an "isometry."

EXAMPLE 12.2

Define $f: \mathcal{R}^2 \to \mathcal{R}^2$ by $f(x,y) = (x, -y)$. This sends each point of the plane into the image one would expect if the x axis were a mirror. (See Fig. 12.1.) It

[†] Preservation of the inner product is such a strong condition that we need not even assume the operator is linear. We can prove it! (See the problems.)

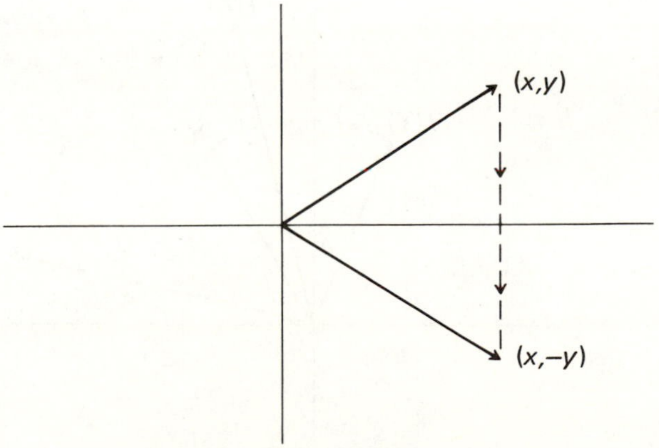

Fig. 12.1 Reflection in the x axis

is called a "reflection in the x axis." (See Prob. 35, Sec. 11.3, for a general definition of "reflection.") To show that it is unitary, let $\mathbf{X} = (x_1,x_2)$ and $\mathbf{Y} = (y_1,y_2)$; then

$$f(\mathbf{X}) \cdot f(\mathbf{Y}) = (x_1,-x_2) \cdot (y_1,-y_2) = x_1y_1 + x_2y_2 = \mathbf{X} \cdot \mathbf{Y}$$

Hence f preserves inner products.

EXAMPLE 12.3

Let $f\colon \mathcal{R}^2 \to \mathcal{R}^2$ be the rotation of the plane described in Example 3.8. Some pencil-pushing is required to show algebraically that $f(\mathbf{X}) \cdot f(\mathbf{Y}) = \mathbf{X} \cdot \mathbf{Y}$. Rather than go through this, observe that it is trivial if $\mathbf{X}$ or $\mathbf{Y}$ is zero, whereas if they are not zero, neither their lengths nor the angle θ between them is changed by the rotation. (See Fig. 12.2.) Hence we can write

$$\mathbf{X} \cdot \mathbf{Y} = \|\mathbf{X}\|\|\mathbf{Y}\| \cos\theta = \|f(\mathbf{X})\|\|f(\mathbf{Y})\| \cos\theta = f(\mathbf{X}) \cdot f(\mathbf{Y})$$

(because θ is the angle between $f(\mathbf{X})$ and $f(\mathbf{Y})$ as well as between $\mathbf{X}$ and $\mathbf{Y}$).

The fact that a unitary operator preserves unit vectors is trivial (as are the other properties listed in Theorem 12.6). What is more interesting is that *any* linear operator which preserves unit vectors is unitary; thus this property is both a necessary and sufficient condition. (It is this fact that is responsible for the term "*unitary*.") Some of the other properties listed in Theorem 12.6 are also equivalent to that of being unitary, as we now prove.

THEOREM 12.7

Let $f\colon S \to S$ be a linear operator on the unitary space S. Each of the following statements is equivalent to each of the others.

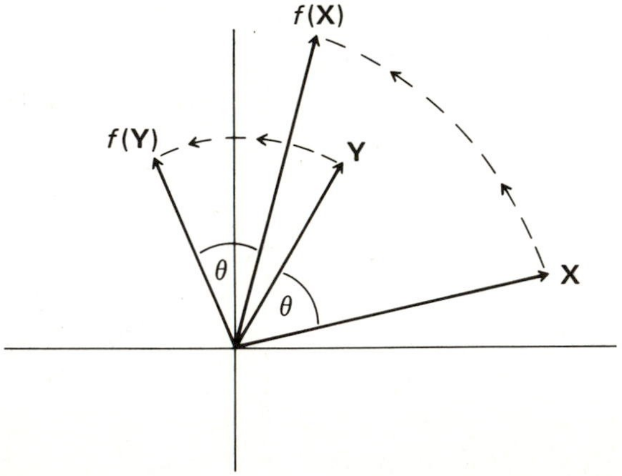

Fig. 12.2 Rotation preserves lengths and angles.

(1) f preserves inner products.

(2) f preserves unit vectors.

(3) f preserves lengths.

(4) f preserves distances.

Proof

The implication (1) $\Rightarrow$ (2) is done in Theorem 12.6. To prove (2) $\Rightarrow$ (3), suppose that f preserves unit vectors; the problem is to show that

$$\|f(x)\| = \|x\| \qquad \text{for all } x \in S$$

If $x = 0$, this is trivial, so assume that x is nonzero. Then $y = x/\|x\|$ is a unit vector and hence $\|f(y)\| = 1$. But then, letting $c = \|x\|$, we have

$$\begin{aligned} \|f(x)\| &= \|f(cy)\| \\ &= \|cf(y)\| \\ &= c\|f(y)\| \qquad \text{(Why?)} \\ &= c \\ &= \|x\| \end{aligned}$$

To show that (3) $\Rightarrow$ (4), suppose that f preserves lengths. Then for all x and y in S,

$$d[f(x), f(y)] = \|f(x) - f(y)\| = \|f(x - y)\| = \|x - y\| = d(x,y)$$

so f preserves distances.

The proof will be complete if we show that (4) $\Rightarrow$ (1), so assume that f preserves distances. To prove that

$$f(x) * f(y) = x * y$$

observe that for all x and y in S,

$$\begin{aligned} \|x - y\|^2 &= (x - y) * (x - y) \\ &= x * x - x * y - y * x + y * y \\ &= \|x\|^2 + \|y\|^2 - (x * y + y * x) \\ &= \|x\|^2 + \|y\|^2 - 2a \qquad (\text{where } x * y = a + bi) \end{aligned}$$

(This is the law of cosines if S is euclidean; see Prob. 39, Sec. 10.3.) Hence

$$2a = \|x\|^2 + \|y\|^2 - \|x - y\|^2$$

The same argument with $f(x)$ and $f(y)$ in place of x and y yields

$$\begin{aligned} 2A &= \|f(x)\|^2 + \|f(y)\|^2 - \|f(x) - f(y)\|^2 \qquad (\text{where } f(x) * f(y) = A + Bi) \\ &= \|f(x)\|^2 + \|f(y)\|^2 - \|f(x - y)\|^2 \end{aligned}$$

Since f preserves distances, it also preserves lengths (obvious?), so this result reduces to

$$2A = \|x\|^2 + \|y\|^2 - \|x - y\|^2$$

Therefore $A = a$. If $\mathfrak{F} = \mathfrak{R}$, we are done (why?), but if $\mathfrak{F} = \mathfrak{C}$, we still have to show that $B = b$. To do this, note that

$$x * (iy) = \bar{i}(x * y) = -i(a + bi) = b - ai$$

and

$$f(x) * f(iy) = f(x) * [if(y)] = \bar{i}[f(x) * f(y)] = -i(A + Bi) = B - Ai$$

Hence we need only replace y by iy in the above argument to conclude that $B = b$. ▮

This theorem enables us to say that a linear operator is unitary if and only if it has any one of the four properties listed. (Thus Examples 12.2 and 12.3 are trivial, since it is geometrically apparent that a reflection or a rotation preserves lengths.) Note, however, that preservation of orthogonality is *not* listed, for although a unitary operator obviously has this property, the property itself is not sufficient to guarantee that a linear operator is unitary. For example, the linear operator $f(\mathbf{X}) = 2\mathbf{X}$ on $\mathfrak{R}^2$ preserves orthogonality, but it is not unitary. (Confirm this!)

For this reason it is unfortunate that unitary operators on euclidean spaces are called "orthogonal"; the term suggests a characterization that is not true. Similarly, an angle-preserving linear operator is not necessarily unitary, although all unitary operators (on a euclidean space) preserve angles. (Why?) Such an operator is called *conformal*. (See Prob. 52, Sec. 10.4.)

The space S in Theorem 12.7 need not be finite-dimensional, as you can see by examining the proof. However, if we impose this condition, then we can state another characterization of unitary operators as follows.

THEOREM 12.8

Let $f: S \to S$ be a linear operator on the finite-dimensional unitary space S. Then f is unitary if and only if $f^* = f^{-1}$.

Proof

Suppose that $f^* = f^{-1}$. Then $f^* \circ f = i$ (the identity map on S) and for all x and y in S, we have

$$f(x) * f(y) = x * f^*[f(y)] = x * i(y) = x * y$$

(See Sec. 12.1.) Hence f is unitary.† Conversely, if f is unitary, then it is invertible (why?), that is, f^{-1} exists. Hence we may write

$$f(x) * y = f(x) * f[f^{-1}(y)] = x * f^{-1}(y) \qquad (f \text{ preserves inner products})$$

for all x and y in S. By the definition of the adjoint of a linear operator, it follows that $f^{-1} = f^*$. ▮

† Note that this part of the argument does not invoke the finite dimension of S.

EXAMPLE 12.4

The reflection $f(x,y) = (x, -y)$ in Example 12.2 is its own inverse, that is, $f^{-1} = f$. (This is obvious geometrically; see Prob. 35, Sec. 11.3, for the same statement about reflections in general.) The reflection is also its own adjoint, that is, $f^* = f$. (Look at its matrix!) Hence $f^* = f^{-1}$, as Theorem 12.8 predicts.

EXAMPLE 12.5

The rotation of the plane through an angle θ has matrix

$$[f] = \begin{bmatrix} \cos\theta & -\sin\theta \\ \sin\theta & \cos\theta \end{bmatrix}$$

Since both the inverse and transpose of this are

$$\begin{bmatrix} \cos\theta & \sin\theta \\ -\sin\theta & \cos\theta \end{bmatrix}$$

we have $[f]^* = [f]^{-1}$, that is, $[f^*] = [f^{-1}]$, from which $f^* = f^{-1}$.

Theorem 12.8 should serve as adequate motivation for our next definition.

UNITARY MATRIX

A square matrix A with entries in $\mathfrak{F}$ is called *unitary* if $A^* = A^{-1}$. When $\mathfrak{F} = \mathfrak{R}$, in which case the defining condition reduces to $A^T = A^{-1}$, we also call A *orthogonal*. (In other words, an orthogonal matrix is a unitary matrix with real entries.†)

THEOREM 12.9

A linear operator on a finite-dimensional unitary space is unitary if and only if its matrix (relative to an orthonormal basis) is unitary.

Proof

Let $f\colon S \to S$ be the operator and let $A = [f]$ be its matrix relative to an orthonormal basis. Then

$$\begin{aligned} f \text{ is unitary} &\Leftrightarrow f^* = f^{-1} && \text{(Theorem 12.8)} \\ &\Leftrightarrow [f^*] = [f^{-1}] && \text{(Theorems 4.1 and 4.2)} \\ &\Leftrightarrow [f]^* = [f]^{-1} && \text{(Why?)} \\ &\Leftrightarrow A^* = A^{-1} && (A = [f]) \\ &\Leftrightarrow A \text{ is unitary} \ \blacksquare \end{aligned}$$

† See Probs. 27 and 28, Sec. 8.4, for a preview of this definition.

QUESTION: Where in this argument do we use the fact that the basis is orthonormal?

THEOREM 12.10

The $n \times n$ matrix A is unitary if and only if its rows [columns] are orthonormal in $\mathfrak{F}^n$.

Proof

The rows of A are orthonormal

$$
\begin{aligned}
&\Leftrightarrow \text{(ith row of A)} * \text{(jth row of A)} = \delta_{ij} \\
&\Leftrightarrow \text{(ith row of A)} \cdot \text{(jth row of $\bar{A}$)} = \delta_{ij} \qquad \text{(Why?)} \\
&\Leftrightarrow \text{(ith row of A)} \cdot \text{(jth column of $\bar{A}^T$)} = \delta_{ij} \\
&\Leftrightarrow A\bar{A}^T = I \qquad \text{(definition of matrix multiplication)} \\
&\Leftrightarrow AA^* = I \qquad \text{(Sec. 4.3)} \\
&\Leftrightarrow A^* = A^{-1} \\
&\Leftrightarrow A \text{ is unitary}
\end{aligned}
$$

This does it as far as the rows are concerned. To prove it for columns, one may imitate the argument for rows, but it is more interesting to write

$$
\begin{aligned}
A \text{ is unitary} &\Leftrightarrow A^T \text{ is unitary} && \text{(Why?)} \\
&\Leftrightarrow \text{the rows of } A^T \text{ are orthonormal} && \text{(just proved)} \\
&\Leftrightarrow \text{the columns of } A \text{ are orthonormal} \ \blacksquare
\end{aligned}
$$

EXAMPLE 12.6

The matrix of the reflection $f(x,y) = (x,-y)$ is

$$A = \begin{bmatrix} 1 & 0 \\ 0 & -1 \end{bmatrix}$$

(relative to the standard basis). The rows of A are $(1,0)$ and $(0,-1)$, which are perpendicular unit vectors. Hence A is unitary.

EXAMPLE 12.7

The matrix of the rotation in Example 12.5 has columns $(\cos\theta, \sin\theta)$ and $(-\sin\theta, \cos\theta)$. These are orthonormal, as you can easily check.

EXAMPLE 12.8

Relative to the standard basis for $\mathcal{R}^3$, the matrix

$$A = \begin{bmatrix} 2/3 & 1/3 & -2/3 \\ 1/3 & 2/3 & 2/3 \\ 2/3 & -2/3 & 1/3 \end{bmatrix} = \frac{1}{3}\begin{bmatrix} 2 & 1 & -2 \\ 1 & 2 & 2 \\ 2 & -2 & 1 \end{bmatrix}$$

represents the linear operator

$$f(x,y,z) = \tfrac{1}{3}(2x + y - 2z,\ x + 2y + 2z,\ 2x - 2y + z)$$

A direct attack on the orthogonality of f is tedious, for we have to show that

$$f(\mathbf{X}) \cdot f(\mathbf{Y}) = \mathbf{X} \cdot \mathbf{Y} \qquad \text{for all } \mathbf{X} \text{ and } \mathbf{Y} \text{ in } \mathcal{R}^3$$

Instead of bothering with this, we conclude that f is orthogonal because the rows [columns] of A are orthonormal (as can be seen by inspection).

EXAMPLE 12.9

The matrix

$$A = \frac{1}{\sqrt{3}} \begin{bmatrix} 1 & 1-i \\ 1+i & -1 \end{bmatrix}$$

is unitary, as you can check by finding A^* and confirming that $AA^* = I$. We may also test this by looking at its rows:

$$\begin{aligned} \|\text{row } 1\|^2 &= (\text{row } 1) * (\text{row } 1) = \tfrac{1}{3}[(1)(1) + (1-i)(1+i)] = 1 \\ \|\text{row } 2\|^2 &= (\text{row } 2) * (\text{row } 2) = \tfrac{1}{3}[(1+i)(1-i) + (-1)(-1)] = 1 \\ & \quad\; (\text{row } 1) * (\text{row } 2) = \tfrac{1}{3}[(1)(1-i) + (1-i)(-1)] = 0 \end{aligned}$$

EXAMPLE 12.10

If

$$A = \begin{bmatrix} \sqrt{2} & i \\ -i & \sqrt{2} \end{bmatrix}$$

then $A^T = A^{-1}$, as you can check. But this is a red herring! If you conclude that A is orthogonal, you must also admit that it is unitary (by definition). Yet the rows of A are not orthonormal.† Of course the explanation is that the term ''orthogonal'' is reserved for matrices with real entries.

We end this section by observing that when S is not finite-dimensional, Theorem 12.8 may not make sense, since f^* doesn't always exist, nor is f necessarily invertible when it is unitary. When f^* does exist, however, we may develop a substitute for Theorem 12.8:

$$f \text{ is unitary if and only if } f^* \circ f = i$$

The ''if'' part of this proposition is already done (see the proof of Theorem 12.8). But the converse is surprisingly subtle; we'll defer it to Sec. 12.5.

† The rows *are* orthonormal if the dot product is used, but this is total confusion. Recall that the definitions of length and distance in complex n-space require the standard inner product; the dot product is not positive-definite.

Problems

16. Which of the following linear operators are unitary?

a. $f: \mathcal{R}^2 \to \mathcal{R}^2$ defined by $f(x,y) = (y,x)$. Describe f geometrically.

b. $f: \mathcal{C}^2 \to \mathcal{C}^2$ defined by $f(x,y) = (-iy,ix)$

c. $f: \mathcal{R}^3 \to \mathcal{R}^3$ defined by $f(\mathbf{X}) = A\mathbf{X}$, where

$$A = \begin{bmatrix} 2 & 1 & -2 \\ 1 & 2 & 2 \\ 2 & -2 & 1 \end{bmatrix}$$

d. $\phi: S \to S$ defined by $\phi(f) = -f$, where S is the euclidean space of real-valued functions continuous on $[-\pi,\pi]$, with inner product

$$f * g = \int_{-\pi}^{\pi} f(x)g(x)\,dx$$

e. $f: S \to S$ defined by $f(x) = x_1$, where x_1 is the orthogonal projection of x on a (proper) subspace of the finite-dimensional unitary space S.

17. Show that a unitary operator preserves unit vectors, orthogonality, lengths, and distances. (This completes the proof of Theorem 12.6.)

18. Explain why an orthogonal linear operator preserves angles. Then give an example of an angle-preserving linear operator which is not orthogonal.

19. We defined a unitary operator to be a linear operator on a unitary space which preserves inner products. Prove that "linear" is redundant in this definition; that is, if $f: S \to S$ preserves inner products, then f *must* be linear. *Hint:* Establish the identities $d[f(x + y), f(x) + f(y)] = 0$ and $d[f(kx), kf(x)] = 0$.

20. In the proof of the implication (4) $\Rightarrow$ (1) in Theorem 12.7 we used the fact that (4) $\Rightarrow$ (3). Prove this.

21. In Theorem 12.8 suppose that we drop the requirement that S is finite-dimensional and merely assume that f^* exists. Why doesn't the proof work anymore?

22. Answer the question following the proof of Theorem 12.9.

23. The unitary map $f: \mathcal{R}^2 \to \mathcal{R}^2$ defined by $f(x,y) = (x,-y)$ has matrices

$$A = \begin{bmatrix} 0 & -1 \\ -1 & 0 \end{bmatrix} \quad \text{and} \quad B = \begin{bmatrix} 1 & -2 \\ 0 & -1 \end{bmatrix}$$

relative to the bases $\{(1,1), (-1,1)\}$ and $\{(1,0), (-1,1)\}$, respectively; note that A is unitary and B is not. How can this be?

24. Prove Theorem 12.10 for columns by imitating the argument for rows given in the text.

25. In the proof of Theorem 12.10 for columns we started by saying that the matrix A is unitary if and only if A^T is unitary. Show that this is correct. *Hint:* See Prob. 27*e*, Sec. 4.3.

26. Prove that if A is a unitary matrix, so is A^*.

27. Show that a unitary operator on a finite-dimensional space has a unitary adjoint:

a. By using Prob. 26;

b. By an argument not involving matrices.

In Sec. 12.4 we'll give an example showing that this statement is not necessarily true in an infinite-dimensional space.

28. Show that the composition of two unitary operators is unitary. Why does it follow that the product of two unitary matrices is unitary?

29. Explain why the set of unitary operators on a finite-dimensional space is a group relative to composition. Is this group abelian?

30. It follows from Prob. 29 that the set of $n \times n$ unitary matrices is a multiplicative group. Show that the set of $n \times n$ orthogonal matrices is a subgroup.

31. Prove that a linear operator on a finite-dimensional unitary space is unitary if and only if it sends every orthonormal basis into an orthonormal basis. *Hint:* For the ''if'' part, show that f preserves lengths by using Theorem 10.9.

32. Prove that the eigenvalues of a unitary operator have absolute value 1.

33. Why does it follow from Prob. 32 that the determinant of a unitary matrix has absolute value 1? What are the possible values of the determinant of an orthogonal matrix? (See Prob. 27, Sec. 8.4, where these matters were brought up before eigenvalues had been mentioned.)

34. Confirm that the eigenvalues and determinant of the unitary matrix

$$\begin{bmatrix} 0 & i \\ i & 0 \end{bmatrix}$$

have absolute value 1.

35. Give an example showing that $|\det A| = 1$ does not imply that A is unitary.

36. Let $f: S \to S$ be a unitary operator on the finite-dimensional space S and suppose that U is a subspace of S invariant under f. [That is, $x \in U \Rightarrow f(x) \in U$; see Prob. 11, Sec. 12.1.]

a. Explain why f maps U onto U.

b. Prove that $U^\perp$ is also invariant under f (and hence that f maps $U^\perp$ onto $U^\perp$).

†12.3 Orthogonal Transformations of the Plane

Let $f: \mathcal{R}^2 \to \mathcal{R}^2$ be an orthogonal linear operator on $\mathcal{R}^2$. What we propose to do in this section is to classify f by looking at the various forms its matrix can have.

Since $\mathbf{E}_1 = (1,0)$ is a unit vector, so is $f(\mathbf{E}_1)$. Hence $f(\mathbf{E}_1)$ is a point of the unit circle centered at the origin, that is,

$$f(\mathbf{E}_1) = (\cos\theta, \sin\theta) \qquad \text{for some } \theta,\ 0 \leq \theta < 2\pi$$

(Crank up your high school trigonometry!) Similarly, $f(\mathbf{E}_2)$ is on the unit circle. Moreover, it is perpendicular to $f(\mathbf{E}_1)$ because $\mathbf{E}_1$ and $\mathbf{E}_2$ are perpendicular and f preserves orthogonality. Hence $f(\mathbf{E}_2)$ is either 90° counterclockwise or 90° clockwise from $f(\mathbf{E}_1)$, which means that

$$f(\mathbf{E}_2) = \left[\cos\left(\theta + \frac{\pi}{2}\right), \sin\left(\theta + \frac{\pi}{2}\right)\right] = (-\sin\theta, \cos\theta)$$

or

$$f(\mathbf{E}_2) = \left[\cos\left(\theta - \frac{\pi}{2}\right), \sin\left(\theta - \frac{\pi}{2}\right)\right] = (\sin\theta, -\cos\theta)$$

(See Fig. 12.3.) The matrix $A = [f]$ (relative to the standard basis for $\mathcal{R}^2$) has $f(\mathbf{E}_1)$ and $f(\mathbf{E}_2)$ for columns; hence

$$A = \begin{bmatrix} \cos\theta & -\sin\theta \\ \sin\theta & \cos\theta \end{bmatrix} \qquad \text{(Case 1)}$$

or

$$A = \begin{bmatrix} \cos\theta & \sin\theta \\ \sin\theta & -\cos\theta \end{bmatrix} \qquad \text{(Case 2)}$$

† This section can be omitted without loss of continuity.

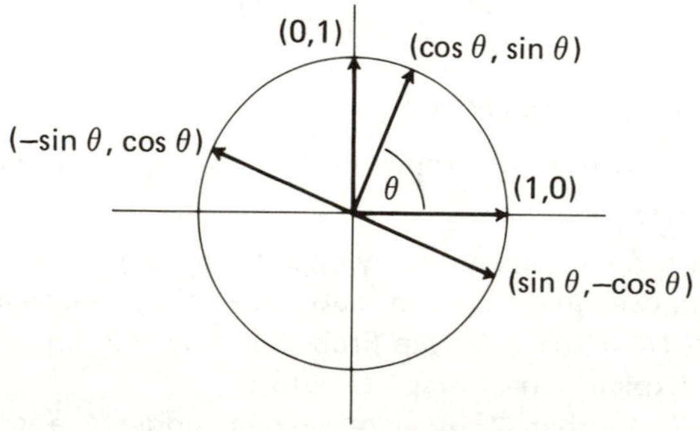

Fig. 12.3 Images of $\mathbf{E}_1$ and $\mathbf{E}_2$ under f

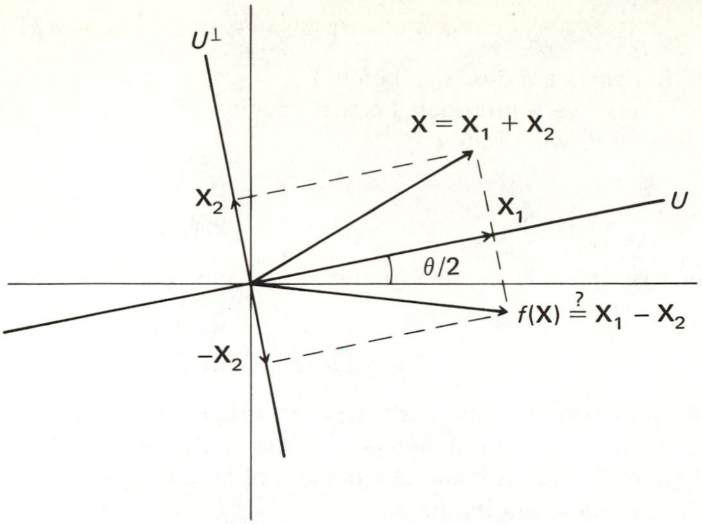

Fig. 12.4 Is *f* a reflection in *U*?

In Case 1 we are dealing with a rotation of the plane through the angle θ. (See Prob. 5, Sec. 4.1.)

In Case 2 we claim that f is a reflection of the plane in the line U through the origin with inclination $\theta/2$. To see why, let $\mathbf{X}$ be any vector in $\mathcal{R}^2$ and write $\mathbf{X} = \mathbf{X}_1 + \mathbf{X}_2$, where $\mathbf{X}_1$ and $\mathbf{X}_2$ are the orthogonal projections of $\mathbf{X}$ on U and $U^\perp$, respectively. The problem is to show that $f(\mathbf{X}) = \mathbf{X}_1 - \mathbf{X}_2$. (See Prob. 35, Sec. 11.3, and look at Fig. 12.4.) Since $f(\mathbf{X}) = f(\mathbf{X}_1 + \mathbf{X}_2) = f(\mathbf{X}_1) + f(\mathbf{X}_2)$, it is enough to prove that $f(\mathbf{X}_1) = \mathbf{X}_1$ and $f(\mathbf{X}_2) = -\mathbf{X}_2$.

To show that $f(\mathbf{X}_1) = \mathbf{X}_1$, observe that $\mathbf{X}_1 = c_1[\cos(\theta/2), \sin(\theta/2)]$ for some scalar c_1. (Why?) The formula for f is

$$f(x,y) = (x\cos\theta + y\sin\theta,\ x\sin\theta - y\cos\theta)$$

so

$$\begin{aligned} f(\mathbf{X}_1) &= c_1 f\left(\cos\frac{\theta}{2}, \sin\frac{\theta}{2}\right) \\ &= c_1\left(\cos\theta\cos\frac{\theta}{2} + \sin\theta\sin\frac{\theta}{2},\ \sin\theta\cos\frac{\theta}{2} - \cos\theta\sin\frac{\theta}{2}\right) \\ &= c_1\left[\cos\left(\theta - \frac{\theta}{2}\right), \sin\left(\theta - \frac{\theta}{2}\right)\right] = c_1\left(\cos\frac{\theta}{2}, \sin\frac{\theta}{2}\right) = \mathbf{X}_1 \end{aligned}$$

Similarly,

$$\mathbf{X}_2 = c_2\left[\cos\left(\frac{\theta}{2} + \frac{\pi}{2}\right), \sin\left(\frac{\theta}{2} + \frac{\pi}{2}\right)\right] = c_2\left(-\sin\frac{\theta}{2}, \cos\frac{\theta}{2}\right)$$

for some scalar c_2. (Why?) Hence

$$f(\mathbf{X}_2) = c_2 f\left(-\sin\frac{\theta}{2}, \cos\frac{\theta}{2}\right) = -\mathbf{X}_2$$

(The details are much as before.)

Thus every orthogonal transformation of the plane is either a rotation through θ, with matrix

$$\begin{bmatrix} \cos\theta & -\sin\theta \\ \sin\theta & \cos\theta \end{bmatrix}$$

or a reflection in the line through the origin with inclination $\theta/2$, with matrix

$$\begin{bmatrix} \cos\theta & \sin\theta \\ \sin\theta & -\cos\theta \end{bmatrix}$$

A quick way to distinguish between these cases is by looking at the determinants, which are 1 and -1, respectively. (See Prob. 33, Sec. 12.2, or Prob. 27, Sec. 8.4; the determinant of an orthogonal matrix is always ± 1.) For example, the matrix

$$A = \begin{bmatrix} \sqrt{3}/2 & 1/2 \\ 1/2 & -\sqrt{3}/2 \end{bmatrix}$$

is orthogonal. (Confirm this!) Since its determinant is -1, the corresponding map f: $\mathcal{R}^2 \to \mathcal{R}^2$ is a reflection in the line through the origin with inclination 15°. (Why?)

Problems

37. Let

$$A = \begin{bmatrix} -1 & 0 \\ 0 & 1 \end{bmatrix}$$

a. Confirm that A is orthogonal.

b. Write A in one of the forms

$$\begin{bmatrix} \cos\theta & -\sin\theta \\ \sin\theta & \cos\theta \end{bmatrix} \quad \text{or} \quad \begin{bmatrix} \cos\theta & \sin\theta \\ \sin\theta & -\cos\theta \end{bmatrix}$$

and name the angle θ ($0 \le \theta < 2\pi$).

c. The result in part *b* classifies A as the matrix of a reflection in a line U through the origin. What is U?

d. Confirm part *c* by finding the formula for $f(x,y)$, where f: $\mathcal{R}^2 \to \mathcal{R}^2$ is the reflection whose matrix is A.

38. Give a geometric description of the linear operator f: $\mathcal{R}^2 \to \mathcal{R}^2$ whose matrix (relative to the standard basis) is

$$A = \begin{bmatrix} 0 & 1 \\ 1 & 0 \end{bmatrix}$$

Then find the formula for $f(x,y)$ and confirm that your description is correct.

39. Suppose that $f: \mathbb{R}^2 \to \mathbb{R}^2$ is orthogonal and let $A = [f]$. If $\det A = 1$, we know that f is a rotation, so assume that $\det A = -1$.

a. If

$$B = \begin{bmatrix} 0 & 1 \\ 1 & 0 \end{bmatrix} \quad \text{and} \quad C = BA$$

explain why C represents a rotation, why $A = BC$, and why this shows that f is a rotation followed by a reflection in the line $y = x$.

b. Illustrate part *a* in the case $f(\mathbf{X}) = A\mathbf{X}$, where

$$A = \frac{1}{2}\begin{bmatrix} 1 & \sqrt{3} \\ \sqrt{3} & -1 \end{bmatrix}$$

by showing that f is orthogonal and by naming a rotation h such that $f = g \circ h$, where g is a reflection in the line $y = x$.

This problem shows that every orthogonal transformation of the plane is either a rotation or a rotation followed by a reflection.

40. Let $f: \mathbb{R}^2 \to \mathbb{R}^2$ be a rotation of the plane through the angle θ, with matrix

$$A = \begin{bmatrix} \cos\theta & -\sin\theta \\ \sin\theta & \cos\theta \end{bmatrix}$$

a. Confirm that

$$A = \begin{bmatrix} \cos\theta & \sin\theta \\ \sin\theta & -\cos\theta \end{bmatrix}\begin{bmatrix} 1 & 0 \\ 0 & -1 \end{bmatrix}$$

noting that each factor of the product represents a reflection, the first in a line through the origin with inclination $\theta/2$, the second in the x axis.

b. Why does this result show that f is a composition of two reflections?

It follows that every orthogonal transformation of the plane is either a reflection or a composition of two reflections. (This generalizes to n-space, every orthogonal operator on $\mathbb{R}^n$ being a composition of at most n reflections.)

41. Let $f: \mathbb{R}^3 \to \mathbb{R}^3$ be the orthogonal map with matrix

$$A = \frac{1}{3}\begin{bmatrix} 2 & 1 & -2 \\ 1 & 2 & 2 \\ 2 & -2 & 1 \end{bmatrix}$$

(See Example 12.8.)

a. What points are left fixed by f? In other words, for what values of $\mathbf{X}$ is $f(\mathbf{X}) = \mathbf{X}$? Show that these points constitute a line in $\mathcal{R}^3$ through the origin. (Note that this is the eigenspace of f associated with the eigenvalue 1; see Prob. 14, Sec. 9.2.)

b. Show that $\det A = 1$. (It can be shown that this implies that f is a rotation; the line in part *a* is the axis of rotation.)

42. The orthogonal matrix

$$A = \frac{1}{3}\begin{bmatrix} 1 & 2 & 2 \\ 2 & 1 & -2 \\ 2 & -2 & 1 \end{bmatrix}$$

represents a reflection in the plane $x - y - z = 0$ in $\mathcal{R}^3$. (See Prob. 36*c*, Sec. 11.3.) Show that $\det A = -1$.

†12.4 An Example in Hilbert Space

In Sec. 12.1 we proved that every linear operator on a finite-dimensional unitary space has an adjoint, while in Sec. 12.2 we observed that a unitary operator is necessarily one-to-one, hence invertible when the space is finite-dimensional. Both these statements leave infinite-dimensional spaces out of account (although the arguments that a unitary map is one-to-one and that an adjoint is unique when it exists do not depend on dimension). In this section we give an example of what can happen in an infinite-dimensional space.

Let $x_1, x_2, x_3, \ldots$ be a sequence of real numbers with the property that $\sum_{i=1}^{\infty} x_i^2$ converges, and label it $\mathbf{X} = (x_1, x_2, x_3, \ldots)$. The set S of all such sequences may be regarded as a generalization of n-space by defining addition and scalar multiplication ''componentwise'' (as in $\mathcal{R}^n$):

$$\mathbf{X} + \mathbf{Y} = (x_1 + y_1, x_2 + y_2, \ldots) \qquad \text{and} \qquad k\mathbf{X} = (kx_1, kx_2, \ldots)$$

Of course it is necessary to show that these definitions make sense. Addition, for example, yields an element of S only if

$$\sum_{i=1}^{\infty}(x_i + y_i)^2 \text{ converges when } \sum_{i=1}^{\infty}x_i^2 \text{ and } \sum_{i=1}^{\infty}y_i^2 \text{ do}$$

Recall from calculus that this is a question of the convergence of the ''partial sums''

$$S_n = \sum_{i=1}^{n}(x_i + y_i)^2 \qquad n = 1, 2, 3, \ldots$$

†The purpose of this section is to provide counterexamples to the generalization of certain theorems to infinite-dimensional spaces. It may be omitted, but you probably ought to read enough of it to get the point (if not the precise details).

For each n, define

$$\mathbf{X}_n = (x_1, \ldots, x_n) \qquad \text{and} \qquad \mathbf{Y}_n = (y_1, \ldots, y_n)$$

Then

$$\begin{aligned} S_n &= \sum_{i=1}^{n} x_i^2 + 2\sum_{i=1}^{n} x_i y_i + \sum_{i=1}^{n} y_i^2 \\ &= \|\mathbf{X}_n\|^2 + 2(\mathbf{X}_n \cdot \mathbf{Y}_n) + \|\mathbf{Y}_n\|^2 \\ &\leq \|\mathbf{X}_n\|^2 + 2|\mathbf{X}_n \cdot \mathbf{Y}_n| + \|\mathbf{Y}_n\|^2 \\ &\leq \|\mathbf{X}_n\|^2 + 2\|\mathbf{X}_n\|\,\|\mathbf{Y}_n\| + \|\mathbf{Y}_n\|^2 \qquad \text{(Cauchy-Schwarz inequality)} \\ &= (\|\mathbf{X}_n\| + \|\mathbf{Y}_n\|)^2 \end{aligned}$$

Since $\Sigma_{i=1}^{\infty} x_i^2$ and $\Sigma_{i=1}^{\infty} y_i^2$ converge, we know that

$$\|\mathbf{X}_n\|^2 = \sum_{i=1}^{n} x_i^2 \qquad \text{and} \qquad \|\mathbf{Y}_n\|^2 = \sum_{i=1}^{n} y_i^2$$

have limits as $n \to \infty$; hence so does $(\|\mathbf{X}_n\| + \|\mathbf{Y}_n\|)^2$. This means that it is bounded; the inequalities

$$0 \leq S_n \leq (\|\mathbf{X}_n\| + \|\mathbf{Y}_n\|)^2 \qquad n = 1, 2, 3, \ldots$$

therefore imply that the sequence $S_1, S_2, S_3, \ldots$ is bounded. The sequence is also monotonic ($S_1 \leq S_2 \leq S_3 \leq \cdots$), so it converges.

Since convergence of $\Sigma_{i=1}^{\infty} x_i^2$ implies convergence of $\Sigma_{i=1}^{\infty} (kx_i)^2$, the definition of scalar multiplication also makes sense. It is tedious to check the other properties of a vector space, so we won't bore you further. But note that

$$\mathbf{E}_1 = (1,0,0, \ldots), \mathbf{E}_2 = (0,1,0, \ldots), \ldots$$

are linearly independent elements of S; hence we are dealing with an infinite-dimensional space.†

To make S a euclidean space, we define the dot product as in $\mathcal{R}^n$, namely

$$\mathbf{X} \cdot \mathbf{Y} = x_1 y_1 + x_2 y_2 + x_3 y_3 + \cdots$$

It is not hard to show that $\Sigma_{i=1}^{\infty} x_i y_i$ converges absolutely and that the dot product has the usual positive-definite, symmetric, and bilinear properties. Hence S is euclidean; it is a concrete example of a class of unitary spaces

† In other words $\{\mathbf{E}_1, \ldots, \mathbf{E}_n\}$ is linearly independent for each n. Thus S has linearly independent subsets of arbitrarily large size and cannot be finite-dimensional.

known as "Hilbert spaces." They are of enormous importance in modern mathematics.[†]

Now (to get to the point of this section), define the map $f: S \to S$ by

$$f(x_1,x_2,x_3, \ldots) = (0,x_1,x_2, \ldots)$$

Then f is linear; since it preserves dot products, it is also orthogonal. (See the problems.) Moreover, f^* exists and is defined by

$$f^*(x_1,x_2,x_3, \ldots) = (x_2,x_3, \ldots)$$

(Confirm this!) *However, f is not invertible*. The vector $(1,0,0, \ldots)$ is not the image of any element of S; that is, f is not onto.

Thus we have a counterexample to the proposal that every unitary operator is invertible. This same map serves as a counterexample to other proposals (see the problems). It should make you cautious about generalizing properties of maps on finite-dimensional spaces.

Problems

(In each of the following problems, S is the Hilbert space described in this section.)

43. Convince yourself that S is a vector space over R.

44. Explain why $\{\mathbf{E}_1, \ldots, \mathbf{E}_n\}$ is linearly independent for each n.

45. Show that the dot product on S is positive-definite, symmetric, and bilinear.

46. Define $f: S \to S$ by $f(x_1,x_2,x_3, \ldots) = (0,x_1,x_2, \ldots)$.

a. Show that f is linear.

b. Prove that f is orthogonal by confirming that it preserves dot products.

c. Use Theorem 12.7 to prove that f is orthogonal by confirming that it preserves lengths.

d. Explain why the adjoint of f is defined by $f^*(x_1,x_2,x_3, \ldots) = (x_2,x_3, \ldots)$.

e. Confirm that $f^* \circ f = i$, while $f \circ f^* \neq i$.

f. Having confirmed that $f^* \circ f = i$, how could you predict (without

[†] David Hilbert (1862–1943) is celebrated for the problems he set before the International Congress of Mathematicians in 1900. These were problems the solution of which in his view would have much to do with the future direction of mathematics. Some are still unsolved today, but the work done on them, as Hilbert foresaw, has been of fundamental importance. See Irving Kaplansky's article on "Mathematics" in the 1971 *Encyclopaedia Britannica Yearbook*. The tenth Hilbert problem was solved in 1970, the seventh (solved in 1934) was significantly generalized. Among other things, we now know (for the first time) what kind of a number $\pi + \log 2$ is; some of the great problems are still constructed of such deceptively simple elements!

checking it out) that $f \circ f^* \neq i$? *Hint:* If $f \circ f^* = f^* \circ f = i$, what would Theorem 5.6 say about f?

Look up the remark following Prob. 51, Sec. 5.4, to the effect that a linear operator on a finite-dimensional space need only have a "one-sided" inverse to be invertible. In other words, if f is given and g can be found satisfying either $f \circ g = i$ or $g \circ f = i$, then f is invertible and $f^{-1} = g$. This problem shows that a one-sided inverse is not enough in general. (Also see Prob. 16, Sec. 5.2.)

g. f is unitary, but $f^* \neq f^{-1}$. Why doesn't this contradict Theorem 12.8?

h. Show that f^* is not orthogonal.

A unitary operator on a finite-dimensional space has a unitary adjoint (Prob. 27, Sec. 12.2). This shows that the corresponding statement for infinite-dimensional spaces is false.

12.5 Another Characterization of Unitary Operators

At the end of Sec. 12.2 we observed that the statement

$$f^* = f^{-1} \quad \Leftrightarrow \quad f \text{ is unitary}$$

does not apply to linear operators on infinite-dimensional unitary spaces. This is because f^* may not exist (although we have not shown you an example of this), nor is f always invertible when it is unitary (an example of this is given in the last section). We proposed as a substitute (in case f^* exists) the statement

$$f^* \circ f = i \quad \Leftrightarrow \quad f \text{ is unitary}$$

The implication

$$f^* \circ f = i \quad \Rightarrow \quad f \text{ is unitary}$$

is proved in the "if" part of Theorem 12.8; we now address ourselves to the much less obvious converse.

Suppose that $f\colon S \to S$ is unitary and assume that f^* exists. To show that $f^* \circ f = i$, let $g = f^* \circ f$ and observe that for all $x \in S$ we have

$$\begin{aligned} g(x) * x &= f^*[f(x)] * x \\ &= f(x) * f(x) && \text{(Sec. 12.1)} \\ &= x * x && \text{(f preserves inner products)} \\ &= i(x) * x \end{aligned}$$

It is tempting to conclude from the identity

$$g(x) * x = i(x) * x$$

that $g = i$. Or, since this identity is equivalent to

$$(g - i)(x) * x = 0$$

it is easy to jump to the conclusion that $g - i = 0$. Recall, for example, the proof of Theorem 12.1, in which the identity

$$x * (g - h)(y) = 0$$

led to $g - h = 0$. It is true that if $\phi: S \to S$ is a linear operator satisfying

$$\phi(x) * y = 0$$

for all x and y, then $\phi = 0$. The reason that this works is that there are two independent variables. Fixing $x \in S$, we still have one left in the statement

$$\phi(x) * y = 0 \qquad \text{for all } y$$

Put $y = \phi(x)$ to get $\phi(x) * \phi(x) = 0$ and hence $\phi(x) = 0$ (because $*$ is positive-definite). Since x (though fixed) was arbitrary, we conclude that ϕ is the zero map.

This argument fails in an identity like

$$\phi(x) * x = 0$$

because there is only one independent variable to work with, and we cannot (in general) conclude that $\phi = 0$.

EXAMPLE 12.11

Let $\phi: \mathcal{R}^2 \to \mathcal{R}^2$ be the rotation of the plane through 90°. Then ϕ is not the zero map, but $\phi(\mathbf{X}) * \mathbf{X} = \phi(\mathbf{X}) \cdot \mathbf{X} = 0$ for all $\mathbf{X} \in \mathcal{R}^2$. (Why?)

Thus the identity

$$(g - i)(x) * x = 0$$

is apparently not going to yield $g - i = 0$ and our attempt to prove that $g = f^* \circ f = i$ when f is unitary has misfired.

However, this is a superficial impression, for despite the above remarks we are going to prove that $g - i$ *is* zero! The reason we can do this is that $\phi = g - i$ is a special kind of map—it is its own adjoint:

$$\phi^* = (g - i)^* = g^* - i^* = (f^* \circ f)^* - i = f^* \circ (f^*)^* - i = f^* \circ f - i = \phi$$

Operators with this property are called "self-adjoint," although we shall more often call them "Hermitian" (also "symmetric" when $\mathcal{F} = \mathcal{R}$); see Sec. 4.3 for the analogous definition for matrices. We'll discuss their properties in the next section, but now we make just one crucial observation.

THEOREM 12.11

Let $\phi: S \to S$ be a Hermitian operator on the unitary space S and suppose that $\phi(x) * x = 0$ for all $x \in S$. Then $\phi = 0$.

Proof

For all x and y in S, we can write

$$\begin{aligned} \phi(x + y) * (x + y) &= 0 \\ \phi(x + y) * x + \phi(x + y) * y &= 0 \\ [\phi(x) + \phi(y)] * x + [\phi(x) + \phi(y)] * y &= 0 \\ \phi(x) * x + \phi(y) * x + \phi(x) * y + \phi(y) * y &= 0 \end{aligned}$$

Since $\phi(x) * x$ and $\phi(y) * y$ are zero, this reduces to the identity

$$\phi(y) * x + \phi(x) * y = 0 \tag{12.1}$$

Now suppose that $\mathfrak{F} = \mathcal{C}$. Then we may replace x by ix in the above argument to obtain the identity

$$\phi(y) * (ix) + \phi(ix) * y = 0$$

or

$$-i[\phi(y) * x] + i[\phi(x) * y] = 0$$

Dividing by i yields the identity

$$-\phi(y) * x + \phi(x) * y = 0 \tag{12.2}$$

Adding Eq. (12.1) and Eq. (12.2) and dividing by 2, we have

$$\phi(x) * y = 0 \qquad \text{for all } x \text{ and } y$$

from which (as we pointed out in the prelude to this theorem) it follows that $\phi = 0$.[†]

This argument cannot be used if $\mathfrak{F} = \mathcal{R}$. (Why?) But in this case we may use the fact that $\phi^* = \phi$ to write

$$\begin{aligned} \phi(y) * x &= y * \phi^*(x) \\ &= y * \phi(x) \\ &= \phi(x) * y \qquad \text{(why?)} \end{aligned}$$

and Eq. (12.1) reduces to

$$\phi(x) * y = 0$$

Again it follows that $\phi = 0$. ∎

[†] It is worth observing that so far we have not used the fact that ϕ is Hermitian. Thus *any* linear operator satisfying the identity $\phi(x) * x = 0$ must be zero if the base field is $\mathcal{C}$. A map like ϕ in Example 12.11 cannot occur. (See the problems for more on this point.)

COROLLARY 12.11a

If S is a unitary space over the complex field and $\phi: S \rightarrow S$ is a linear operator satisfying $\phi(x) * x = 0$ for all $x \in S$, then $\phi = 0$.

Proof

See the footnote in the proof of Theorem 12.11.

THEOREM 12.12

Let $f: S \rightarrow S$ be a linear operator on the unitary space S and suppose that f^* exists. Then f is unitary if and only if $f^* \circ f = i$.

Proof

The "if" part is done in the proof of Theorem 12.8. To do the converse (which ground to a halt earlier), suppose that f is unitary and let $g = f^* \circ f$. Then (as we have shown)

$$g(x) * x = i(x) * x \qquad \text{for all } x \in S$$

that is,

$$\phi(x) * x = 0 \qquad \text{for all } x \in S$$

where $\phi = g - i$. Since ϕ is Hermitian (already shown), it follows from Theorem 12.11 that $\phi = 0$, that is, $g = f^* \circ f = i$. ▮

It is not apparent why this substitute for the statement

$$f^* = f^{-1} \iff f \text{ is unitary}$$

should read as it does. One might well expect that if

$$f^* \circ f = i \iff f \text{ is unitary}$$

then the condition $f \circ f^* = i$ ought to serve as well. Of course if *each* condition is equivalent to unitary, we are back to Theorem 12.8, since

$$f^* = f^{-1} \iff f^* \circ f = f \circ f^* = i$$

In view of the example in the last section (in which we defined a unitary operator whose adjoint exists but whose inverse doesn't), this is too much to expect.

EXAMPLE 12.12

Let S be the Hilbert space described in the last section. The linear operator

$$f(x_1,x_2,x_3, \ldots) = (0,x_1,x_2, \ldots)$$

is unitary and has adjoint

$$f^*(x_1,x_2,x_3, \ldots) = (x_2,x_3, \ldots)$$

Theorem 12.12 says that $f^* \circ f = i$; to confirm this, we observe that for every $\mathbf{X} = (x_1, x_2, x_3, \ldots)$ in S we have

$$(f^* \circ f)(\mathbf{X}) = f^*(0, x_1, x_2, \ldots) = (x_1, x_2, \ldots) = \mathbf{X}$$

But $f \circ f^* \neq i$, because f^{-1} does not exist. To confirm this, we need only note that

$$(f \circ f^*)(\mathbf{E}_1) = (f \circ f^*)(1,0,0,\ldots) = f(0,0,\ldots) = (0,0,\ldots) \neq \mathbf{E}_1$$

Problems

47. Let $\phi: \mathcal{C}^2 \to \mathcal{C}^2$ be the linear operator defined by the same formula as the 90° rotation of the plane, namely $\phi(x,y) = (-y,x)$.

a. How can you be sure (without checking it out) that the identity $\phi(\mathbf{X}) * \mathbf{X} = 0$ in Example 12.11 is not true on the domain $\mathcal{C}^2$?

b. Confirm part *a* by naming a vector $\mathbf{X} \in \mathcal{C}^2$ for which $\phi(\mathbf{X}) * \mathbf{X} \neq 0$.

This should clarify the footnote in the proof of Theorem 12.11, where we said that a map like ϕ in Example 12.11 cannot occur when the base field is $\mathcal{C}$. The only linear operator on a *complex* unitary space satisfying the identity $\phi(x) * x = 0$ is the zero map.

48. Let $\phi: \mathcal{R}^2 \to \mathcal{R}^2$ be the rotation of the plane through 90°.

a. Explain why $\phi(\mathbf{X}) * \mathbf{X} = 0$ for all $\mathbf{X} \in \mathcal{R}^2$.

b. How can you be sure (without checking it out) that $\phi^* \neq \phi$?

c. Find ϕ^* and confirm that $\phi^* \neq \phi$.

49. In the proof of Theorem 12.11 we separated the cases $\mathcal{F} = \mathcal{R}$ and $\mathcal{F} = \mathcal{C}$.

a. Why can't the argument given when $\mathcal{F} = \mathcal{C}$ be used when $\mathcal{F} = \mathcal{R}$?

b. The argument when $\mathcal{F} = \mathcal{R}$ utilizes the identity

$$\phi(y) * x = y * \phi^*(x) = y * \phi(x) = \phi(x) * y$$

to reduce Eq. (12.1) to $\phi(x) * y = 0$; this implies that $\phi = 0$. Why doesn't this work when $\mathcal{F} = \mathcal{C}$?

50. Explain why Theorem 12.8 is a corollary of Theorem 12.12. *Hint:* For the "only if" part, consult the remark following Prob. 51, Sec. 5.4.

12.6 Self-adjoint Operators

The subject of this chapter, as its title says, is the class of linear operators on a given unitary space S.† When S is finite-dimensional, there is a fascinating analogy between $L(S)$ and the class $\mathcal{C}$ of complex numbers.

† Recall from Sec. 3.5 that this class is denoted by $L(S)$, or simply L when S is understood.

To see what we mean, note first that each map in L has an adjoint (Theorem 12.2). It is therefore legitimate to consider the function which sends each element of L into its adjoint, namely

$$\phi: L \to L \qquad \text{defined by } \phi(f) = f^*$$

According to the formulas

$$(f + g)^* = f^* + g^* \quad \text{and} \quad (kf)^* = \bar{k}f^* \qquad \text{(Theorem 12.5)}$$

ϕ has the "almost-linear" properties

$$\phi(f + g) = \phi(f) + \phi(g) \qquad \text{and} \qquad \phi(kf) = \bar{k}\phi(f)$$

Ordinarily this would be a cause of distress, but now we propose to exploit the resemblance of this function to the function which sends each element of $\mathcal{C}$ into its conjugate, namely

$$\psi: \mathcal{C} \to \mathcal{C} \qquad \text{defined by } \psi(z) = \bar{z}$$

The almost-linear properties

$$\psi(u + v) = \psi(u) + \psi(v) \quad \text{and} \quad \psi(ku) = \bar{k}\psi(u) \qquad \text{(Prob. 7, Sec. 3.1)}$$

are the same; moreover, both functions are "involutions." That is, apply them twice and you're back where you started. To be precise, we have

$$(\phi \circ \phi)(f) = \phi(f^*) = (f^*)^* = f^* \qquad \text{(Theorem 12.4)}$$

and

$$(\psi \circ \psi)(z) = \psi(\bar{z}) = \bar{\bar{z}} = z$$

that is, ϕ^2 and ψ^2 are the identity maps on L and $\mathcal{C}$, respectively.

Another parallel may be drawn by noting the behavior of complex numbers of unit length under ψ and the behavior of unitary operators under ϕ:

If $|z| = 1$, the conjugate of z is its reciprocal, that is, $\psi(z) = \bar{z} = z^{-1}$. (Why?)

If f is unitary, its adjoint is its inverse, that is, $\phi(f) = f^* = f^{-1}$.

Evidently there is a conspiracy. We might as well push the analogy further and see where it goes. Consider, for example, the set of "real" complex numbers (numbers of the form $z = a + bi$ with $b = 0$). They may be characterized in terms of the map ψ by observing that

$$z \text{ is real if and only if } \psi(z) = \bar{z} = z$$

Why not call a linear operator "real" if $\phi(f) = f^* = f$?

LECTOR *Because you've already called it Hermitian.*

AUCTOR *Do you subscribe to Occam's Razor?*

LECTOR *Not if it's a magazine.*

AUCTOR *It's a philosophic principle to the effect that categories should not be needlessly proliferated.*

LECTOR *I'll buy that.*

AUCTOR *In that case we ought to call Hermitian operators real.*

LECTOR *Personally I prefer self-adjoint.*

AUCTOR *In any case they act like real numbers in $\mathcal{C}$.*

A complex number $z = a + bi$ is "pure imaginary" ($a = 0$) if and only if $\psi(z) = \bar{z} = -z$. (Confirm this![†]) The corresponding class of linear operators, defined by the property $\phi(f) = f^* = -f$, is identified by the prefix "skew." Just as every complex number is the sum of a real part and a pure imaginary part, every linear operator (whose adjoint exists) can be expressed as the sum of a self-adjoint part and a skew part. Before stating this formally, however, we pause for definitions.

SELF-ADJOINT AND SKEW OPERATORS

Let $f: S \to S$ be a linear operator on the unitary space S and suppose that f^* exists. If $f^* = f$, we call f *self-adjoint* or *Hermitian* (also *symmetric* when $\mathfrak{F} = \mathfrak{R}$).[‡] If $f^* = -f$, then f is said to be *skew-Hermitian* (also *skew-symmetric* when $\mathfrak{F} = \mathfrak{R}$).

THEOREM 12.13

Every linear operator having an adjoint can be written in exactly one way as the sum of a Hermitian operator and a skew-Hermitian operator.

Proof

Let f be the linear operator. There is at most one way to write $f = G + H$, where G and H are Hermitian and skew-Hermitian, respectively. Suppose that such operators exist. Then $f^* = G^* + H^* = G - H$; solving the equations

$$G + H = f \qquad G - H = f^*$$

for G and H, we find that

$$G = \tfrac{1}{2}(f + f^*) \qquad \text{and} \qquad H = \tfrac{1}{2}(f - f^*)$$

Hence if we are going to name G and H, we have no other choice but these.

[†] In checking this out, you should note that we are including $0 + 0i$ among the pure imaginaries as well as calling it real. Since it is the intersection of the "real axis" and the "imaginary axis" in the complex plane (Prob. 37, Sec. 11.3), there is no reason we cannot agree to this. It just sounds strange.

[‡] See Sec. 4.3, where we said that a square matrix A is "Hermitian" if $A^* = A$. When A has real entries, this reduces to $A^T = A$, in which case we also called A "symmetric."

The question of existence is settled by naming $G = \frac{1}{2}(f + f^*)$ and $H = \frac{1}{2}(f - f^*)$ and confirming that they perform as advertised. Clearly $f = G + H$; moreover (by Theorems 12.4 and 12.5), we have

$$G^* = \tfrac{1}{2}(f^* + f) = G \qquad \text{and} \qquad H^* = \tfrac{1}{2}(f^* - f) = -H$$

Hence G is Hermitian and H is skew-Hermitian. ▮

COROLLARY 12.13a

If the base field is $\mathcal{C}$, then $f = g + ih$, where g and h are unique Hermitian operators.

Proof

Use Theorem 12.13 to write $f = G + H$, where G and H are unique Hermitian and skew-Hermitian operators, respectively. Let $g = G$ and $h = -iH$ (legitimate because the base field is $\mathcal{C}$). Then $H = ih$ and we have $f = g + ih$, where g is Hermitian because G is and h is Hermitian because

$$h^* = (-iH)^* = iH^* = -iH = h \;\blacksquare$$

Of course we cannot write $f = g + ih$ when the base field is $\mathcal{R}$, but must content ourselves with Theorem 12.13. The analogy between L and $\mathcal{C}$ is not as impressive in this case.† When S is a finite-dimensional space over $\mathcal{C}$, however, each $f \in L$ has the form

$$f = (\text{Hermitian operator}) + i(\text{Hermitian operator})$$

just as each $z \in \mathcal{C}$ has the form

$$z = (\text{real number}) + i(\text{real number})$$

EXAMPLE 12.13

Define $f: \mathcal{C}^2 \to \mathcal{C}^2$ by $f(x,y) = (ix, x - iy)$. In the proof of Theorem 12.13 we observed that if $f = G + H$, where G and H are Hermitian and skew-Hermitian, respectively, then

$$G = \tfrac{1}{2}(f + f^*) \qquad \text{and} \qquad H = \tfrac{1}{2}(f - f^*)$$

Since the matrix of f relative to the standard basis is

$$A = \begin{bmatrix} i & 0 \\ 1 & -i \end{bmatrix}$$

† All analogies limp! Otherwise they would not be analogies, but exact correspondences.

the matrix of f^* is

$$A^* = \begin{bmatrix} -i & 1 \\ 0 & i \end{bmatrix}$$

The matrices of G and H are therefore

$$B = \frac{1}{2}(A + A^*) = \frac{1}{2}\begin{bmatrix} 0 & 1 \\ 1 & 0 \end{bmatrix} \quad \text{and} \quad C = \frac{1}{2}(A - A^*) = \frac{1}{2}\begin{bmatrix} 2i & -1 \\ 1 & -2i \end{bmatrix}$$

from which

$$G(x,y) = \tfrac{1}{2}(y,x) \quad \text{and} \quad H(x,y) = \tfrac{1}{2}(2ix - y,\ x - 2iy)$$

You can see by inspection that $f = G + H$ and that B and C are Hermitian and skew-Hermitian, respectively. It follows from Theorem 12.14 that G and H are too (as Theorem 12.13 predicts).

To illustrate Corollary 12.13a, we put

$$g(x,y) = G(x,y) = \tfrac{1}{2}(y,x)$$

and

$$h(x,y) = -iH(x,y) = \tfrac{1}{2}(2x + iy,\ -ix - 2y)$$

Then $f = g + ih$ and g and h are Hermitian because their matrices are. (Confirm this!)

LECTOR *Have you ever defined what a skew-Hermitian matrix is?*

AUCTOR *No. By now I expect you to have enough imagination to fill in some of these things yourself.*

LECTOR *Well, that's very kind, but I've developed a sort of affection for your little boxes.*

AUCTOR *All right, here's another.*

HERMITIAN AND SKEW-HERMITIAN MATRICES

Let A be a square matrix with entries in $\mathfrak{F}$. If $A^* = A$, we say that A is *Hermitian* (also *symmetric* if $\mathfrak{F} = \mathfrak{R}$, in which case $A^T = A$). If $A^* = -A$, then A is called *skew-Hermitian* (also *skew-symmetric* if $\mathfrak{F} = \mathfrak{R}$, in which case $A^T = -A$).

EXAMPLE 12.14

The matrix

$$A = \begin{bmatrix} 0 & -3 & 1 \\ 3 & 0 & 5 \\ -1 & -5 & 0 \end{bmatrix}$$

is skew-symmetric, as you can check. The linear operator f: $\mathcal{R}^3 \to \mathcal{R}^3$ whose matrix (relative to the standard basis) is A is defined by

$$f(x,y,z) = (-3y + z,\ 3x + 5z,\ -x - 5y)$$

It is probably not too surprising that f is also skew-symmetric, that is, $f^* = -f$. (Confirm this!) The next theorem (which we have already used in Example 12.13) is a statement of this sort of thing in general.

THEOREM 12.14

A linear operator on a finite-dimensional unitary space is Hermitian (skew-Hermitian) if and only if its matrix relative to an orthonormal basis is Hermitian (skew-Hermitian).

Proof

Let f be the linear operator and let $A = [f]$ be its matrix relative to an orthonormal basis. If A is Hermitian, we have

$$\begin{aligned} [f^*] &= [f]^* && \text{(Corollary 12.3a)} \\ &= A^* \\ &= A && \text{(definition of Hermitian matrix)} \\ &= [f] \end{aligned}$$

from which $f^* = f$. Hence f is Hermitian. The rest of the proof is left for the problems. ▮

COROLLARY 12.14a

A linear operator on a finite-dimensional euclidean space is symmetric (skew-symmetric) if and only if its matrix relative to an orthonormal basis is symmetric (skew-symmetric).

Returning to the analogy between $L(S)$ and $\mathcal{C}$ (at the beginning of this section), note that since the sum of real numbers in $\mathcal{C}$ is real, we might reasonably expect the sum of Hermitian operators (or matrices) to be Hermitian.

This is, in fact, the case, as you can check. But the product of Hermitian operators is not necessarily Hermitian. (See the problems.) For although Hermitian operators act like real numbers in $\mathcal{C}$, and the product of real numbers is real, the resemblance between the functions

$$\phi(f) = f^* \quad \text{and} \quad \psi(z) = \bar{z}$$

does not extend to multiplication. If f and g are linear operators in L, then

$$\begin{aligned}\phi(f \circ g) &= (f \circ g)^* \\ &= g^* \circ f^* \qquad \text{(Theorem 12.5)} \\ &= \phi(g) \circ \phi(f) \\ &\neq \phi(f) \circ \phi(g)\end{aligned}$$

(except in special cases), while if u and v are complex numbers, then

$$\psi(uv) = \overline{uv} = \bar{u}\,\bar{v} = \psi(u)\psi(v)$$

Thus, when considered as mappings on *rings* (Sec. 5.3), ϕ is not an isomorphism and ψ is.

Note that it is the failure of composition of maps to be commutative that is responsible for this; hence we may expect some improvement when the maps commute. (See Prob. 60.)

Problems

51. Let $f: S \to S$ be a linear operator on the unitary space S. Explain the following.

a. f is Hermitian if and only if $f(x) * y = x * f(y)$ for all x and y.

b. f is skew-Hermitian if and only if $f(x) * y = -x * f(y)$ for all x and y.

52. Let f be a linear operator on a complex unitary space and suppose that f^* exists.

a. Prove that f is Hermitian if and only if $f(x) * x$ is real for all x. *Hint:* For the "if" part, use Corollary 12.11a to show that $\phi = f^* - f$ is the zero map.

b. Prove that f is skew-Hermitian if and only if $f(x) * x$ is pure imaginary for all x.

53. Prove that a linear operator ϕ on a complex unitary space is Hermitian if and only if $i\phi$ is skew-Hermitian.

54. The conjugate of a real number in $\mathcal{C}$ is real. Prove that the adjoint of a Hermitian operator is Hermitian. (See Prob. 37, Sec. 4.3, for the same statement in connection with matrices.)

55. A pure imaginary number z has the property that z^2 is real. Prove that a skew-Hermitian operator f has the property that f^2 is Hermitian.

56. A linear operator is said to be "normal" if it commutes with its adjoint, that is, $f \circ f^* = f^* \circ f$. Explain why Hermitian and skew-Hermitian operators are normal, and why unitary operators on a finite-dimensional space are normal. (Why the restriction in the unitary case?)

57. Show that the linear operator $f\colon \mathcal{C}^2 \to \mathcal{C}^2$ defined by $f(x,y) = (x + iy,\ ix + y)$ is normal, but neither Hermitian, skew-Hermitian, nor unitary.

Problems 56 and 57 show that the class of normal operators on a finite-dimensional space includes Hermitian, skew-Hermitian, and unitary operators, but is not confined to these. (See Sec. 13.3 for more on normal operators.)

58. Let f be a linear operator on a unitary space. Then f is *unitary* if $f^* \circ f = i$, *self-adjoint* if $f^* = f$, and is called an *involution* if $f^2 = i$. Prove that any two of these properties imply the third.

59. Confirm that the maps $f\colon \mathcal{R}^2 \to \mathcal{R}^2$ and $g\colon \mathcal{R}^2 \to \mathcal{R}^2$ defined by $f(x,y) = (x,2y)$ and $g(x,y) = (2y,\ 2x + y)$ are Hermitian, but that neither $f \circ g$ nor $g \circ f$ is.

60. Prove that the composition of two Hermitian operators is Hermitian if and only if the operators commute. Why does it follow that the product of two Hermitian matrices is Hermitian if and only if the matrices commute?

61. Prove that an invertible Hermitian operator on a finite-dimensional space has a Hermitian inverse. Why does it follow that a nonsingular Hermitian matrix has a Hermitian inverse?

62. Let U be a subspace of the finite-dimensional euclidean space S. The map $f\colon S \to S$ defined by $f(x) = x_1 - x_2$ (where x_1 and x_2 are the orthogonal projections of x on U and $U^\perp$, respectively) is called a "reflection" in U. (See Prob. 35, Sec. 11.3.)

a. Prove that f is Hermitian by showing that $f(x) * y = x * f(y)$ for all x and y. (See Prob. 51.)

b. Use Prob. 58 to prove that f is Hermitian.

63. Suppose that $f: S \to S$ is Hermitian or skew-Hermitian. Prove that $\ker f = (\text{rng } f)^{\perp}$. *Hint:* See Prob. 12, Sec. 12.1.

64. Prove that an orthogonal projection (Sec. 11.3) is Hermitian. *Hint:* Use Prob. 51.

65. Let A be the matrix of a linear operator f relative to an orthonormal basis. Finish the proof of Theorem 12.14 by showing the following.

a. f is Hermitian $\Rightarrow A^* = A$

b. f is skew-Hermitian $\Leftrightarrow A^* = -A$

66. Let A be an $n \times n$ matrix with entries in $\mathfrak{F}$.

a. Explain why there are unique Hermitian and skew-Hermitian matrices B and C such that $A = B + C$.

b. If $\mathfrak{F}$ is the complex field, show that $A = B + iC$, where B and C are (unique) Hermitian matrices.

c. Illustrate parts *a* and *b* in the case

$$A = \begin{bmatrix} 1 & i \\ 2 & -1 \end{bmatrix}$$

d. What does the statement in part *b* amount to if $n = 1$?

67. The diagonal entries of a Hermitian matrix are real (Prob. 36, Sec. 4.3). Show that the diagonal entries of a skew-Hermitian matrix are pure imaginary. Why does it follow that the diagonal entries of a skew-symmetric matrix are zero? (See Example 12.14 for an illustration.)

68. If n is odd, prove that the determinant of an $n \times n$ skew-Hermitian matrix is pure imaginary (zero if the matrix is skew-symmetric). *Hint:* See Prob. 8, Sec. 8.2, and Prob. 24, Sec. 8.4.)

Review Quiz

True or false?

1. A rotation of the plane is a unitary operator.

2. A unitary operator on $\mathfrak{F}^n$ sends the standard basis into an orthonormal basis.

3. If A is an $n \times n$ matrix with real entries, then $(A\mathbf{X}) \cdot \mathbf{Y} = \mathbf{X} \cdot (A^T\mathbf{Y})$ for all $\mathbf{X}$ and $\mathbf{Y}$ in $\mathfrak{R}^n$.

4. The linear operator on $\mathcal{R}^2$ whose matrix is

$$\frac{1}{2}\begin{bmatrix} -1 & \sqrt{3} \\ \sqrt{3} & 1 \end{bmatrix}$$

is a reflection in the line $y = \sqrt{3}x$.

5. The class of self-adjoint operators on a unitary space is an abelian group relative to addition.

6. The diagonal entries of a skew-symmetric matrix are zero.

7. Let $f\colon \mathcal{R}^2 \to \mathcal{R}^2$ be an orthogonal map whose matrix has determinant -1. Then the set of points left fixed by f is a straight line through the origin.

8. Two symmetric matrices which commute have a symmetric product.

9. If f is a linear operator with the property that $f^* \circ f = i$, then f is unitary.

10. Orthogonal projections are unitary operators.

11. If f is a linear operator on a unitary space and $f(x) * y = -x * f(y)$ for all x and y, then f is skew-Hermitian.

12. An orthogonal linear operator on $\mathcal{R}^2$ maps the unit circle into itself.

13. If f is a linear operator on a unitary space and f^* exists, then $f - f^*$ is skew-Hermitian.

14. If A is a 2×2 matrix with determinant 1, the linear operator on $\mathcal{R}^2$ whose matrix is A is a rotation.

15. The adjoint of a skew-Hermitian operator is skew-Hermitian.

16. The determinant of a Hermitian matrix is real.

17. If f and g are linear operators on a complex unitary space and $f(x) * x = g(x) * x$ for all x, then $f = g$.

18. The matrix

$$\begin{bmatrix} 1 & 1 \\ -1 & 1 \end{bmatrix}$$

is orthogonal.

19. The adjoint of $f(x,y) = (x - y, x + y)$ is $f^*(x,y) = (x + y, y - x)$.

20. An orthogonal linear operator is invertible.

21. Given the linear operator f on the unitary space S, there is at most one linear operator $g\colon S \to S$ satisfying the identity $f(x) * y = x * g(y)$.

22. A rotation of the plane through 90° is Hermitian.

23. The only linear operator on a unitary space which is both Hermitian and skew-Hermitian is the zero map.

24. The determinant of a skew-symmetric matrix is 0.

25. If $f\colon \mathcal{C}^2 \rightarrow \mathcal{C}^2$ is the linear operator whose matrix is

$$A = \begin{bmatrix} 1 & i \\ -1 & 0 \end{bmatrix}$$

then the matrix of f^* is

$$A^T = \begin{bmatrix} 1 & -1 \\ i & 0 \end{bmatrix}$$

26. If A is a matrix representation of the Hermitian operator f, then A is Hermitian.

27. The map $f\colon \mathcal{R}^2 \rightarrow \mathcal{R}^2$ defined by $f(x,y) = (x + y, x + 3y)$ is self-adjoint.

28. Let $f\colon S \rightarrow S$ be a linear operator on the unitary space S and suppose that $x * y = 0 \Rightarrow f(x) * f(y) = 0$. Then f is unitary.

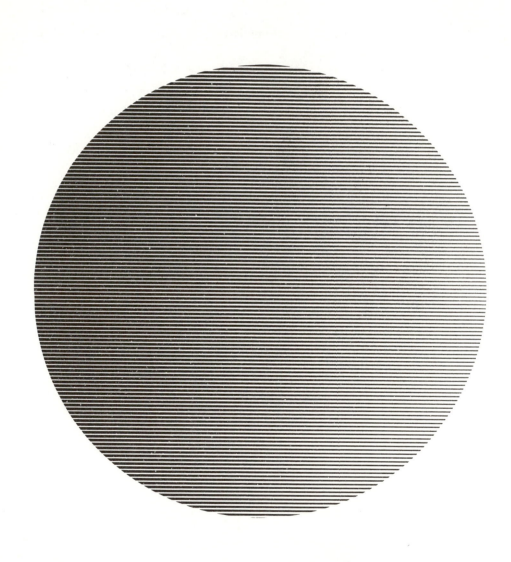

CHAPTER 13

Spectral Theory

This chapter is a continuation of the last, the emphasis now being on the eigenvalues and eigenvectors of certain types of linear operators on a unitary space. When the space is finite-dimensional, we shall be interested in finding a "spectral basis" associated with the operator, that is, a set of orthonormal eigenvectors spanning the space. The fact that this can always be done when the operator is self-adjoint is the first of several "spectral theorems" we'll prove. Another is the statement that every Hermitian matrix is "unitarily similar" to a diagonal matrix; that is, if A is Hermitian (symmetric in the real case), there exists a unitary matrix P such that $B = PAP^*$ is diagonal. In the real case this is sometimes called the "principal axis theorem," because of its relation to the problem of rotation of axes in analytic geometry.

When these ideas are extended to "normal" operators and matrices, we'll have reached our last spectral theorem, and the end of this book. The generalization of this theorem to infinite-dimensional spaces, in which spectral theory achieves its true climax, is one of the most interesting and important developments of modern mathematics.

13.1 Spectral Theorems for Self-adjoint Operators

THEOREM 13.1

The eigenvalues of a self-adjoint operator (if any) are real.

Proof

Let $f\colon S \to S$ be the operator and suppose that r is an eigenvalue of f. Then $f(x) = rx$ for some nonzero x. Since

$$r(x * x) = (rx) * x = f(x) * x = x * f^*(x) = x * f(x) = x * (rx) = \bar{r}(x * x)$$

and since $x * x \neq 0$, we have $\bar{r} = r$. Therefore r is real. ∎

Note that Theorem 13.1 says nothing about the existence of eigenvalues; as we saw in Chap. 9, some maps don't have any. (Also see Example 13.2.) However (this is a crucial point in the theory), self-adjoint operators on finite-dimensional spaces always do, as we now show.

THEOREM 13.2

The characteristic polynomial of a self-adjoint operator on an n-dimensional space has n real roots.[†]

Proof

Let $f: S \rightarrow S$ be the operator. The fundamental theorem of algebra (Prob. 30, Sec. 9.3) guarantees that the characteristic polynomial of f has n roots in $\mathcal{C}$. If we could be sure these are eigenvalues of f, the proof would be over, since Theorem 13.1 says that the eigenvalues of f are real. But unless the base field for S is $\mathfrak{F} = \mathcal{C}$ we cannot say this right off; linear operators on real vector spaces don't always have eigenvalues. (See Example 9.6.)

There is a way around this, however. Let A be the matrix of f relative to an orthonormal basis (guaranteed to exist by Corollary 10.12a). Then A is Hermitian; by definition, its characteristic polynomial is the characteristic polynomial of f. Regarding A as a matrix with entries in $\mathcal{C}$ (which we may always do whether the base field for S is $\mathcal{R}$ or $\mathcal{C}$), we know that its eigenvalues are the above roots. Let $g: \mathcal{C}^n \rightarrow \mathcal{C}^n$ be the map whose matrix relative to the standard basis for $\mathcal{C}^n$ is A. Then g is self-adjoint; by Theorem 9.8 we know it has the same eigenvalues as A. But then Theorem 13.1 says that these eigenvalues are real! Hence they *are* eigenvalues of f after all, and the proof is complete. ▮

COROLLARY 13.2a

A self-adjoint operator on an n-dimensional space has n real eigenvalues.

COROLLARY 13.2b

An $n \times n$ Hermitian matrix has n real eigenvalues. (In particular, an $n \times n$ symmetric matrix has n real eigenvalues.)

THEOREM 13.3

The characteristic polynomial of a Hermitian operator (or matrix) has real coefficients.

Proof

By the above corollaries, there are n real eigenvalues, say $r_1, \ldots, r_n$. If $p(x) = a_0 + a_1x + \cdots + a_nx^n$ is the characteristic polynomial, then

$$p(x) = a_n(x - r_1) \cdots (x - r_n) \qquad \text{(Prob. 30, Sec. 9.3)}$$

Since $a_n = (-1)^n$ (why?) and $r_1, \ldots, r_n$ are real, $p(x)$ has real coefficients. ▮

[†] As usual, count multiple roots as though they were distinct.

EXAMPLE 13.1

Theorem 13.3 is true even though the matrix may not have real entries. The characteristic polynomial of the Hermitian matrix

$$\begin{bmatrix} 1 & -i \\ i & 2 \end{bmatrix}$$

for example, is

$$\begin{vmatrix} 1 - r & -i \\ i & 2 - r \end{vmatrix} = r^2 - 3r + 1$$

As guaranteed by Corollary 13.2b, its eigenvalues are real, namely $\frac{1}{2}(3 \pm \sqrt{5})$.

EXAMPLE 13.2

Self-adjoint operators on infinite-dimensional spaces do not necessarily have any eigenvalues. Let S be the euclidean space of real-valued functions continuous on $[0,1]$, with inner product

$$f * g = \int_0^1 f(x)g(x)\,dx$$

Define ϕ: $S \rightarrow S$ by

$$\phi(f) = h$$

where

$$h(x) = xf(x) \qquad \text{for all } x \in [0,1]$$

Since

$$\phi(f) * g = \int_0^1 xf(x)g(x)\,dx = f * \phi(g)$$

we know that ϕ is self-adjoint. (Why?) But there are no eigenvalues. Now suppose there were a scalar r, and an element $f \in S$ (other than the zero function), such that $\phi(f) = rf$. Then for all $x \in [0,1]$, we have

$$xf(x) = rf(x) \qquad \text{that is} \quad (x - r)f(x) = 0$$

Hence $f(x) = 0$ for all $x \neq r$. Indeed, since f is continuous, $f(r) = 0$, too. Therefore f is the zero function, which is impossible.

THEOREM 13.4 Spectral Theorem for Self-adjoint Operators

Let f: $S \rightarrow S$ be a self-adjoint operator on the finite-dimensional space S. Then S has an orthonormal basis consisting of eigenvectors of f.

Proof

Induction on $n = \dim S$. By Corollary 13.2a we know that f has n real eigenvalues; let r_1 be one of these and let v_1 be a unit eigenvector associated with r_1. If $n = 1$, we are done, since $\{v_1\}$ is already an orthonormal basis for S. Hence assume that $n > 1$ and that the theorem is true for self-adjoint operators on $(n - 1)$-dimensional spaces.

Let $U = \langle v_1 \rangle$ and $V = U^{\perp}$; then V is $(n - 1)$-dimensional. (Why?) The idea of the proof is to restrict f to the domain V and apply the induction hypothesis to obtain an orthonormal basis $\{v_2, \ldots, v_n\}$ for V consisting of eigenvectors of f. Then $\{v_1, \ldots, v_n\}$ is the basis we seek.

However, this is delicate because the map $g: V \to S$ defined by $g(x) = f(x)$ for each $x \in V$ (the restriction of f to V) is not, on the face of it, a linear operator, but just a linear map. It is essential to be able to write it $g: V \to V$; that is, we must show that g maps V into V, not just V into S. To do this, note that U is invariant under f, that is, $x \in U \Rightarrow f(x) \in U$. (Why?) It follows from Prob. 11, Sec. 12.1, that V is invariant under f^*. But f is self-adjoint, so $f^* = f$ and we conclude that *V is invariant under f, too*. Since $g = f$ on V, we have $x \in V \Rightarrow g(x) \in V$.

Now it is clear sailing. We observe that g is self-adjoint because f is; then we apply the induction hypothesis to g to obtain an orthonormal basis $\{v_2, \ldots, v_n\}$ for V consisting of eigenvectors of g. Each of these is also an eigenvector of f and $\{v_1, \ldots, v_n\}$ is an orthonormal set. (Why?) Hence $\{v_1, \ldots, v_n\}$ is an orthonormal basis for S consisting of eigenvectors of f. ∎

COROLLARY 13.4a Spectral Theorem for Hermitian Matrices

Let A be an $n \times n$ Hermitian matrix with entries in $\mathfrak{F}$ (a symmetric matrix if $\mathfrak{F} = \mathfrak{R}$). Then $\mathfrak{F}^n$ has an orthonormal basis consisting of eigenvectors of A.

EXAMPLE 13.3

Define $f: \mathfrak{R}^3 \to \mathfrak{R}^3$ by $f(\mathbf{X}) = A\mathbf{X}$, where

$$A = \begin{bmatrix} 2 & 0 & 1 \\ 0 & 1 & 0 \\ 1 & 0 & 2 \end{bmatrix}$$

Then f (and A) are Hermitian (in fact symmetric); the spectral theorem says that $S = \mathfrak{R}^3$ has an orthonormal basis consisting of eigenvectors of f. As a matter of fact, we have already found the orthogonal eigenvectors $(0,1,0)$, $(-1,0,1)$, $(1,0,1)$ in Example 9.11; to produce the desired orthonormal basis, we need only normalize these:

$$\left\{(0,1,0), \frac{1}{\sqrt{2}}(-1,0,1), \frac{1}{\sqrt{2}}(1,0,1)\right\}$$

Note that in this case we have also found an orthonormal basis of eigenvectors of A. (See Theorem 9.10.)

SPECTRAL BASIS

The set of eigenvalues of a linear operator (or matrix) is called its *spectrum*. An orthonormal basis consisting of eigenvectors is called a *spectral basis* associated with the operator (or matrix).

Using this terminology, we may paraphrase the spectral theorems by saying that it is always possible to find a spectral basis associated with a self-adjoint operator on a finite-dimensional space (or with a Hermitian matrix). Recall from Sec. 9.4 (Theorem 9.12) that the existence of a basis of eigenvectors is equivalent to the statement that the matrix can be diagonalized (and is similar to a diagonal matrix). Hence every Hermitian matrix (in particular every symmetric matrix) can be diagonalized.

But we have shown more than that: we have come up with an *orthonormal* basis of eigenvectors; therefore the similarity relation we are talking about is *unitary*. To be precise, we make the following definition.

UNITARY SIMILARITY

If the square matrices A and B are related by an equation of the form $B = PAP^*$, where P is unitary ($P^* = P^{-1}$), we say that A is *unitarily similar* to B. In the real case, when P is orthogonal ($P^T = P^{-1}$), we also call A *orthogonally similar* to B.

THEOREM 13.5

Every Hermitian matrix is unitarily similar to a diagonal matrix.

Proof

Let A be the matrix and define $f\colon \mathfrak{F}^n \to \mathfrak{F}^n$ by $[f]_\alpha = A$, where α is the standard basis for $\mathfrak{F}^n$. Let β be the spectral basis for $\mathfrak{F}^n$ guaranteed by Corollary 13.4a. Then

$$B = [f]_\beta = \text{diag}\,(r_1, \ldots, r_n)$$

where $r_1, \ldots, r_n$ are the eigenvalues of A. Moreover, $B = PAP^{-1}$, where P is the matrix relating α to β (Theorem 6.10). Since α and β are orthonormal, P is unitary (see the problems); hence $P^{-1} = P^*$. ∎

COROLLARY 13.5a

Every symmetric matrix is orthogonally similar to a diagonal matrix.

In other words, if A is symmetric, there exists an orthogonal matrix P such that $B = PAP^T$ is diagonal. However, one must read the literature carefully on this point. Some writers use the term "symmetric" to mean any matrix equal to its transpose (even when the entries are not real). The theory we have developed for real symmetric and complex Hermitian matrices simultaneously does not apply to complex symmetric matrices; this is because A^T and A^* are not the same in the complex case.†

EXAMPLE 13.4

The symmetric matrix

$$A = \begin{bmatrix} 2 & 0 & 1 \\ 0 & 1 & 0 \\ 1 & 0 & 2 \end{bmatrix}$$

is orthogonally similar to the diagonal matrix

$$B = \begin{bmatrix} 1 & 0 & 0 \\ 0 & 1 & 0 \\ 0 & 0 & 3 \end{bmatrix}$$

by virtue of the relation $B = PAP^T$, where

$$P = \frac{1}{\sqrt{2}} \begin{bmatrix} 0 & \sqrt{2} & 0 \\ -1 & 0 & 1 \\ 1 & 0 & 1 \end{bmatrix}$$

is orthogonal. We found P as in Example 9.11, by using the orthonormal eigenvectors

$$(0,1,0) \qquad \frac{1}{\sqrt{2}}(-1,0,1) \qquad \frac{1}{\sqrt{2}}(1,0,1)$$

(Example 13.3) as columns of the matrix

$$P^{-1} = \frac{1}{\sqrt{2}} \begin{bmatrix} 0 & -1 & 1 \\ \sqrt{2} & 0 & 0 \\ 0 & 1 & 1 \end{bmatrix}$$

and then computing its inverse P. (Note that the "computation" of the inverse is trivial, since it is just the transpose. Why?)

The spectral theorem in the real case says that if the operator is symmetric, there is a spectral basis. The next theorem is a converse of this.

† See the footnote in the definition of symmetric matrices, Sec. 4.3.

THEOREM 13.6

Let $f: S \to S$ be a linear operator on the euclidean space S and suppose that there is a spectral basis associated with f. Then f is symmetric.

Proof

Let $\{v_1, \ldots, v_n\}$ be the spectral basis. Since the v_j are eigenvectors of f, and since S is euclidean, there are real scalars $r_1, \ldots, r_n$ such that

$$f(v_j) = r_j v_j \qquad j = 1, \ldots, n$$

The matrix of f relative to this basis is $A = \text{diag}\,(r_1, \ldots, r_n)$, while the matrix of f^* is

$$A^* = \text{diag}\,(\bar{r}_1, \ldots, \bar{r}_n) = \text{diag}\,(r_1, \ldots, r_n) = A$$

Hence $f^* = f$; that is, f is symmetric. ▮

Real Hermitian (symmetric) operators (on a finite-dimensional space) are therefore *characterized* by the fact that a spectral basis exists:

f is symmetric $\Leftrightarrow$ S has an orthonormal basis consisting of eigenvectors of f

Equivalently, in the case of an $n \times n$ matrix A, we have:

A is symmetric $\Leftrightarrow$ $\mathcal{R}^n$ has an orthonormal basis consisting of eigenvectors of A

If the base field is $\mathcal{C}$, however, the existence of a spectral basis is not sufficient to guarantee that f is Hermitian. Then the scalars $r_1, \ldots, r_n$ in the proof of Theorem 13.6 need not be real and $A^* = \text{diag}\,(\bar{r}_1, \ldots, \bar{r}_n)$ is not necessarily the same as A.

We can still salvage something in this case. Since diagonal matrices commute,

$$AA^* = A^*A$$

and hence

$$f \circ f^* = f^* \circ f$$

Linear operators which commute with their adjoint are called *normal*. (See Prob. 56, Sec. 12.6.) Thus we have shown the following.

THEOREM 13.7

Let $f: S \to S$ be a linear operator on the unitary space S. If S has a spectral basis associated with f, then f is normal.

It is a remarkable fact that *this goes the other way in the complex case*. In other words, if f is a normal operator on a finite-dimensional space over $\mathcal{C}$, then a spectral basis exists. The proof of this fact will be given in Sec. 13.3.

Thus *symmetric* operators on a euclidean space and *normal* operators on a (complex) unitary space are precisely those classes of linear operators for which a spectral basis exists. To quote Hillaire Belloc (in *The Path to Rome*), we have just about reached the Grand Climacteric of the book.

LECTOR *What is that in a book?*

AUCTOR *Why, it is the point where the reader has caught on, enters into the book and desires to continue reading it.*

LECTOR *It comes earlier in some books than in others.*

Problems

1. The eigenvalues of a Hermitian operator are real (Theorem 13.1). Prove that the eigenvalues of a skew-Hermitian operator are pure imaginary (zero if the operator is skew-symmetric).
2. Confirm Prob. 1 in the case of the map $f\colon \mathfrak{F}^3 \to \mathfrak{F}^3$ defined by $f(\mathbf{X}) = A\mathbf{X}$, where

$$A = \begin{bmatrix} 0 & -3 & 1 \\ 3 & 0 & 5 \\ -1 & -5 & 0 \end{bmatrix}$$

(Note that if $\mathfrak{F} = \mathfrak{R}$, there is only one eigenvalue, while if $\mathfrak{F} = \mathfrak{C}$, there are three.)
3. In the euclidean space of real-valued functions continuous on $[-\pi,\pi]$ let T be the subspace generated by $\phi_1(x) = (1/\sqrt{\pi}) \cos x$ and $\phi_2(x) = (1/\sqrt{\pi}) \sin x$. Define $D\colon T \to T$ by $D(f) = f'$.
   a. Apply Prob. 51*b*, Sec. 12.6, to show that D is skew-symmetric. *Hint:* Use integration by parts.
   b. $\{\phi_1,\phi_2\}$ is an orthonormal basis for T (Example 10.7). Confirm that the matrix of D relative to this basis is skew-symmetric. (This illustrates Corollary 12.14a.)
   c. [BOOBY TRAP] What can you say about the eigenvalues of D? (Note that if you say they are $\pm i$, you are contradicting Prob. 1!)
4. Fill in the reasoning in the proof of Theorem 13.4 as follows.
   a. Why is V $(n - 1)$-dimensional?
   b. Why is U invariant under f?
   c. Why is the map $g\colon V \to V$ self-adjoint?
   d. Why are the vectors $v_2, \ldots, v_n$ eigenvectors of f?
   e. Why is $\{v_1, \ldots, v_n\}$ an orthonormal set?
5. Let $f\colon S \to S$ be a self-adjoint operator.
   a. Prove that eigenvectors associated with distinct eigenvalues are or-

thogonal. (This simplifies the problem of finding a spectral basis. Only when orthonormal eigenvectors associated with repeated eigenvalues are sought do we have to worry about orthogonality; it is automatic in the case of distinct eigenvalues.)

b. Why does it follow that the eigenspaces of f are mutually orthogonal subspaces of S?

c. Assuming that S is finite-dimensional, let $T_1, \ldots, T_m$ be the eigenspaces of f. If $f_j\colon S \to S$ is the orthogonal projection of S on T_j, where $j = 1, \ldots, m$, prove that $f_j \circ f_k = 0$ when $j \neq k$.

6. Show that unitary similarity is an equivalence relation on $M_n(\mathfrak{F})$; that is, show that it is reflexive, symmetric, and transitive. (See Theorem 6.6 for the same statement in connection with ordinary similarity.)

7. Let α be an orthonormal basis for the unitary space S and suppose that β is any basis for S. Prove that the matrix $P = [i]_{\alpha\beta}$ is unitary if and only if β is orthonormal. *Hint:* Let $\alpha = \{u_1, \ldots, u_n\}$ and $\beta = \{v_1, \ldots, v_n\}$. To do the "if" part (which fills in a gap in the proof of Theorem 13.5), let $P = [p_{ij}]$ and make use of the fact that

$$u_t = \sum_{s=1}^{n} p_{st} v_s$$

to show that

$$u_i * u_j = \sum_{k=1}^{n} p_{ki}\bar{p}_{kj}$$

For the "only if" part, let $Q = P^* = [q_{ij}]$ and show that

$$v_i * v_j = \sum_{k=1}^{n} q_{ki}\bar{q}_{kj}$$

8. Prove that two matrices are unitarily similar if and only if they represent the same linear operator on a unitary space (each relative to an orthonormal basis). See Theorem 6.7 for the same statement in connection with ordinary similarity. The proof is the same except that now we have to worry about orthonormal bases.

9. Let A be an $n \times n$ matrix with real entries. Show that if an orthogonal matrix P exists such that $B = PAP^T$ is diagonal, then A is symmetric. (This is the converse of Corollary 13.5a.)

10. Check that each of the following matrices is Hermitian, find a spectral basis associated with it, name a unitary matrix P such that $B = PAP^*$ is diagonal, and find B.

a. $A = \begin{bmatrix} 1 & 0 \\ 0 & 2 \end{bmatrix}$ b. $A = \begin{bmatrix} 2 & 1 \\ 1 & 2 \end{bmatrix}$ c. $A = \begin{bmatrix} 1 & i \\ -i & 1 \end{bmatrix}$

d. $A = \begin{bmatrix} 1 & 1 & 1 \\ 1 & 1 & 1 \\ 1 & 1 & 1 \end{bmatrix}$ e. $A = \begin{bmatrix} 1 & 1 & 1 \\ 1 & 2 & 0 \\ 1 & 0 & 2 \end{bmatrix}$ f. $A = \begin{bmatrix} 1 & 0 & i \\ 0 & 1 & -1 \\ -i & -1 & 0 \end{bmatrix}$

†13.2 An Application to Geometry

The various *quadric surfaces* studied in analytic geometry (ellipsoids, paraboloids, hyperboloids, cones, certain cylinders) are all represented by second-degree equations in x, y, and z, as you no doubt recall. One learns a number of standard forms as an aid to identification of these surfaces; the equation

$$\frac{x^2}{a^2} + \frac{y^2}{b^2} + \frac{z^2}{c^2} = 1$$

for example, represents an ellipsoid with center at the origin and "semidiameters" a, b, and c. One also learns how to translate the axes to reduce certain equations to standard form when they would otherwise be hard to analyze. The equation

$$x^2 + y^2 - 2x + z - 3 = 0$$

for example, may be written in the form

$$(x - 1)^2 + y^2 = -(z - 4)$$

The translation

$$\overline{x} = x - 1 \qquad \overline{y} = y \qquad \overline{z} = z - 4$$

reduces this to the equation of a circular paraboloid,

$$\overline{x}^2 + \overline{y}^2 = -\overline{z}$$

In plane analytic geometry (where second-degree equations in x and y represent various *conics*), one also learns how to rotate the axes to get rid of the xy term; this simplifies the equation to an easily recognizable form. Our purpose in this section is to do the same thing in 3-space, where the

†*Although this section may be omitted, it is worth looking through—just to appreciate an ingenious application of spectral theory to a hard problem in geometry.*

possibilities for rotation are obviously more complicated. Spectral theory for symmetric operators turns out to be precisely the needed tool!

LECTOR *Is this the Grand Climacteric of the book?*
AUCTOR *This is fun and games.*

Consider the general second-degree equation in x, y, and z,

$$ax^2 + by^2 + cz^2 + 2dxy + 2exz + 2fyz + px + qy + rz = k$$

where $a, b, c, \ldots$ are real.† The second-degree part of this, namely

$$ax^2 + by^2 + cz^2 + 2dxy + 2exz + 2fyz$$

is what makes the graph hard to identify; the problem is to remove the xy, xz, and yz terms by an appropriate rotation of axes. Of course this will also change the coefficients in $px + qy + rz$, but not its degree (as you will see). Hence we confine our attention for the present to equations of the form

$$ax^2 + by^2 + cz^2 + 2dxy + 2exz + 2fyz = k$$

The left-hand side of this is a so-called "quadratic form" in x, y, and z; with it we associate the matrix

$$A = \begin{bmatrix} a & d & e \\ d & b & f \\ e & f & c \end{bmatrix}$$

LECTOR *Why?*
AUCTOR *An adequate explanation would require a digression.*
LECTOR *I haven't noticed much reluctance to digress in this book.*
AUCTOR *As Pythagoras said about the diagonal of the unit square, I am beginning to worry about its length.*
LECTOR *That's irrational at this late stage.*
AUCTOR *It was irrational from the beginning.*

Note that A is symmetric, with the coefficients of x^2, y^2, and z^2 as diagonal entries; hence it is easy to remember. If $\mathbf{X} = (x,y,z)$, our equation may be written

$$(A\mathbf{X}) \cdot \mathbf{X} = k$$

as you can check.

Since A is symmetric, we know from the last section that its eigenvalues are real, and that $\mathcal{R}^3$ has a basis β whose elements are orthonormal eigen-

† We insert 2 in the coefficients of the mixed product terms for later convenience.

vectors of A. The orthogonal matrix P relating α to β (where α is the standard basis) has the property that

$$B = PAP^T = \text{diag}\,(r_1,r_2,r_3)$$

where r_1, r_2, and r_3 are the eigenvalues of A associated with the elements of β.

For each $\mathbf{X} \in \mathcal{R}^3$, write $\mathbf{X}' = \beta(\mathbf{X})$, the coordinate vector of $\mathbf{X}$ relative to β. Since

$$P\alpha(\mathbf{X}) = \beta(\mathbf{X}) \qquad \text{(Theorem 6.8)}$$

and $\alpha(\mathbf{X}) = \mathbf{X}$ (why?), we have

$$P\mathbf{X} = \mathbf{X}' \qquad \text{for all } \mathbf{X} \in \mathcal{R}^3$$

Since $P^{-1} = P^T$, we obtain

$$\mathbf{X} = P^T\mathbf{X}' \qquad \text{for all } \mathbf{X} \in \mathcal{R}^3$$

Hence the equation of our quadric surface,

$$(A\mathbf{X}) \cdot \mathbf{X} = k$$

may be written

$$(AP^T\mathbf{X}') \cdot (P^T\mathbf{X}') = k$$

or, moving P^T "across the dot" (see Prob. 9, Sec. 12.1),

$$(PAP^T\mathbf{X}') \cdot \mathbf{X}' = k$$

But $PAP^T = B = \text{diag}\,(r_1,r_2,r_3)$, so this reads

$$(B\mathbf{X}') \cdot \mathbf{X}' = k$$

or

$$r_1x'^2 + r_2y'^2 + r_3z'^2 = k$$

where $\mathbf{X}' = (x',y',z')$. This is the equation of our quadric surface in the x', y', z' coordinate system; the xy, xz, yz terms are gone!

Since P is orthogonal, we know that $\det P = \pm 1$; it can be shown that if $\det P = 1$, the change of basis from α to β is a rotation of axes, while if $\det P = -1$, a reflection is also involved. In any case, the change to a spectral basis is precisely what is needed to simplify the equation of the quadric surface. (In effect, we have changed from x, y, z axes to axes which are lined up with the principal axes of the surface; hence the spectral theorem in this context is often called the *principal axis theorem*.)

For example, suppose that the surface is defined by

$$2x^2 + y^2 + 2z^2 + 2xz = 1$$

The symmetric matrix A associated with the quadratic form $2x^2 + y^2 + 2z^2 + 2xz$ is

$$A = \begin{bmatrix} 2 & 0 & 1 \\ 0 & 1 & 0 \\ 1 & 0 & 2 \end{bmatrix}$$

with eigenvalues $r_1 = r_2 = 1$, $r_3 = 3$. (See Example 9.11.) Hence there is a rotation of axes (or a rotation followed by a reflection) which transforms the equation to

$$x'^2 + y'^2 + 3z'^2 = 1$$

This is an ellipsoid of revolution, a fact which is not apparent from an inspection of the original equation.

If we are interested in the details of the change of basis which simplified our equation, we compute a spectral basis $\beta = \{\mathbf{V}_1, \mathbf{V}_2, \mathbf{V}_3\}$, say

$$\mathbf{V}_1 = (0,1,0) \qquad \mathbf{V}_2 = \frac{1}{\sqrt{2}}(-1,0,1) \qquad \mathbf{V}_3 = \frac{1}{\sqrt{2}}(1,0,1)^\dagger$$

Since $P^T = P^{-1}$ is the matrix relating β to α, its columns are $\mathbf{V}_1$, $\mathbf{V}_2$, $\mathbf{V}_3$ (Prob. 28, Sec. 6.3). Hence

$$P^T = \frac{1}{\sqrt{2}} \begin{bmatrix} 0 & -1 & 1 \\ \sqrt{2} & 0 & 0 \\ 0 & 1 & 1 \end{bmatrix}$$

and

$$P = \frac{1}{\sqrt{2}} \begin{bmatrix} 0 & \sqrt{2} & 0 \\ -1 & 0 & 1 \\ 1 & 0 & 1 \end{bmatrix}$$

(Also see Example 13.4.) You can check that $B = PAP^T = \text{diag}\,(1,1,3)$; since $\det P = 1$, the change of basis is a rotation of axes.‡

As a matter of fact (see Sec. 6.3 for a discussion of this in $\mathcal{R}^2$), we may change our point of view and regard P as the matrix of a rotation of 3-space (rather than the matrix of a change of basis). Let $f\colon \mathcal{R}^3 \to \mathcal{R}^3$ be defined by $f(\mathbf{X}) = P\mathbf{X}$. The points $\mathbf{X} = (x,y,z)$ left fixed by f (see Prob. 41, Sec. 12.3) are the solutions of $f(\mathbf{X}) = \mathbf{X}$, namely

$$\{(x,y,z) \mid x = t,\ y = t,\ z = (1 + \sqrt{2})t\}$$

This is the line through the origin containing the vector $(1, 1, 1 + \sqrt{2})$; it is the axis of rotation. Alternatively (returning to our original point of view)

† In general, this may require Gram-Schmidt orthogonalization as well as the techniques of diagonalization described in Sec. 9.4. We want not merely a basis of eigenvectors, but an *orthonormal* basis. (See Example 13.3.) In other words, we are interested in orthogonal similarity, not just similarity. However, eigenvectors associated with distinct eigenvalues are automatically orthogonal. (See Prob. 5, Sec. 13.1.)

‡ This may always be arranged! Had we chosen $\mathbf{V}_1 = (0,-1,0)$, for example, $\det P$ would have been -1. Just reverse its direction to get $\det P = 1$.

we may say that $\mathcal{R}^3$ stays put; the coordinate axes, however, are rotated around this line to a new position. In other words, the given quadric surface doesn't move, but its equation changes because the x, y, z coordinate system has been replaced by the x', y', z' system.

The only remaining question is what to do with the first-degree part of the equation

$$ax^2 + by^2 + cz^2 + 2dxy + 2eyz + 2fyz + px + qy + rz = k$$

The change of basis from α *to* β is represented by $P\mathbf{X} = \mathbf{X}'$, or (equivalently) by $\mathbf{X} = P^T\mathbf{X}'$. This gives x, y, z in terms of x', y', z', namely

$$\begin{aligned} x &= q_{11}x' + q_{12}y' + q_{13}z' \\ y &= q_{21}x' + q_{22}y' + q_{23}z' \\ z &= q_{31}x' + q_{32}y' + q_{33}z' \end{aligned}$$

where $P^T = Q = [q_{ij}]$. It is clear from the linearity of these equations that substitution in $px + qy + rz$ does not change its degree, but merely its coefficients. Our new equation therefore takes the form

$$r_1x'^2 + r_2y'^2 + r_3z'^2 + p'x' + q'y' + r'z' = k$$

and a translation of axes will produce a standard form. The crucial part of the discussion is the change to a spectral basis.

Problems

11. Discuss the graph in $\mathcal{R}^2$ of the general second-degree equation

$$ax^2 + bxy + cy^2 + dx + ey + f = 0$$

as follows.

a. Let

$$A = \begin{bmatrix} a & b/2 \\ b/2 & c \end{bmatrix}$$

be the symmetric matrix associated with the quadratic form $ax^2 + bxy + cy^2$. Confirm that $(A\mathbf{X}) \cdot \mathbf{X} = ax^2 + bxy + cy^2$, where $\mathbf{X} = (x,y)$.

b. Show that $r_1r_2 = -\frac{1}{4}(b^2 - 4ac)$, where r_1 and r_2 are the eigenvalues of A. *Hint:* Don't use brute force! See Probs. 29 and 30, Sec. 9.3.

c. The same reasoning as in the text shows that a rotation of axes will transform $ax^2 + bxy + cy^2$ to $r_1x'^2 + r_2y'^2$; the original equation takes the form

$$r_1x'^2 + r_2y'^2 + d'x' + e'y' + f = 0$$

Use part b to explain why (except for degenerate cases) this represents a hyperbola, parabola, or ellipse depending on whether $b^2 - 4ac$ is positive, zero, or negative, respectively. (This is a well-known theorem in analytic geometry.)

12. Use spectral theory to reduce each of the following equations to recognizable form. Name the matrix (with determinant 1) relating the original basis to the new basis and find the angle through which the coordinate system is rotated.

a. $2x^2 + \sqrt{3}xy + y^2 = 1$

b. $2xy - x - y = 1$

13. Reduce $x^2 + y^2 + z^2 + 2xy + 2xz + 2yz = 4$ to the form

$$r_1x'^2 + r_2y'^2 + r_3z'^2 = 4$$

and note that the "quadric surface" degenerates to the union of two parallel planes. What are the equations of these planes in the original x, y, z system?

14. What kind of quadric surface is represented by $x^2 + 2y^2 + 2z^2 + 2xy + 2xz = 1$?

15. If the coordinate system is rotated to reduce $5y^2 + 4z^2 + 4xy + 8xz + 12yz = 3$ to standard form, what is the axis of rotation?

13.3 Normal Operators and the Spectral Theorem

NORMAL OPERATORS AND MATRICES

Let $f: S \to S$ be a linear operator on the unitary space S. If f^* exists and commutes with f, that is, if

$$f \circ f^* = f^* \circ f$$

we say that f is *normal*. Similarly, an $n \times n$ matrix A is said to be *normal* if $AA^* = A^*A$.

As we saw in Probs. 56 and 57, Sec. 12.6, Hermitian and skew-Hermitian operators are normal, as are unitary operators on a finite-dimensional space, but not every normal operator is one of these types. Thus the class of normal operators on a finite-dimensional space includes the special classes we have studied, together with others.

In this section we shall prove that every normal operator on a *complex* finite-dimensional space has orthonormal eigenvectors spanning the space. Or equivalently, every normal matrix is unitarily similar to a diagonal matrix (provided that we are willing to work with complex numbers). These spectral theorems make normal operators (and matrices) especially nice to deal with. Moreover (Theorem 13.7) the *only* operators with this property are normal. Since many of the applications of linear algebra involve normal matrices, they have been studied intensively. Some of the results obtained are either false or undecided for operators in general; in the latter case, there is clearly room for research.

THEOREM 13.8

Let $f: S \to S$ be a linear operator on the unitary space S. Then f is normal if and only if $\|f(x)\| = \|f^*(x)\|$ for all $x \in S$.

Proof

Suppose that $\|f(x)\| = \|f^*(x)\|$ for all $x \in S$. To prove that $f \circ f^* = f^* \circ f$, observe that

$$\begin{aligned}(f \circ f^*)(x) * x &= f[f^*(x)] * x \\ &= f^*(x) * f^*(x) \\ &= \|f^*(x)\|^2 \\ &= \|f(x)\|^2 \\ &= f(x) * f(x) \\ &= f^*[f(x)] * x \\ &= (f^* \circ f)(x) * x\end{aligned}$$

Hence

$$\phi(x) * x = 0 \qquad \text{for all } x \in S$$

where $\phi = f \circ f^* - f^* \circ f$. Since ϕ is Hermitian (confirm this!), it follows from Theorem 12.11 that $\phi = 0$, that is, $f \circ f^* = f^* \circ f$. Thus f is normal. The converse is easy. ∎

THEOREM 13.9

If r is an eigenvalue of the normal operator f, then $\bar{r}$ is an eigenvalue of f^*. Moreover, the eigenspace of f associated with r is identical to the eigenspace of f^* associated with $\bar{r}$.

Proof

Observe first that $f - ri$ is normal because f is. (See the problems.) Let x be any element of S satisfying $f(x) = rx$. To prove that $f^*(x) = \bar{r}x$, we show that $d[f^*(x), \bar{r}x] = 0$:

$$\begin{aligned} d[f^*(x),\bar{r}x] &= \|f^*(x) - \bar{r}x\| \\ &= \|(f^* - \bar{r}i)(x)\| \\ &= \|(f - ri)^*(x)\| \\ &= \|(f - ri)(x)\| \qquad \text{(Theorem 13.8 applied to } f - ri) \\ &= \|f(x) - rx\| \\ &= d[f(x),rx] \\ &= 0 \end{aligned}$$

This shows that $\bar{r}$ is an eigenvalue of f^* and that $U \subset V$, where U and V are the eigenspaces of f and f^* associated with r and $\bar{r}$, respectively. The proof is completed by showing that $V \subset U$, which is easy. (Just reverse the above steps.) ▮

COROLLARY 13.9a

Each eigenvector of a normal operator f is also an eigenvector of f^*.

THEOREM 13.10

Eigenvectors of a normal operator associated with distinct eigenvalues are orthogonal.

Proof

Suppose that r_1 and r_2 are distinct eigenvalues of the normal operator f, and let x_1 and x_2 be corresponding eigenvectors. Then

$$f(x_1) = r_1x_1 \qquad \text{and} \qquad f(x_2) = r_2x_2$$

Since

$$\begin{aligned} r_1(x_1 * x_2) &= (r_1x_1) * x_2 \\ &= f(x_1) * x_2 \\ &= x_1 * f^*(x_2) \\ &= x_1 * (\bar{r}_2x_2) \\ &= r_2(x_1 * x_2) \end{aligned}$$

we have $(r_1 - r_2)(x_1 * x_2) = 0$. However $r_1 - r_2$ is nonzero, so $x_1 * x_2 = 0$; hence x_1 and x_2 are orthogonal. ▮

COROLLARY 13.10a

Eigenvectors of a Hermitian (or skew-Hermitian) operator associated with distinct eigenvalues are orthogonal.

COROLLARY 13.10b

Eigenvectors of a unitary operator (on a finite-dimensional space) associated with distinct eigenvalues are orthogonal.

It is worth noting that Corollary 13.10a for Hermitian operators was suggested in Prob. 5, Sec. 13.1. The proof there was simpler because $f^* = f$ and the eigenvalues are real:

$$r_1(x_1 * x_2) = f(x_1) * x_2 = x_1 * f(x_2) = r_2(x_1 * x_2)$$

THEOREM 13.11 Spectral Theorem for Normal Operators

Let $f\colon S \to S$ be a normal operator on the complex finite-dimensional space S. Then S has an orthonormal basis consisting of eigenvectors of f.

Proof

Only minor alterations are needed in the proof of the spectral theorem for Hermitian operators (Sec. 13.1). Since S is a complex space, we know that f has n eigenvalues, where $n = \dim S$; let r_1 be one of these and let v_1 be a unit eigenvector associated with r_1. If $n = 1$, we are done, since $\{v_1\}$ is already an orthonormal basis for S. Hence assume that $n > 1$ and that the theorem is true for normal operators on $(n - 1)$-dimensional spaces. Since v_1 is an eigenvector of f, it is also an eigenvector of f^* (Corollary 13.9a), so $U = \langle v_1 \rangle$ is invariant under f^*. But then $U^\perp$ is invariant under $(f^*)^* = f$ (Prob. 11, Sec. 12.1). Let $g\colon U^\perp \to U^\perp$ be the restriction of f to the domain $U^\perp$. Then g is normal; since $\dim U^\perp = n - 1$, the induction hypothesis yields an orthonormal basis $\{v_2, \ldots, v_n\}$ for $U^\perp$ consisting of eigenvectors of g. Each of these is also an eigenvector of f and $\{v_1, \ldots, v_n\}$ is an orthonormal set. Hence it is an orthonormal basis for S consisting of eigenvectors of f. ∎

COROLLARY 13.11a Spectral Theorem for Normal Matrices

Let A be an $n \times n$ normal matrix with entries in $\mathcal{C}$.† Then $\mathcal{C}^n$ has an orthonormal basis consisting of eigenvectors of A.

THEOREM 13.12

Every normal matrix is unitarily similar to a diagonal matrix (provided that we are willing to work with complex numbers).

Proof

This proceeds exactly like the proof of Theorem 13.5 except that $\mathcal{F}^n$ should be replaced by $\mathcal{C}^n$.

Now we have completed our characterization of linear operators and matrices in the context of spectral theory. We may summarize as follows.

† *Every* matrix with numerical entries can be considered as having entries in $\mathcal{C}$, because $\mathcal{R} \subset \mathcal{C}$. We use the phrase "with entries in $\mathcal{C}$" to emphasize the fact that the theorem may require us to go into complex n-space to find eigenvectors.

Let S be a finite-dimensional unitary space over $\mathfrak{F}$. The class of linear operators on S with which a spectral basis can be associated is *normal* or *symmetric* depending on whether $\mathfrak{F} = \mathcal{C}$ or $\mathfrak{F} = \mathcal{R}$. Equivalently, the class of matrices in $M_n(\mathfrak{F})$ which can be diagonalized by a unitary similarity transformation is *normal* or *symmetric* depending on whether $\mathfrak{F} = \mathcal{C}$ or $\mathfrak{F} = \mathcal{R}$.

THEOREM 13.13

Let $f\colon S \to S$ be a normal operator on the complex finite-dimensional space S. Then

a. f is Hermitian if and only if its eigenvalues are real.

b. f is skew-Hermitian if and only if its eigenvalues are pure imaginary.

c. f is unitary if and only if its eigenvalues have absolute value 1.

Proof

The "only if" part of each statement has been done before (Theorem 13.1 and Prob. 1, Sec. 13.1, and Prob. 32, Sec. 12.2). To prove the "if" part, note that since f is normal there is an orthonormal basis relative to which its matrix is $B = \text{diag}\,(r_1, \ldots, r_n)$, where $r_1, \ldots, r_n$ are its eigenvalues. But B is Hermitian if the eigenvalues are real, skew-Hermitian if they are pure imaginary, and unitary if they have absolute value 1. (See the problems.) Hence f is, too. ▮

COROLLARY 13.13a

An $n \times n$ normal matrix with entries in $\mathcal{C}$ is Hermitian, skew-Hermitian, unitary if and only if its eigenvalues are real, pure imaginary, of absolute value 1, respectively.

This gives some idea of the structure of the class of normal operators (or matrices) in terms of eigenvalues. Thinking of the eigenvalues as points in the complex plane, we may say that an operator (or matrix) is Hermitian, skew-Hermitian, unitary if and only if its eigenvalues all lie on the real axis, imaginary axis, unit circle, respectively. When an eigenvalue lies somewhere else, the operator (or matrix) is not one of these special types.

We end this section, and the book, with a version of the spectral theorem which can be generalized to Hilbert spaces. It is fair to call this result the central theorem of linear algebra; it is an appropriate launching pad for future study of the subject.

THEOREM 13.14 The Spectral Theorem

If $f\colon S \to S$ is a normal [symmetric] operator on the complex [real] finite-dimensional space S and if $r_1, \ldots, r_m$ are its distinct eigenvalues, then

1. $S = T_1 \oplus \cdots \oplus T_m$, where T_j is the eigenspace associated with r_j;

2. $f = r_1 f_1 + \cdots + r_m f_m$, where f_j: $S \rightarrow S$ is the orthogonal projection of S on T_j;
3. $f_1 + \cdots + f_m = i$;
4. $f_j \circ f_k = 0$ if $j \neq k$.

Proof

Regardless of the base field, f has n eigenvalues, where $n = \dim S$.† By the spectral theorem for Hermitian operators (Sec. 13.1) or the same theorem for normal operators (in the present section) we know that S has an orthonormal basis consisting of eigenvectors of f. It follows from Theorem 9.14 that

$$\dim T_j = k_j \qquad \text{where } k_j \text{ is the multiplicity of } r_j, \quad j = 1, \ldots, m$$

Since the r_j are distinct, the sum of the T_j is a direct sum (Prob. 9*b*, Sec. 9.2) and we have

$$\dim (T_1 \oplus \cdots \oplus T_m) = k_1 + \cdots + k_m = n$$

(See Corollary 11.2a.) Therefore

$$S = T_1 \oplus \cdots \oplus T_m \qquad \text{(Theorem 2.17)}$$

and each $x \in S$ can be written (uniquely) in the form

$$x = x_1 + \cdots + x_m \qquad \text{where } x_j \in T_j, \quad j = 1, \ldots, m$$

Since eigenvectors associated with distinct eigenvalues are orthogonal (why?), the eigenspaces $T_1, \ldots, T_m$ are mutually orthogonal. Hence each T_j has the sum of the others for its orthogonal complement (Theorem 11.7). It follows that the orthogonal projection of S on T_j is the map f_j: $S \rightarrow S$ defined by $f_j(x) = x_j$. But then

$$f(x) = \sum_{j=1}^{m} f(x_j) = \sum_{j=1}^{m} r_j x_j = \sum_{j=1}^{m} r_j f_j(x)$$

which means that

$$f = r_1 f_1 + \cdots + r_m f_m$$

Moreover, since

$$(f_1 + \cdots + f_m)(x) = f_1(x) + \cdots + f_m(x) = x_1 + \cdots + x_m = x = i(x)$$

we have

$$f_1 + \cdots + f_m = i$$

† This is always the case if $\mathfrak{F} = \mathfrak{C}$, whereas if $\mathfrak{F} = \mathfrak{R}$, it follows from the fact that f is symmetric (Corollary 13.2a).

Finally, if $j \neq k$, we have

$$(f_j \circ f_k)(x) = f_j[f_k(x)] = f_j(x_k) = 0$$

so $f_j \circ f_k$ is the zero map. ∎

SPECTRAL DECOMPOSITION

The expression

$$f = r_1 f_1 + \cdots + r_m f_m$$

in which f is written as a linear combination of orthogonal projections on distinct eigenspaces, is called a *spectral decomposition* of f.

Thus we may summarize the spectral theorem by saying that every normal [symmetric] operator on a complex [real] finite-dimensional space has a spectral decomposition. It is this sort of thing (as opposed to its matrix interpretation) which generalizes to Hilbert spaces; we end the book with that thought in mind.

LECTOR *Have you no last exhortation, no charge to the reader, no call to arms?*

AUCTOR *The final exam is next week.*

LECTOR *That's not very helpful.*

AUCTOR *I tell you naught for your comfort,*
Yea, naught for your desire,
Save that the sky grows darker yet
And the sea rises higher.

LECTOR *And I have miles to go before I sleep.*

Problems

16. Let A be the matrix (relative to an orthonormal basis) of the linear operator $f: S \to S$. Prove that f is normal if and only if A is normal.

17. Give an example showing that the product of two normal matrices is not necessarily normal.

18. Prove that the inverse of a nonsingular normal matrix is normal. Do nonsingular $n \times n$ normal matrices constitute a group relative to multiplication?

19. Every linear operator whose adjoint exists can be written $f = G + H$, where G is Hermitian and H is skew-Hermitian (Theorem 12.13). Prove that f is normal if and only if G and H commute.

20. Suppose that $f: S \to S$ is normal.

a. Show that $\|f(x)\| = \|f^*(x)\|$ for all $x \in S$. (This completes the proof of Theorem 13.8.)

b. Show that $f - ri$ is normal, where r is any scalar and i is the identity map on S. (This justifies the opening move in the proof of Theorem 13.9.)

c. Prove that if r is an eigenvalue of f and $f^*(x) = \bar{r}x$, then $f(x) = rx$. (This shows that $V \subset U$ in the proof of Theorem 13.9.)

21. Prove that if U is an eigenspace of the normal operator f, then $U^\perp$ is invariant under f.

22. Let $f: S \to S$ be a normal operator and suppose that U is an eigenspace of f. If $g: U^\perp \to U^\perp$ is the restriction of f to $U^\perp$, why is g normal? (This explains an assertion in the proof of Theorem 13.11.)

23. Prove that a normal operator and its adjoint have the same kernel.

24. Prove that if an $n \times n$ normal matrix has a zero eigenvalue of multiplicity k, its rank is $n - k$. Then show that this is false for the nonnormal matrix

$$\begin{bmatrix} 0 & 1 \\ 0 & 0 \end{bmatrix}$$

25. Prove that a normal operator f is unitary if and only if $f^* = f^{-1}$. (In a finite-dimensional space we need not assume that f is normal; see Theorem 12.8. But in general we cannot say that $f^* = f^{-1}$ when f is merely unitary. Why?)

26. Explain why a diagonal matrix with entries in $\mathcal{C}$ is Hermitian, skew-Hermitian, unitary depending on whether its diagonal entries are real, pure imaginary, of absolute value 1, respectively. (This completes the proof of Theorem 13.13.)

27. Confirm that

$$A = \begin{bmatrix} 1 & i \\ i & 1 \end{bmatrix}$$

is normal, and find:

a. A unitary matrix P such that PAP^* is diagonal

b. A spectral decomposition of the normal operator $f(\mathbf{X}) = A\mathbf{X}$

28. Find a spectral decomposition of the symmetric operator $f(\mathbf{X}) = A\mathbf{X}$, where

$$A = \begin{bmatrix} 1 & 1 & 1 \\ 1 & 2 & 0 \\ 1 & 0 & 2 \end{bmatrix}$$

29. Let $f: S \to S$ be a normal [symmetric] operator on the complex [real] finite-dimensional space S, and let $r_1, \ldots, r_m$ be its distinct eigenvalues.

a. Explain why the eigenspace T_j associated with r_j has an orthonormal basis consisting of k_j eigenvectors of f, where k_j is the multiplicity of r_j.

b. Why is the union of these bases a spectral basis associated with f?

c. If $f = r_1f_1 + \cdots + r_mf_m$ is a spectral decomposition of f, prove that $f^2 = r_1^2f_1 + \cdots + r_m^2f_m$. *Hint:* See Prob. 34*a*, Sec. 11.3.

d. Prove that if f is invertible, then $f^{-1} = r_1^{-1}f_1 + \cdots + r_m^{-1}f_m$.

30. Prove that an idempotent normal operator on a complex finite-dimensional space is Hermitian. *Hint:* See Prob. 64, Sec. 12.6, and Prob. 12*a*, Sec. 9.2.

31. Prove that a normal [symmetric] operator on a complex [real] finite-dimensional space is idempotent if and only if its eigenvalues are 0 or 1.

32. Suppose that A is an $n \times n$ normal matrix which is nilpotent, that is, $A^k = 0$ for some positive integer k. (See Prob. 53, Sec. 9.6.) Prove that $A = 0$. *Hint:* Let P be a unitary matrix such that $B = PAP^*$ is diagonal. Then $B^k = PA^kP^*$. (Why?)

Review Quiz

True or false?

1. The sum of two $n \times n$ normal matrices is normal.

2. The matrix

$$\begin{bmatrix} 1 & 2 & 0 & -1 \\ 2 & 5 & 1 & 1 \\ 0 & 1 & -3 & 0 \\ -1 & 1 & 0 & 5 \end{bmatrix}$$

has four real eigenvalues.

3. Every orthogonal operator on a finite-dimensional space has orthonormal eigenvectors spanning the space.

4. Eigenvectors associated with distinct eigenvalues of an orthogonal matrix are orthogonal.

5. The eigenvalues of a reflection of the plane in a line through the origin have absolute value 1.

6. If A is a symmetric matrix, there is an orthogonal matrix P such that PAP^T is diagonal.

7. An $n \times n$ matrix is Hermitian if and only if its eigenvalues are real.

8. Every unitary operator is normal.
9. If r is an eigenvalue of

$$A = \begin{bmatrix} 0 & -1 & -2 \\ 1 & 0 & 5 \\ 2 & -5 & 0 \end{bmatrix}$$

then $\bar{r}$ is an eigenvalue of A^T.

10. The characteristic polynomial of

$$\begin{bmatrix} 1 & -i & 0 \\ i & 2 & i \\ 0 & -i & 0 \end{bmatrix}$$

has real coefficients.

11. Unitary similarity is an equivalence relation.
12. A square matrix can be diagonalized by a unitary similarity transformation if and only if it is normal.

Appendix 1

A1.1 The Dual Space of a Vector Space

Associated with every vector space over $\mathfrak{F}$ is the vector space $L(S,\mathfrak{F})$ of linear maps from S to $\mathfrak{F}$.[†] If S is finite-dimensional, Theorem 3.10 says that

$$\dim L(S,\mathfrak{F}) = (\dim S)(\dim \mathfrak{F}) = \dim S$$

so $L(S,\mathfrak{F})$ and S are isomorphic (Theorem 3.18). The relation between them has been exploited in several profound ways by modern mathematicians.

DUAL SPACE

Let S be a vector space over $\mathfrak{F}$. The vector space $S^* = L(S,\mathfrak{F})$ is called the *dual space* of S; its elements (linear maps from S to $\mathfrak{F}$) are called *linear functionals*.

The reason for the special name is that a map from S to $\mathfrak{F}$ has *numerical values* (real or complex depending on whether $\mathfrak{F}$ is $\mathcal{R}$ or $\mathcal{C}$). In other words, we are mapping S into its own field of scalars, rather than into some other more general space T.

EXAMPLE A1.1

If $S = \mathcal{R}^3$, the dual space is $S^* = L(\mathcal{R}^3,\mathcal{R})$. A typical element is the linear map $f\colon \mathcal{R}^3 \to \mathcal{R}$ defined by $f(x,y,z) = 2x + y - z$, an ordinary real-valued (linear) function of three variables. Letting $\mathbf{X} = (x,y,z)$, we observe that

$$f(\mathbf{X}) = \mathbf{A}\cdot\mathbf{X} \qquad \text{where } \mathbf{A} = (2,1,-1)$$

a formula which is completely described just by naming $\mathbf{A}$. That is, the map is *characterized* by the vector $\mathbf{A}$.

More generally, every element of S^* is of the form

$$f(x,y,z) = ax + by + cz = \mathbf{A}\cdot\mathbf{X} \qquad \text{where } \mathbf{A} = (a,b,c)$$

[†] We regard $\mathfrak{F}$ as a vector space over itself, that is, $\mathfrak{F} = \mathfrak{F}^1$ with dimension 1. See Prob. 2, Sec. 2.1.

This is because a linear map from $\mathcal{R}^3$ to $\mathcal{R}^1$ must send (x,y,z) into a scalar expression linear in x, y, z. (See the remark following Example 3.4 and its proof in Prob. 3, Sec. 4.1.) Hence every linear functional on $\mathcal{R}^3$ has a unique vector "representative" (the vector **A** in the above formula). We'll generalize this idea in the next section.

Now suppose that S is finite-dimensional, with basis $\{u_1, \ldots, u_n\}$. There is a natural choice of basis for S^* corresponding to the given basis for S; for each $j = 1, \ldots, n$, define $f_j \colon S \to \mathfrak{F}$ by

$$f_j(x) = x_j \qquad \text{where } x = \sum_{j=1}^{n} x_j u_j \text{ is any element of } S$$

In other words, f_j is the "projection" which maps each element of S into the jth component of its coordinate vector $(x_1, \ldots, x_n)$ relative to the given basis. Obviously $f_1, \ldots, f_n$ are linear functionals; since they are linearly independent in S^* (confirm this!), they constitute a basis for S^*. (Why?)

DUAL BASIS

Let S be a vector space over $\mathfrak{F}$ with basis $\{u_1, \ldots, u_n\}$. For each $j = 1, \ldots, n$, define the linear functional $f_j \in S^*$ by

$$f_j(x) = x_j \qquad \text{where } x = \sum_{j=1}^{n} x_j u_j \text{ is any element of } S$$

Then $\{f_1, \ldots, f_n\}$ is called the *dual basis* of $\{u_1, \ldots, u_n\}$.

EXAMPLE A1.2

If $\{\mathbf{E}_1, \mathbf{E}_2, \mathbf{E}_3\}$ is the standard basis for $S = \mathcal{R}^3$, the dual basis is $\{f_1, f_2, f_3\}$, where the f_j are the linear functionals on $\mathcal{R}^3$ defined by

$$f_1(x,y,z) = x \qquad f_2(x,y,z) = y \qquad f_3(x,y,z) = z$$

Each $f \in S^*$ can be written (uniquely) as a linear combination of f_1, f_2, and f_3, say, $f = af_1 + bf_2 + cf_3$. Hence each $f \in S^*$ has the form $f(x,y,z) = ax + by + cz$, an observation we made in Example A1.1 without appealing to the dual basis.

Problems

1. Let S be a vector space over $\mathfrak{F}$ with basis $\{u_1, \ldots, u_n\}$.
   a. Show that every linear functional on S has the form $f(x) = \mathbf{A} \cdot \mathbf{X}$, where $\mathbf{A}$ is the vector with components $a_j = f(u_j)$, $j = 1, \ldots, n$, and $\mathbf{X}$ is the coordinate vector of x relative to the given basis.
   b. Why does it follow that every linear functional on n-space is of the form $f(x_1, \ldots, x_n) = a_1x_1 + \cdots + a_nx_n$?
2. Let S be a vector space over $\mathfrak{F}$ with basis $\{u_1, \ldots, u_n\}$. For each $j = 1, \ldots, n$, define $f_j: S \to \mathfrak{F}$ by

$$f_j(x) = x_j \qquad \text{where } x = \sum_{j=1}^{n} x_j u_j \text{ is any element of } S$$

   a. Explain why f_j is a linear functional on S, hence an element of S^*.
   b. Show that $f_1, \ldots, f_n$ are linearly independent in S^*.
   c. Why does it follow that $\{f_1, \ldots, f_n\}$ is a basis for S^*?
3. Let S be the vector space of real-valued functions differentiable on $[0,1]$. Which of the following maps are elements of S^*?
   a. $\phi: S \to \mathfrak{R}$ defined by $\phi(f) = f(0)$
   b. $I: S \to \mathfrak{R}$ defined by $I(f) = \int_0^1 f(x)\,dx$
   c. $D: S \to \mathfrak{R}$ defined by $D(f) = f'(0)$
4. Let S be the vector space of $n \times n$ matrices with entries in $\mathfrak{F}$. The "trace" of a matrix $A = [a_{ij}]$ is $\operatorname{tr} A = \sum_{j=1}^{n} a_{jj}$ (the sum of the diagonal entries). Is the function $f(A) = \operatorname{tr} A$ an element of S^*? What about the function $g(A) = \det A$?
5. What is the dual basis of the basis $\{E_{11}, E_{12}, E_{21}, E_{22}\}$ for the space of 2×2 matrices?
6. A basis for $S = \mathfrak{R}^2$ is $\{(1,1), (-1,1)\}$. What is the dual basis for S^*?
7. According to Prob. 1*b*, every linear functional on $\mathfrak{F}^n$ is of the form $f(x_1, \ldots, x_n) = a_1x_1 + \cdots + a_nx_n$. Prove this by using the dual basis of the standard basis.

A1.2 The Representation Theorem

In the last section we proved that every linear functional on n-space is of the form

$$f(x_1, \ldots, x_n) = a_1x_1 + \cdots + a_nx_n = \mathbf{A} \cdot \mathbf{X} \qquad \text{where } \mathbf{A} = (a_1, \ldots, a_n)$$

It is natural to say that the vector **A** "represents" the map f, for if we know **A**, then f is determined. The role played by the dot product in this representation is obvious; our next definition is designed to generalize the idea.

SCALAR PRODUCT

Let S be a vector space over $\mathfrak{F}$ and suppose that $\circ$ is a scalar-valued product on S which is symmetric and bilinear, that is,

$$x \circ y = y \circ x \qquad \text{and} \qquad (ax + by) \circ z = a(x \circ z) + b(y \circ z)$$

for all x, y, and z in S and all scalars a and b. Then $\circ$ is called a *scalar product*. A *nondegenerate* scalar product has the additional property that if $x \circ y = 0$ for all x, then $y = 0$.

EXAMPLE A1.3

The dot product on $\mathfrak{F}^n$

$$\mathbf{X} \cdot \mathbf{Y} = \sum_{j=1}^{n} x_j y_j$$

is symmetric and bilinear; hence it is a scalar product. As we remarked in Sec. 10.1, it is not positive-definite unless $\mathfrak{F} = \mathfrak{R}$, since $\mathbf{X} \cdot \mathbf{X} = 0$ does not imply that $\mathbf{X} = \mathbf{0}$. It is, however, nondegenerate, because

$$\mathbf{X} \cdot \mathbf{Y} = 0 \text{ for all } \mathbf{X} \quad \Rightarrow \quad \mathbf{Y} = \mathbf{0} \qquad \text{(Why?)}$$

EXAMPLE A1.4

A "degenerate" scalar product (Prob. 22, Sec. 10.2) is

$$\mathbf{X} \circ \mathbf{Y} = (x_1 - x_2)(y_1 - y_2)$$

where $\mathbf{X} = (x_1, x_2)$ and $\mathbf{Y} = (y_1, y_2)$ are elements of $\mathfrak{R}^2$. For if $\mathbf{Y} = (1,1)$, then $\mathbf{X} \circ \mathbf{Y} = 0$ for all $\mathbf{X}$, yet $\mathbf{Y}$ is nonzero.

REPRESENTATIVE OF A LINEAR FUNCTIONAL

Let S be a vector space over $\mathfrak{F}$ with a nondegenerate scalar product $\circ$. If $f: S \rightarrow \mathfrak{F}$ is a linear functional and there exists an element $a \in S$ with the property that $f(x) = a \circ x$ for all $x \in S$, we say that *a represents f* relative to $\circ$.

Such a representative is unique if it exists (see the problems), but unless S is finite-dimensional we cannot guarantee its existence. Note, too, that the bilinearity of $\circ$ is essential (otherwise the function $a \circ x$ would not be linear). That is why we use the dot product on n-space (rather than the standard inner product) and why we have introduced the idea of scalar product as a substitute for inner product (which is not bilinear when $\mathfrak{F} = \mathcal{C}$).

THE REPRESENTATION THEOREM

Let S be a finite-dimensional space over $\mathfrak{F}$ with a nondegenerate scalar product $\circ$. Then every linear functional $f\colon S \to \mathfrak{F}$ has a unique vector representative relative to $\circ$; that is, the formula defining f is $f(x) = a \circ x$ for some unique $a \in S$.

Proof

Define the map $p\colon S \to S^*$ by $p(a) = f$, where $a \in S$ and $f\colon S \to \mathfrak{F}$ is the linear functional $f(x) = a \circ x$. Then p is linear (confirm this!) and one-to-one (because $\circ$ is nondegenerate). Since $\dim S = \dim S^*$, it is also onto. Both the existence and uniqueness of a representative $a \in S$ for each $f \in S^*$ follow. ∎

This theorem doesn't seem to say much when we look at a linear functional like $f(x,y,z) = 2x + y - z$. The vector representing f relative to the dot product is simply $\mathbf{A} = (2,1,-1)$, as we observed in the last section. For a less obvious illustration, consider the following.

EXAMPLE A1.5

Let $F\colon \mathcal{R}^2 \to \mathcal{R}$ be a function that is continuous and has continuous partial derivatives F_1 and F_2. In Example 3.10 we pointed out that the "differential" of F at a point P, usually written

$$dF = F_1(x,y)\,dx + F_2(x,y)\,dy$$

is really a linear map $dF\colon \mathcal{R}^2 \to \mathcal{R}$ defined by

$$(dF)(x_1,x_2) = F_1(P)x_1 + F_2(P)x_2$$

According to the representation theorem, there is a unique vector $\mathbf{A}$ in $\mathcal{R}^2$ which represents dF relative to the dot product, that is,

$$(dF)(\mathbf{X}) = \mathbf{A}\cdot\mathbf{X} \qquad \text{for all } \mathbf{X} \in \mathcal{R}^2$$

You can see what $\mathbf{A}$ is by looking at the formula defining dF; it is the vector $(F_1(P),F_2(P))$. In calculus this is called the "gradient" of F at P, usually written $\nabla F(P)$. Thus

$$(df)(\mathbf{X}) = \nabla F(P)\cdot\mathbf{X} \qquad \text{for all } \mathbf{X} \in \mathcal{R}^2$$

When S is not finite-dimensional the question of representing a linear functional on S by a vector in S is much deeper. For example, let S be the vector space of real-valued functions continuous on $[-\pi,\pi]$, with nondegenerate scalar product

$$f * g = \int_{-\pi}^{\pi} f(x)g(x)\, dx$$

If $\phi\colon S \to \mathcal{R}$ is a linear functional (see Prob. 3 in Sec. A1.1 for the sort of thing we have in mind), it is not clear whether there is a unique function $a \in S$ such that $\phi(f) = a * f$ for all $f \in S$. One result in this direction is the famous Riesz representation theorem, which states that if S is enlarged to the space of all functions whose squares are integrable (this includes continuous functions) and if ϕ has a finite "norm," then the answer is yes.† However, this is not as simple as it sounds, for the scalar product on S is now defined in terms of the "Lebesgue" integral, not the ordinary Riemann integral of elementary calculus. The Lebesgue theory of integration was developed shortly after the turn of the century to answer questions of this sort arising in Fourier analysis and elsewhere.

Problems

8. Prove that the dot product on $\mathfrak{F}^n$ is nondegenerate, that is,

$$\mathbf{X} \cdot \mathbf{Y} = 0 \text{ for all } \mathbf{X} \quad \Rightarrow \quad \mathbf{Y} = \mathbf{0}$$

Hint: Let $\mathbf{Y} = (y_1, \ldots, y_n)$ and take $\mathbf{X} = (\bar{y}_1, \ldots, \bar{y}_n)$.

9. Explain why the positive-definite property of an inner product is stronger than the nondegenerate property, that is,

$$\text{Positive-definite} \quad \Rightarrow \quad \text{nondegenerate}$$

but the reverse implication is invalid.

10. Let S be a vector space over $\mathfrak{F}$ with a nondegenerate scalar product $\circ$, and suppose that $a \in S$. Define the map $f\colon S \to \mathfrak{F}$ by $f(x) = a \circ x$ for all $x \in S$.

a. Show that f is linear, hence an element of S^*.

b. Explain why f would not be linear if S were a unitary space over $\mathfrak{F} = \mathcal{C}$ and its inner product were used in place of $\circ$.

c. Define the map $p\colon S \to S^*$ by $p(a) = f$. Why is p linear?

d. Prove that p is one-to-one. *Hint:* Use the fact that $\circ$ is nondegenerate to show that ker p is the zero subspace of S.

This shows that the vector representative of a linear functional on S is unique if it exists. However, unless p is onto (hence an isomorphism between S and S^*), we cannot conclude that every $f \in S^*$ has a repre-

† See, for example, N. I. Akhiezer and I. M. Glazman, *Theory of Linear Operators in Hilbert Space* (translated from the Russian by Nestell), New York: Ungar, 1961. The "norm" of a linear functional is something we have not defined.

sentative $a \in S$. There is no problem when $\dim S = \dim S^*$; it is when this cannot be invoked that representation theory gets deep.

11. Define $F: \mathcal{R}^2 \to \mathcal{R}$ by $F(x,y) = e^{xy}$. Show that the vector representing the linear functional dF is $e^{xy}(y,x)$.

12. What vector in $\mathcal{R}^3$ represents the differential of $F(x,y,z) = x^2 + y^2 + z^2$?

13. Let S be the vector space of 2×2 matrices with entries in $\mathfrak{F}$.

a. Show that the product $\circ$ defined by $A \circ B = a_{11}b_{11} + a_{12}b_{12} + a_{21}b_{21} + a_{22}b_{22}$, where $A = [a_{ij}]$ and $B = [b_{ij}]$, is a nondegenerate scalar product on S. Is it positive-definite?

b. Define $f: S \to \mathfrak{F}$ by $f(A) = \operatorname{tr} A$. (See Prob. 4 in Sec. A1.1.) What element of S represents f relative to $\circ$?

14. Let S be the euclidean space of linear combinations of $\cos x$ and $\sin x$, with inner product

$$f * g = \int_{-\pi}^{\pi} f(x)g(x)\,dx$$

Why is $*$ a nondegenerate scalar product on S? What element of S represents the linear functional $D: S \to \mathcal{R}$ defined by $D(f) = f'(0)$?

A1.3 Orthogonality Revisited

In Chap. 11 we proved that if U is a subspace of the finite-dimensional unitary space S, then

$$\dim U + \dim U^\perp = \dim S$$

where $U^\perp$ is the orthogonal complement of U. When $S = \mathfrak{F}^n$, this reads

$$\dim U + \dim U^\perp = n$$

with the understanding that orthogonality is defined in terms of the dot product when $\mathfrak{F} = \mathcal{R}$ and in terms of the standard inner product when $\mathfrak{F} = \mathcal{C}$.

Our purpose in this section is to prove that the dot product can be used in both cases, a fact which is useful in proving that the row rank and column rank of a matrix are equal.† The key to the argument is the dual space.

† See Theorem 7.12 and Prob. 33, Sec. 7.5, also Prob. 23, Sec. 11.2. The theorem has been proved for matrices with real entries, but not in general.

ANNIHILATOR

Let U be a (nonempty) subset of the vector space S. The set of maps in S^* which send every element of U into 0 is called the *annihilator* of U, denoted by U^0.

EXAMPLE A1.6

Let U be the x axis in $S = \mathcal{R}^3$. The annihilator of U is the set of linear functionals $f(x,y,z) = ax + by + cz$ with the property that

$$f(x,0,0) = 0 \qquad \text{for all } x$$

Since the identity $f(x,0,0) = ax = 0$ implies that $a = 0$, U^0 is the set of maps of the form $f(x,y,z) = by + cz$, that is,

$$U^0 = \{f \in S^* \mid f = bf_2 + cf_3\} = \langle f_2,f_3 \rangle$$

where $\{f_1,f_2,f_3\}$ is the dual basis of $\{\mathbf{E}_1,\mathbf{E}_2,\mathbf{E}_3\}$. (See Example A1.2.) Hence U^0 is a subspace of S^*, a fact we ask you to prove in general in the problems.

Note that since $\dim U = 1$ and $\dim U^0 = 2$ in this example, we have

$$\dim U + \dim U^0 = \dim S$$

This is no accident, as we now show.

THEOREM A1.1

If U is a subspace of the finite-dimensional space S and U^0 is its annihilator in the dual space, then $\dim U + \dim U^0 = \dim S$.

Proof

Assume that U is neither the zero subspace of S nor all of S. (Otherwise the theorem is trivial.) Then U has a basis $\{u_1, \ldots, u_m\}$; build this up to a basis $\{u_1, \ldots, u_n\}$ for S and let $\{f_1, \ldots, f_n\}$ be the dual basis for S^*. We assert that $\{f_{m+1}, \ldots, f_n\}$ is a basis for U^0. If so, the proof is complete.

Note that $f_{m+1}, \ldots, f_n$ are elements of U^0, because each maps u_j into 0 if $j = 1, \ldots, m$. (Why?) Since they are linearly independent, we need only show that they span U^0. Take any $f \in U^0$. Then there are scalars $c_1, \ldots, c_n$ such that $f = \sum_{i=1}^{n} c_i f_i$. But $f(u_j) = 0$, $j = 1, \ldots, m$, so

$$\sum_{i=1}^{n} c_i f_i(u_j) = \sum_{i=1}^{n} c_i \delta_{ij} = c_j = 0 \qquad j = 1, \ldots, m$$

Hence

$$f = \sum_{i=m+1}^{n} c_i f_i \; \blacksquare$$

Now recall from the last section that the dot product on $\mathfrak{F}^n$ is nondegenerate, symmetric, and bilinear. These properties are inadequate for the definition of length and distance in complex n-space, as we saw in Sec. 10.1. But they are sufficient for a new definition of orthogonality. Let us agree (for our present purposes only) that

$$\mathbf{X} \text{ and } \mathbf{Y} \text{ are } \textit{orthogonal} \text{ if } \mathbf{X} \cdot \mathbf{Y} = 0^\dagger$$

If U is a subspace of $\mathfrak{F}^n$, we'll use the symbol $U^\perp$ for the subspace of $\mathfrak{F}^n$ consisting of vectors orthogonal to every vector in U. (Again this is not the same as $U^\perp$ defined in $\mathfrak{F}^n$ as a unitary space, unless $\mathfrak{F} = \mathfrak{R}$. We are using the same symbol because the idea is the same in terms of our new definition of orthogonality.)

THEOREM A1.2

If U is a subspace of $\mathfrak{F}^n$ and

$$U^\perp = \{\mathbf{Y} \in \mathfrak{F}^n \mid \mathbf{X} \cdot \mathbf{Y} = 0 \text{ for all } \mathbf{X} \in U\}$$

then $\dim U + \dim U^\perp = n$.

Proof

Let $S = \mathfrak{F}^n$ and define the map $p\colon S \to S^*$ as in the proof of the representation theorem in the last section: $p(\mathbf{A}) = f$, where $\mathbf{A} \in S$ and $f\colon S \to \mathfrak{F}$ is the linear functional $f(\mathbf{X}) = \mathbf{A} \cdot \mathbf{X}$. Then (Prob. 10, Sec. A1.2) p is linear and one-to-one. The key to the proof is that p sends every element of $U^\perp$ into U^0, the annihilator of U. If $\mathbf{A} \in U^\perp$, then $p(\mathbf{A}) = f$ is a linear functional with the property that

$$f(\mathbf{X}) = \mathbf{A} \cdot \mathbf{X} = 0 \qquad \text{for every } \mathbf{X} \in U$$

Restricting p to the domain $U^\perp$, we have a linear one-to-one map $p\colon U^\perp \to U^0$; if this map is onto, then $\dim U^\perp = \dim U^0$ and the equation

$$\dim U + \dim U^0 = \dim S$$

from Theorem A1.1 reads

$$\dim U + \dim U^\perp = n \qquad (!)$$

To prove that p maps $U^\perp$ onto U^0, take any $f \in U^0$. The problem is to name a vector $\mathbf{A} \in U^\perp$ such that $p(\mathbf{A}) = f$. But this is easy! According to the representation theorem in the last section, f has a unique representative $\mathbf{A} \in S$ relative to the dot product; this means that $f(\mathbf{X}) = \mathbf{A} \cdot \mathbf{X}$ for all $\mathbf{X}$. Since $f \in U^0$, we have $\mathbf{A} \cdot \mathbf{X} = 0$ for all $\mathbf{X} \in U$, so $\mathbf{A} \in U^\perp$. ∎

THEOREM A1.3

The row rank and column rank of a matrix are equal.

† Remember that in $\mathfrak{F}^n$ as a unitary space $\mathbf{X}$ and $\mathbf{Y}$ are orthogonal if $\mathbf{X} * \mathbf{Y} = 0$, where $*$ is the standard inner product. We are abandoning this definition for the moment, at least when $\mathfrak{F} = \mathfrak{C}$.

Proof

Let $A \in M_{m,n}(\mathfrak{F})$ and define the standard map $f: \mathfrak{F}^n \to \mathfrak{F}^m$ by $f(\mathbf{X}) = A\mathbf{X}$. Since ker f consists of those vectors in $\mathfrak{F}^n$ whose dot product with each of the rows of A is 0, we have

$$\ker f = [R(A)]^\perp$$

where $R(A)$ is the row space of A and the orthogonal complement is understood as in this section.

The range of f is the set of vectors $A\mathbf{X} \in \mathfrak{F}^m$ obtained when $\mathbf{X}$ takes on values in $\mathfrak{F}^n$. But $A\mathbf{X}$ (a matrix product having one column) is a linear combination of the columns of A (Theorem 7.1). Hence the range of f is the set of all linear combinations of the columns of A, that is,

$$\text{rng } f = C(A)$$

the column space of A. Sylvester's law of nullity therefore yields

$$\dim [R(A)]^\perp + \dim C(A) = n$$

On the other hand, Theorem A1.2 says that

$$\dim R(A) + \dim [R(A)]^\perp = n$$

It follows that

$$r(A) = \dim R(A) = \dim C(A) = c(A)$$

where $r(A)$ and $c(A)$ are the row rank and column rank of A, respectively. ▮

Problems

15. Let U be a (nonempty) subset of the vector space S. Prove that its annihilator U^0 is a subspace of S^*.

16. If S is finite-dimensional and U is the zero subspace of S or all of S, explain why $\dim U + \dim U^0 = \dim S$.

17. Let U be the xy plane in $S = \mathcal{R}^3$. Find U^0 and confirm the formula $\dim U + \dim U^0 = \dim S$.

18. Prove that if U and V are subspaces of the vector space S, then

$$(U + V)^0 = U^0 \cap V^0$$

(See Prob. 20, Sec. 11.2, for a similar formula concerning orthogonal complements.)

19. Confirm the formula in Prob. 18 if U and V are the x axis and xy plane, respectively, in $S = \mathcal{R}^3$.

20. Let S be a finite-dimensional space which is the direct sum of subspaces U and V. Prove that the dual space is the direct sum of the annihilators U^0 and V^0. *Hint:* Use Corollary 11.2a.

21. Let U be a subspace of $\mathfrak{F}^n$ and let $U^\perp = \{\mathbf{Y} \in \mathfrak{F}^n | \mathbf{X} \cdot \mathbf{Y} = 0 \text{ for all } \mathbf{X} \in U\}$. Explain why $U^\perp$ and the annihilator of U are isomorphic.

Appendix 2

A2.1 Triangulation of a Matrix

TRIANGULAR MATRIX

An $n \times n$ matrix $A = [a_{ij}]$ is said to be *upper triangular* if the entries below the main diagonal are all zero, that is, if $a_{ij} = 0$ whenever $i > j$. If the entries above the main diagonal are zero, that is, if $a_{ij} = 0$ whenever $i < j$, we say that A is *lower triangular*. The unqualified term *triangular* will mean upper triangular (since our discussion will be confined to these matrices and can be duplicated for lower triangular matrices).

EXAMPLE A2.1

In the matrix

$$A = \begin{bmatrix} 1 & -1 & 0 \\ 0 & 2 & 1 \\ 0 & 0 & 3 \end{bmatrix}$$

we have $a_{21} = a_{31} = a_{32} = 0$; a glance at the main diagonal and the entries above it will show why A is called upper triangular.

The object of this section is to show that every matrix representation of a linear operator can be "triangulated" (provided that the base field is taken to be $\mathfrak{F} = \mathfrak{C}$). In other words, every square matrix is similar to an upper triangular matrix. In view of the failure of diagonalization in some cases, this is a refreshing result. It is also important, since triangular matrices are almost as convenient as diagonal matrices for many purposes.

There are several ways to prove this theorem. We follow Serge Lang's approach via "fans," for it is not only elegant in its own right, but leads to a relatively simple proof of the Cayley-Hamilton theorem. (See Sec. 9.6.)

INVARIANT SUBSPACE

Let $f: S \rightarrow S$ be a linear operator and suppose that T is a subspace of S which f maps into itself, that is,

$$x \in T \quad \Rightarrow \quad f(x) \in T$$

Then T is said to be *invariant* under f.†

EXAMPLE A2.2

Define $f: \mathcal{R}^3 \rightarrow \mathcal{R}^3$ by $f(x,y,z) = (x + y + z, 2x, 0)$ and let T be the xy plane. Then T is invariant under f, since for every $(x,y,0) \in T$ we have $f(x,y,0) = (x + y, 2x, 0) \in T$. The eigenspaces of f, namely

$$T_1 = \langle(1,1,0)\rangle \qquad T_2 = \langle(1,-2,0)\rangle \qquad T_3 = \langle(0,1,-1)\rangle$$

are also invariant under f. However, note that T is not an eigenspace; instead, $T = T_1 \oplus T_2$.

FAN OF A LINEAR OPERATOR

Let $f: S \rightarrow S$ be a linear operator on the n-dimensional space S $(n > 0)$. A *fan* of f is a telescoping sequence of invariant subspaces, that is, a sequence $T_1, \ldots, T_n$ such that

$$T_1 \subset T_2 \subset \cdots \subset T_n$$

each T_k is invariant under f, and

$$\dim T_1 = 1, \dim T_2 = 2, \ldots, \dim T_n = n$$

Note that $T_n = S$. (Why?) A *fan basis* associated with the fan $T_1, \ldots, T_n$ is a basis $\{v_1, \ldots, v_n\}$ for S with the property that $\{v_1, \ldots, v_k\}$ is a basis for T_k, where $k = 1, \ldots, n$.

EXAMPLE A2.3

If $f: \mathcal{R}^3 \rightarrow \mathcal{R}^3$ is the orthogonal projection of $S = \mathcal{R}^3$ on the xy plane, a fan of f is the sequence T_1, T_2, T_3, where T_1 is the x axis, T_2 is the xy plane, and $T_3 = \mathcal{R}^3$. A fan basis associated with this fan is $\{\mathbf{E}_1, \mathbf{E}_2, \mathbf{E}_3\}$.

While Example A2.3 is simple enough, it is not obvious (in general) that an arbitrary linear operator has a fan. Assuming, however, that $T_1, \ldots, T_n$

† See Prob. 8, Sec. 9.2, where we asked you to explain why each eigenspace of f is invariant under f.

is a fan of f, it is easy to construct a fan basis. We may let v_1 be any nonzero element of T_1, then build up $\{v_1\}$ to a basis $\{v_1, v_2\}$ for T_2, and so on until we have built up the basis $\{v_1, \ldots, v_n\}$ for $T_n = S$. (See Theorem 2.7.) The significance of such a basis for our present purposes is shown by the following.

THEOREM A2.1

The matrix of a linear operator relative to a fan basis is upper triangular.

Proof

Let $f: S \to S$ be the given map and suppose that $T_1, \ldots, T_n$ is a fan of f, the corresponding fan basis being $\{v_1, \ldots, v_n\}$. Let j be any one of the integers $1, \ldots, n$. Since $v_j \in T_j$ and T_j is invariant under f, we know that $f(v_j) \in T_j$. But $\{v_1, \ldots, v_j\}$ is a basis for T_j, so there are scalars $a_{1j}, \ldots, a_{jj}$ such that

$$f(v_j) = a_{1j}v_1 + \cdots + a_{jj}v_j$$

Letting j take on the values $1, \ldots, n$, we find that

$$\begin{aligned} f(v_1) &= a_{11}v_1 \\ f(v_2) &= a_{12}v_1 + a_{22}v_2 \\ &\cdots\cdots\cdots\cdots\cdots\cdots \\ f(v_n) &= a_{1n}v_1 + a_{2n}v_2 + \cdots + a_{nn}v_n \end{aligned}$$

Hence the matrix of f relative to $\{v_1, \ldots, v_n\}$ is

$$A = \begin{bmatrix} a_{11} & a_{12} & \cdots & a_{1n} \\ & a_{22} & \cdots & a_{2n} \\ \bigcirc & & \ddots & \vdots \\ & & & a_{nn} \end{bmatrix}$$

where the $\bigcirc$ means that all entries below the main diagonal are zero. Thus A is upper triangular. ▮

Note that the first equation in the above list, $f(v_1) = a_{11}v_1$, implies that v_1 is an eigenvector of f. The vectors $v_1, \ldots, v_n$ in a fan basis are a generalization of eigenvectors spanning S (which lead to a *diagonal* matrix, as we saw in Sec. 9.4). While such eigenvectors do not always exist, a fan basis does (at least when the base field is $\mathfrak{F} = \mathfrak{C}$), as we now show.

THEOREM A2.2

Every linear operator on a (nonzero) finite-dimensional vector space over $\mathfrak{C}$ has a fan.

Proof

Let $f: S \to S$ be the linear operator. We use induction on n, where $n = \dim S$. If $n = 1$, we need only take $T_1 = S$ and we have a fan already. (Why?) Hence assume that $n > 1$ and that the theorem is true for linear operators on $(n - 1)$-dimensional spaces over $\mathcal{C}$. Since the base field is $\mathcal{C}$, we know that f has at least one eigenvector v_1. Let $T_1 = \langle v_1 \rangle$ and use Theorem 11.4 to write $S = U \oplus V$, where $U = T_1$ and V is some $(n - 1)$-dimensional subspace of S.

Now although $U = T_1$ is invariant under f (it is a subspace of an eigenspace!), there is no reason to suppose that V is. However we can find a map under which V *is* invariant by using an idea similar to orthogonal projection. Let $g: S \to S$ and $h: S \to S$ be the functions defined by

$$g(x) = x_1 \qquad \text{and} \qquad h(x) = x_2$$

where $x = x_1 + x_2$ is the unique expression of $x \in S$ as a sum of $x_1 \in U$ and $x_2 \in V$.†

The maps g and h are linear and their sum is the identity map on S. (Confirm this!) Moreover, V is invariant under $h \circ f$ (why?), so if we restrict $h \circ f$ to the domain V, we are dealing with a linear operator $h \circ f: V \to V$ on an $(n - 1)$-dimensional space. According to the induction hypothesis, $h \circ f$ has a fan, say $W_1, \ldots, W_{n-1}$; this is a sequence of subspaces of V which are invariant under $h \circ f$, with

$$W_1 \subset W_2 \subset \cdots \subset W_{n-1} \qquad \text{and} \qquad \dim W_k = k \qquad k = 1, \ldots, n-1$$

We use it to build up T_1 to a fan of f, as follows. Let

$$T_2 = U \oplus W_1,\ T_3 = U \oplus W_2,\ \ldots,\ T_n = U \oplus W_{n-1}$$

(Why are these *direct* sums?) It is obvious that

$$T_1 \subset T_2 \cdots \subset T_n \qquad \text{and} \qquad \dim T_k = k \qquad k = 1, \ldots, n$$

To prove that $T_1, \ldots, T_n$ is a fan of f, we need only show that each T_k is invariant under f, that is, $x \in T_k \Rightarrow f(x) \in T_k$.

Take any $x \in T_k$, where k is one of the integers $2, \ldots, n$. (If $k = 1$, there is nothing to prove.) Since $T_k = U \oplus W_{k-1}$, we may write

$$x = y + z \qquad \text{where } y \in U \text{ and } z \in W_{k-1}$$

To show that $f(x) \in T_k$, observe that $f(x) = f(y) + f(z)$, where $f(y) \in U$ because U is invariant under f. Since $U = T_1 \subset T_k$, we have $f(y) \in T_k$, and all that remains is to show that $f(z) \in T_k$. (T_k is a subspace of S, hence closed relative to addition.)

Since $g + h$ is the identity map on S, we can write

$$f = i \circ f = (g + h) \circ f = (g \circ f) + (h \circ f)$$

from which $f(z) = g[f(z)] + h[f(z)]$. However g sends everything into U and $U \subset T_k$, and therefore $g[f(z)] \in T_k$. Moreover, $z \in W_{k-1}$ and W_{k-1} is invariant

† If S happens to be unitary and $V = U^{\perp}$, then g and h are the orthogonal projections of S on U and $U^{\perp}$, respectively.

under $h \circ f$, so $h[f(z)] \in W_{k-1}$. Since $W_{k-1} \subset T_k$ (why?), we have $h[f(z)] \in T_k$. Hence $f(z) \in T_k$. ∎

COROLLARY A2.2a

Let f: $S \to S$ be a linear operator, where S is a (nonzero) finite-dimensional vector space over $\mathcal{C}$. Then S has a basis relative to which the matrix of f is upper triangular.

COROLLARY A2.2b

Every $n \times n$ matrix is similar to an upper triangular matrix (provided that we are willing to accept complex entries); that is, every square matrix can be triangulated.

Note that if the base field is $\mathcal{R}$, triangulation is not always possible, because it requires the existence of an eigenvalue (why?) and not every matrix with real entries has a real eigenvalue. (See Example 9.13.)

EXAMPLE A2.4

Let $f(\mathbf{X}) = A\mathbf{X}$, where

$$A = \begin{bmatrix} 3 & 0 & 1 \\ 2 & 2 & 1 \\ -6 & -1 & -2 \end{bmatrix}$$

In Prob. 32g, Sec. 9.4, we found the triple eigenvalue $r = 1$, the associated eigenspace being $T_1 = \langle \mathbf{V}_1 \rangle$, where $\mathbf{V}_1 = (1,0,-2)$. Hence A cannot be diagonalized. As in the proof of Theorem A2.2, write $U = T_1$ and $S = U \oplus V$, where $S = \mathcal{R}^3$ and V is 2-dimensional. A convenient choice for V is the yz plane; since we can write each element $\mathbf{X} = (x,y,z)$ of S as

$$\mathbf{X} = \mathbf{X}_1 + \mathbf{X}_2$$

where

$$\mathbf{X}_1 = (x,0,-2x) \in U \quad \text{and} \quad \mathbf{X}_2 = (0,\ y,\ 2x + z) \in V$$

the map h in the proof is defined by

$$h(x,y,z) = (0,\ y,\ 2x + z)$$

Then $h \circ f$ is defined by

$$\begin{aligned}(h \circ f)(x,y,z) &= h(3x + z,\ 2x + 2y + z,\ -6x - y - 2z) \\ &= (0,\ 2x + 2y + z,\ -y)\end{aligned}$$

As you can see, V is invariant under $h \circ f$.

Regarding $h \circ f$ as a linear operator on V, we need only find an eigenvector to construct a fan W_1, W_2. (Why?) For this purpose, we compute the matrix of $h \circ f$ relative to the basis $\{\mathbf{E}_2, \mathbf{E}_3\}$ for V:

$$(h \circ f)(\mathbf{E}_2) = (0,2,-1) = 2\mathbf{E}_2 - \mathbf{E}_3$$
$$(h \circ f)(\mathbf{E}_3) = (0,1,0) = \mathbf{E}_2$$

The matrix is

$$\begin{bmatrix} 2 & 1 \\ -1 & 0 \end{bmatrix}$$

with double eigenvalue $r = 1$ and corresponding eigenvector $(1,-1)$. In the context of V as a subspace of $\mathcal{R}^3$ with basis $\{\mathbf{E}_2, \mathbf{E}_3\}$, this is the vector $\mathbf{V}_2 = \mathbf{E}_2 - \mathbf{E}_3 = (0,1,-1)$; we may now name the fan

$$W_1 = \langle \mathbf{V}_2 \rangle \qquad W_2 = V$$

of $h \circ f$ and use it to build up T_1 to a fan of f. Let

$$T_2 = U \oplus W_1 = \langle \mathbf{V}_1 \rangle \oplus \langle \mathbf{V}_2 \rangle = \langle \mathbf{V}_1, \mathbf{V}_2 \rangle \qquad \text{and} \qquad T_3 = U \oplus W_2 = \mathcal{R}^3$$

A fan basis is $\{\mathbf{V}_1, \mathbf{V}_2, \mathbf{V}_3\}$, where $\mathbf{V}_3$ is any vector not in T_2; we choose $\mathbf{V}_3 = (0,0,1)$ and compute

$$f(\mathbf{V}_1) = \mathbf{V}_1$$
$$f(\mathbf{V}_2) = -\mathbf{V}_1 + \mathbf{V}_2$$
$$f(\mathbf{V}_3) = \mathbf{V}_1 + \mathbf{V}_2 + \mathbf{V}_3$$

The matrix of f relative to the fan basis is

$$\begin{bmatrix} 1 & -1 & 1 \\ 0 & 1 & 1 \\ 0 & 0 & 1 \end{bmatrix}$$

Thus although A cannot be diagonalized, it *can* be triangulated—in this case even without resorting to complex numbers. You should not, however, regard this procedure as a practical one; the idea is to throw some light on the proof of Theorem A2.2.

LECTOR *I suppose you've seen the motto "Fiat Lux" of certain colleges.*

AUCTOR *Let there be light.*

LECTOR *Precisely. And do you know the only difference between these colleges and God?*

AUCTOR *I'm afraid to ask.*

LECTOR *Oh, come on.*

AUCTOR *All right, what is the only difference between these colleges and God?*

LECTOR *When God said "Let there be light," there was light.*

Problems

1. Let U be the set of $n \times n$ upper triangular matrices with entries in $\mathfrak{F}$.
   a. Show that U is a subspace of $M_n(\mathfrak{F})$. Can you find its dimension?
   b. Prove that U is closed relative to matrix multiplication and that if $A = [a_{ij}]$ and $B = [b_{ij}]$ are upper triangular, the diagonal entries of AB are $a_{ii}b_{ii}$, where $i = 1, \ldots, n$.
   c. Give an example showing that upper triangular matrices do not necessarily commute.
   d. Explain why U is a ring with unit element.
   e. Explain why U is a linear algebra.
2. Suppose that A and B are $n \times n$ upper triangular matrices. What are the eigenvalues of $A + B$ and AB? *Hint:* See Prob. 37, Sec. 9.4.
3. Prove that the determinant of a triangular matrix is the product of its diagonal entries.
4. Let f: $S \to S$ be a linear operator. Prove that the sum and intersection of two subspaces invariant under f are also invariant under f.
5. Let f: $S \to S$ be a linear operator and suppose that T is a 1-dimensional subspace invariant under f. Prove that T is a subspace of an eigenspace of f.
6. Define f: $\mathfrak{R}^2 \to \mathfrak{R}^2$ by $f(x,y) = (x + 2y, -2x - 3y)$. (See Example 9.12, where we proved that the matrix of f cannot be diagonalized.)
   a. Name a fan of f, and a corresponding fan basis.
   b. Find the triangular matrix of f relative to the fan basis named in part *a*.
   c. Explain why (as in this case) a 2×2 matrix with real entries and a real eigenvalue can be triangulated without appealing to complex numbers.
7. Let

$$A = \begin{bmatrix} 0 & -1 \\ 1 & 0 \end{bmatrix}$$

   a. As the matrix of a linear operator on $\mathfrak{R}^2$, A represents a rotation of the plane through 90°. Why does this imply that A cannot be triangulated using only real numbers?
   b. According to our theory, *every* square matrix can be triangulated using complex numbers. Triangulate A. (Note that, in fact, A can be diagonalized!)
   c. Explain why (as in this case) a 2×2 matrix with real entries and no real eigenvalues cannot be triangulated over $\mathfrak{R}$ but can be diagonalized over $\mathfrak{C}$.

8. Triangulate the matrix

$$A = \begin{bmatrix} 3 & 2 & 1 \\ -1 & 0 & -1 \\ 1 & 1 & 2 \end{bmatrix}$$

(See Prob. 32*h*, Sec. 9.4, where you showed that A cannot be diagonalized.)

9. Let g and h be the maps defined in the proof of Theorem A2.2.
a. Explain why g and h are linear.
b. Why is $g + h$ the identity map on S?
c. Why is V invariant under $h \circ f$?

10. Prove Corollary A2.2a.

11. Prove Corollary A2.2b.

A2.2 The Cayley-Hamilton Theorem

In Prob. 29, Sec. 9.3, it is shown that every 2×2 matrix satisfies its characteristic equation, while in Sec. 9.6 we proved the same thing for $n \times n$ matrices that can be diagonalized. In this section we complete the argument.

THE CAYLEY-HAMILTON THEOREM

Let A be an $n \times n$ matrix with entries in $\mathfrak{F}$ and let p be its characteristic polynomial. Then $p(A) = 0$.

Proof

For the moment suppose that we have proved the theorem when the base field is $\mathfrak{F} = \mathcal{C}$. Then it is also true when $\mathfrak{F} = \mathcal{R}$, for in that case A has real entries and its characteristic equation has real coefficients. Regarding these entries and coefficients as complex numbers, we can apply the theorem already proved to conclude that $p(A) = 0$.

Next, assuming that $\mathfrak{F} = \mathcal{C}$, suppose that the theorem is known to be true for upper triangular matrices. Then it is true in general, since any $n \times n$ matrix with entries in $\mathcal{C}$ is similar to an upper triangular matrix (Corollary A2.2b in the last section) and similar matrices have the same characteristic polynomial. (For an elucidation of this point, see Prob. 49, Sec. 9.6.) Hence we may assume without loss of generality that A is an upper triangular matrix.

Let $f\colon \mathcal{C}^n \to \mathcal{C}^n$ be the standard map $f(\mathbf{X}) = A\mathbf{X}$ and, for each $k = 1, \ldots, n$, let T_k be the subspace of $\mathcal{C}^n$ generated by the standard basis vectors $\mathbf{E}_1, \ldots, \mathbf{E}_k$. Then $T_1, \ldots, T_n$ is a fan of f, the corresponding fan basis being the standard basis; that is, each T_k is invariant under f, has basis $\{\mathbf{E}_1, \ldots, \mathbf{E}_k\}$, and $T_1 \subset \cdots \subset T_n$. (Confirm this!) Since

$$f(\mathbf{E}_k) = a_{1k}\mathbf{E}_1 + \cdots + a_{kk}\mathbf{E}_k$$

we see that

$$f(\mathbf{E}_k) - a_{kk}\mathbf{E}_k \in T_{k-1} \qquad k = 1, \ldots, n$$

(Take T_0 to be the zero subspace to make this sensible when $k = 1$.) In other words,

$$(A - a_{kk}I)\mathbf{E}_k \in T_{k-1} \qquad k = 1, \ldots, n \tag{A2.1}$$

Now $a_{11}, \ldots, a_{nn}$ are the eigenvalues of A (Prob. 37, Sec. 9.4), so the characteristic polynomial of A is

$$p(r) = c(r - a_{11}) \cdots (r - a_{nn})^{\dagger}$$

It follows that

$$p(A) = c(A - a_{11}I) \cdots (A - a_{nn}I)$$

We now use induction to prove that if k is any one of the integers $1, \ldots, n$, then

$$B_k\mathbf{X} = \mathbf{0} \qquad \text{for all } \mathbf{X} \in T_k \tag{A2.2}$$

where

$$B_k = (A - a_{11}I) \cdots (A - a_{kk}I)$$

is the product of the first k factors of $p(A)$. If this can be done, then taking $k = n$ yields $p(A)\mathbf{X} = \mathbf{0}$ for all $\mathbf{X} \in \mathcal{C}^n$, which implies that $p(A) = 0$. (Why?)

The first case of Eq. (A2.2) reads

$$B_1\mathbf{X} = (A - a_{11}I)\mathbf{X} = \mathbf{0} \qquad \text{for all } \mathbf{X} \in T_1$$

which is true because $T_1 = \langle \mathbf{E}_1 \rangle$ and Eq. (A2.1) says that $(A - a_{11}I)\mathbf{E}_1 = \mathbf{0}$. Hence assume that $1 < k \leq n$ and suppose that the $k - 1$ case of Eq. (A2.2) is true. To prove the k case, take any $\mathbf{X} \in T_k$ and write

$$\mathbf{X} = \mathbf{Y} + \mathbf{Z}$$

where

$$\mathbf{Y} \in T_{k-1} \qquad \text{and} \qquad \mathbf{Z} = a\mathbf{E}_k$$

for some scalar a. (Why is this possible?) We have to show that

$$B_k\mathbf{X} = B_k(\mathbf{Y} + \mathbf{Z}) = B_k\mathbf{Y} + B_k\mathbf{Z} = \mathbf{0}$$

† Actually $c = (-1)^n$, but its value is immaterial.

But this is easy! The induction hypothesis says that B_{k-1} times any element of T_{k-1} is $\mathbf{0}$. Since $B_k = B_{k-1}(A - a_{kk}I)$, we need only observe that $(A - a_{kk}I)\mathbf{Y}$ and $(A - a_{kk}I)\mathbf{Z}$ are in T_{k-1}. To confirm that they are, note that $\mathbf{Y} \in T_{k-1}$ and T_{k-1} is invariant under $f(\mathbf{X}) = A\mathbf{X}$. Hence

$$(A - a_{kk}I)\mathbf{Y} = A\mathbf{Y} - a_{kk}\mathbf{Y} \in T_{k-1}$$

Moreover, Eq. (A2.1) says that

$$(A - a_{kk}I)\mathbf{Z} = a(A - a_{kk}I)\mathbf{E}_k \in T_{k-1} \blacksquare$$

Bibliography

These are books from which I have learned something. The list is highly personal, drawn from books I have encountered as a student, used as a teacher, or consulted as a writer. They are uneven in quality, but each has something to offer the reader who wishes to browse in linear algebra and its applications.

Akhiezer, N. I. and Glazman, I. M., *Theory of Linear Operators in Hilbert Space* (trans. from the Russian by Nestell), New York: Ungar, 1961.

Birkhoff, G., and MacLane, S., *A Survey of Modern Algebra* (3d ed.), New York: Macmillan, 1965.

Finkbeiner, D. T., *Introduction to Matrices and Linear Transformations* (2d ed.), San Francisco: Freeman, 1966.

Fisher, R. C., *An Introduction to Linear Algebra,* Encino, Calif.: Dickenson, 1970.

Gillett, P., *Linear Mathematics,* Boston: Prindle, Weber and Schmidt, 1970.

Halmos, P. R., *Finite-Dimensional Vector Spaces* (2d ed.), New York: Van Nostrand, 1958.

Hoffman, K., and Kunze, R., *Linear Algebra* (2d ed.), Englewood Cliffs, N.J.: Prentice-Hall, 1971.

Krause, E., *Introduction to Linear Algebra,* New York: Holt, Rinehart and Winston, 1970.

Kreider, D. L., Kuller, R. G., Ostberg, D. R., and Perkins, F. N., *An Introduction to Linear Analysis,* Reading, Mass.: Addison-Wesley, 1966.

Lang, S., *Linear Algebra,* Reading, Mass.: Addison-Wesley, 1966.

Lipschutz, S., *Linear Algebra,* New York: (Schaum) McGraw-Hill, 1968.

Mal'cev, A. I., *Foundations of Linear Algebra* (trans. from the Russian by T. C. Brown, ed. by J. B. Roberts), San Francisco: Freeman, 1963.

Marcus, M., and Minc, H., *Introduction to Linear Algebra,* New York: Macmillan, 1965.

Nomizu, K., *Fundamentals of Linear Algebra,* New York: McGraw-Hill, 1966.

Schneider, H., and Barker, G. P., *Matrices and Linear Algebra,* 2d ed., New York: Holt, Rinehart and Winston, 1973.

Stewart, F. M., *Introduction to Linear Algebra,* New York: Van Nostrand, 1963.

Stoll, R. R., *Linear Algebra and Matrix Theory,* New York: McGraw-Hill, 1952.

Selected Answers and Hints

These are offered as a check on numerical errors and a help when you are stuck. But a too hasty retreat to this section is bad strategy for anyone trying to develop his mathematical powers. Moreover, the style is often terse, even cryptic—not to be taken as a model of how a lucid argument should read.

Section 1.1

1. If P, Q, R are the tips of $\mathbf{X}$, $\mathbf{Y}$, $\mathbf{X} + \mathbf{Y}$, respectively, then $d(0,P) = d(Q,R)$ and $d(0,Q) = d(P,R)$, where d is the usual distance function.

3. $\mathbf{X}$ is arbitrary; take it to be $(0,0)$.

5. A "coordinate-free" argument (useful later on):

$$\mathbf{X} + \mathbf{Y} = \mathbf{0} \quad \Rightarrow \quad (-\mathbf{X}) + (\mathbf{X} + \mathbf{Y}) = (-\mathbf{X}) + \mathbf{0} \quad \Rightarrow \quad \mathbf{Y} = -\mathbf{X}$$

12. For a coordinate-free argument, see the proof of Theorem 2.4. In the present context, it is easier to write

$$\begin{aligned} c\mathbf{X} = \mathbf{0} \quad &\Leftrightarrow \quad (cx_1,cx_2) = (0,0) \\ &\Leftrightarrow \quad cx_1 = 0 \text{ and } cx_2 = 0 \\ &\Leftrightarrow \quad c = 0 \text{ or } x_1 = x_2 = 0 \\ &\Leftrightarrow \quad c = 0 \text{ or } \mathbf{X} = \mathbf{0} \end{aligned}$$

13. Assuming that length and direction have their usual meaning (they are not formally defined until Chap. 10), it is easy to confirm that $\|c\mathbf{X}\| = |c|\ \|\mathbf{X}\|$ (where the double bars mean length in $\mathcal{R}^2$). The direction of $\mathbf{X}$ is determined by the angle θ $(0 \le \theta < 2\pi)$ satisfying $x_1 = \|\mathbf{X}\| \cos \theta$, $x_2 = \|\mathbf{X}\| \sin \theta$. If ϕ is the angle determining the direction of $c\mathbf{X}$, show that $\phi = \theta$ when $c > 0$, whereas ϕ and θ differ by π when $c < 0$.

14. The usual laws of addition and multiplication in $\mathcal{R}$.

Section 1.2

16. $\mathbf{X} = c_1\mathbf{U}_1 + c_2\mathbf{U}_2 \iff c_1 = x_1$ and $c_2 = x_2 - x_1$

17. The only elements of $\mathcal{R}^2$ that can be written in terms of $\mathbf{U}_1$ and $\mathbf{U}_2$ are the points of the line $y = -2x$ (and each of these can be written in terms of $\mathbf{U}_1$ and $\mathbf{U}_2$ in more than one way).

18. There is only one way to write $\mathbf{0}$ as a linear combination of $\mathbf{U}_1$ and $\mathbf{U}_2$, namely $\mathbf{0} = 0\mathbf{U}_1 + 0\mathbf{U}_2$.

20. Linear dependence is equivalent to the falsity of the implication

$$c_1\mathbf{U}_1 + c_2\mathbf{U}_2 = \mathbf{0} \Rightarrow c_1 = c_2 = 0$$

21. $2\mathbf{U}_1 + \mathbf{U}_2 = \mathbf{0}$

22. $\mathbf{U}_2 = c\mathbf{U}_1 \Rightarrow c\mathbf{U}_1 + (-1)\mathbf{U}_2 = \mathbf{0}$, a linear dependence relation because $-1 \neq 0$. Similarly, $\mathbf{U}_1 = c\mathbf{U}_2 \Rightarrow$ linear dependence. Conversely, suppose that $c_1\mathbf{U}_1 + c_2\mathbf{U}_2 = \mathbf{0}$, where c_1 and c_2 are not both zero. If $c_1 \neq 0$, then $\mathbf{U}_1 = c\mathbf{U}_2$; if $c_1 = 0$, then $c_2 \neq 0$ and $\mathbf{U}_2 = c\mathbf{U}_1$.

23. *a*. If they were linearly dependent, one would be a scalar multiple of the other (Prob. 22) and they could not span $\mathcal{R}^2$.

b. As in the text,

$$\begin{aligned} &\text{Linear independence of } \mathbf{U}_1 = (a_{11},a_{21}) \text{ and } \mathbf{U}_2 = (a_{12},a_{22}) \\ &\qquad \Rightarrow \quad a_{11}a_{22} - a_{12}a_{21} \neq 0 \\ &\qquad \Rightarrow \quad \mathbf{U}_1 \text{ and } \mathbf{U}_2 \text{ span } \mathcal{R}^2 \end{aligned}$$

24. If they were, $\mathbf{U}_1$ and $\mathbf{U}_2$ would be, too. But then $\mathbf{U}_1$ and $\mathbf{U}_2$ would span $\mathcal{R}^2$ (Prob. 23). As in the text, this implies that $\mathbf{U}_1, \ldots, \mathbf{U}_m$ are linearly *dependent,* a contradiction.

25. $c_1 = 1$, $c_2 = -2$, $c_3 = 3$

26. If $\mathbf{X} = (x_1,x_2)$, then $\mathbf{X} = t\mathbf{U}_1 + (-2t + x_2)\mathbf{U}_2 + (3t - x_1 - x_2)\mathbf{U}_3$, where t is arbitrary.

27. 5 ways.

28. The "only if" part begins with the assumption that (0,0) is the only solution. To prove that $ad - bc \neq 0$, assume the contrary. Then $ad = bc$ and the following cases arise.

(1) If $ab \neq 0$, the system reduces to $ax + by = 0$, $k(ax + by) = 0$, where $k = c/a = d/b$. There are infinitely many solutions (all the solutions of $ax + by = 0$).

(2) If $ab = 0$, then $a = 0$ or $b = 0$.

a. If $a = 0$, then $bc = 0$ and hence $b = 0$ or $c = 0$. If $b = 0$, the system reduces to $0x + 0y = 0$, $cx + dy = 0$ and again there are infinitely many solutions. If $c = 0$, the system becomes $0x + by = 0$, $0x + dy = 0$ and every pair $(x,0)$ is a solution.

b. If $b = 0$, reason as in (*a*).

(*Ed. note:* This is a proof by brute force. Better methods will show up later.)

Section 1.3

32. The closure law of scalar multiplication.

34. *a*. $\mathbf{U}_4 = 0\mathbf{U}_1 + (-1)\mathbf{U}_2 + 2\mathbf{U}_3$

b. The assumption that it can leads to the false equation $-1 = 0$.

c. Every $\mathbf{X} = (x_1,x_2,x_3)$ in $\mathcal{R}^3$ can be written $\mathbf{X} = c_1\mathbf{U}_1 + c_2\mathbf{U}_2 + c_3\mathbf{U}_3 + c_4\mathbf{U}_4$, where $c_1 = -x_2$, $c_2 = t + x_1 + 2x_2$, $c_3 = -2t - \frac{1}{2}(x_1 + 2x_2 - x_3)$, $c_4 = t$ (t being arbitrary).

d. Take $t = 0$ in part *c*.

e. The only elements of $\mathcal{R}^3$ that can be written in terms of $\mathbf{U}_1$ and $\mathbf{U}_2$ are the points of the plane $x + 2y - z = 0$.

f. Only $\{\mathbf{U}_1,\mathbf{U}_2,\mathbf{U}_3\}$.

35. $c_1 = 0$, $c_2 = 1$, $c_3 = -2$, $c_4 = 1$

37. If no element of $\mathcal{F}^n$ can be written as a linear combination of $\mathbf{U}_1, \ldots, \mathbf{U}_m$ in more than one way, then $\mathbf{0}$ cannot be. Hence

$$c_1\mathbf{U}_1 + \cdots + c_m\mathbf{U}_m = \mathbf{0} \quad \Rightarrow \quad c_1 = \cdots = c_m = 0$$

Conversely, assume that $\mathbf{U}_1, \ldots, \mathbf{U}_m$ are linearly independent and suppose that $\mathbf{X} = a_1\mathbf{U}_1 + \cdots + a_m\mathbf{U}_m$ and $\mathbf{X} = b_1\mathbf{U}_1 + \cdots + b_m\mathbf{U}_m$. Then $(a_1 - b_1)\mathbf{U}_1 + \cdots + (a_m - b_m)\mathbf{U}_m = \mathbf{0}$, which implies that $a_j = b_j$, $j = 1, \ldots, m$.

41. Relabel the elements (if necessary) so that $\mathbf{U}_1 = \mathbf{U}_2$. Then

$$1\mathbf{U}_1 + (-1)\mathbf{U}_2 + 0\mathbf{U}_3 + \cdots + 0\mathbf{U}_m = \mathbf{0}$$

which shows that $\mathbf{U}_1, \ldots, \mathbf{U}_m$ are linearly dependent.

43. *a*. $z \in \mathcal{R} \Rightarrow z = a + bi$, where $a \in \mathcal{R}$ and $b = 0 \Rightarrow \bar{z} = z$. Conversely, $\bar{z} = z \Rightarrow a - bi = a + bi \Rightarrow b = 0 \Rightarrow z \in \mathcal{R}$. In these circumstances $|z| = |a|$, the ordinary absolute value of a real number.

d. If $z = a + bi$, then $z + \bar{z} = 2a \le 2\sqrt{a^2 + b^2} = 2|z|$, with equality $\Leftrightarrow a = \sqrt{a^2 + b^2} \Leftrightarrow a \ge 0$ and $b = 0 \Leftrightarrow z = a \ge 0$.

e. If $u = a + bi$ and $v = c + di$, then $uv = (ac - bd) + (ad + bc)i$ and $\bar{u}\,\bar{v} = (ac - bd) - (ad + bc)i$. Hence $\overline{uv} = \bar{u}\,\bar{v}$.

f. $u/v = u\bar{v}/|v|^2$. Since $|v|^2$ is real, one need only perform the multiplication $u\bar{v}$. If $u = 2 + i$, $v = 2 - i$, then $u/v = \frac{3}{5} + \frac{4}{5}i$.

44. Each element $\mathbf{X} = (x_1,x_2)$ in $\mathcal{C}^2$ can be written $\mathbf{X} = c_1\mathbf{U}_1 + c_2\mathbf{U}_2$ in exactly one way, by taking $c_1 = i(x_1 - x_2)$ and $c_2 = -x_1 + (1 - i)x_2$.

45. Yes.

Review Quiz

1. T	**7.** T	**12.** F	**17.** T
2. F	**8.** T	**13.** T	**18.** T
3. F	**9.** T	**14.** F	**19.** F
4. F	**10.** T	**15.** T	**20.** T
5. F	**11.** Well, maybe. But see page 1.	**16.** F	**21.** T
6. T			

Section 2.1

1. $\{1,i\}$

2. *b*. $\{1\}$

4. $\left\{\begin{bmatrix}1 & 0\\0 & 0\end{bmatrix}, \begin{bmatrix}0 & 1\\0 & 0\end{bmatrix}, \begin{bmatrix}0 & 0\\1 & 0\end{bmatrix}, \begin{bmatrix}0 & 0\\0 & 1\end{bmatrix}\right\}$

5. $\left\{\begin{bmatrix}1 & 0\\0 & 1\end{bmatrix}, \begin{bmatrix}0 & 1\\-1 & 0\end{bmatrix}\right\}$

Section 2.2

12. $1(0) = 0$

13. Linearly independent elements are always distinct.

14. *a*. $1u_1 + (-1)u_2 = 0$
b. The "set" $\{u_1,u_2\}$ is really $\{u_1\}$.

15. Label the elements of the given set $u_1, \ldots, u_n$ in such a way that the subset consists of $u_1, \ldots, u_m$, $1 \leq m \leq n$. Then

$$\sum_{j=1}^{m} c_j u_j = 0 \Rightarrow \sum_{j=1}^{n} c_j u_j = 0 \quad \text{where each } c_j \text{ after } c_m \text{ (if any) is } 0$$
$$\Rightarrow \text{Each } c_j \text{ is } 0 \quad j = 1, \ldots, n$$
$$\Rightarrow c_1 = \cdots = c_m = 0$$

16. Corollary 2.6b: (1) Label the elements $u_1, \ldots, u_m$ so that the one that is 0 is u_m. Then $u_m = 0u_1 + \cdots + 0u_{m-1}$, which, by Theorem 2.6, makes $u_1, \ldots, u_m$ linearly dependent. (2) Label so that $u_1 = u_2$. Then $u_1 = 1u_2 + 0u_3 + \cdots + 0u_m$.

Section 2.3

17. *b.* $c_1 = 1, c_2 = 1, c_3 = -2, c_4 = 0$

18. *a.* Let x be a point of I for which $W(x) \neq 0$. Then

$$\begin{aligned} c_1\phi_1 + c_2\phi_2 = 0 \quad &\Rightarrow \quad c_1\phi_1' + c_2\phi_2' = 0 \\ &\Rightarrow \quad \begin{cases} \phi_1(x)c_1 + \phi_2(x)c_2 = 0 \\ \phi_1'(x)c_1 + \phi_2'(x)c_2 = 0 \end{cases} \\ &\Rightarrow \quad c_1 = c_2 = 0 \end{aligned}$$

c. The implications

$$W \neq 0 \quad \Rightarrow \quad \text{linear independence}$$

and

$$\text{Linear dependence} \quad \Rightarrow \quad W = 0$$

are equivalent.

19. $W(x) = (r_2 - r_1)e^{(r_1+r_2)x} \neq 0$

21. *a.* $W(x) = 2 \neq 0$

22. *c.* $\phi(x) = 2 \cos x - \sin x$

23. *b.* $\phi(x) = \cosh x + 3 \sinh x$

Section 2.4

25. *a.* 2 *b.* 1 *c.* 4 *d.* 2 *e.* 1

26. *c.* $\mathbf{V}_3$ **28.** mn

29. Label the larger set $\{v_1, \ldots, v_m\}$, where $m > n$. Each $x \in S$ can be written as $x = \Sigma_{j=1}^{n} c_j v_j = \Sigma_{j=1}^{m} c_j v_j$, where each c_j after c_n is 0.

33. If $\{u_1, \ldots, u_m\}$ and $\{v_1, \ldots, v_n\}$ are both bases, then $m \leq n$ because the first is linearly independent and the second is a spanning set. Reverse their roles to get $m \geq n$. Then $m = n$.

34. Assuming that $r > 0$, we know that S has a basis consisting of r elements. Their linear independence implies (by Theorem 2.8) that any spanning set must have at least r elements.

Section 2.5

35. If $r = 1$, then $\{u_1\}$ is linearly independent $\Leftrightarrow$ $u_1 \neq 0$ $\Leftrightarrow$ $\{u_1\}$ is a spanning set.

38. A basis $\{u_1, \ldots, u_m\}$ is a linearly independent set. It is also a spanning set, so for every $x \in S$ the elements $u_1, \ldots, u_m, x$ are linearly dependent because x is a linear combination of $u_1, \ldots, u_m$.

39. If $n = 1$, the spanning set $\{v_1\}$ is linearly independent because $v_1 \neq 0$. Hence it is a basis.

40. A basis $\{v_1, \ldots, v_n\}$ is a spanning set. It is also linearly independent, so that its spanning power is destroyed if any element is left out (this element cannot be written as a linear combination of the others). If $n = 1$, this argument fails, but a 1-element basis is clearly a minimal spanning set.

41. *a.* Neither property holds. *b.* Both properties hold. *c.* Neither property holds.

42. Omit (1,1,1,1) and (0,0,1,1).

Section 2.6

43. *a.* Straight line; not a subspace.
b. Plane; subspace with basis $\{(1,2,0), (-1,0,2)\}$.
c. Straight line; subspace with basis $\{(1,1)\}$.
d. Origin; subspace with no basis.
e. Straight line; subspace with basis $\{(1,2,0)\}$.
f. Straight line; subspace with basis $\{(2,-1)\}$.

46. 1. Theorem 2.16 says nothing, since S is not finite-dimensional.

49. The x axis and y axis in $\mathcal{R}^2$ are subspaces, but their union is not.

51. His use of Theorem 2.7 requires T to be finite-dimensional, a fact that is not known.

53. *a.* $U^\perp = S$. *b.* $U^\perp = \{(x,y,z) \mid (1,-1,2) \cdot (x,y,z) = 0\}$
c. Same as in part *b*. *e.* The zero subspace
f. $U^\perp$ contains the zero element, so it is nonempty. If $\mathbf{Y}$ and $\mathbf{Z}$ are in $U^\perp$ and a and b are scalars, then for every $\mathbf{X} \in U$, $\mathbf{X} \cdot (a\mathbf{Y} + b\mathbf{Z}) = a(\mathbf{X} \cdot \mathbf{Y}) + b(\mathbf{X} \cdot \mathbf{Z}) = a0 + b0 = 0$. Hence $a\mathbf{Y} + b\mathbf{Z} \in U^\perp$.

Review Quiz

1. F	8. T	15. T	22. T
2. F	9. F	16. T	23. F
3. T	10. F	17. F	24. F
4. F	11. F	18. F	25. T
5. T	12. T	19. T	26. T
6. T	13. F	20. T	27. F
7. F	14. T	21. F	

Section 3.1

3. *a.* $f(0) = f(0 + 0) = f(0) + f(0)$ **4.** $f(1 + 1) \neq f(1) + f(1)$

5. $f(\pi,0) \neq 2f(\pi/2, 0)$ **6.** They all are. **7.** *b.* $f(ku) = \bar{k}f(u)$

9. *a.* $A^* = \begin{bmatrix} 1 - i & 2 \\ 1 & -i \end{bmatrix}$ *d.* Not unless $\mathfrak{F} = \mathfrak{R}$.

11. *a.* 5 *b.* 0 *c.* 1

12. $\mathbf{X} \cdot \mathbf{X} = x_1^2 + \cdots + x_n^2 \geq 0$. But in $\mathfrak{C}^2$ take $\mathbf{X} = (0,i)$ to see that $\mathbf{X} \cdot \mathbf{X}$ may be negative.

15. $f(x,y,z) = 2x - y + 3z$. Yes.

16. *b.* $g' = f$, so g has derivatives of all orders on $(-\infty,\infty)$. Since $g(0) = 0$, $g \in T$.

17. For each $\mathbf{X} = \mathbf{X}_1 + \mathbf{X}_2$, where $\mathbf{X}_1 \in U$ and $\mathbf{X}_2 \in U^\perp$, we have $(f + g)(\mathbf{X}) = f(\mathbf{X}) + g(\mathbf{X}) = \mathbf{X}_1 + \mathbf{X}_2 = \mathbf{X}$.

18. If U is the zero subspace, then $f = 0$ and $g = i$. If $U = S$, then $f = i$ and $g = 0$.

19. *a.* The pair $\frac{1}{13}(9x - 6y, -6x + 4y)$ satisfies the equation $2x + 3y = 0$.
b. $g(x,y) = \frac{1}{13}(4x + 6y, 6x + 9y)$

20. *b.* $g(x,y,z) = \frac{1}{3}(x - y - z, -x + y + z, -x + y + z)$
c. The triple $\frac{1}{3}(x - y - z, -x + y + z, -x + y + z)$ satisfies the equations $x = t$, $y = -t$, $z = -t$.

Section 3.2

22. *a.* The line $x - 2y = 0$ in $\mathfrak{R}^2$, with dimension 1.
b. The line in $\mathfrak{R}^3$ with parametric equations $x = t$, $y = t$, $z = -3t$, dimension 1.
c. The origin in $\mathfrak{R}^2$, dimension 0.
d. The origin in $\mathfrak{R}^1$, dimension 0.

23. *a.* $\mathfrak{R}^1$ *b.* $\mathfrak{R}^2$ *c.* $\mathfrak{R}^2$
d. The line in $\mathfrak{R}^3$ with parametric equations $x = 2t$, $y = 0$, $z = -t$, dimension 1.

25. $\ker f = \{0\}$, $\text{rng } f = \mathfrak{C}$

26. *a.* $\ker f$ is the zero subspace of S, $\text{rng } f = \mathfrak{F}^4$.
b. $\ker f$ is the zero subspace of S, $\text{rng } f = S$.

28. $\ker D$ is the set of constant functions in S; its dimension is 1. The law of nullity says nothing, since S is not finite-dimensional.

29. $\ker I$ consists of the zero function alone; $\text{rng } I = T$ (because if $g \in T$, the function $f = g'$ is sent into g by I).

30. $\ker f = \{(x,y) \mid 3x - 2y = 0\} = U^{\perp}$; $\operatorname{rng} f = U$.

31. $\ker f = \{(x,y,z) \mid x = -t,\ y = t,\ z = t\} = U^{\perp}$; $\operatorname{rng} f = U$.

32. *a.* Each $\mathbf{X} \in S$ can be written (uniquely) in the form $\mathbf{X} = \mathbf{X}_1 + \mathbf{X}_2$, where $\mathbf{X}_1 \in U$ and $\mathbf{X}_2 \in U^{\perp}$. Hence $\mathbf{X} \in \ker f \iff f(\mathbf{X}) = \mathbf{0} \iff \mathbf{X}_1 = \mathbf{0} \iff \mathbf{X} \in U^{\perp}$.

b. Since f sends $\mathbf{X}$ into $\mathbf{X}_1 \in U$, $\operatorname{rng} f \subset U$. To prove the reverse inclusion, let $\mathbf{X} \in U$ and write $\mathbf{X} = \mathbf{X}_1 + \mathbf{X}_2$, where $\mathbf{X}_1 = \mathbf{X} \in U$ and $\mathbf{X}_2 = \mathbf{0} \in U^{\perp}$. Then $f(\mathbf{X}) = \mathbf{X}$ and hence $\mathbf{X} \in \operatorname{rng} f$.

33. $\dim T = \dim(\operatorname{rng} f) = \dim S - \dim(\ker f) \leq \dim S$

34. If $f(x) = y$, then [since $f(x_p) = y$] we have $f(x) - f(x_p) = 0$, that is, $f(x - x_p) = 0$. Hence $x - x_p \in \ker f$, say $x - x_p = x_h$. Then $x = x_h + x_p$. Conversely,

$$x = x_h + x_p\ (x_h \in \ker f) \implies f(x) = f(x_h) + f(x_p) = f(x_p) = y$$

35. *a.*
$$\begin{aligned} L(c_1\phi_1 + c_2\phi_2) &= (c_1\phi_1 + c_2\phi_2)'' + (c_1\phi_1 + c_2\phi_2) \\ &= c_1(\phi_1'' + \phi_1) + c_2(\phi_2'' + \phi_2) \\ &= c_1L(\phi_1) + c_2L(\phi_2) \end{aligned}$$

d. $L(\phi_p) = g \implies g \in \operatorname{rng} L$

f. $\phi(x) = 3\cos x - 2\sin x + x^2 - 2$

Section 3.3

36. $f^{-1}(x) = \frac{1}{3}(x + 2)$

38. *a.* Onto, but not one-to-one. *b.* Onto, but not one-to-one. *c.* Invertible. *d.* One-to-one, but not onto.

39. *a.* $f^{-1}(x,y) = (3x - y, -2x + y)$ *b.* $f^{-1} = f$ *c.* $f^{-1} = f$

41. No.

42. $I^{-1} = D$, where $D: T \to S$ is defined by $D(g) = g'$.

43. $x \in \ker f \implies f(x) = 0 \implies f(x) = f(0) \implies x = 0$

45. Use the law of nullity.

47. c. Since $f^{-1}: T \to S$ is an isomorphism (Theorem 3.5), the argument is the same as in part *b*.

Section 3.4

50. $(-1,5,0)$; $f = i$

51. $(-\frac{5}{2}, \frac{17}{22}, \frac{2}{11})$

52. $(\frac{1}{2}, -\frac{1}{2})$; $(-\frac{1}{2}, \frac{1}{2})$

54. Suppose that $x = \sum_{j=1}^{n} x_j u_j \in \ker f$. Then $f(x) = (x_1, \ldots, x_n) = (0, \ldots, 0)$, so $x = 0$.

55. The identity map on S is an isomorphism, so $S \approx S$. To prove the symmetric law, use Theorem 3.5.

56. We could if S and U were known to have the same dimension.

58. Reflexive and transitive.

59. The map sending $0 \in S$ into $0 \in T$ is an invertible linear map.

Section 3.5

60. *a.*
$$\begin{aligned}(f + g)(ax + by) &= f(ax + by) + g(ax + by)\\ &= af(x) + bf(y) + ag(x) + bg(y)\\ &= a(f + g)(x) + b(f + g)(y)\end{aligned}$$

61. $f(x,y) = (x + 2y, y, -x + 5y)$

62. *a.* The map which sends the polynomial $ax^2 + bx + c$ into the polynomial $-ax^2 - bx + (2a - c)$, that is, $f(\phi) = (2a - c)\phi_1 - b\phi_2 - a\phi_3$.
b. $1 - x^2$ *c.* Yes, since it is one-to-one (the kernel is zero).

63. Suppose that $f = a_{11}f_{11} + a_{12}f_{12} + a_{13}f_{13} + a_{21}f_{21} + a_{22}f_{22} + a_{23}f_{23} = 0$ (the zero map from $\mathcal{R}^3$ to $\mathcal{R}^2$). Then

$$f(\mathbf{E}_1) = (a_{11},a_{21}) = (0,0) \qquad f(\mathbf{E}_2) = (a_{12},a_{22}) = (0,0)$$
$$f(\mathbf{E}_3) = (a_{13},a_{23}) = (0,0)$$

so all the coefficients are 0.

65. $f_{11}(x,y) = (x,0)$; $f_{12}(x,y) = (y,0)$; $f_{21}(x,y) = (0,x)$; $f_{22}(x,y) = (0,y)$.

66. Let $w_j = f(u_j)$ and apply Theorem 3.11 to conclude that if $g(u_j) = w_j$, then $g = f$. The converse is trivial.

Review Quiz

1. T
2. T
3. F
4. T
5. The James Sylvester who discovered the law of nullity was born in 1814.
6. T
7. T
8. F
9. F
10. T
11. T
12. T
13. T
14. F
15. F
16. F
17. T
18. T
19. F
20. T
21. T
22. T
23. F
24. T
25. F
26. T
27. T
28. T
29. T
30. T
31. F
32. T
33. T
34. T
35. T
36. F
37. T

Section 4.1

1. *a.* $\begin{bmatrix} 1 & -2 & 1 \\ 0 & 2 & 0 \end{bmatrix}$ *b.* $\begin{bmatrix} 1 & -1 \\ 2 & 1 \end{bmatrix}$ *c.* $[2 \quad 1 \quad -1]$ *d.* $\begin{bmatrix} 2 \\ -1 \\ 1 \end{bmatrix}$

e. $\begin{bmatrix} 1 & 0 & 0 \\ 0 & 1 & 0 \\ 0 & 0 & 1 \end{bmatrix}$ *f.* $\begin{bmatrix} 0 & 0 & 0 \\ 0 & 0 & 0 \end{bmatrix}$ *g.* $[3]$

3. Suppose that f is defined as in the problem. Then if $\mathbf{P} = (p_1, \ldots, p_n)$ and $\mathbf{Q} = (q_1, \ldots, q_n)$ are vectors in $\mathfrak{F}^n$, we have $f(\mathbf{P}) = (r_1, \ldots, r_m)$ and $f(\mathbf{Q}) = (s_1, \ldots, s_m)$, where $r_i = \sum_{j=1}^{n} a_{ij}p_j$ and $s_i = \sum_{j=1}^{n} a_{ij}q_j$, for $i = 1, \ldots, m$. On the other hand, $f(\mathbf{P} + \mathbf{Q}) = f(p_1 + q_1, \ldots, p_n + q_n) = (t_1, \ldots, t_m)$, where $t_i = \sum_{j=1}^{n} a_{ij}(p_j + q_j) = r_i + s_i$, for $i = 1, \ldots, m$. Hence $f(\mathbf{P} + \mathbf{Q}) = f(\mathbf{P}) + f(\mathbf{Q})$. Similarly, $f(k\mathbf{P}) = kf(\mathbf{P})$ for every scalar k, so f is linear.

Conversely, if f is linear, let A be its matrix relative to the standard bases, and let a_{ij} be the entry in the ith row and jth column of A. Then if $\mathbf{E}_j$ is the jth element of the standard basis for $\mathfrak{F}^n$, we have $f(\mathbf{E}_j) = (a_{1j}, \ldots, a_{mj})$, for $j = 1, \ldots, n$. Given $\mathbf{X} = (x_1, \ldots, x_n)$ in $\mathfrak{F}^n$, write $\mathbf{X} = \sum_{j=1}^{n} x_j\mathbf{E}_j$, from which

$$f(\mathbf{X}) = \sum_{j=1}^{n} x_j f(\mathbf{E}_j) = \sum_{j=1}^{n} x_j(a_{1j}, \ldots, a_{mj})$$

$$= \sum_{j=1}^{n} (a_{1j}x_j, \ldots, a_{mj}x_j) = (y_1, \ldots, y_m)$$

where

$$y_i = \sum_{j=1}^{n} a_{ij}x_j \qquad i = 1, \ldots, m$$

Hence f has the form described in the problem.

4. In the solution of Prob. 3 it is clear that if f is defined this way, its matrix relative to the standard bases has a_{ij} in its ith row and jth column. Hence the ith row has entries $a_{i1}, \ldots, a_{in}$.

5. $[f] = \begin{bmatrix} \cos\theta & -\sin\theta \\ \sin\theta & \cos\theta \end{bmatrix}$, $[f^{-1}] = \begin{bmatrix} \cos\theta & \sin\theta \\ -\sin\theta & \cos\theta \end{bmatrix}$

6. *a.* $\begin{bmatrix} 0 & 1 \\ -1 & 0 \end{bmatrix}$ *b.* $\begin{bmatrix} 0 & 0 \\ 0 & 0 \end{bmatrix}$

7. The columns of A get interchanged.

8. $[-3 \quad 2 \quad 3]$

9. *b.* If $z = a + bi$ is plotted as the point (a,b) in the complex plane, then $f(z) = a - bi$ is the point $(a,-b)$.

10. *a.* $\begin{bmatrix} 1 & 0 & 0 & 0 \\ 0 & 1 & 0 & 0 \\ 0 & 0 & 1 & 0 \\ 0 & 0 & 0 & 1 \end{bmatrix}$ *b.* $\begin{bmatrix} 1 & 0 & 0 & 0 \\ 0 & 0 & 1 & 0 \\ 0 & 1 & 0 & 0 \\ 0 & 0 & 0 & 1 \end{bmatrix}$

11. Since $u_j = \Sigma_{i=1}^n \delta_{ij}u_i$, $f(u_j) = (\delta_{1j}, \ldots, \delta_{nj}) = \mathbf{E}_j$. Hence the jth column of $[f]$ is $\mathbf{E}_j$.

12. *a.* $f(x,y,z) = (2x - 3y + z, x + 5y)$ *b.* $f(x,y) = (ix, x + iy)$
c. $f(x,y,z) = x + y + 2z$ *d.* $f(x) = 2x$

14. $i(u_j) = u_j = \sum_{i=1}^{n} \delta_{ij}u_i$ **15.** $\begin{bmatrix} 0 & 1 \\ 1 & 0 \end{bmatrix}$

16. *a.* One basis for U is $\beta = \{(1,1,0), (1,0,1)\}$. If α is the standard basis for $\mathcal{R}^3$, then

$$[f]_{\alpha\beta} = \frac{1}{3}\begin{bmatrix} 1 & 2 & -1 \\ 1 & -1 & 2 \end{bmatrix}$$

b. $\frac{1}{3}\begin{bmatrix} 2 & 1 & 1 \\ 1 & 2 & -1 \\ 1 & -1 & 2 \end{bmatrix}$

17. *a.* $A = \begin{bmatrix} 1 & -1 & 1 \\ 0 & 3 & 0 \end{bmatrix}$; $B = \begin{bmatrix} 0 & 1 & 2 \\ 1 & -1 & 0 \end{bmatrix}$

b. $A + B = \begin{bmatrix} 1 & 0 & 3 \\ 1 & 2 & 0 \end{bmatrix}$; $2A = \begin{bmatrix} 2 & -2 & 2 \\ 0 & 6 & 0 \end{bmatrix}$

c. $(f + g)(x,y,z) = (x + 3z, x + 2y)$; $(2f)(x,y,z) = (2x - 2y + 2z, 6y)$

18. *a.* $A = \begin{bmatrix} 3 & 0 & 1 \\ -1 & 0 & 1 \\ 0 & 2 & 0 \\ 0 & 1 & 1 \end{bmatrix}$; $B = \begin{bmatrix} 1 & 0 \\ 1 & -2 \\ 1 & 1 \end{bmatrix}$ *b.* $C = \begin{bmatrix} 4 & 1 \\ 0 & 1 \\ 2 & -4 \\ 2 & -1 \end{bmatrix}$

Section 4.2

22. $[\phi] = I_6$, the 6×6 identity matrix. (See Prob. 14, Sec. 4.1.)

24. *b.* $AB = \begin{bmatrix} 4 & -5 & 13 \\ 1 & 3 & -1 \end{bmatrix}$

25. *a.* $f(x,y,z) = (2x - z, 3x + 5y + z)$

Section 4.3

27. *d.* Let $B = A^* = \bar{A}^T = A^T$. Then $(A^*)^* = B^* = \bar{B}^T = (\bar{\bar{A}}^T)^T = (A^T)^T = A$. More directly, compare the i,j entries of B^* and A.

29. ϕ is linear by Theorem 4.5. It is one-to-one because $\phi(A) = \phi(B) \Rightarrow A^T = B^T \Rightarrow A = B$. It is automatically onto. (Why?)

31. f is onto because if $A \in M_{n,m}(\mathfrak{F})$, then $f(A^*) = A$. It is one-to-one by the same reasoning as in Prob. 29. However it is not linear (unless $\mathfrak{F} = \mathfrak{R}$).

33. First note that $\mathbf{X} * \mathbf{Y}$ and $\mathbf{Y} * \mathbf{X}$ are conjugates. (Why?) Hence the i,j entry of A^T is $a_{ji} = \mathbf{U}_j * \mathbf{U}_i = \overline{\mathbf{U}_i * \mathbf{U}_j} = \bar{a}_{ij}$, which is the i,j entry of $\bar{A}$.

34. $A^T = \bar{A} \Leftrightarrow \overline{A^T} = \bar{\bar{A}} \Leftrightarrow \bar{A}^T = A \Leftrightarrow A^* = A$

35. $\begin{bmatrix} 1 & -i \\ i & 0 \end{bmatrix}$ **36.** $A^* = A \Rightarrow \bar{a}_{ii} = a_{ii} \Rightarrow a_{ii}$ is real

39. *a.* $f^*(x,y) = (x + 3y, -x + y)$ *c.* $g^*(x,y) = (x, -iy)$

Review Quiz

1. T	**4.** T	**7.** F	**10.** T	**13.** F
2. T	**5.** T	**8.** F	**11.** T	**14.** T
3. T	**6.** F	**9.** T	**12.** T	**15.** T

Section 5.1

1. $(f \circ g)(x,y) = (2x + y, 6x + 2y, -4x - y)$; $g \circ f$ is undefined.

2. $f^2(x,y) = (x \cos 2\theta - y \sin 2\theta, x \sin 2\theta + y \cos 2\theta)$

3. *a.* $f \circ g = g \circ f = 0$ (the zero map) *b.* $f^n = f$ and $g^n = g$

6. Linearity is needed in part *a*.

9. *c.* $\cos x$ and $\sin x$ are solutions of $L(y) = 0$, so the terms $c_4 \cos x + c_5 \sin x$ drop out when the trial solution is substituted in $L(y) = g$.

d. The substitution yields the identity $(2c_3 + c_1) + c_2x + c_3x^2 = 0 + 0x + 1x^2$. The coefficients of these linear combinations of $1, x, x^2$

must be the same (Theorem 2.5), that is, $c_1 + 2c_3 = 0$, $c_2 = 0$, and $c_3 = 1$, from which $c_1 = -2$, $c_2 = 0$, and $c_3 = 1$.

10. *b*. Use the distributive laws and part *a*.

Section 5.2

12. $f \circ u = f$ for every $f \Rightarrow i \circ u = i \Rightarrow u = i$

14. $f^2 = i$; $f^{-1} = f$

15. $f \circ g = i_T$, $g \circ f = i_S$; $f^{-1} = g$; $g^{-1} = f$.

16. *a*. $D \circ I = i_S$, $I \circ D = i_T$, $I^{-1} = D$, $D^{-1} = I$.

b. $D \circ I = i$, but $I \circ D \neq i$. Instead $I \circ D$ sends $g \in S$ into $g + C$, where $C = -g(0)$. In other words, differentiation reverses integration, but integration reverses differentiation only in the sense that the original function is recovered to within an arbitrary constant. As for the invertibility of I and D, I is not onto and D is not one-to-one:

No function $f \in S$ exists such that $I(f) = g$, where $g(x) = 1$.
If $f(x) = 0$ and $g(x) = 1$, then $D(f) = D(g)$, but $f \neq g$.

19. *a*. $(f \circ g)(x,y) = (2x + y, 3x + y)$ *b*. $(f \circ g)^{-1}(x,y) = (y - x, 3x - 2y)$
c. No.

20. The set doesn't contain multiplicative inverses of all its nonzero elements. For example, there is no integer b satisfying $2b = 1$.

22. $L(S)$ is not a commutative ring, since in general $f \circ g \neq g \circ f$; nor does every nonzero linear operator on S have an inverse.

Section 5.3

23. *a*. $A = \begin{bmatrix} 1 & 1 & -2 \\ 0 & 3 & 1 \end{bmatrix}$, $B = \begin{bmatrix} 1 & 1 \\ 2 & 0 \\ 0 & -1 \end{bmatrix}$, $C = \begin{bmatrix} 3 & 3 \\ 6 & -1 \end{bmatrix}$,

$$D = \begin{bmatrix} 1 & 4 & -1 \\ 2 & 2 & -4 \\ 0 & -3 & -1 \end{bmatrix}$$

28. Use the distributive laws and the formula $(kA)B = k(AB) = A(kB)$ from Prob. 24*c*, Sec. 4.2.

29. $(A + B)^2 = (A + B)(A + B) = (A + B)A + (A + B)B =$
$A^2 + BA + AB + B^2 = A^2 + AB + AB + B^2$.
Since $M_n(\mathfrak{F})$ is a vector space, $AB + AB = 1(AB) + 1(AB) = (1 + 1)AB = 2AB$.

31. Define $f: \mathfrak{F}^n \to \mathfrak{F}^m$ by $[f] = A$. Writing $S = \mathfrak{F}^n$ and $T = \mathfrak{F}^m$, we have $AI_n = [f][i_S] = [f \circ i_S] = [f] = A$ and $I_m A = [i_T][f] = [i_T \circ f] = [f] = A$.

34. *b*. $[f][f^{-1}] = [f \circ f^{-1}] = [i] = I$

35. *a*. $B = \begin{bmatrix} 5 & -3 \\ -3 & 2 \end{bmatrix}$ *b*. $f^{-1}(x,y) = (5x - 3y, -3x + 2y)$

38. Because $f(AB) = f(B)f(A)$, which in general is different from $f(A)f(B)$.

39. $f: \mathfrak{F} \to M_1(\mathfrak{F})$ defined by $f(a) = [a]$ (the 1×1 matrix having entry $a \in \mathfrak{F}$).

40. *a*. What must be checked is that $\mathcal{C}'$ is closed relative to addition and multiplication, that the identity elements 0 and I belong to $\mathcal{C}'$, that each element of $\mathcal{C}'$ has an additive inverse in $\mathcal{C}'$, and that multiplication is commutative. The remaining properties of a ring are inherited from $M_2(\mathcal{R})$.

Section 5.4

43. *a*. Suppose that

$$B = \begin{bmatrix} x & u \\ y & v \end{bmatrix}$$

satisfies $AB = I$. Then x and y must satisfy $4x + 2y = 1$ and $6x + 3y = 0$, which is impossible.

c. The equation $A\mathbf{X} = \mathbf{0}$ is the system $4x + 2y = 0$, $6x + 3y = 0$, which has solution set $\{(x,y) \mid x = t, y = -2t\}$.

44. If A^{-1} exists, let $g: T \to S$ be the linear map defined by $[g] = A^{-1}$ (relative to the given bases). Then $[f \circ g] = [f][g] = AA^{-1} = I = [i_T]$, from which $f \circ g = i_T$. Similarly, $g \circ f = i_S$. It follows from Theorem 5.6 that f is invertible. The converse is proved as in the "if" part of Theorem 5.13.

45. $f^{-1}(x,y,z) = \frac{1}{3}(6x + 2y - 3z, y, -3x - y + 3z)$

52. *b*. If $ad - bc = 0$, the matrix equation in the hint leads to

$$\begin{aligned}(ad - bc)x &= d \\ (ad - bc)y &= -c\end{aligned} \quad \text{and} \quad \begin{aligned}(ad - bc)u &= -b \\ (ad - bc)v &= a\end{aligned}$$

Hence $a = b = c = d = 0$, that is, $A = 0$ and the equation $AB = I$ is impossible.

54. The formula is trivial when $r = 1$. Assuming that it holds when $r = k - 1$ $(k > 1)$, we have

$$\begin{aligned}(A_1 \cdots A_k)^{-1} &= [(A_1 \cdots A_{k-1})A_k]^{-1} \\ &= A_k^{-1}(A_1 \cdots A_{k-1})^{-1} && \text{(Theorem 5.14)} \\ &= A_k^{-1}(A_{k-1}^{-1} \cdots A_1^{-1}) && \text{(induction hypothesis)} \\ &= A_k^{-1} \cdots A_1^{-1}\end{aligned}$$

Hence the formula is correct for every positive integer r.

55. $AB = [i]_{\alpha^\beta}[i]_{\beta^\alpha} = [i \circ i]_{\beta^\beta} = [i]_\beta = I$

56. $A = \dfrac{1}{3}\begin{bmatrix} 2 & 1 \\ -1 & 1 \end{bmatrix}$

$B = \begin{bmatrix} 1 & -1 \\ 1 & 2 \end{bmatrix}$

57. Suppose that the ith row of the $n \times n$ matrix A is zero. Then for every $n \times n$ matrix B, the ith row of AB is zero (by Prob. 33, Sec. 5.3). Thus no B exists satisfying $AB = I$, since the rows of I are nonzero. Hence A is singular.

58. *c*. The equation $A\mathbf{X} = \mathbf{0}$ has nonzero solutions $(1,-2)$ and $(-1,2)$. Use these as columns of B.

61. *a*. If A is nonzero, then a and b are not both zero, so $\det A = a^2 + b^2 \neq 0$. Hence

$$A^{-1} = \frac{1}{a^2 + b^2}\begin{bmatrix} a & -b \\ b & a \end{bmatrix} \in \mathcal{C}'$$

b. Since $b = 0$, a is nonzero; hence

$$A^{-1} = \frac{1}{a^2}\begin{bmatrix} a & 0 \\ 0 & a \end{bmatrix} = \begin{bmatrix} a^{-1} & 0 \\ 0 & a^{-1} \end{bmatrix} \in \mathcal{R}'$$

Review Quiz

1. T	8. T	15. F	22. T
2. T	9. T	16. T	23. F
3. T	10. T	17. F	24. T
4. F	11. F	18. T	25. F
5. T	12. F	19. T	26. T
6. F	13. T	20. T	27. F
7. T	14. T	21. T	

Section 6.1

1. *a*. An easy way to confirm that P is nonsingular (finding its inverse is hard!) is to observe that the only solution of the equation $P\mathbf{X} = \mathbf{0}$ (equivalent to the system $x = 0$, $y = 0$, $x - y + z = 0$) is $(0,0,0)$. Q is nonsingular because det Q is nonzero.
   b. If such a basis existed, its first element (x,y) would have to satisfy $f(x,y) = (1,0,0)$, that is, $x + y = 1$, $2x = 0$, $x - y = 0$. But this system is inconsistent.

2. Let γ and δ be the standard bases for $S = \mathcal{R}^3$ and $T = \mathcal{R}^2$, respectively. Then one choice of P and Q is

$$P = [i_T]_{\beta^\delta} = \begin{bmatrix} 1 & 1 \\ -1 & 1 \end{bmatrix} \quad \text{and} \quad Q = [i_S]_{\gamma^\alpha} = \begin{bmatrix} 1 & -1 & 0 \\ 0 & 1 & 0 \\ 0 & -1 & 1 \end{bmatrix}$$

4. $\alpha = \{(2,1), (-1,1)\}$

5. Define $f\colon \mathfrak{F}^n \to \mathfrak{F}^n$ by $[f] = P$ (relative to β). If $\mathbf{U}_j = f(\mathbf{E}_j)$, then (as in the proof of Theorem 6.3) $\alpha = \{\mathbf{U}_1, \ldots, \mathbf{U}_n\}$ is a basis for $\mathfrak{F}^n$ and $P = [i]_{\alpha^\beta}$. The definition of P as the matrix of f says that its jth column is the coordinate vector of $f(\mathbf{E}_j)$ relative to β. Since β is the standard basis, this is $f(\mathbf{E}_j)$ itself. Hence the columns of P are $\mathbf{U}_1, \ldots, \mathbf{U}_n$.

10. *c*. Given that $B \in M$, let $A = P^{-1}BQ^{-1}$. Then $f(A) = B$.

Section 6.2

11. $f(x,y) = (x,-y)$; $\alpha = \{(1/\sqrt{2})(1,-1), (1/\sqrt{2})(1,1)\}$, $\beta = \{(1,0), (0,1)\}$

14. Since $B = PAQ$, where

$$P = I \quad \text{and} \quad Q = \begin{bmatrix} 0 & 1 \\ 1 & 0 \end{bmatrix}$$

are nonsingular, A and B are equivalent. However they are not similar; for suppose that there is a nonsingular matrix

$$P = \begin{bmatrix} a & b \\ c & d \end{bmatrix}$$

such that $B = PAP^{-1}$. Then $BP = PA$, that is,

$$\begin{bmatrix} c & d \\ 0 & 0 \end{bmatrix} = \begin{bmatrix} a & 0 \\ c & 0 \end{bmatrix}$$

from which $a = c = d = 0$. But then

$$P = \begin{bmatrix} 0 & b \\ 0 & 0 \end{bmatrix}$$

which is singular.

16. $P = [i]_{\alpha\beta} = \begin{bmatrix} 0 & 1 & 0 \\ 0 & 0 & 1 \\ -1 & 0 & 0 \end{bmatrix}$

18. *a.* Given the nonsingular matrix P, let $B = PAP^{-1}$. Then $A \approx B$, which implies that $B = A$. Hence $A = PAP^{-1}$ and $AP = PA$.

21. Given that $B = PAP^{-1}$, use induction on r to prove that $B^r = PA^rP^{-1}$.

23. $AB = A(BA)A^{-1}$

24. *a.* The i,i entry of AB is $\Sigma_{j=1}^{n} a_{ij}b_{ji}$, so tr $AB = \Sigma_{i=1}^{n} \Sigma_{j=1}^{n} a_{ij}b_{ji}$. Interchange the sums to get tr BA.

d. 0

Section 6.3

26. *b.* If $x = \Sigma_{j=1}^{n} x_j u_j$, then $\alpha(x) = \mathbf{X} = (x_1, \ldots, x_n)$. Since $f(x) = \Sigma_{j=1}^{n} x_j f(u_j)$, we have

$$\beta[f(x)] = \sum_{j=1}^{n} x_j \beta[f(u_j)] = \sum_{j=1}^{n} x_j \text{ (jth column of A)} = A\mathbf{X} = A\alpha(x)$$

27. *a.* $P = \begin{bmatrix} 1 & 0 \\ 0 & -1 \end{bmatrix}$; $P^{-1} = P$

b. $P = \begin{bmatrix} 1 & 1 & 0 \\ 1 & 0 & 1 \\ 0 & 1 & 1 \end{bmatrix}$; $P^{-1} = \frac{1}{2}\begin{bmatrix} 1 & 1 & -1 \\ 1 & -1 & 1 \\ -1 & 1 & 1 \end{bmatrix}$

c. $P = \begin{bmatrix} 0 & -1 \\ 1 & 0 \end{bmatrix}$; $P^{-1} = \begin{bmatrix} 0 & 1 \\ -1 & 0 \end{bmatrix}$

33. *a.* $P = \frac{1}{7}\begin{bmatrix} 3 & 1 \\ -1 & 2 \end{bmatrix}$, $Q = \begin{bmatrix} 1 & 0 & -1 \\ 0 & 1 & -2 \\ 0 & 0 & 7/5 \end{bmatrix}$, $B = \begin{bmatrix} 1 & 0 & 0 \\ 0 & 1 & 0 \end{bmatrix}$

b. $P = \frac{1}{3}\begin{bmatrix} 0 & 3 & 0 \\ 0 & 1 & -1 \\ 3 & -6 & 0 \end{bmatrix}$, $Q = \begin{bmatrix} 1 & 0 \\ 0 & 1 \end{bmatrix}$, $B = \begin{bmatrix} 1 & 0 \\ 0 & 1 \\ 0 & 0 \end{bmatrix}$

c. $P = \begin{bmatrix} 1 & 0 \\ 0 & 1 \end{bmatrix}$, $Q = \begin{bmatrix} 1 & -1 \\ -1 & 2 \end{bmatrix}$, $B = \begin{bmatrix} 1 & 0 \\ 0 & 1 \end{bmatrix}$

34. a. $P = \frac{1}{2}\begin{bmatrix} 1 & 1 \\ 1 & -1 \end{bmatrix}$, $P^{-1} = \begin{bmatrix} 1 & 1 \\ 1 & -1 \end{bmatrix}$, $B = \begin{bmatrix} 1 & 0 \\ 0 & 3 \end{bmatrix}$

b. $P = \begin{bmatrix} 0 & -1 & -1 \\ 0 & 0 & 1 \\ 1 & 1 & 0 \end{bmatrix}$, $P^{-1} = \begin{bmatrix} 1 & 1 & 1 \\ -1 & -1 & 0 \\ 0 & 1 & 0 \end{bmatrix}$, $B = \begin{bmatrix} 1 & 0 & 0 \\ 0 & 2 & 1 \\ 0 & 0 & 2 \end{bmatrix}$

c. $P = \frac{1}{2}\begin{bmatrix} 0 & 2 & 0 \\ -1 & 0 & 1 \\ 1 & 0 & 1 \end{bmatrix}$, $P^{-1} = \begin{bmatrix} 0 & -1 & 1 \\ 1 & 0 & 0 \\ 0 & 1 & 1 \end{bmatrix}$, $B = \begin{bmatrix} 1 & 0 & 0 \\ 0 & 1 & 0 \\ 0 & 0 & 3 \end{bmatrix}$

Review Quiz

1. T	**7.** T	**13.** F
2. T	**8.** T	
3. F	**9.** F	
4. T	**10.** T	
5. T	**11.** F	
6. T	**12.** F	

Section 7.1

1. a. $\mathbf{X} = (-\frac{1}{5}, -\frac{1}{10})$ b. $\mathbf{X} = (\frac{5}{9}, -\frac{5}{9}, -\frac{2}{9})$

2. a. The solution set is empty. b. $\{(x,y) \mid y = 2x + 3\}$

3. a. The result called for is

$$\begin{bmatrix} 1 & 0 & 0 & \frac{1}{6} & -\frac{2}{3} \\ 0 & 1 & 0 & -\frac{1}{3} & \frac{1}{3} \\ 0 & 0 & 1 & -\frac{7}{6} & \frac{8}{3} \end{bmatrix}$$

4. b. $S_0 = \langle(1,-2,-7,-6)\rangle$; $\dim S_0 = 1$.

6. a. The result called for is

$$\begin{bmatrix} 1 & 2 & 0 & 0 & -\frac{1}{5} \\ 0 & 0 & 1 & 0 & \frac{23}{5} \\ 0 & 0 & 0 & 1 & \frac{8}{5} \end{bmatrix}$$

b. $S_0 = \langle(-2,1,0,0)\rangle$; $\mathbf{X}_h = t(-2,1,0,0) = (-2t,t,0,0)$.
c. $\mathbf{X}_p = (-\frac{1}{5},0,\frac{23}{5},\frac{8}{5})$; $\mathbf{X} = \mathbf{X}_h + \mathbf{X}_p = (-2t - \frac{1}{5},t,\frac{23}{5},\frac{8}{5})$.

Section 7.2

7. *a.* $[B|\mathbf{D}] = \begin{bmatrix} 1 & 0 & 0 & -\frac{1}{2} & 1 \\ 0 & 1 & \frac{1}{2} & \frac{3}{4} & 0 \\ 0 & 0 & 0 & 0 & 0 \end{bmatrix}$

(1) $\mathbf{U}_1 = (0,-1,2,0)$, $\mathbf{U}_2 = (2,-3,0,4)$ (2) $\mathbf{X}_p = (1,0,0,0)$
(3) $\mathbf{X} = t_1\mathbf{U}_1 + t_2\mathbf{U}_2 + \mathbf{X}_p = (2t_2 + 1, -t_1 - 3t_2, 2t_1, 4t_2)$

b. $[B|\mathbf{D}] = \begin{bmatrix} 1 & 0 & 0 & \frac{8}{9} \\ 0 & 1 & 0 & \frac{1}{9} \\ 0 & 0 & 1 & \frac{5}{9} \end{bmatrix}$

(3) $\mathbf{X} = \mathbf{X}_p = (\frac{8}{9}, \frac{1}{9}, \frac{5}{9})$, the only solution.

c. $B = \begin{bmatrix} 1 & -1 & 0 & 3 & -1 \\ 0 & 0 & 1 & -3 & 1 \\ 0 & 0 & 0 & 0 & 0 \\ 0 & 0 & 0 & 0 & 0 \end{bmatrix}$

(1) $\mathbf{U}_1 = (1,1,0,0,0)$, $\mathbf{U}_2 = (-3,0,3,1,0)$, $\mathbf{U}_3 = (1,0,-1,0,1)$
(2) $\mathbf{O}$ (3) $\mathbf{X} = t_1\mathbf{U}_1 + t_2\mathbf{U}_2 + t_3\mathbf{U}_3 = (t_1 - 3t_2 + t_3, t_1, 3t_2 - t_3, t_2, t_3)$

d. $[B|\mathbf{D}] = \begin{bmatrix} 1 & 0 & -\frac{1}{8} & -\frac{3}{16} \\ 0 & 1 & -\frac{3}{8} & \frac{15}{16} \end{bmatrix}$

(1) $\mathbf{U} = (1,3,8)$ (2) $\mathbf{X}_p = (-\frac{3}{16}, \frac{15}{16}, 0)$
(3) $\mathbf{X} = t\mathbf{U} + \mathbf{X}_p = (t - \frac{3}{16}, 3t + \frac{15}{16}, 8t)$

e. $[B|\mathbf{D}] = \begin{bmatrix} 1 & 0 & 5 \\ 0 & 1 & 3 \\ 0 & 0 & 0 \\ 0 & 0 & 0 \end{bmatrix}$ (3) $\mathbf{X} = \mathbf{X}_p = (5,3)$, the only solution.

8. $[B|\mathbf{D}] = \begin{bmatrix} 1 & 0 & 0 & \frac{1}{7} \\ 0 & 1 & 0 & \frac{5}{7} \\ 0 & 0 & 1 & -\frac{6}{7} \\ 0 & 0 & 0 & -\frac{33}{7} \end{bmatrix}$

9. *a.* By Prob. 33*b*, Sec. 5.3, the product of A and the jth column of A^{-1} is the jth column of $AA^{-1} = I$.

b. $[I|P] = \begin{bmatrix} 1 & 0 & 0 & \frac{5}{12} & \frac{1}{6} & \frac{1}{4} \\ 0 & 1 & 0 & -\frac{1}{4} & \frac{1}{2} & \frac{1}{4} \\ 0 & 0 & 1 & -\frac{1}{12} & \frac{1}{6} & -\frac{1}{4} \end{bmatrix}$

10. $A^{-1} = \frac{1}{3}\begin{bmatrix} 6 & 2 & -3 \\ 0 & 1 & 0 \\ -3 & -1 & 3 \end{bmatrix}$

11. One can reduce $[A|I]$ to

$$\begin{bmatrix} 1 & 0 & 2 & 1 & 2/5 & 0 \\ 0 & 1 & -1 & 0 & 1/5 & 0 \\ 0 & 0 & 0 & -3 & -7/5 & 1 \end{bmatrix}$$

the last row of which shows the inconsistency of the systems $A\mathbf{X} = \mathbf{E}_1$, $A\mathbf{Y} = \mathbf{E}_2$, $A\mathbf{Z} = \mathbf{E}_3$ (where $\mathbf{X},\mathbf{Y},\mathbf{Z}$ are as in Prob. 9). Hence $[A|I]$ cannot be reduced to $[I|P]$ by row operations.

12. *b.* $(1) \leftrightarrow (2)$, $(1) \to c^{-1}(1)$, $(2) \to b^{-1}(2)$, $(1) \to (1) - c^{-1}d(2)$.

Section 7.3

15. $P = \frac{1}{6}\begin{bmatrix} 1 & 1 & 1 \\ -2 & 0 & 2 \\ -1 & -3 & 1 \end{bmatrix}$ **19.** *a.* $\begin{bmatrix} 1 & 0 & 0 \\ 0 & 1 & 0 \\ 0 & 0 & 1/2 \end{bmatrix}$ *b.* $\begin{bmatrix} 1 & 4 & 0 \\ 0 & 1 & 0 \\ 0 & 0 & 1 \end{bmatrix}$

21. *a.* Let $A = [i,j]$ and $B = I_m - E_{ii} - E_{jj} + E_{ij} + E_{ji}$. If p is neither i nor j, then $b_{p*} = \mathbf{E}_p - \mathbf{0} - \mathbf{0} + \mathbf{0} + \mathbf{0} = \mathbf{E}_p = a_{p*}$, while $b_{i*} = \mathbf{E}_i - \mathbf{E}_i - \mathbf{0} + \mathbf{E}_j + \mathbf{0} = \mathbf{E}_j = a_{i*}$ and $b_{j*} = \mathbf{E}_j - \mathbf{0} - \mathbf{E}_j + \mathbf{0} + \mathbf{E}_i = \mathbf{E}_i = a_{j*}$. Hence $A = B$.

Section 7.4

30. *d.* Let P be the nonsingular matrix found in part *a* and let β be the standard basis for $T = \mathcal{R}^3$. By Theorem 6.3 there is a basis γ for $\mathcal{R}^3$ such that $P = [i_T]_{\beta\gamma}$. Then since $A = [f]_{\alpha\beta}$, we have $B = PA = [f]_{\alpha\gamma}$.

e. Suppose that there were such a basis for $S = \mathcal{R}^2$. Then $B = AQ$, where $Q = [i_S]_{\delta\alpha}$, contradicting part *c*.

Section 7.5

31. $\{a_{1*},a_{3*}\}$; $r(A) = 2$ **32.** $\{a_{*1},a_{*2}\}$; $c(A) = 2$

42. Suppose that $A \to B$. Then $B = PA$, where P is nonsingular, and hence $r(B) = r(A)$, that is, $\dim R(B) = \dim R(A)$. This result implies that $R(B) = R(A)$ because $R(B)$ is a subspace of $R(A)$. (The rows of B are linear combinations of the rows of A because B is obtained from A by row operations.) However, A and B need not have the same column space. See Prob. 30, Sec. 7.4, where we have $A \to B$ but $C(A) \neq C(B)$.

45. $\langle(-2,0,1)\rangle$

46. Sylvester's law shows that ker f is the zero space.

47. $r(A) = 3$; nullity $= 0$. **48.** The nullity is positive.

Section 7.6

54. (1) $\Leftrightarrow$ (2) because $\mathbf{X} = (x_1, \ldots, x_n)$ is a solution of $A\mathbf{X} = \mathbf{C}$ if and only if $\mathbf{C} = x_1 a_{*1} + \cdots + x_n a_{*n}$ (Theorem 7.1). (2) $\Leftrightarrow$ (3) because $[A|\mathbf{C}]$ and A have the same rank if and only if the extra column $\mathbf{C}$ does not affect the dimension of the column space when A is augmented, that is, if and only if $\mathbf{C}$ is an element of the column space of A. (1) $\Leftrightarrow$ (4) by Prob. 5*b*, Sec. 7.1.

55. Let A and B be $n \times n$. Then

$$A \text{ nonsingular and } r(AB) = n \quad \Rightarrow \quad B = A^{-1}(AB) \text{ is nonsingular}$$

Therefore

$$A \text{ nonsingular and } B \text{ singular} \quad \Rightarrow \quad r(AB) < n$$

56. *a*. If A could be reduced to I by row operations, I would be its row echelon form. (There is only one by Theorem 7.21.)

57. Write P as a product of elementary matrices.

60. Use row operations to reduce A to its row echelon form

$$\begin{bmatrix} 1 & 0 & -1 & 10/7 \\ 0 & 1 & 1 & -4/7 \\ 0 & 0 & 0 & 0 \end{bmatrix}$$

Then use column operations to transform this to B. Hence A is equivalent to B. But it is not row-equivalent to B; if it were, B would be its row echelon form.

61. Two $m \times n$ matrices with the same rank r are either both zero (if $r = 0$), or else each is equivalent to the same canonical form (if $r > 0$). In any case, they are equivalent.

Review Quiz

1. T	**9.** T	**16.** F	**23.** F
2. T	**10.** T	**17.** T	**24.** T
3. F	**11.** F	**18.** F	**25.** F
4. F	**12.** T	**19.** F	**26.** T
5. T	**13.** T	**20.** T	**27.** T
6. T	**14.** F	**21.** T	**28.** F
7. F	**15.** T	**22.** T	**29.** T
8. T			

Section 8.1

3. Expand by the second column to get -30.

Section 8.2

8. $\det cA = c^n \det A$ 12. *a.* 70 *b.* 112

Section 8.4

24. *a.* Expand $\det A^T$ by the jth column and $\det A$ by the jth row.
26. 0
27. *b.* ± 1
28. The determinants are -1, 1, $(1/\sqrt{2})(-1 + i)$, 1; in each case, the absolute value is 1.

Section 8.5

32. *a.* $(\operatorname{adj} A)^{-1} = A/\det A$ 34. $A^{-1} = -\frac{1}{8}\begin{bmatrix} 3 & -7 & 8 \\ -6 & 6 & -8 \\ 5 & -9 & 8 \end{bmatrix}$

Review Quiz

1. T	5. F	9. T	13. F
2. T	6. F	10. T	14. T
3. F	7. T	11. T	15. T
4. F	8. F	12. T	16. T

Section 9.1

1. *b.* If $A = \operatorname{diag}(a_1, \ldots, a_n)$ is any element of D, then $A = a_1A_1 + \cdots + a_nA_n$ (and this expression is unique).
2. Write $A = [a_i\delta_{ij}]$ and $B = [b_i\delta_{ij}]$. The i,j entry of AB is

$$\sum_{k=1}^{n} (a_i\delta_{ik})(b_k\delta_{kj}) = a_ib_i\delta_{ij}$$

Section 9.2

10. *a.* -1; $\mathcal{R}^2$ *b.* 1 and -1; $\{(x,y) \mid y = x\}$ and $\{(x,y) \mid y = -x\}$
c. 1 and 0; xy plane and z axis

d. Since ϕ_1 and ϕ_2 are linearly independent, with $\phi_1' = \phi_1$ and $\phi_2' = 2\phi_2$,

$$\begin{aligned} &r \text{ is an eigenvalue} \iff D(\phi) = r\phi && \text{for some } \phi \neq 0 \\ &\iff \phi' = r\phi,\ \phi = c_1\phi_1 + c_2\phi_2 && c_1 \text{ and } c_2 \text{ not both } 0 \\ &\iff c_1\phi_1' + c_2\phi_2' = rc_1\phi_1 + rc_2\phi_2 && c_1 \text{ and } c_2 \text{ not both } 0 \\ &\iff c_1(1-r)\phi_1 + c_2(2-r)\phi_2 = 0 && c_1 \text{ and } c_2 \text{ not both } 0 \\ &\iff r = 1 \text{ or } r = 2 \end{aligned}$$

The eigenspaces are (respectively)

$$\{\phi \in S \mid D(\phi) = \phi\} = \{\phi \in S \mid \phi(x) = ce^x\} = \langle \phi_1 \rangle$$

and

$$\{\phi \in S \mid D(\phi) = 2\phi\} = \{\phi \in S \mid \phi(x) = ce^{2x}\} = \langle \phi_2 \rangle$$

11. *a.* Write $\mathbf{X} = \mathbf{X}_1 + \mathbf{X}_2$, $\mathbf{X}_1 \in U$, $\mathbf{X}_2 \in U^\perp$. If $\mathbf{X} \in U$, then $\mathbf{X}_1 = \mathbf{X}$ and $\mathbf{X}_2 = \mathbf{0}$, so $f(\mathbf{X}) = \mathbf{X}$. If $\mathbf{X} \in U^\perp$, then $\mathbf{X}_1 = \mathbf{0}$ and $\mathbf{X}_2 = \mathbf{X}$, so $f(\mathbf{X}) = \mathbf{0}$.

b. If 1 is to be an eigenvalue, then $f(\mathbf{X}) = \mathbf{X}$ for some nonzero $\mathbf{X}$. Hence U cannot be the zero space. If 0 is an eigenvalue, then $f(\mathbf{X}) = 0\mathbf{X} = \mathbf{0}$ for some nonzero $\mathbf{X}$, so $U^\perp$ cannot be the zero space, that is, U cannot be all of S.

c. Suppose that $\mathbf{X}$ is an eigenvector which is neither in U nor in $U^\perp$. Since $f(\mathbf{X}) = r\mathbf{X}$ for some r, and since $\mathbf{X} = \mathbf{X}_1 + \mathbf{X}_2$ ($\mathbf{X}_1 \in U$, $\mathbf{X}_2 \in U^\perp$), we have $\mathbf{X}_1 = r(\mathbf{X}_1 + \mathbf{X}_2)$, from which $(1-r)\mathbf{X}_1 - r\mathbf{X}_2 = \mathbf{0}$. Since neither $\mathbf{X}_1$ nor $\mathbf{X}_2$ can be zero, they are linearly independent (being perpendicular). Hence $1 - r = 0$ and $-r = 0$. This is impossible.

12. *a.*
$$\begin{aligned} r \text{ is an eigenvalue} &\Rightarrow f(x) = rx \text{ for some } x \neq 0 \\ &\Rightarrow f^2(x) = f(rx) = rf(x) = r^2x \\ &\Rightarrow f(x) = r^2x \quad \text{(because } f^2 = f\text{)} \\ &\Rightarrow r(1-r)x = 0 \\ &\Rightarrow r = 0 \text{ or } r = 1 \quad \text{(because } x \neq 0\text{)} \end{aligned}$$

14. *a.*
$$\begin{aligned} f(\mathbf{X}) = r\mathbf{X} &\Rightarrow (A - rI)\mathbf{X} = \mathbf{0} \\ &\Rightarrow \det(A - rI) = 0 \quad \text{(if } \mathbf{X} \text{ is to be nonzero)} \\ &\Rightarrow -\tfrac{1}{3}(r-1)(3r^2 - 2r + 3) = 0 \\ &\Rightarrow r = 1 \quad \text{(since } r \text{ is to be real)} \end{aligned}$$

The eigenspace is $\langle(1,1,0)\rangle$.

Section 9.3

19. If $A = \operatorname{diag}(a_1, \ldots, a_n)$, the jth column of A is $A\mathbf{E}_j = a_j\mathbf{E}_j$, so $\mathbf{E}_j$ is an eigenvector. (*Ed. note:* This also shows that a_j is an eigenvalue, which is another way to do Prob. 16.)

20. The jth column of $P^{-1}B$ is $P^{-1}(r_j\mathbf{E}_j) = r_j\mathbf{V}_j$, where $\mathbf{V}_j$ is the jth column of P^{-1}. The jth column of AP^{-1} is $A\mathbf{V}_j$ (Prob. 33, Sec. 5.3), so $A\mathbf{V}_j = r_j\mathbf{V}_j$.

Since P^{-1} is nonsingular, each $\mathbf{V}_j$ is nonzero, so they are eigenvectors of A.

21. Let $\mathbf{V}_1, \ldots, \mathbf{V}_n$ be the given eigenvectors, and $r_1, \ldots, r_n$ the corresponding eigenvalues. If $B = PAP^{-1}$, the jth column of B is

$$\begin{aligned} P(j\text{th column of } AP^{-1}) &= PA(j\text{th column of } P^{-1}) \\ &= PA\mathbf{V}_j \\ &= r_jP\mathbf{V}_j \\ &= r_jP(j\text{th column of } P^{-1}) \\ &= r_j\mathbf{E}_j \qquad j = 1, \ldots, n \end{aligned}$$

Hence $B = \text{diag}\,(r_1, \ldots, r_n)$.

24. The eigenvalues of A are $r_1 = r_2 = 1$. If $A \approx B$, where B is diagonal, then $B = \text{diag}\,(r_1, r_2) = I$. Since $B = PAP^{-1}$ for some nonsingular P, we have $A = P^{-1}BP = I$. But $A \neq I$!

26. An example of the kind called for is to be found in Prob. 24.

27. Show that $A^{-1}\mathbf{X} = r^{-1}\mathbf{X}$ for some nonzero $\mathbf{X}$.

29. *b*. The product of the roots of a quadratic polynomial $ax^2 + bx + c$ is c/a. (*Ed. note:* We are regarding A as a matrix with entries in $\mathcal{C}$, which is always legitimate when no base field is specified, since $\mathcal{R} \subset \mathcal{C}$.)
c. Compute $p(A) = A^2 - (\text{tr}\,A)A + (\det A)I$ to get

$$\begin{bmatrix} 0 & 0 \\ 0 & 0 \end{bmatrix}$$

30. *a*. Use induction on n.

Section 9.4

31. *a*. 1, with multiplicity n. *b*. None (the characteristic polynomial has no real roots). *c*. ± 1 *d*. -1, with multiplicity 2. *e*. ± 1 *f*. 0, 1 (with multiplicity 2). *g*. 1, 2 (*Ed. note:* Compare the simplicity of this to the argument without matrices in Prob. 10*d*, Sec. 9.2.)

32. In each of the following, basis vectors for the eigenspace are listed after the eigenvalue.
a. $r_1 = 1$, $\mathbf{V}_1 = (1,1)$; $r_2 = 3$, $\mathbf{V}_2 = (1,-1)$
b. $r_1 = 2 + i$, $\mathbf{V}_1 = (1 + i, 2)$; $r_2 = 2 - i$, $\mathbf{V}_2 = (1 - i, 2)$
c. $r_1 = r_2 = 2$, $\mathbf{V}_1 = (1,1)$
d. $r_1 = 6$, $\mathbf{V}_1 = (1,1,-2)$; $r_2 = 2$, $\mathbf{V}_2 = (1,1,0)$; $r_3 = -4$, $\mathbf{V}_3 = (-1,1,0)$
e. $r_1 = 1$, $\mathbf{V}_1 = (1,-1,1)$; $r_2 = r_3 = 2$, $\mathbf{V}_2 = (-1,1,0)$, $\mathbf{V}_3 = (-1,0,1)$
f. $r_1 = -1$, $\mathbf{V}_1 = (0,0,1)$; $r_2 = r_3 = 2$, $\mathbf{V}_2 = (0,3,-2)$
g. $r_1 = r_2 = r_3 = 1$, $\mathbf{V}_1 = (1,0,-2)$
h. $r_1 = 1$, $\mathbf{V}_1 = (1,-1,0)$; $r_2 = r_3 = 2$, $\mathbf{V}_2 = (1,-1,1)$

33. Yes in (*a*), (*b*), (*d*), (*e*); no in the others.

34. In each of the following, the answers given are not the only ones possible (although B is unique except for the order of the diagonal entries).

a. $P = \frac{1}{2}\begin{bmatrix} 1 & 1 \\ 1 & -1 \end{bmatrix}$, $B = \text{diag}\,(1,3)$

b. $P = \frac{1}{4}\begin{bmatrix} -2i & 1+i \\ 2i & 1-i \end{bmatrix}$, $B = \text{diag}\,(2+i, 2-i)$

d. $P = \frac{1}{2}\begin{bmatrix} 0 & 0 & -1 \\ 1 & 1 & 1 \\ -1 & 1 & 0 \end{bmatrix}$, $B = \text{diag}\,(6,2,-4)$

e. $P = \begin{bmatrix} 1 & 1 & 1 \\ 1 & 2 & 1 \\ -1 & -1 & 0 \end{bmatrix}$, $B = \text{diag}\,(1,2,2)$

36. First prove (by induction on n) that the determinant of an $n \times n$ lower triangular matrix is the product of its diagonal entries.

37. Use the transpose and appeal to Prob. 36.

38. Apply Theorem 9.13 to eigenvectors of the map $D\colon S \to S$ defined by $D(\phi) = \phi'$. (*Ed. note:* Compare this proof with the one ordinarily found in textbooks on differential equations. This involves the Wronskian of $e^{r_1 x}, \ldots, e^{r_n x}$, as in Probs. 19 and 21, Sec. 2.3, and is relatively complicated.)

40. Let A be the matrix and f the corresponding standard map. Then $r_1, \ldots, r_n$ are eigenvalues of f; by Corollary 9.13*a*, $\mathfrak{F}^n$ has a basis of eigenvectors of f (which are also eigenvectors of A). By Theorem 9.12, A is similar to a diagonal matrix (whose eigenvalues are those of A). By Prob. 6, Sec. 9.1, we may assume that this is diag $(r_1, \ldots, r_n)$.

41. If r is the only eigenvalue of f, then $k = n$ and the statement is trivial. Hence assume that there are others. Relabel $v_1, \ldots, v_n$ (if necessary) so that the k elements of β associated with r are $v_1, \ldots, v_k$. The problem is to show that every eigenvector x associated with r can be written as a linear combination of $v_1, \ldots, v_k$. First write $x = \sum_{j=1}^n x_j v_j$. Since $f(x) = rx$, we have $\sum_{j=1}^n x_j f(v_j) = \sum_{j=1}^n r x_j v_j$. But $f(v_j) = rv_j$ for $j = 1, \ldots, k$, while if $j > k$ we have $f(v_j) = r_j v_j$, where r_j is an eigenvalue of f distinct from r. The above equation now reads

$$\sum_{j=1}^{k} r x_j v_j + \sum_{j=k+1}^{n} r_j x_j v_j = \sum_{j=1}^{k} r x_j v_j + \sum_{j=k+1}^{n} r x_j v_j$$

from which

$$\sum_{j=k+1}^{n} (r_j - r)x_j v_j = 0$$

This implies $(r_j - r)x_j = 0$ and hence $x_j = 0$, $j = k+1, \ldots, n$. Therefore $x = \sum_{j=1}^{k} x_j v_j$.

42. Since $u_1 + \cdots + u_m = 0$, the u_i are linearly dependent. If they were all nonzero, they would be eigenvectors of f associated with distinct eigenvalues, hence linearly independent (by Theorem 9.13). Therefore at least one of them is 0; relabel (if necessary) so that $u_m = 0$. Then $u_1 + \cdots + u_{m-1} = 0$. Repeat the argument to conclude that they are all 0.

43. *b.* Deduce the rank of $B - rI$ from Theorem 9.5.

Section 9.5

45. *a.* $\Phi(x) = c_1 e^x(-2,1) + c_2 e^{4x}(1,1)$
b. $\Phi(x) = c_1 e^x(1,-1,1) + c_2 e^{2x}(-1,1,0) + c_3 e^{2x}(-1,0,1)$

46. $\Phi(x) = (2e^x - e^{4x}, -e^x - e^{4x})$

Section 9.6

48. *a.* $p(A)\mathbf{V}_i = c(A - r_1 I) \cdots (A - r_n I)\mathbf{V}_i$. Look at the effect on $\mathbf{V}_i$ of the factors $A - r_n I, \ldots, A - r_1 I$ (in that order). The typical factor is $A - r_j I$, where j takes on the values $n, \ldots, 1$ (in that order). As long as $j > i$, the effect of this factor is to transform $\mathbf{V}_i$ into $(r_i - r_j)\mathbf{V}_i$. The scalar $r_i - r_j$ may be moved to the left, leaving $\mathbf{V}_i$ to be multiplied by the next factor in the list; this continues until $j = i$. Since $(A - r_i I)\mathbf{V}_i = \mathbf{0}$, we get $\mathbf{0}$ at this point.
b. Since the $\mathbf{V}_i$ are linearly independent, the null space of $p(A)$ is n-dimensional. Hence the rank of $p(A)$ is 0 and $p(A)$ must be the zero matrix.

50. *a.* $A^{-1} = \dfrac{1}{17}(3I - A) = \dfrac{1}{17}\begin{bmatrix} 2 & 3 \\ -5 & 1 \end{bmatrix}$

b. $A^{-1} = 3I - 3A + A^2 = \begin{bmatrix} -3 & -1 & -2 \\ -2 & 0 & -1 \\ 10 & 3 & 6 \end{bmatrix}$

51. *a.* Use Prob. 22, Sec. 9.3.

Review Quiz

1. F	7. T	13. F	19. T
2. F	8. T	14. T	20. F
3. T	9. T	15. T	21. T
4. T	10. F	16. F	22. T
5. T	11. T	17. T	23. F
6. T	12. F	18. T	24. T

Section 10.1

2. $2 + 7i$ 3. *a.* $14 - 6i$ *b.* $\mathbf{X} * \mathbf{X} = 4$ *c.* $\mathbf{X} * \mathbf{X} = 5$

5. $\sqrt{7}$ 7. *a.* $-\dfrac{1}{\sqrt{2}}$ *b.* $\dfrac{1}{6}$

9. $$|\mathbf{X} \cdot \mathbf{Y}| > \|\mathbf{X}\|\,\|\mathbf{Y}\| \implies (x_1y_1 + x_2y_2 + x_3y_3)^2 > (x_1^2 + x_2^2 + x_3^2)(y_1^2 + y_2^2 + y_3^2)$$
$$\implies (x_1y_2 - x_2y_1)^2 + (x_1y_3 - x_3y_1)^2 + (x_2y_3 - x_3y_2)^2 < 0$$

which is impossible in $\mathcal{R}$.

12. *a.* $\sqrt{30}$ *b.* $\sqrt{17}$

17. *c.* Assume that $t(\mathbf{X},\mathbf{Y}) < d(\mathbf{X},\mathbf{Y})$, square, and obtain a contradiction.

19. *a.* xy, the ordinary product in $\mathcal{R}$.
 b. $x^2 \geq 0$, with equality $\iff$ $x = 0$; the commutative law in $\mathcal{R}$; the distributive laws in $\mathcal{R}$.
 c. $\sqrt{x \cdot x} = |x|$, the ordinary absolute value in $\mathcal{R}$.
 d. 0 or π. *e.* $|xy| = |x|\,|y|$ *f.* $|x + y| \leq |x| + |y|$
 g. $d(x,y) = |x - y|$, the usual distance on the number scale between x and y.

Section 10.2

20. *a.* $f * f = \int_a^b f(x)^2\,dx \geq 0$. Obviously this is 0 if $f = 0$, but the converse is not so apparent. Consult any good calculus book to see why $\int_a^b f(x)^2\,dx = 0 \implies f = 0$, provided that f is continuous. This is not necessarily true of integrable functions that are not continuous. For example, if $f(x) = 0$ when $0 \leq x < 1$ and $f(1) = 1$, then $\int_0^1 f(x)^2\,dx = 0$, but f is not the zero function on [0,1].

21. *a.* 2 *b.* 3 *c.* $1/\sqrt{30}$

29. *c.* $$\frac{e^{ix}}{e^{iy}} = \frac{\cos x + i \sin x}{\cos y + i \sin y} \cdot \frac{\cos y - i \sin y}{\cos y - i \sin y} = \cos (x - y) + i \sin (x - y)$$

d. If $p = q$, then $f * g = 2\pi$.

Section 10.3

34. *b.* If $\theta = 0$ or $\theta = \pi$, then $|\cos \theta| = 1$, from which $|x * y| = \|x\| \, \|y\|$. Letting $z = y - kx$, where $k = (y * x)/\|x\|^2$, we find that $z * z = 0$, so $z = 0$, that is, $y = kx$. Moreover,

$$k = \frac{\|x\| \, \|y\| \cos \theta}{\|x\|^2} = \frac{\|y\|}{\|x\|} \cos \theta$$

so $k > 0$ if $\theta = 0$ and $k < 0$ if $\theta = \pi$.

35. Let $k = \|v\|$. Then

$$u = \frac{v}{\|v\|} \quad \Rightarrow \quad v = ku \quad \Rightarrow \quad \|v\| = |k| \, \|u\| = \|v\| \, \|u\| \cdot \quad \Rightarrow \quad \|u\| = 1$$

Moreover, $k > 0$; if the space is euclidean, u and v have the same direction by Prob. 34.

36. Proceeding from the hint, it is easy to show that equality holds if $y = kx$. Conversely, suppose that $|x * y| = \|x\| \, \|y\|$. Let $z = y - kx$, where $k = (y * x)/\|x\|^2$, and show that $z * z = 0$.

37. Proceeding from the hint, suppose that $\|x + y\| = \|x\| + \|y\|$. Since the inequalities in the proof of Theorem 10.2 are now equalities, we have $x * y + \overline{x * y} = 2|x * y|$ and $|x * y| = \|x\| \, \|y\|$. The first of these implies that $x * y$ is a nonnegative real number (Prob. 43*d*, Sec. 1.3), while the second implies that $y = kx$, where $k = (y * x)/\|x\|^2$ (as in Prob. 36). Hence $k \geq 0$, because $y * x = \overline{x * y} = x * y \geq 0$.

Section 10.4

44. Try $\mathbf{X} = (1,0)$ and $\mathbf{Y} = (i,0)$. The "only if" part of Prob. 43 is still correct, however.

46. *b.* $\mathbf{X} = -\tfrac{5}{2}\mathbf{V}_1 + \tfrac{17}{22}\mathbf{V}_2 + \tfrac{2}{11}\mathbf{V}_3$

47. *c.* $u_1(x) = 1$, $u_2(x) = 2\sqrt{3}(x - \tfrac{1}{2})$

d. $f = -2u_1 + (1/\sqrt{3})u_2$;
$g = \tfrac{1}{2}u_1 + (\sqrt{3}/6)u_2$

48. *a.* $a_0 = 1$, while all the other coefficients are 0. Hence the Fourier series collapses to 1.

b. $a_1 = 1$, while all the other coefficients are 0. Hence the Fourier series collapses to $\cos x$.

c. $a_m = 0$, $m = 0, 1, 2, \ldots$, while $b_m = (-1)^{m+1}(2/m)$, $m = 1, 2, \ldots$. Hence the Fourier series is $2(\sin x - \frac{1}{2}\sin 2x + \frac{1}{3}\sin 3x - \cdots)$.
(*Ed. note:* Naturally one wonders whether this converges to the original function. In parts *a* and *b* this is trivial, but not here.)

49. *c*. $x * y = \sum_{j=1}^{n} x_j \bar{y}_j = \mathbf{X} * \mathbf{Y}$, where $\mathbf{X} = (x_1, \ldots, x_n)$ and $\mathbf{Y} = (y_1, \ldots, y_n)$. (*Ed. note:* This is Parseval's identity, stated in the next section.)

50. *e*. $A = \text{diag}\ (6,210,35)$

51. *d*. First show that if $\mathbf{X} = (x_1,x_2)$ is any element of $\mathcal{R}^2$, then $\mathbf{X} = (2x_1 - 3x_2)\mathbf{U}_1 + (-x_1 + 2x_2)\mathbf{U}_2$.
e. -8 *f*. $\sqrt{5}$

53. *c*. The matrix is

$$\frac{1}{\sqrt{5}}\begin{bmatrix} 1 & 2 \\ -2 & 1 \end{bmatrix}$$

of the form

$$\begin{bmatrix} \cos\theta & -\sin\theta \\ \sin\theta & \cos\theta \end{bmatrix}$$

which represents a rotation.

Section 10.5

54. $(-\frac{5}{2}, \frac{17}{22}, \frac{2}{11})$

55. *a*. $(-2, 1/\sqrt{3})$ and $(\frac{1}{2}, \sqrt{3}/6)$ *b*. $-\frac{5}{6}$ *c*. $\sqrt{39}/3$ and $\sqrt{3}/3$ *d*. $\sqrt{57}/3$

56. $3 + i$

63. Bessel's inequality simply says that if $\mathbf{X} = (x_1,x_2,x_3)$, then $x_1^2 + x_2^2 \leq x_1^2 + x_2^2 + x_3^2$.

65. *b*. Use Prob. 61.
c. Use Prob. 61. Or, by brute force, compute

$$\int_0^1 e^x\,dx = e - 1 \qquad \int_0^1 (2x - 1)e^x\,dx = 3 - e$$

$$\int_0^1 e^{2x}\,dx = \tfrac{1}{2}(e^2 - 1)$$

The inequality to be verified then reads $(e - 1)^2 + 3(3 - e)^2 < \frac{1}{2}(e^2 - 1)$, or (after simplifying) $(7e - 19)(e - 3) < 0$. Since the solution set of the inequality $(7x - 19)(x - 3) < 0$ is the interval $I = (^{19}/_7, 3)$ and since $e \in I$, the desired inequality is correct. (*Ed. note:* It is a close shave! $^{19}/_7 \approx 2.714$, while $e \approx 2.718$.)

66. Apply Bessel's inequality in Prob. 60 to the space $\langle u_1, \ldots, u_n \rangle$ to conclude that the partial sums are bounded by $\|x\|^2$.

67. *a.* $1/\sqrt{2\pi}, (1/\sqrt{\pi})\cos x, (1/\sqrt{\pi})\sin x, (1/\sqrt{\pi})\cos 2x, (1/\sqrt{\pi})\sin 2x, \ldots$

b. $\sqrt{2\pi}\, a_0, \sqrt{\pi}\, a_1, \sqrt{\pi}\, b_1, \sqrt{\pi}\, a_2, \sqrt{\pi}\, b_2, \ldots$

68. *a.* $a_0 = 1$, while all the other coefficients are 0. Parseval's equality reads $2 = (1/\pi) \int_{-\pi}^{\pi} dx$, which is correct.

b. $a_1 = 1$, while all the other coefficients are 0. Parseval's equality reads $1 = (1/\pi) \int_{-\pi}^{\pi} \cos^2 x\, dx$, which is correct.

69. $a_m = 0, m = 0, 1, 2, \ldots$, while $b_m = (-1)^{m+1}(2/m), m = 1, 2, \ldots$. Parseval's equality therefore reduces to

$$b_1^2 + b_2^2 + \cdots = \frac{1}{\pi}\int_{-\pi}^{\pi} x^2\, dx \qquad \text{or} \qquad \sum_{k=1}^{\infty} \frac{1}{k^2} = \frac{\pi^2}{6}$$

Section 10.6

70. $\left\{\frac{1}{2}(1,1,1,1), \frac{1}{2\sqrt{3}}(3,-1,-1,-1), \frac{1}{\sqrt{2}}(0,-1,0,1),\right.$
$\left.\frac{1}{\sqrt{6}}(0,-1,2,-1)\right\}$

71. One answer is $\left\{\frac{1}{3}(1,-2,2), \frac{1}{\sqrt{5}}(2,1,0), \frac{1}{3\sqrt{5}}(-2,4,5)\right\}$.

72. $\frac{1}{\sqrt{2}}(0,0,1,-1)$ or $\frac{1}{\sqrt{2}}(0,0,-1,1)$

73. $v_1(x) = 1$, $v_2(x) = x - \frac{1}{2}$

74. $v_1(x) = 1$, $v_2(x) = x$, $v_3(x) = x^2 - \frac{1}{3}$, $v_4(x) = x^3 - \frac{3}{5}x$. The first four Legendre polynomials are $P_0(x) = 1$, $P_1(x) = x$, $P_2(x) = \frac{3}{2}(x^2 - \frac{1}{3})$, $P_3(x) = \frac{5}{2}(x^3 - \frac{3}{5}x)$.

75. $\left\{\frac{1}{\sqrt{2}}(i,1,0), \frac{1}{\sqrt{6}}(-i,1,2)\right\}$

81. Yes

82. The Gram-Schmidt process yields $\mathbf{V}_1 = (1,1,0,-1)$, $\mathbf{V}_2 = (0,-1,0,-1)$, $\mathbf{V}_3 = (0,0,0,0)$.

Review Quiz

1. F	9. T	17. T	25. T	33. T
2. T	10. F	18. T	26. T	34. T
3. F	11. F	19. T	27. T	35. T
4. T	12. T	20. F	28. T	36. T
5. F	13. T	21. T	29. F	37. T
6. F	14. F	22. T	30. T	38. T
7. T	15. T	23. F	31. F	39. T
8. T	16. T	24. T	32. F	40. T

Section 11.1

3. $T_1 + T_2 = \langle \mathbf{U}_1, \mathbf{U}_2, \mathbf{U}_3 \rangle$

5. $T_1 + T_2 = T_1$

8. $T_1 + T_2 = \{(x,y,z) \mid 2x - y + z = 0\}$, the plane spanned by $(1,3,1) \in T_1$ and $(-2,1,5) \in T_2$.

9. *a.* Neither U nor V is finite-dimensional because they contain the functions $1, x^2, x^4, \ldots$ and $x, x^3, x^5, \ldots$, respectively.

b. The only function that is both even and odd is the zero function, because $f(-x) = f(x) = -f(x) \Rightarrow f(x) = 0$. Hence $U \cap V = \{0\}$.

d. $e^x = \cosh x + \sinh x$

11. The second subspace may be taken to be any line through the origin distinct from U.

Section 11.2

13. *a.* The origin. (Note that U is not a subspace.) *b.* The xy plane. *c.* The line $y = x$. *d.* $\langle(1,1,1)\rangle$, the line through the origin perpendicular to U.

15. Suppose that $S = U + V$, where S is unitary and U and V are orthogonal subspaces. To prove that $V = U^\perp$, we show $U^\perp \subset V$ (the reverse inclusion being trivial). Let $y \in U^\perp$ and use the fact that $S = U + V$ to write $y = y_1 + y_2$, where $y_1 \in U$ and $y_2 \in V$. Then $y_1 * y_1 = y_1 * (y - y_2) = y_1 * y - y_1 * y_2 = 0 - 0 = 0$, so $y_1 = 0$. Hence $y = y_2 \in V$.

This shows that V is the orthogonal complement of U. To prove that U is the orthogonal complement of V, interchange U and V in the above argument. [*Ed. note:* One might approach this second part by writing $V = U^\perp \Rightarrow V^\perp = (U^\perp)^\perp = U$. However, we shall establish that $(U^\perp)^\perp = U$ only when U is finite-dimensional. See Prob. 26, Sec. 11.3.]

17. $\langle \phi_2 \rangle$

18. *a.* $U^\perp$ is the subspace generated by $v(x) = x$.

19. *c.* V is the intersection of the plane $U^\perp$ and the plane T. See Fig. 11.4; $U^\perp$ is the plane shown, while T is the plane containing the lines U and V. If T is regarded as an isomorphic copy of $\mathcal{R}^2$, then U and V are lines in 2-space, each being the orthogonal complement of the other in this space.

20. The inclusion $(U + V)^\perp \subset U^\perp \cap V^\perp$ is easy. To prove the reverse inclusion, suppose that $y \in U^\perp \cap V^\perp$. Then if $x = x_1 + x_2$ is any element of $U + V$ (where $x_1 \in U$ and $x_2 \in V$), we have $x * y = x_1 * y + x_2 * y = 0 + 0 = 0$. Hence $y \in (U + V)^\perp$.

21. *b.* Let $V = \langle u_{m+1}, \ldots, u_n \rangle$. Then $S = U \oplus V$ by the same argument as in Theorem 11.4. Since U and V are orthogonal, Theorem 11.7 implies that $V = U^\perp$.

Another argument is to use Theorem 10.7 to write each $y \in S$ in the form $y = \Sigma_{i=1}^n y_i u_i$, where $y_i = y * u_i$, $i = 1, \ldots, n$. Since $y \in U^\perp \iff y * u_i = 0$, $i = 1, \ldots, m$ (Theorem 11.5), we have $y \in U^\perp \iff y = \Sigma_{i=m+1}^n y_i u_i \iff y \in V$. Hence $V = U^\perp$.

24. The first is $\langle (1, -i) \rangle$, while the second is $\langle (1, i) \rangle$. In each case, the dimension is 1.

Section 11.3

28. $\mathbf{X}_1 = \frac{1}{3}(8,1,7)$, $\mathbf{X}_2 = \frac{4}{3}(1,-1,-1)$, $\|\mathbf{X}_2\| = 4/\sqrt{3}$

29. $\mathbf{X}_1 = \frac{1}{3}(5 - i,\ 1 - i,\ 2 + 4i)$

31. *a.* $f(x,y,z) = (x,y,0)$
b. $f(x,y) = \frac{1}{13}(9x - 6y,\ -6x + 4y)$
c. $f(x,y,z) = \frac{1}{3}(2x + y + z,\ x + 2y - z,\ x - y + 2z)$

33. *b.* $g(x,y,z) = (0,0,z)$; $g(x,y) = \frac{1}{13}(4x + 6y,\ 6x + 9y)$; $g(x,y,z) = \frac{1}{3}(x - y - z,\ -x + y + z,\ -x + y + z)$

34. *b.* Let S be the space and f the operator. Since rng f is finite-dimensional, the projection theorem says that every $x \in S$ can be written $x = x_1 + x_2$, where $x_1 \in \text{rng } f$ and $x_2 \in (\text{rng } f)^\perp$. However $(\text{rng } f)^\perp = \ker f$, so $f(x) = f(x_1)$. Since $x_1 \in \text{rng } f$, there is a $y \in S$ such that $f(y) = x_1$. Hence $f(x) = f^2(y) = f(y) = x_1$, which means that f is the orthogonal projection of S on rng f.

35. *b.* The identity map. *c.* $f(x) = -x$ *f.* $f^{-1} = f$

36. *a.* $f(x,y) = (y,x)$ *b.* $f(x,y) = \frac{1}{13}(5x - 12y,\ -12x - 5y)$
c. $f(x,y,z) = \frac{1}{3}(x + 2y + 2z,\ 2x + y - 2z,\ 2x - 2y + z)$

37. *c.* $U^\perp = \langle i \rangle$. The orthogonal projections of $x = a + bi$ on U and $U^\perp$ are $x_1 = a$ and $x_2 = bi$. The orthogonal projections of $\mathbf{X} = (a,b)$ on the coordinates axes in $\mathcal{R}^2$ are $f(x_1) = (a,0)$ and $f(x_2) = (0,b)$.

Section 11.4

38. Use Theorem 10.7 to write $x = \Sigma_{i=1}^{n} (x * u_i)u_i$. By Theorem 11.9 the sum of the first m terms is the orthogonal projection of x on U. Hence its orthogonal projection on $U^{\perp}$ is $x_2 = \Sigma_{i=m+1}^{n}(x * u_i)u_i$. Theorem 10.9 (applied to the space spanned by $u_{m+1}, \ldots, u_n$) yields $\|x_2\| = (\Sigma_{i=m+1}^{n} |x * u_i|^2)^{1/2}$.

39. A basis for $\mathcal{R}^3$ whose first element spans the given line is $\{(2,1,-1), (0,1,0), (0,0,1)\}$. The Gram-Schmidt process converts this to the orthonormal basis $\{\mathbf{U}_1,\mathbf{U}_2,\mathbf{U}_3\}$, where

$$\mathbf{U}_1 = \frac{1}{\sqrt{6}}(2,1,-1) \qquad \mathbf{U}_2 = \frac{1}{\sqrt{30}}(-2,5,1) \qquad \mathbf{U}_3 = \frac{1}{\sqrt{5}}(1,0,2)$$

The distance from $\mathbf{X} = (2,-1,1)$ to the line is then

$$\sqrt{(\mathbf{X}\cdot\mathbf{U}_2)^2 + (\mathbf{X}\cdot\mathbf{U}_3)^2} = \frac{4}{\sqrt{3}}$$

40. The square of the distance from $(2,-1,1)$ to an arbitrary point $(x,y,z) = (2t,t,-t)$ of the given line is $f(t) = 2(3t^2 - 2t + 3)$. The minimum value of f is $f(1/3) = 16/3$, so the desired distance is $4/\sqrt{3}$.

41. $\sqrt{15}/3$

42. Let $\mathbf{X} = (x_0,y_0)$ and suppose that $\{\mathbf{U}_1,\mathbf{U}_2\}$ is an orthonormal basis for $\mathcal{R}^2$ whose second element is $\mathbf{U}_2 = \dfrac{1}{\sqrt{a^2 + b^2}}(a,b)$. Then $\mathbf{U}_1$ spans the given line; by Theorem 11.11 the distance from $\mathbf{X}$ to the line is

$$d = \sqrt{(\mathbf{X}\cdot\mathbf{U}_2)^2} = \frac{|ax_0 + by_0|}{\sqrt{a^2 + b^2}}$$

(*Ed. note:* It is unnecessary to specify $\mathbf{U}_1$!)

43. Proceed as in Prob. 42 by noting that a unit vector perpendicular to the plane is

$$\frac{1}{\sqrt{a^2 + b^2 + c^2}}(a,b,c)$$

44. *a.* 1 *b.* $\cos x$

c. $2\left[\sin x - \dfrac{1}{2}\sin 2x + \dfrac{1}{3}\sin 3x - \cdots + \dfrac{(-1)^{n+1}}{n}\sin nx\right]$

Review Quiz

1. F	5. T	9. T	13. T
2. T	6. T	10. F	14. T
3. F	7. T	11. F	15. T
4. T	8. T	12. T	16. F

Section 12.1

2. *a.* $f^*(x,y,z) = (x + 2y + z, x, -x - z)$
b. $f^* = f$ *c.* $f^*(x,y) = (x, 4x - y)$

8. *c.* Using matrices relative to an orthonormal basis, we have $[(f \circ g)^*] = [f \circ g]^* = ([f][g])^* = [g]^*[f]^* = [g^*][f^*] = [g^* \circ f^*]$, from which $(f \circ g)^* = g^* \circ f^*$.

11. $y \in U^\perp \Rightarrow x * y = 0$ for all $x \in U \Rightarrow f(x) * y = 0$ for all $x \in U \Rightarrow x * f^*(y) = 0$ for all $x \in U \Rightarrow f^*(y) \in U^\perp$.

15. *a.* $f^*(x,y,z) = (x + 2y + z, x - z)$

Section 12.2

16. Parts *c* and *e* are not unitary, the others are.

21. It fails in the "only if" part, where the hypothesis that f is unitary does not imply that f^{-1} exists (even though f^* may). For an example, see Sec. 12.4.

36. *a.* Since f maps U into U, we may restrict its domain to U and consider it as a linear operator on U. Being unitary, it is one-to-one, hence onto.
b. Let $y \in U^\perp$. We know from part *a* that for each $x \in U$ there is a $z \in U$ such that $f(z) = x$. Hence $x * f(y) = f(z) * f(y) = z * y = 0$, which means $f(y) \in U^\perp$.

Section 12.3

37. *b.* $A = \begin{bmatrix} \cos \pi & \sin \pi \\ \sin \pi & -\cos \pi \end{bmatrix}$
c. The y axis.
d. $f(x,y) = (-x,y)$

38. A reflection in the line $y = x$, namely $f(x,y) = (y,x)$.

39. *a.* $\det C = (\det B)(\det A) = 1$, so C (being orthogonal because it is the product of orthogonal matrices) represents a rotation.

b. Define $h: \mathcal{R}^2 \to \mathcal{R}^2$ by

$$[h] = C = BA = \begin{bmatrix} \cos 30^\circ & -\sin 30^\circ \\ \sin 30^\circ & \cos 30^\circ \end{bmatrix}$$

Then h is a rotation through 30° and $f = g \circ h$, where $g: \mathcal{R}^2 \to \mathcal{R}^2$ is the reflection defined by $[g] = B$.

41. *a*. $\{\mathbf{X} = (x,y,z) \mid x = t,\ y = t,\ z = 0\} = \langle(1,1,0)\rangle$

Section 12.4

46. *d*. Define $g: S \to S$ by $g(x_1,x_2,x_3, \ldots) = (x_2,x_3, \ldots)$. Then $f(\mathbf{X}) \cdot \mathbf{Y} = (0, x_1, x_2, \ldots) \cdot (y_1, y_2, y_3, \ldots) = x_1y_2 + x_2y_3 + \cdots = \mathbf{X} \cdot g(\mathbf{Y})$. Hence $g = f^*$.

h. $f^*(\mathbf{E}_1) = \mathbf{0}$, so f^* is not one-to-one (as it must be if it is to be orthogonal).

Section 12.5

47. *a*. If it were, Corollary 12.11a would imply that ϕ is the zero map.

b. $\mathbf{X} = (i,1)$

48. *c*. $\phi(x,y) = (-y,x)$, while $\phi^*(x,y) = (y,-x)$.

49. *a*. When $\mathfrak{F} = \mathcal{C}$, we replaced x by ix in Eq. (12.1). When $\mathfrak{F} = \mathcal{R}$, the scalar i is not available.

b. The step $y * \phi(x) = \phi(x) * y$ is incorrect in a complex space.

Section 12.6

51. *b*. The identity $f(x) * y = -x * f(y)$ is equivalent to $f(x) * y = x * (-f)(y)$, which in turn is equivalent to $f^* = -f$.

52. *b*. For the "if" part, suppose that $f(x) * x$ is pure imaginary for all x and let $\phi = f^* + f$. Argue that $\phi = 0$ as in part *a*, using the fact that z is pure imaginary if and only if $\bar{z} = -z$.

57. Let

$$A = [f] = \begin{bmatrix} 1 & i \\ i & 1 \end{bmatrix}$$

Since

$$A^* = \begin{bmatrix} 1 & -i \\ -i & 1 \end{bmatrix}$$

we have $AA^* = 2I = A^*A$, so $f \circ f^* = f^* \circ f$ and f is normal. However A^* is neither A, $-A$, nor A^{-1}.

58. One of the three arguments is as follows. Suppose that $f^* \circ f = i$ and $f^2 = i$. Postmultiplying each side of $f^* \circ f = i$ by f yields $f^* = f$.

66. *a*. Define $f: \mathfrak{F}^n \to \mathfrak{F}^n$ by $[f] = A$ (relative to an orthonormal basis). By Theorem 12.13, $f = G + H$, where G and H are Hermitian and skew-Hermitian, respectively. Let $B = [G]$ and $C = [H]$. Then B is Hermitian and C is skew-Hermitian, and $A = B + C$.

Uniqueness of B and C is not obvious from this argument, for although G and H are unique (once f is specified), there is more than one way to name f (since different orthonormal bases may be used). To avoid this difficulty, one may give a direct argument imitating the proof of Theorem 12.13; that is, show that the only choices are $B = \frac{1}{2}(A + A^*)$ and $C = \frac{1}{2}(A - A^*)$.

c. To illustrate part *a*, compute

$$B = \frac{1}{2}\begin{bmatrix} 2 & 2+i \\ 2-i & -2 \end{bmatrix} \quad \text{and} \quad C = \frac{1}{2}\begin{bmatrix} 0 & -2+i \\ 2+i & 0 \end{bmatrix}$$

Then $A = B + C$, $B^* = B$, and $C^* = -C$. To illustrate part *b*, compute B as before, but take

$$C = -\frac{i}{2}(A - A^*) = \frac{1}{2}\begin{bmatrix} 0 & 1+2i \\ 1-2i & 0 \end{bmatrix}$$

Then $A = B + iC$, $B^* = B$, and $C^* = C$.

Review Quiz

1. T	**8.** T	**15.** T	**22.** F
2. T	**9.** T	**16.** T	**23.** T
3. T	**10.** F	**17.** T	**24.** F
4. T	**11.** T	**18.** F	**25.** F
5. T	**12.** T	**19.** T	**26.** F
6. T	**13.** T	**20.** F	**27.** T
7. T	**14.** F	**21.** T	**28.** F

Section 13.1

2. If $\mathfrak{F} = \mathfrak{R}$, the only eigenvalue is 0, while if $\mathfrak{F} = \mathfrak{C}$, the eigenvalues are $0, \pm i\sqrt{35}$.

3. *a*. Let $f = a_1\phi_1 + a_2\phi_2$ and $g = b_1\phi_1 + b_2\phi_2$ be any elements of T. Then

$$D(f) * g = f' * g = \int_{-\pi}^{\pi} f'(x)g(x)\,dx = f(x)g(x)\Big|_{-\pi}^{\pi} - \int_{-\pi}^{\pi} f(x)g'(x)\,dx$$

Since

$$f(\pm\pi) = -\frac{a_1}{\sqrt{\pi}} \quad \text{and} \quad g(\pm\pi) = -\frac{b_1}{\sqrt{\pi}}$$

we find that $f(x)g(x)\Big|_{-\pi}^{\pi} = 0$ and hence that $D(f) * g = -f * D(g)$.

b. $[D] = \begin{bmatrix} 0 & 1 \\ -1 & 0 \end{bmatrix}$ *c*. It has none.

5. *a*. Show that if x_1 and x_2 are eigenvectors associated with distinct eigenvalues r_1 and r_2, then $r_1(x_1 * x_2) = r_2(x_1 * x_2)$.

c. Let x be any element of S and write $x = x_1 + x_2$, where $x_1 \in T_k$ and $x_2 \in T_k{}^{\perp}$. Then $f_k(x) = x_1$. But x_1 is orthogonal to T_j, that is, $x_1 \in T_j^{\perp}$, which means that $f_j(x_1) = 0$. Hence $(f_j \circ f_k)(x) = f_j(x_1) = 0$.

10. *a*. $\{\mathbf{E}_1, \mathbf{E}_2\}$; $P = I$; $B = A$

b. $\left\{\frac{1}{\sqrt{2}}(1,1), \frac{1}{\sqrt{2}}(1,-1)\right\}$; $P = \frac{1}{\sqrt{2}}\begin{bmatrix} 1 & 1 \\ 1 & -1 \end{bmatrix}$; $B = \text{diag}\,(3,1)$

c. $\left\{\frac{1}{\sqrt{2}}(1,-i), \frac{1}{\sqrt{2}}(1,i)\right\}$; $P = \frac{1}{\sqrt{2}}\begin{bmatrix} 1 & i \\ 1 & -i \end{bmatrix}$; $B = \text{diag}\,(2,0)$

d. $\left\{\frac{1}{\sqrt{3}}(1,1,1), \frac{1}{\sqrt{2}}(1,0,-1), \frac{1}{\sqrt{6}}(1,-2,1)\right\}$;

$$P = \frac{1}{6}\begin{bmatrix} 2\sqrt{3} & 2\sqrt{3} & 2\sqrt{3} \\ 3\sqrt{2} & 0 & -3\sqrt{2} \\ \sqrt{6} & -2\sqrt{6} & \sqrt{6} \end{bmatrix}; \; B = \text{diag}\,(3,0,0)$$

e. $\left\{\frac{1}{\sqrt{3}}(1,1,1), \frac{1}{\sqrt{2}}(0,1,-1), \frac{1}{\sqrt{6}}(2,-1,-1)\right\}$;

$$P = \frac{1}{6}\begin{bmatrix} 2\sqrt{3} & 2\sqrt{3} & 2\sqrt{3} \\ 0 & 3\sqrt{2} & -3\sqrt{2} \\ 2\sqrt{6} & -\sqrt{6} & -\sqrt{6} \end{bmatrix}; \; B = \text{diag}\,(3,2,0)$$

f. $\left\{\frac{1}{\sqrt{3}}(1,i,-i), \frac{1}{\sqrt{2}}(1,-i,0), \frac{1}{\sqrt{6}}(-i,1,2)\right\}$;

$$P = \frac{1}{6}\begin{bmatrix} 2\sqrt{3} & -2i\sqrt{3} & 2i\sqrt{3} \\ 3\sqrt{2} & 3i\sqrt{2} & 0 \\ i\sqrt{6} & \sqrt{6} & 2\sqrt{6} \end{bmatrix}; \; B = \text{diag}\,(2,1,-1)$$

Section 13.2

11. *c.* From part *b* it is clear that r_1 and r_2 either have opposite signs, one of them is zero, or they have the same sign, depending on whether $b^2 - 4ac$ is positive, zero, or negative, respectively.

12. *a.* $\frac{5}{2}x'^2 + \frac{1}{2}y'^2 = 1$ (representing an ellipse); $P = \begin{bmatrix} \cos 30° & \sin 30° \\ -\sin 30° & \cos 30° \end{bmatrix}$

(*Ed. note:* See the discussion of this in Sec. 6.3. *P* is not being considered as the matrix of a rotation of the plane; instead the axes are rotated.)

b. The quadratic part of the new equation is $x'^2 - y'^2$; $P = \begin{bmatrix} \cos 45° & \sin 45° \\ -\sin 45° & \cos 45° \end{bmatrix}$. Since

$$(x,y) = \mathbf{X} = P^T\mathbf{X}' = \frac{1}{\sqrt{2}}\begin{bmatrix} 1 & -1 \\ 1 & 1 \end{bmatrix}(x',y') = \frac{1}{\sqrt{2}}(x' - y', x' + y')$$

the equations for x and y in terms of x' and y' are

$$x = \frac{1}{\sqrt{2}}(x' - y') \qquad y = \frac{1}{\sqrt{2}}(x' + y')$$

Substitution in the linear part of $2xy - x - y$ yields $-x - y = -\sqrt{2}\,x'$; the new equation is $x'^2 - y'^2 - \sqrt{2}\,x' = 1$ (representing a hyperbola).

13. $3x'^2 = 4$. This represents the parallel planes with equations $x' = 2/\sqrt{3}$ and $x' = -2/\sqrt{3}$; their original equations are most easily discovered by writing the given equation in the form $(x + y + z)^2 = 4$. Hence they are $x + y + z = 2$ and $x + y + z = -2$.

14. $3x'^2 + 2y'^2 = 1$, representing an elliptical cylinder.

15. $4x'^2 - z'^2 = 1$; $P = \frac{1}{3}\begin{bmatrix} 1 & 2 & 2 \\ 2 & -2 & 1 \\ 2 & 1 & -2 \end{bmatrix}$

The set of points left fixed by the map $P(\mathbf{X}) = P\mathbf{X}$ is $\{(x,y,z) \mid x = 2t,\ y = t,\ z = t\}$, the axis of rotation.

Section 13.3

17. The matrices

$$A = \begin{bmatrix} 0 & 1 \\ 1 & 0 \end{bmatrix} \quad \text{and} \quad B = \begin{bmatrix} 1 & 1 \\ 1 & 0 \end{bmatrix}$$

are Hermitian (hence normal). But

$$C = AB = \begin{bmatrix} 1 & 0 \\ 1 & 1 \end{bmatrix}$$

is not normal.

21. Use Prob. 11, Sec. 12.1.

22. Prob. 21 says that $U^\perp$ is invariant under f, so the restriction of f to $U^\perp$ is the linear operator $g\colon U^\perp \to U^\perp$. It seems reasonable to suppose that $g^*\colon U^\perp \to U^\perp$ is the restriction of f^* to $U^\perp$; then $g \circ g^* = g^* \circ g$ because $f \circ f^* = f^* \circ f$. However, this is not as trivial as it looks. The reason it works is that $U^\perp$ is invariant under f^* (because U is invariant under f). Hence if $h\colon U^\perp \to U^\perp$ is the restriction of f^* to $U^\perp$, we have (for all x and y in $U^\perp$)

$$g(x) * y = f(x) * y = x * f^*(y) = x * h(y)$$

It follows from this that $g^* = h$.

24. Let A be the matrix, and consider it as an element of $M_n(\mathbb{C})$. By Theorem 13.12 there is a unitary matrix P such that $B = PAP^* = \text{diag}\,(r_1, \ldots, r_n)$, where $r_1, \ldots, r_n$ are the eigenvalues of A. Since B has exactly $n - k$ nonzero diagonal entries, its rank is $n - k$ (Theorem 9.5). Hence the rank of A is $n - k$.

27. *a.* $P = \dfrac{1}{\sqrt{2}}\begin{bmatrix} 1 & 1 \\ -1 & 1 \end{bmatrix}$

b. $f = (1 + i)f_1 + (1 - i)f_2$, where $f_1\colon \mathbb{C}^2 \to \mathbb{C}^2$ and $f_2\colon \mathbb{C}^2 \to \mathbb{C}^2$ are defined by $f_1(x,y) = \frac{1}{2}(x + y)(1,1)$ and $f_2(x,y) = \frac{1}{2}(x - y)(1,-1)$.

28. $f = 3f_1 + 2f_2 + 0f_3$, where the $f_j\colon \mathbb{R}^3 \to \mathbb{R}^3$ are defined by $f_1(x,y,z) = \frac{1}{3}(x + y + z)(1,1,1)$, $f_2(x,y,z) = \frac{1}{2}(y - z)(0,1,-1)$, $f_3(x,y,z) = \frac{1}{6}(2x - y - z)(2,-1,-1)$.

29. *a.* Gram-Schmidt theory (Sec. 10.6) guarantees that T_j has an orthonormal basis. Since dim $T_j = k_j$ (see the proof of Theorem 13.14), this basis has k_j elements.

b. Each basis is orthonormal; elements of distinct bases are orthogonal by Theorem 13.10 [or its first corollary]. Hence the union is an orthonormal set. Since $k_1 + \cdots + k_m = \dim S$, this set is a basis.

Review Quiz

1. F	4. T	7. F	10. T
2. T	5. T	8. F	11. T
3. F	6. T	9. T	12. T

Appendix A1.1

3. They all are.

5. $\{f_{11}, f_{12}, f_{21}, f_{22}\}$, where $f_{ij}: S \to \mathfrak{F}$ is defined by $f_{ij}(A) = a_{ij}$ (the i,j entry of A).

6. $\{f_1, f_2\}$, where $f_1: \mathfrak{R}^2 \to \mathfrak{R}$ and $f_2: \mathfrak{R}^2 \to \mathfrak{R}$ are defined by $f_1(x,y) = \frac{1}{2}(y + x)$ and $f_2(x,y) = \frac{1}{2}(y - x)$.

Appendix A1.2

10. *c*. Suppose that a and b are elements of S and that c and d are scalars. The linear functionals $f = p(a)$ and $g = p(b)$ are defined by $f(x) = a \circ x$ and $g(x) = b \circ x$, while $h = p(ca + db)$ is defined by

$$\begin{aligned} h(x) = (ca + db) \circ x = c(a \circ x) + d(b \circ x) &= cf(x) + dg(x) \\ &= (cf + dg)(x) \end{aligned}$$

Hence $h = cf + dg$, that is, $p(ca + db) = cp(a) + dp(b)$.

12. $\nabla F(P) = 2P$, where P is the point in $\mathfrak{R}^3$ at which the differential is computed.

13. *a*. The product $\circ$ is the same as the dot product on $\mathfrak{F}^4$. Hence it is a nondegenerate scalar product, but not positive-definite (unless $\mathfrak{F} = \mathfrak{R}$).

b. $\begin{bmatrix} 1 & 0 \\ 0 & 1 \end{bmatrix}$

14. Every inner product on a euclidean space is a nondegenerate scalar product. The representative of D relative to $*$ is the function a defined by $a(x) = (1/\pi)\sin x$.

Appendix A1.3

17. $U^0 = \langle f_3 \rangle$, where f_3 is the linear functional defined by $f_3(x,y,z) = z$ (the third element of the dual basis of the standard basis).

19. If $\{f_1, f_2, f_3\}$ is the dual basis of the standard basis, then $U^0 = \langle f_2, f_3 \rangle$ and

$V^0 = \langle f_3 \rangle$. Hence $U^0 \cap V^0 = V^0$. On the other hand, $U + V = V$, so $(U + V)^0 = V^0$.

20. By Corollary 11.2a, $\dim S = \dim U + \dim V$. However, $\dim S = \dim U + \dim U^0$ and $\dim S = \dim V + \dim V^0$, so $\dim U^0 = \dim V$ and $\dim V^0 = \dim U$. Therefore $\dim U^0 + \dim V^0 = \dim V + \dim U = \dim S = \dim S^*$. Also, $\dim U^0 + \dim V^0 = \dim (U^0 \oplus V^0)$ because $U^0 \cap V^0 = (U \oplus V)^0 = S^0$, the zero subspace of S^*. Hence $S^* = U^0 \oplus V^0$.

21. They have the same dimension.

Appendix A2.1

1. *a.* To find $\dim U$, let E_{ij} be the $n \times n$ matrix with i,j entry 1 and 0s elsewhere. If $i \leq j$, then $E_{ij} \in U$; these matrices constitute a basis for U. For each $j = 1, \ldots, n$, there are j choices of the integer i satisfying $i \leq j$, so the number of elements in this basis is $1 + 2 + \cdots + n = \frac{1}{2}n(n + 1) = \dim U$.

b. Suppose that A and B are elements of U. The i,j entry of $C = AB$ is $c_{ij} = \sum_{k=1}^{n} a_{ik}b_{kj}$. Since $a_{ik} = 0$ when $k < i$ and $b_{kj} = 0$ when $j < k$, the terms of this sum are all zero unless there is an integer k satisfying $i \leq k \leq j$. When $i > j$, no such integer exists, so $c_{ij} = 0$. Hence $C \in U$. When $i = j$, the only integer k satisfying $i \leq k \leq j$ is $k = i$, so $c_{ii} = a_{ii}b_{ii}$.

c. $A = \begin{bmatrix} 1 & 0 \\ 0 & 0 \end{bmatrix}$ and $B = \begin{bmatrix} 0 & 1 \\ 0 & 0 \end{bmatrix}$

6. *a.* T_1, T_2, where $T_1 = \langle(-1,1)\rangle$, $T_2 = \mathbb{R}^2$. A fan basis is $\{\mathbf{V}_1,\mathbf{V}_2\}$, where $\mathbf{V}_1 = (-1,1)$ and $\mathbf{V}_2$ is any element of $\mathbb{R}^2$ not parallel to $\mathbf{V}_1$, say $\mathbf{V}_2 = (0,1)$.

b. $\begin{bmatrix} -1 & -2 \\ 0 & -1 \end{bmatrix}$

7. *b.* $A \approx \begin{bmatrix} i & 0 \\ 0 & -i \end{bmatrix}$

c. The characteristic polynomial is quadratic, with real coefficients but no real roots. The roots in $\mathbb{C}$ are therefore conjugate imaginary numbers, hence distinct. This means the matrix can be diagonalized over $\mathbb{C}$.

8. The eigenvalues are $r_1 = 1$, $r_2 = 2$ (multiplicity 2), with corresponding eigenvectors $\mathbf{V}_1 = (1,-1,0)$, $\mathbf{V}_2 = (1,-1,1)$. A fan of the map $f(\mathbf{X}) = A\mathbf{X}$ is therefore $T_1 = \langle \mathbf{V}_1 \rangle$, $T_2 = \langle \mathbf{V}_1, \mathbf{V}_2 \rangle$, $T_3 = \mathbb{R}^3$. A fan basis is $\{\mathbf{V}_1, \mathbf{V}_2, \mathbf{V}_3\}$, where $\mathbf{V}_3$ is any vector not in T_2. Choosing $\mathbf{V}_3 = (1,1,0)$, we find that

$$A \approx \begin{bmatrix} 1 & 0 & 1 \\ 0 & 2 & 2 \\ 0 & 0 & 2 \end{bmatrix}$$

9. *c*. Let x be any element of V and write $f(x) = y_1 + y_2$, where $y_1 \in U$ and $y_2 \in V$. Then $(h \circ f)(x) = y_2 \in V$.

Index

A page reference in italics indicates a formal definition or prominent mention of an item with several references.

Abel, Niels (1802–1829), 26n, 287n
Abelian group, 26
Absalom and Achitophel (Dryden), 195
Absolute value
of a complex number, 22
of a real number, 18
Addition
closure law, 3, *25*
of functions, 28
of linear maps, 97
of matrices, 28, *118*
of vectors in *n*-space, 2, *15*
in a vector space, 25
Adjoint
of a linear map, 128, *388–389,* 396
of a linear operator, inverse of, 395
of a matrix, 261
see also Self-adjoint
Angle, *308,* 323
Annihilator
of a subspace, 460
of a vector, 135
Associative law
of composition, 133, 135
of matrix multiplication, 146
in a vector space, 25–26
Augmented matrix, 196

Ballad of The White Horse, The (Chesterton), 449
Basis, 10, 17, *34*
equivalent definition of, 35
infinite, 45n
ordered, 92
standard, 17
Belloc, Hilaire (1870–1953), xii, 170, 436
Bessel's inequality, 343–344
Bilinearity
of composition, 149
of the dot product, 69, 73, 305
of the inner product on a euclidean space, 316
of matrix multiplication, 151
Boldface type for vectors in *n*-space, 27
Boole, George (1815–1864), 357n
Boolean formula, 357
Bunyakovsky, Viktor (1804–1889), 308

$\mathfrak{C}$, 2
Canonical form, 172, 227n, 235
Cardano, Girolamo (1501–1576), 287
Cartesian plane, 1
Cauchy, Augustin-Louis (1789–1857), 308
Cauchy-Schwarz inequality, 308, *320*
Cayley-Hamilton theorem, 284, *298, 471*
Characteristic equation, 279, 280
Characteristic polynomial, 279, 280
Characteristic value, 272
Characteristic vector, 272
Circular definition, 64n
Circular reasoning, 31
Closure law
of addition, 3, *25*
of composition, 133
of matrix multiplication, 145
of scalar multiplication, 26
Cofactor, 241
Column
equivalence, 186, *221*
operation, 219
rank, 174, *223*
space, 223
vector, 110n, 144

Commutative group, 26
Commutative law, 25
Commute, 148n
Compatible matrices, 145
Complex numbers, 21–22
  absolute value, 22
  as a subfield of M_2 ($\mathcal{R}$), 152–153, 163–164
  as a vector space over $\mathcal{R}$, 32
Complex plane, 115, 374–375
Complex-valued function, 314n
Composition
  distributive law of, 133–134
  inverse of a, 139
  of linear maps, 131
Conformal map, 334, 400
Conjugate
  of a complex number, 22
  of a matrix, 73, *125*
  transpose, 73, *125*
Conjugate-bilinear law, 306, *316*
Conjugate-symmetric law, 305, *316*
Consistent linear system, 201
Coordinate vector, 92
Coordinate vector map, 93, *179*
Cosines, *see* Law of cosines
Cramer, Gabriel (1704–1752), 192
Cramer's rule, 88, 192, *259*

De Morgan, Augustus (1806–1871), 369
De Morgan's law, 369
Descartes, René (1596–1650), 1
Determinant
  of the conjugate transpose of a matrix, 257
  of a diagonal matrix, 269
  of the identity matrix, 243
  of the inverse of a matrix, 257
  of a linear operator, 257
  of a linear system, 11
  of a matrix, 239
  of a nonsingular matrix, 254
  of an orthogonal matrix, 257
  of a product of matrices, 256
  of similar matrices, 257
  of a singular matrix, 255
  of the transpose of a matrix, 257
  of a triangular matrix, 470
  of a unitary matrix, 257
Diagonal entries
  of a diagonal matrix, 268
  of a matrix, 128
Diagonalization of a matrix, 286

Diagonal matrix, 58, *267*
  determinant of, 269
Differential equations, 40
  solution of, 40n
  system of, 295
Differential of a function, 70, 457
Dimension, 11, *47*
Direction, 323
Direct sum, 355
Distance
  between a point and a subspace, 377
  between points, 309, *318*
  function, 309–310, *322*
Distributive law
  of composition, 133–134
  of matrix multiplication, 147
  in a vector space, 26
Dot product, 59, 69, *304*
  on $\mathcal{R}^n$, 73
  symmetric law, 73, 305
Dual basis, 454
Dual space, 453

Eigenspace, 272, 280
Eigenvalue, 272, 280
Eigenvector, 272, 280
Elementary matrix, 208–209
  inverse of, 213
Equality in $\mathcal{R}^2$, 3n
Equivalence class, 95
Equivalence relation, 95
Equivalent matrices, 169
Equivalent systems of equations, 17n, 194n
Euclidean space, 316
Euler, Leonhard (1707–1783), 317n
Euler's formula, 317
Even function, 362n

Factor theorem, 284
Fan, 465
Fan basis, 465
Ferrari, Lodovico (1522–1565), 287
Field, 140
Finite-dimensional, 33
Fourier, Jean-Baptiste (1768–1830), 314n
Fourier analysis, 381, 383
Fourier coefficient, *329,* 330–331
Fourier series, 330–331, 381, 383
Fundamental theorem of algebra, 284

Galois, Evariste (1811–1832), 287n
Gauss, Karl (1777–1855), 153, 284

General solution, 198
Generators, 56n
Gradient, 70, 457
Gram-Schmidt orthogonalization, 347
Group, *26*
  of invertible linear operators, 153
  of nonsingular matrices, 158
  of orthogonal matrices, 405
  of unitary matrices, 405

Hamlet (Shakespeare), 26, 195, 260, 339
Hardy, Godfrey (1877–1947), 392n
Heredity principle, 53
Hermite, Charles (1822–1901), 126
Hermitian matrix, 126, 421
  spectral theorem, 432
Hermitian operator, 419
Hermitian product, 317n
Hilbert, David (1862–1943), 412n
Hilbert problems, 412n
Hilbert space, 412
Homogeneous linear equation, 82
Homogeneous linear system, 198, *202*
Hyperbolic functions, 41

Idempotent, *278,* 374
Identity law, 25–26
Identity map, 66
Identity matrix, 116, *148*
  determinant of, 243
"If and only if," 13n
Image, 64
Infinite basis, 45n
Infinite-dimensional, 29
Injective, 85
Inner product, *316*
  of functions, 314, 318
  space, 317n
Invariant, 172
  subspace, 277, *465*
Inverse
  of the adjoint of a linear operator, 395
  of a composition, 139
  of the conjugate transpose of a matrix, 163
  of a diagonal matrix, 269
  of an elementary matrix, 213
  law, 25
  of a map, 84, *87*
  of a matrix, 151, *156,* 217, 262
  of a matrix product, 157, 158
  one-sided, 161, 413
  of the transpose of a matrix, 163
  of a vector, 25, *30*
Invertible map, 87
Involution, 418
Isometry, 397
Isomorphic, 87, 94
Isomorphism, 84, *87,* 150
  as an equivalence relation, 94
  between linear algebras, 150
  between rings, 149

Johnson, Samuel (1709–1784), 64n

Kernel, 74–75
Kronecker, Leopold (1823–1891), 18n, 19
Kronecker's delta, 18

Latent value, 272
Latent vector, 272
Law of cosines, 324
Law of nullity, 78
Laws of Thought, The (Boole), 357n
Leading entry, 201
Lebesgue integral, 458
Legendre polynomials, 351
Length, 307
Linear algebra
  informally described, 71, 107
  as a mathematical system, 134, 136, *149*
Linear combination, 7, *33,* 56
Linear dependence, 10, 15, *34*
Linear equation, homogenous, 82
Linear extension, 101–102
Linear functional, 453
Linear independence, 10, 15, *34*
Linear map(s), 65
  adjoint of, 128, *388–389,* 396
  composition of, 131
  nullity of, 80
  null space of, 75
Linear operator(s), 65
  determinant of, 257
  orthogonal, 334, *396*
  ring of, 134
  trace of, 178
Linear space, 26
Linear system, homogeneous, 198, *202*
Linear transformation, 65
Lord of the Rings, The (Tolkien), 63
Lower triangular matrix, 293, *464*

Main diagonal, 116, 267
Mallory, George (1886–1924), 304
Man Who Was Thursday, The (Chesterton), 91
Mapping, 64
Mathematical research, 381n
Matrix (matrices)
  addition of, 28, *118*
  adjoint of, 261
  augmented, 196
  of a change of basis, 180
  compatible, 145
  conjugate of, 73, *125*
  determinant of, 239
  diagonal entries of, 128
  diagonalization of, 286
  elementary, 208–209
  equivalent, 169
  informally mentioned, 28, 49, 67
  inverse of, 151, *156*, 217, 262
  of a linear map, 109
  of a linear system, 191, 193
  lower triangular, 293, *464*
  $n \times n$, ring of, 148
  nonsingular, 156, 254
  normal, 443
  notation, 117, 209, 247
  nullity of, 227
  orthogonal, 257, 334, *396*, 405
  scalar multiplication of, 28, *118*
  symmetric, 124, 421
  trace of, 178
  triangular, 293–294, *464*
  unitary, 257, *401*, 405
  upper triangular, 294, *464*
  zero, 118
Maximal linearly independent set, 51
Metric, 310
Metric space, 310
Minimal spanning set, 51
Minimax principle, 51
Minkowski inequality, 309, *321*
Minor, 241
Multiple roots, 280n
Multiplication
  of linear maps, 131
  of matrices, 121, *145*
  scalar, 26
Multiplicity, 291n

Nilpotent, 300
Nondegenerate scalar product, 456
Nonsingular, 156
Norm, 307, *317*
Normalize, 324
Normal matrix, 443
  spectral theorem, 446
Normal operator, 424, 435, *443*
  spectral theorem, 446
Normal set, 326
Normal vector, 326
n-space, 2, 15
  boldface type for vectors in, 27
Nullity
  of a linear map, 80
  of a matrix, 227
  Sylvester's law of, 78
Null space
  of a linear map, 75
  of a matrix, 227

Occam's razor, 418–419
Odd function, 362n
One-to-one, 85
One-to-one correspondence, 87
Onto, 84
Ordered basis, 92
Orthogonal complement, 59, 70, *364*
Orthogonality relations, 325
Orthogonal linear operator, 334, *396*
Orthogonal matrix, 257, 334, *396*
Orthogonal projection, 70, *371–372*
Orthogonal set, 325
Orthogonal similarity, 433
Orthogonal subspaces, 362
Orthogonal vectors, 59, 310, *325*
Orthonormal, 326

Parallelogram law, 2
Parameter, 17
Parseval's equality, 344
Parseval's identity, 336
Particular solution, 198
Path to Rome, The (Belloc), xii, 170, 436
Polar forms, 324
Positive-definite law
  of the distance function, 309–310, *322*
  of the dot product on $\mathcal{R}^n$, 73
  of the inner product, 316
  of the standard inner product, 307
Postmultiplication, 175
Premultiplication, 175
Principal axis theorem, 440
Projection theorem, 59, 70, *370*

Proper subspace, 369
Proper value, 272
Proper vector, 272
Pythagorean theorem, 313, 332

Quadratic form, 439
Quadric surface, 438

ℛ, 1
Range, 63, 64, *76*
Rank
 of a linear map, 80
 of a matrix, 174, *224,* 461
Recursive definition, 239
Reflection, 273, *374*
Representation theorem, 457
Representative of a linear functional, 456
Research, *see* Mathematical research
Riemann integral, 458
Riesz representation theorem, 458
Ring, *134*
 commutative, 140
 of linear operators, 134
 of $n \times n$ matrices, 148
 with unit element, 137
Rotation
 of axes, 69n, 183, 440
 of the plane, 68
 of 3-space, 441
Row
 echelon form, 202
 equivalence, 186, *216*
 operation, 207–208
 rank, 174, *223*
 space, 223
 vector, 110n, 144
Russell, Bertrand (1872–1970), 233n, 357n

Scalar, 4, 14, *26*
Scalar multiplication
 of functions, 28
 of linear maps, 97
 of matrices, 28, *118*
 of vectors in n-space, 3, *15*
 in a vector space, 26
Scalar product, 319, *456*
Schwarz, Hermann (1843–1921), 308
Self-adjoint, 419
 spectral theorem, 431
Set notation, 21n, 34
Similarity, 175
 orthogonal, 433
Singular, 156
Skew-Hermitian matrix, 421
Skew-Hermitian operator, 419
Skew-symmetric matrix, 421
Skew-symmetric operator, 419
Solution
 of a differential equation, 40n
 of a linear system, 160
 of a system of differential equations, 295
Span, 8, *33*
Spanning set, 33
Spectral basis, 433
Spectral decomposition, 449
Spectral theorem, 447–448
 for Hermitian matrices, 432
 for normal matrices, 446
 for normal operators, 446
 for self-adjoint operators, 431
Spectrum, 433
Standard basis, 17
Standard inner product, 124, *304–305, 339–340*
 positive-definite law, 307
Standard map, 158, *179*
Stopping by Woods on a Snowy Evening (Frost), 449
Subset notation, 58n
Subspace(s), 53
 annihilator of, 460
 distance between point and a, 377
 invariant, 277, *465*
 orthogonal, 362
 proper, 369
 sum of, 355
Subtraction, 5, *30*
Surjective, 84
Swinburne, Algernon (1837–1909), 304
Sylvester, James (1814–1897), 74
Sylvester's law of nullity, 78
Symmetric law
 of the distance function, 310, *322*
 of the dot product, 73, 305
Symmetric matrix, 124, 421
Symmetric operator, 419

Taxicab metric, 313
Thomas Aquinas (1225–1274), 172
Trace
 of a linear operator, 178
 of a matrix, 178
Transpose, 68, *122*
 inverse, of a matrix, 163

Trent's Last Case (Bentley), 91
Triangle inequality, 309, *322*
Triangular matrix, 293–294, *464*
Triangulation of a matrix, 294, *464*

Undetermined coefficients, 135
Unitary matrix, 257, *401*
Unitary operator, 396
Unitary similarity, 433
Unitary space, 316
Unit element, 137
Unit vector, 324, *326*
Upper triangular matrix, 294, *464*

Vacuous, 36, 57
Vandermonde, Alexis (1735–1796), 42n
Vandermonde's determinant, 42n
Vector(s), 26
  annihilator of, 135
  characteristic, 272
  column, 110n, 144
  coordinate, 92
  inverse of, 25, *30*
  normal, 326
  orthogonal, 59, 310, *325*
  proper, 272
  unit, 324, *326*
  zero, 3n, 25, 30
Vector space, 4–5, *25–26*

Wronski, Jozef (1778–1853), 38
Wronskian, 38, 41

Zero
  function, 29
  map, 66
  matrix, 118
  space, 37
  vector, 3n, 25, 30

BCDEFGHIJ–VB–798765